AF552754

MAIZE RESEARCH IN INDIA

About the Editors

S.L. Jat, Scientist (Sr. Scale), ICAR-Indian Institute of Maize Research (IIMR), Ludhiana

Dr. S.L. Jat, Ph.D. Agronomy from ICAR-IARI with IARI Merit Medal and working as a Scientist at ICAR-IIMR, New Delhi since 2010. After doing training of 5 weeks at CIMMYT, Mexico he intensified studies on conservation agriculture (CA)-based sustainable intensification of maize systems. He is also associated with teaching and guiding of the PG students at ICAR-IARI, New Delhi. Dr. Jat has leading a project funded by ICAR-NASF on microbiome and soil health indicators in CA. He is also leading the outreach activities of the ICAR-IIMR and travelled across India for creating awareness and providing management tools for alien invasive pest fall armyworm. Beside crop production technology, he also worked in a team for development of 9 maize hybrids including baby corn and pop corn. Dr Jat has more than 85 research/review papers in his credit with goggle scholar citation of 793 and h-index of 17. Dr Jat is recipient of FAI Golden Jubilee Award (2019), MTAI Young Scientist Award (2020), FAI-Dhiru Moraraji Memorial Award (2016), Appreciation Certificate from Govt of Afghanistan (2016 and 2018), FAI-Shreeram Puruskar (2010 & 2013), IIMR best Scientist Award (2019 and 2020) etc.

Chikkappa G. Karjagi, Scientist (Sr.Scale), ICAR-IIMR, Ludhiana

Dr. Chikkappa Gangadhar Karjagi is actively involved in maize research especially the application of molecular tools and techniques in maize improvement at ICAR-Indian Institute of Maize Research (formerly Directorate of Maize Research). He has developed a protocol for *In-planta* transformation in maize. He has also studied the genetics of inheritance to *Sesamia inferans* in maize. He is also associated in development of single cross maize hybrids DHM121 (BH41009) and also the extension of area of cultivation of maize single cross hybrid DHM117 (BH 40625) in Bihar during both kharif and rabi season. In his Ph.D. research he has developed several NILs of improved Swarna (Golden Swarna, an Indian rice cultivar) through MABB, which can accumulate total carotenoids as high as 16 ìg/g of endosperm, with 85% of pro vitamin A and also studied inheritance of transgene (*psy+crtI*) for which he was honoured with Jawaharlal Nehru Award for outstanding Doctoral Thesis Research in Agricultural and Allied

Sciences in the disciplines of Crop Science by ICAR. Presently he is involved in improvement of nutritional value of maize through marker assisted introgression of gene(s) determining quality traits namely *opaque2*, *lpa2*, and *crtRB1/lcyE*, in elite Indian germplasm.

Bhupender Kumar, Scientist (Sr. Scale), ICAR-IIMR, Ludhiana

Dr. Bhupender Kumar has completed his B.Sc. Ag. from CSKHPKV Palampur, H.P, and M.Sc. & Ph.D. in Genetics from ICAR-IARI, New Delhi, is working as scientist in Maize breeding at ICAR-Indian Institute of Maize Research (IIMR), Ludhiana from 2011. Since joining, he is intensely working on basic and applied science of improving maize for high yield, abiotic and biotic stresses tolerance. He has developed and released six single cross hybrids of maize, many of them have been commercialized through PPP and are very much popular among the farmers. Besides, he has developed and registered two genetic stocks at NBPGR. Research results of his team/group of researchers have been well documented in over 40 research papers in peer reviewed high-impact factor journals. Dr. Bhupender Kumar has several awards and recognitions to his credit including NAAS Associateship (2020), ICAR- IIMR Best Scientist recognition (2019), Young Scientist award of MTAI (2020), DST-SERB early career research award (2018), Young Scientist Award from society for scientific development in agriculture and technology (2015) and Appreciation awards (2015 & 2017) from ICAR-IIMR, Ludhiana.

Suby S.B., Scientist (Sr. Scale), ICAR-IIMR, Ludhiana

Dr. Suby S.B., is an entomologist working in integrated pest management of maize at ICAR-IIMR since October 2009. She listens to the communication between plants and insects and believes in eco-friendly ways to bring equilibrium in agro-ecology. One of her technology was granted patent and one is under review. She has published more than 20 research papers in national and international journals. She was graduated in Agriculture from Kerala Agricultural University with first rank and was a topper in Msc and PhD Entomology from ICAR-Indian Agricultural Research Institute, New Delhi.

C.M. Parihar, Sr. Scientist, ICAR-IARI, New Delhi

Dr. C.M. Parihar (Ph.D. Agronomy, ICAR-IARI, New Delhi), is a Senior Scientist in Indian Council of Agricultural Research (ICAR), New Delhi. After obtaining his Ph.D. devoted more than one decade to intensely working on basic and applied science in natural resource management and specifically on conservation agriculture (CA)-based sustainable intensification and teaching of IARI PG students. Dr. Parihar has conducted comprehensive research on soil health governing parameters under CA. Research results of his team/group of researchers have been well documented in over 65 peer reviewed high-impact journal articles. Dr. Parihar has several awards and recognitions to his credit including Fertilizer Association of India (FAI)-Golden Jubilee Award (2019) for Excellence in Fertilizer Use Research, FAI-Golden Jubilee Award (2017) for Efficient Nutrient Management, ICAR-Indian Institute of Maize Research (IIMR)-Outstanding Researcher Award (2015), ICAR-IIMR-Young Scientist Award (2015), Indian Society of Agronomy (ISA)-PS Deshmukh Young Agronomist Award (2014), ICAR-Junior Research Fellowship (JRF), Gold Medal in M.Sc. (Agronomy), etc.

Meena Sekhar, Principal Scientist, ICAR-NBPGR, New Delhi

Dr. Meena Shekhar is Principal scientist in Division of Plant Quarantine, ICAR-National Bureau of Plant Genetic Resources, New Delhi, started her career from Indian Institute of maize Research in 1984. Working with co-ordinated maize improvement, she has very long experience in maize diseases especially in Post flowering stalk (PFSR) of maize. She worked on host pathogen interaction between PFSR pathogen and maize (host). She developed disease management module and resistant lines for PFSR disease. She also worked on Prevention and Management of Mycotoxin specially aflatoxin contamination in maize, Genes Pyramiding for Resistance to Turcicum Leaf Blight and Polysora Rust in Maize. She investigated rapid Technique, developed/validated for the first time in maize to identify the toxic and nontoxic isolates of *Aspergillus flavus* through Ammonia vapour test. Dr. Shekhar has published several papers in reputed Journals/Book chapters/Books/ Popular Magazines. She has several national and International awards and recognitions to her credit including Distinguished Scientist, Lifetime achievement award in Plant Pathology and Fellow of Indian Phytopathological Society (FPSI 2018).

Sujay Rakshit, Director, ICAR-IIMR, Ludhiana

Dr. Sujay Rakshit did his PhD in Gentics and Plant Breeding from ICAR-Indian Agricultural Research Institute, New Delhi (India) and received Best Ph.D. Student Medal and IARI Merit Medal in 1998. Presently he is a Director of ICAR-Indian Institute of Maize Research PAU Campus, Ludhiana since March 2017. Dr Rakshit is a recipient of prestigious NE Borlaug Fellowship and BOYSCAST Fellowship and was a postdoctoral fellow at Iwate Biotechnology Research Center, Japan (2003-05). He also served as faculty member of the Division of Genetics, IARI, New Delhi and guided 3 PhD students. Dr Rakshit had developed AICRP on Sorghum and Maize automation systems. Dr Rakshit and his team developed and released 4 hybrid maize cultivars, one forage sorghum variety and registered 20 genetic stocks at ICAR-NBPGR. Dr Rakshit has authored research paper (65), review/technical article (28) and bulletins (20). He is a recipient of many awards and honors including Fellow, Indian Society of Genetics & Plant Breeding, New Delhi (2014), President, Maize Technologists Association of India (2018), President, Agriculture and Forestry Section, Indian Science Congress, Kolkata (2019), Associated Editor, Indian Journal of Genetics & Plant Breeding (2016), Indian Science Congress Association Young Scientist Award (1999), Jawaharlal Nehru Award (1999) for outstanding post-graduate research and UNESCO Fellowship in Biotechnology (1998).

Vinay Mahajan, Ex-Director IIMR and Pr. Scientist, ICAR-NBPGR, New Delhi

Dr Vinay Mahajan did PhD Plant Breeding from PAU, Ludhiana and MBA from IGNOU, New Delhi is presently working as Principal Scientist at ICAR-NBPGR, New Delhi. Dr Mahajan served as Head (Crop Improvement Division) at ICAR-VPKAS, Almora (2003-10), in-charge of ICAR-IIMR, Ludhiana (June 2014 to Feb 2016) and acting Director ICAR-Indian Institute of Maize Research, New Delhi (March 2016 to March 2017).

Dr Mahajan has two patents on hybrid wheat and developed 45 cultivars in crops like maize hybrids/composites (18), lentil (5), horse gram (4), wheat (3), soybean (3), field pea (2) and one each in barley, rajmash, arhar, toria and groundnut. In addition, 19 genetic stocks of maize, wheat, barnyard millet and soybean also registered with ICAR-NBPGR. Dr Mahajan was honoured with ICAR Team Award Triennium 1994-96 and recognized by ICARDA and ICRISAT in 2009 and 2010, respectively. Other scientific recognitions were Bioved Fellowship Award 2007, 20th Century Crop Research Award 2000 and visited Syria, Mexico, China and Thailand. He had published >95 research publications, 4 review articles, 4 books, 14 extension bulletins, 20 book chapters and >15 invited lectures were to the credit of Dr Mahajan.

MAIZE RESEARCH IN INDIA
RETROSPECT AND PROSPECT

Editors

S.L. Jat

Chikkappa G. Karjagi

Bhupender Kumar

Suby S.B.

C.M. Parihar

Meena Sekhar

Sujay Rakshit

Vinay Mahajan

NEW INDIA PUBLISHING AGENCY

New Delhi – 110 034

NEW INDIA PUBLISHING AGENCY
101, Vikas Surya Plaza, CU Block, LSC Market
Pitam Pura, New Delhi – 110 034, India
Email: info@nipabooks.com
Web: www.nipabooks.com

For customer assistance, please contact

Phone: + 91-11-27 34 17 17 Fax: + 91-11- 27 34 16 16
E-Mail: feedbacks@nipabooks.com

ISBN : 978-93-89992-00-7

Composed and Designed by NIPA.

सत्यमेव जयते

त्रिलोचन महापात्र, पीएच.डी.
एफ एन ए, एफ एन ए एस सी, एफ एन ए ए एम
सचिव एवं महानिदेशक

भारत सरकार
कृषि अनुसंधान और शिक्षा विभाग एवं
भारतीय कृषि अनुसंधान परिषद्
कृषि एवं कल्याण मंत्रालय
कृषि भवन, नई दिल्ली–110 001

GOVERNMENT OF INDIA
Department of Agricultural Research & Education
and
Indian Council of Agricultural Research
Ministry of Agriculture and Farmers Welfare
Krishi Bhavan, New Delhi-110 001
Tel.: 23382696; 23386711 Fax : 91-11-23381773
E-mail: dg_icar@nic.in

TRILOCHAN MOHAPATRA, Ph.D.
FNA, FNASc, FAAS
SECRETARY & FDIRECTOR GENERAL

Foreword

All India Coordinated Research Project on Maize, the first co-ordinated agricultural research programme in India has completed six decades of its service to the nation by generating technologies to sustain maize production in different agro-ecologies. Since its inception, AICRP on Maize has released and notified 265 cultivars for enhancing productivity, production and quality of maize. Changing utilization pattern of maize has lead to crop intensification and greater spread of its cultivation to non-traditional areas. Nonetheless, there is a great demand of maize in industry and with better export potential, maize can be expanded further.

AICRP on Maize is celebrating its diamond Jubilee this year and to earmark this occasion, this book on 'Maize research in India: Retrospect and Prospect' is relevant and appreciable. I am sure, the book will be of much use to the researchers as well as policy makers.

T. MAHOPATRA

Dated the 29th March, 2017
New Delhi

Preface

The Diamond Jubilee of All India Coordinated Research Project on Maize, the first of its kind in Indian agricultural research is being celebrated this year. The coordinated efforts led to the release of the first set of double-cross hybrids, *viz*., Ganga 1, Ganga 101, Ranjit and Deccan for commercial cultivation in India in 1961. Since then 144 hybrids and 121 composites of maize have been released from public sector through the AICRP on Maize network. The movement of research gradually led to evolution of the project to a full-fledged institute, ICAR-Indian Institute of Maize Research, a single umbrella to co-ordinate the maize research in India. The work carried out by the Indian maize community so far has created a far-reaching impact on the maize scenario of the country as evidenced by the ever-increasing role in Indian economy in the recent years. In this context, it has been out humble efforts to bring forth this journey in form of this book.

During 1950-51, India produced only 1.73 million tonnes (mt) of maize, which has reached 27.72 mt in 2018-19. This has been possible due to increase in area as well as productivity. India with 3.01 t/ha productivity, is far behind the global average of 5.92 t/ha. This gap is largely attributed to the abiotic stresses, especially drought in the rainfed environment, which constitutes 75% of maize production area in India. Apart from this, pest and diseases, which are highly influenced by the abiotic stress, causes unexpected havoc in maize. Climate smart maize technologies may play a key role in achieving the yield potential of cultivars churned out by the maize breeders each year. This will help in crop intensification and make the crop more amenable to the wide geographic and climatic regimes of the country.

It is estimated that by 2025, India would require 50 mt maize grains, of which 32 mt will be required in the feed sector, 15 mt in the industrial sector, 2 mt as food, and 1 mt for seed and miscellaneous purposes. Moreover, the annual export potential of maize stands at 10 mt. An annual growth rate of 7-8% would be required to achieve this target. This opens up the opportunity to strategically deploy all the resources to optimize maize technologies developed over the six decades to meet the challenges. This also prompts to consider the future course of maize research with the shift in geography and utilization pattern.

Single cross hybrid is the single most important technology, which has revolutionized maize research across the globe. However, low adoption of this technology coupled with the lack of proper nutrient management, the lack of right post-harvest handling and gaps in outreach of technologies have led to slow progress. Among field crops, maize has maximum industrial make over. Indian maize research lags behind other global players in this aspect. Strengthening of diversification of germplasm base and heterotic grouping of germplasm will equip us to meet the technological challenges in improving and diversifying the crop for food, feed, fodder and industry. Exploiting the progresses in biotechnology, genomics and information technology will bring efficiency and bridge the time gap in achieving the technology and policy interventions envisaged in the future journey of maize. The gap exist in post harvest processing, value addition, pricing and marketing needs to be addressed with policy interventions and technological back up.

We hope this exercise and the vision embodied in it will pave the way ahead for a smooth maize revolution in India, which is already underway.

Editors

Contents

1

Maize Research and Development in Uttarakhand

R.K. Khulbe[1], N.K. Singh[2], Dibakar Mahanta[1], Rajashekara H.[1] J. Stanley[1] and A. Pattanayak[1]*

[1]ICAR-Vivekanand Parvatiya Krishi Anusandhan Sansthan, Almora, Uttarakhand, India
[2] GB Pant University of Agriculture and Technology, Pantnagar, Uttarakhand, India
**Corresponding Author's Email:rkkhulbe@gmail.com*

Maize is an important crop of Uttarakhand and is cultivated in about 22 thousand ha in both hills and plains. The crop is grown for food, feed and fodder and forms an integral constituent of all major cropping systems prevalent in the state. *Kharif* (June-September) is the main grown season in the hills, whereas in the foothills and *tarai* plains, the crop is grown during *zaid* (Feb-May) as well. The crop is grown broadly for two different end products – grain and green ears. In areas where maize forms staple diet, the crop is grown for grain, whereas in areas adjoining/en route popular tourist destinations and towns, the crop is grown mainly for green ears. The crop is also gaining importance as an industrial crops following establishment of maize processing industry in the state.

The major maize growing districts of the state are Dehradun, Nainital, Pithoragarh, Pauri, Tehri and Almora (Fig. 1), which together comprise about 90% of total

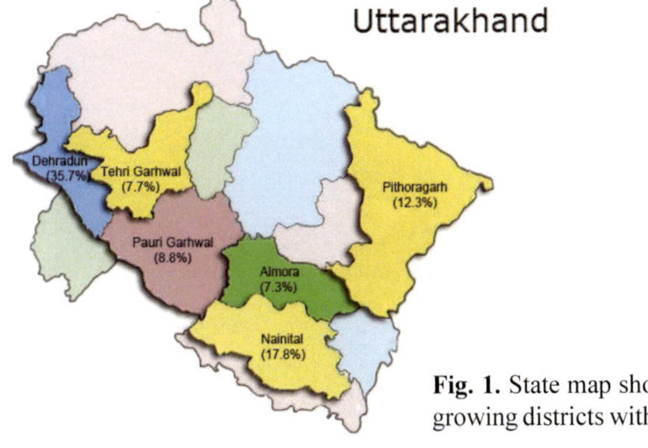

Fig. 1. State map showing major maize growing districts with acreage share.

maize area in the state. Dehradun district alone accounts for over one-third of the total area (35.7%) followed by Nainital (17.8%) and Pithoragarh (12.3%). Out of thirteen districts in the state, two are situated in *tarai* plains (Haridwar and Udham Singh Nagar), eight in hills (Pithoragarh, Almora, Pauri, Tehri, Uttarkashi, Bageshwar, Chamoli, Rudraprayag and Champawat), while two have both hill and plains areas (Dehradun and Nainital). 63% of total maize area in the state lies in the hills with the remaining 37% being in the plains.

The area under maize crop in the state has witnessed a steady decline over the last 15 years. Compared to the starting year (2000-01) the area stands reduced by about 25 per cent in 2015 (Fig. 2).

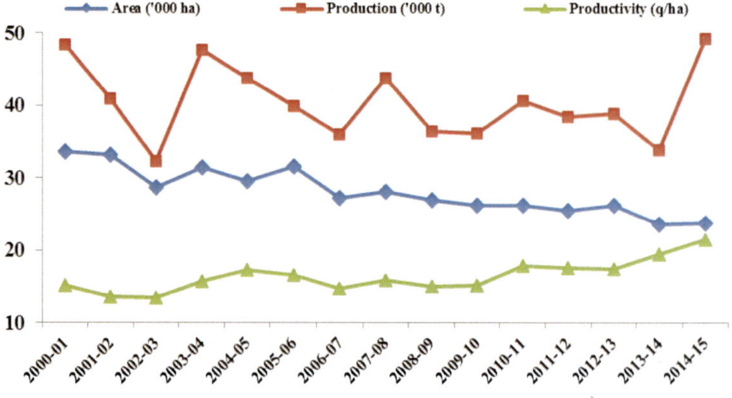

Fig. 2. Maize APY trend in Uttarakhand.

The highest reduction in area, compared to the starting year (2001), has been in Nainital (50%) followed by Dehradun (27.9%) and Pithoragarh (14.1%) (Fig. 3).

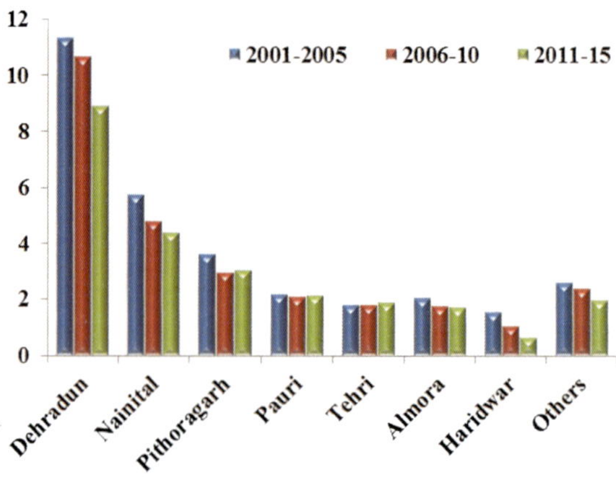

Fig. 3. District-wise area under maize crop in Uttarakhand ('000 ha).

Maize production in the state in the last 15 years exhibits a rather erratic pattern, the major districts (except Dehradun) though show a clear increasing trend over the three 5-year blocks (Fig. 4).

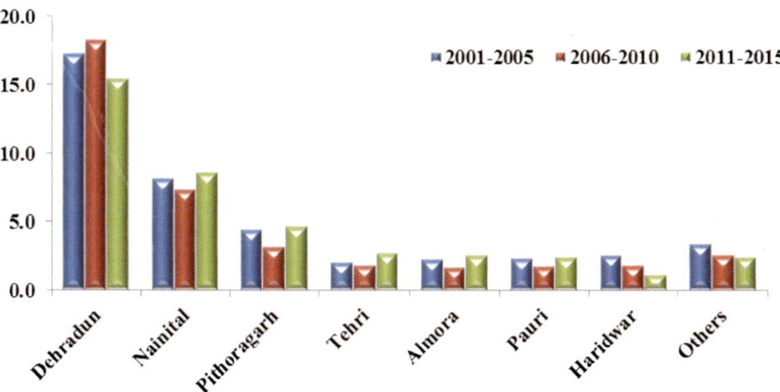

Fig. 4. District-wise maize production in Uttarakhand ('000 t).

In contrast to the area and production trend, maize productivity in the state has increased progressively. The overall increase of 23.8 per cent in productivity has been contributed mainly by Nainital (39.8%), Pithoragarh (29.9%) and Dehradun (13.5%) (Fig 5). The *zaid* yields have also increased by up to 40 per cent. The increase in productivity is driven mainly by increase in area under hybrids (in plains) and improved OPVs (in hills) and adoption of improved crop management practices.

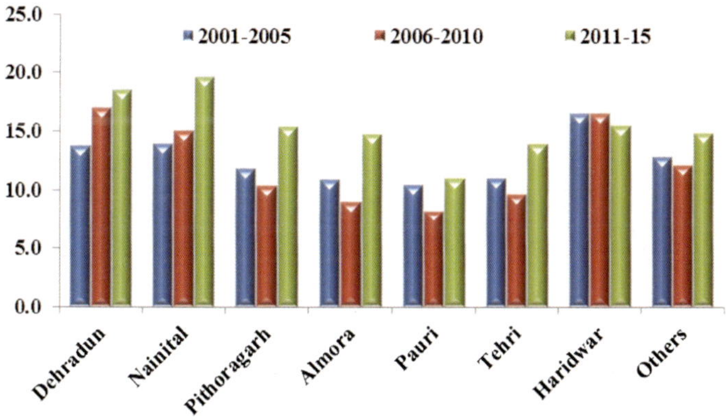

Fig. 5. District-wise maize productivity in Uttarakhand (q/ha).

Diminishing returns from the crop owing chiefly to damage by wild animals and inadequate marketing support are among major causes forcing farmers to switch to alternate crops. Lack of regular demand from industry has been another important reason for declining interest in maize cultivation in the past. However, with the changing scenario in the state in terms of increasing demand of maize due to establishment of processing industries, poultry and cattle feed industries, increased tourism, availability of high yielding hybrids, established market and other enabling factors, the area, production and productivity are expected to increase significantly in the near future over commonly grown wheat, rice and soybean.

Major industries utilizing maize in the state with their daily/monthly/ yearly capacity

The state has a good number of small and medium-scale poultry and cattle feed units, and processing of maize for industrial use has also witnessed an increase in the recent years (Table 1). The collective annual requirement of the industry is estimated at 10 lakh t, 20 times the average maize production in the state. The industrial demand at present is being met by other maize growing states. However, the growing industrial demand is expected to boost maize cultivation in the state, particularly in the plains, where the crop can be grown throughout the year and high yields (up to 60-70 q/ha) yield are achievable.

Table 1. Major maize industries in the state.

Industry	Firms	Maize requirement per year (t)	Products
Food industry	Murliwala Agrotech, Rudrapur and Sitarganj (US Nagar)	54750	Ingredients for mid-day meal
Feed industry	Many small and medium-scale units in Jaspur, Kashipur and Dehradun	100375	Poultry and cattle feed
Processing industry	Roquette Riddhi Siddhi Pvt. Ltd. Rudrapur (US Nagar)	438000	Starch, glucose, liquid glucose, pharma grade liquid glucose, glucose syrup
	Gujrat Ambuja Export Ltd. Sitarganj (US Nagar)	273750	Dextrose monohydrate, dextrin, liquid glucose, malto dextrin, sorbitol, maize starch for foods and pharmaceuticals, maize gluten, maize germ

Fresh food processing industry in the state, that consumes huge volumes of vegetable crops annually, has yet to begin processing specialty corn (sweet corn and baby corn). Establishment of specialty corn processing units can bring about a turnaround in the economy of farmers in the state, as has been witnessed in Haryana.

Major cropping systems and production ecologies

Among *Kharif* cereals, maize ranks third in area and production and first in productivity in Uttarakhand. Maize is grown under rainfed condition in Uttarakhand hills. Maize-wheat is most popular cropping system in Uttarakhand hills followed by maize-lentil and maize-toria. In some areas maize-toria-wheat is also popular. Maize is also followed by vegetable crops such as garden pea and onion in irrigated valleys. Indeterminate rajamsh is most common as a mixed crop with maize.

The major maize-based cropping system in plain areas of Uttarakhand are maize-wheat, maize (green ear)-tomato-potato, maize (green ear)-tomato-onion and rice-vegetable pea-maize. Paired row maize (45/90 cm) is advantageous for intercropping of urdbean in 2+2 row ratio as compared to inter-cropping of one row of urdbean between two rows of maize (67.5cm apart). There is saving of 30 kg N/ha in maize with urdbean intercropping. Furrow placement of fertilizer gave the maximum maize grain equivalent yield and earned more net return than broadcast application. During *Kharif* maize is largely cultivated as rainfed crop. However, irrigation is required for early planted maize and during the dry spells in *Kharif*. During the winter and spring season, maize is cultivated as an irrigated crop.

AICRP centre/s genesis and mandate

ICAR-VPKAS Almora (Voluntary centre)

ICAR-VPKAS Almora is a pioneer maize breeding centre that ushered in hybrid maize development programme in the country. The institute is credited with the development of first maize hybrid (VL Makka 42) in the country which led the way for development of hybrid maize varieties by Indian public sector institutions. The research mandates of the institute are:

1. Genetic enhancement in maize for early and extra early maturity, disease resistance and yield
2. Development of short duration composites and hybrids of normal and specialty corn suitable to hill agriculture
3. Maintenance and augmentation of maize germplasm
4. Identification of resistant sources against important disease of maize in NW-hills and their management
5. To improve the profitability and economic returns of maize farmers by developing matching agro-technology.

The centre has released a total of 31 varieties (10 composites and 21 hybrids) in different consumer segments for different agro-ecologies of the country. The institute of late has gained prominence on account of release of first MAS-derived QPM hybrid (Vivek QPM 9), which has been commercialized with six private seed sector companies and is estimated to be grown over a significant area in different parts of the country. The breeding material generated by the centre has strengthened extra- and early maize breeding and QPM programmes of various AICRP centres. Current focal research areas of the institute include development of biofortified maize hybrids using combination of molecular breeding tools and doubled haploid technology.

GBPUA&T, Pantnagar (AICRP centre)

Govind Ballabh Pant University of Agriculture & Technology, Pantnagar, the first agricultural University established in 1960, is known for ushering in Green Revolution in India and is branded as the Harbinger of Green Revolution. AICRP maize is the oldest coordinated project implemented at the University in the year 1961. During its long journey, the University has been credited with the development of many high yielding and popular composite varieties and hybrids adopted and appreciated by the farmers across different states. With the changing scenario of maize utilization, advantage of hybrids and also with the development of seed production and supply networks, the university has changed its priority from composites to hybrid cultivars. So far, the University has developed 11 composites and 3 hybrids in addition to pools and populations of maize. The mandates of AICRP centre Pantnagar are:

1. To develop short duration composites and hybrids
2. To develop and diversify germplasm
3. To participate in the yield evaluation trial during *Kharif* and *Rabi.*
4. Maintenance and seed production of released hybrid and composites.
5. Development of agronomic practices to enhance productivity and profitability of maize.
6. Screening of germplasm and test entries against BLSB and ESR and development of IDM practices.
7. Survey and surveillance of diseases in the state.
8. Transfer of improved maize cultivars and improved cultural practices through training and demonstration.

Key production problems in state for maize

Limited area under maize hybrids: While most of area in the plains is under hybrids, the area under hybrids in hills, which account for about 67% of the total area, is negligible. The reasons vary from unavailability of seed of public sector hybrids, unsuitability of private sector hybrids for hill conditions to high cost of the hybrid seed. FLDs conducted in different hill regions of the state convincingly demonstrate the superiority of public bred hybrid varieties over the local cultivars, yet the area under maize hybrids has failed to spread chiefly due to unavailability of seed to the farmers. The government seed agencies responsible for production of seed of popular hybrids are reluctant to produce maize hybrid seed as it requires greater expertise and efforts compared of seed production of self-pollinated crops. The initiative has been launched by ICAR-VPKAS in the Jaunsar tribal area of the state. Enabling government policies and collective efforts by the institutes and government seed agencies can ensure easy access maize farmers to hybrid seed at affordable prices, leading to progressive increase in the area under hybrid maize cultivation.

Traditional practices of maize cultivation: Maize in Uttarakhand hills is mostly grown under organic condition following traditional practices. Maize is a heavy feeder crop, especially for nitrogen, and requires adequate supply of nutrients to express its potential. Meeting nutrient requirement of maize through FYM is difficult due to inadequate availability of FYM and its high cost. Chemical fertilizers are often not available in the local market. Similarly, chemical measures of disease and insect-control are rarely used in hills, resulting in heavy losses at times. A number of new production technologies have been developed for maize, especially on plant geometry, plant population, nutrient management, however, the farmers have to yet to adopt them.

Lack of organized procurement: Unlike crops such as wheat and rice, government declared MSP and procurement mechanism for maize lacks in the state, causing difficulty to the farmers in disposal of their produce. Though local mandis exist in and around major maize growing areas, arbitrary fixation of procurement prices by the middlemen results in loss to the farmers. Intervention by the government in the form of declaration of MSP and establish of procurement mechanism in the lines of wheat and rice would eliminate the uncertainty about disposal of the produce.

Crop losses by wildlife: Among the major reasons for decline in maize area, particularly in hills, is the loss caused to the crop by monkeys and wild boars, and in some specific areas, by birds. Though the crop remains vulnerable to damage right from the vegetative stage, the maximum damage is caused during grain filling stages. The rampant damage caused by the wild animals to the laboriously nurtured crops acts as a huge discouragement to the farmers and forces them to switch to alternate crops. Since dealing with wildlife is a complex issue, a comprehensive strategy needs to be formulated at higher government levels and a balanced approach adopted to tackle the wildlife menace in maize crop.

AICRP technologies

Popular land races with their peculiar characteristics for their popularity

The state of Uttarakhand is fairly rich in maize land races diversity. The salient characteristics of popular land races/local cultivars evaluated at ICAR-VPKAS Almora are given in Table 2.

Table 2. Salient features of popular local cultivars of Uttarakhand.

Name of cultivar	Salient characteristics
Bhimtal Local	Extra-early, white flint, 190 cm, used mainly for green ears
Amritpur Local	Early, white flint, 200 cm, used mainly for green ears
Dhiari Local	Early, yellow flint, 185 cm, highly susceptible to TLB
Champawat Local	Medium, yellow flint, 200 cm, used both for green years and grain
Chauda Local	Medium, yellow flint, 210 cm, used both for green years and grain
Kankutia Local (White)	Late, white flint, 235 cm, used both for green years and grain
Kankutia Local (Yellow)	Late, yellow flint, 240 cm, used both for green years and grain
Dobans Local	Medium, yellow flint, 220 cm, used both for green years and grain
Hanol Local	Medium, yellow flint, 195 cm, used both for green years and grain
Dhanpau Local (Yellow)	Medium, yellow flint, 220 cm, used both for green years and grain
Dhanpau Local (White)	Late, white flint, 225 cm, used both for green years and grain
Tehra Local	Medium, orangish-yellow flint, 200 cm, used both for green years and grain
Kwanu Local (Yellow)	Late, yellow flint, 280 cm, good ear size, bolder seed, used both for green years and grain
Kwanu Local (Orange)	Late, orange flint, 260 cm, good yield, used both for green years and grain
Kwanu Local (Orangish yellow)	Late, orangish-yellow flint, good yield, 265 cm, used both for green years and grain

Coverage of cultivars

In Uttarakhand, around 63 per cent maize area is in the hills where predominantly improved OPVs and local cultivars are grown. The remaining 37 per cent area lies in the plains is covered mainly by hybrids with OPVs and local cultivars being grown only in specific pockets for household consumption. Though precise and reliable figures are difficult to obtain as maize seed market in the state is very diffused, the total area under hybrids may be estimated at about 25 per cent of the total maize area in the state. The hybrids in cultivation are mostly from private seed sector and are both single- and 3-way crosses. Mostly different hybrids are preferred for *Kharif* and *zaid* seasons. A small demand exists for sweet corn hybrids also. The reason for smaller presence of public sector hybrids despite availability of good public bred hybrids is the lack of expertise, more so, the reluctance on part of government seed agencies to undertake maize hybrid seed production.

Cultivars released for state by AICRP

A total of 41 cultivars are recommended for cultivation in the state of Uttarakhand. These include 19 single-cross hybrids, 4 double-cross hybrids and 18 composites. Of the 41 cultivars, three are sweetcorn cultivars, one is popcorn and the rest 37 are normal maize cultivars. The details of the cultivars are given in Table 3.

Table 3. Maize cultivars recommended for cultivation in the state.

Type	Cultivar	Year	Release by	Developed by	Characteristics
Double-cross hybrid	VL Makka 54	1962	CVRC	ICAR-VPKAS, Almora	Tall, thick, sturdy stem; yellow, semi-flint grains; maturity 110-120 days; yield 70-80 q/ha
Double-cross hybrid	VL Makka 42	1962 1988	SVRC CVRC	ICAR-VPKAS, Almora	Medium, healthy stem; bright orange flint grains; maturity 85-90 days; yield 35-40 q/ha
Composite	VL Makka 1	1973	SVRC	ICAR-VPKAS, Almora	Medium in height, light green narrow leaves, orange flintish grains; maturity 100-105 days; yield 30-35 q/ha
Composite	Protina	1971	CVRC	GBPUAT, Pantnagar	Medium; yield 40 q/ha; grains with 11% protein & 40% lysine
Composite	Tarun	1977	CVRC	GBPUAT, Pantnagar	Medium height; early; orange yellow flint kernel; moderately tolerant to foliar diseases and insect pests; yield 40-45 q/ha
Composite	Navin	1979	SVRC	GBPUAT, Pantnagar	Medium tall; early; densely branched tassel; moderately tolerant; yield 40-45 q/ha.
Double-cross hybrid	Him 128	1979	CVRC	ICAR-VPKAS, Almora	Tall; sturdy stem; white semi-flint grains; maturity 110-120 days; yield 70-80 q/ha
Composite	Shweta	1980	SVRC	GBPUAT, Pantnagar	Medium tall; early; sparse tassel; white bold semi-flint kernel; yield 45-50 q/ha.
Composite	VL Makka 16	1980	SVRC	ICAR-VPKAS, Almora	Medium stout stem; dark green leaves; yellowish orange flint grains; maturity 95-100 days; yield 35-40 q/ha.

Composite	VL Amber popcorn	1981	SVRC	ICAR-VPKAS, Almora	Small; hard, bright orange flint grain; maturity 100-105 days; tolerant to TLB; excellent popping quality; yield 20-25 q/ha.
Composite	Kanchan	1982	SVRC	GBPUAT, Pantnagar	Medium height; extra-early; yellow flint kernel; yield 30-35 q/ha.
Composite	D-765	1984	CVRC	GBPUAT, Pantnagar	Medium short; early; sparse foliage; tolerant to lodging and moisture stress; yellow semi-flint kernel; yield 30-35 q/ha.
Composite	VL Makka 41	1986	SVRC	ICAR-VPKAS, Almora	Short healthy stalk; medium green leaves; yellow flint round grains; maturity 80-85 days; yield 30-35 q/ha.
Composite	Surya	1988	CVRC	GBPUAT, Pantnagar	Early; yellow flint kernel; Tolerance to stalk rot and N responsive.
Composite	VL Makka 88	1988	CVRC	ICAR-VPKAS, Almora	Medium healthy stalks; medium dark green leaves; orangish yellow flint grains; maturity 85-90 days; yield 35-40 q /ha.
Double top cross hybrid	Him 129	1997	CVRC	ICAR-VPKAS, Almora	Medium in height; medium dark green leaves; yellow flint medium grains; maturity 85-90 days; yield 40-45 q /ha.
Single-cross hybrid	Vivek Hybrid 5	1999	SVRC	ICAR-VPKAS, Almora	Yellow; semi-flint; flat bold grains; maturity 85-90 days; yield 45-50 q /ha.
Composite	Gaurav	1999	CVRC	GBPUAT, Pantnagar	Early; yellow flint kernel; Resistance to major foliar diseases, tolerance to SB
Single-cross hybrid	Vivek Hybrid 9	2000	CVRC	ICAR-VPKAS, Almora	Yellow semi-flint grain; maturity 85-90 days; tolerant to TLB and MLB; yield 50-55 q /ha.
Composite	Vivek Sankul Makka 11	2003	SVRC	ICAR-VPKAS, Almora	Orange-yellow flint grains; maturity 95-100 days; tolerant to TLB; yield 40-45 q/ha.

Single-cross hybrid	Vivek Maize Hybrid 15	2004	CVRC	ICAR-VPKAS, Almora	Yellow flint grain; maturity 85-90 days; yield 45-50 q/ha.
Composite	VL Babycorn 1	2004	CVRC	ICAR-VPKAS, Almora	Bright yellow medium flat grain; maturity 90-95 days; tolerant to TLB; yield 12-13 q/ha (babycorn)
Single-cross hybrid	Pusa Extra Early Hybrid Makka 5	2004	CVRC	ICAR-IARI, New Delhi	Yellow-orange flint grain; extra early maturity; tolerant to TLB, MLB and ESR; yield 50 q/ha.
Composite	Win Orange Sweet Corn	2005	CVRC	IIMR WNC, Hyderabad	Medium maturity; tolerant to TLB, MLB and ESR; yield 65 q/ha.
Single-cross hybrid	Vivek Maize Hybrid 21	2006	CVRC	ICAR-VPKAS, Almora	Yellow semi-flint grain; maturity 85-90 days; yield 45-50 q /ha.
Single-cross hybrid	Vivek Maize Hybrid 23	2006	SVRC	ICAR-VPKAS, Almora	Yellow flint grain; maturity 90-95 days; moderately tolerant to TLB; yield 45-50 q/ha.
Single-cross hybrid	Pratap Makka 4	2006	CVRC	MPUA&T, Udaipur	White semi-flint grain; medium maturity; moderately resistant to stem borer; yield 45 q/ha.
Single-cross hybrid	Vivek Maize Hybrid 25	2007	CVRC	ICAR-VPKAS, Almora	Yellow semi-dent grain; maturity 85-90 days; tolerant to TLB; yield 50-55 q/ha.
Single-cross hybrid	Vivek QPM 9	2007 2008	SVRC CVRC	ICAR-VPKAS, Almora	Yellow semi-flint grain; Tryptophan 0.83%; Lysine 4.19%; Fe 37ppm; Zn 29 ppm; maturity 85-90 days; tolerant to TLB and MLB; yield 50-55 q/ha.
Single-cross hybrid	Pant Sankar Makka 1	2007	SVRC	GBPUAT, Pantnagar	Early; medium plant height; semi-flint yellow kernel
Composite	Vivek Sankul Makka 31	2008	SVRC	ICAR-VPKAS, Almora	Yellow, semi-flint grain; maturity 85-90 days; yield 35-40 q/ha.
Single-cross hybrid	Vivek Maize Hybrid 33	2008	CVRC	ICAR-VPKAS, Almora	Yellow dent grain; maturity 85-90 days; yield 55-60 q /ha.

Composite	Bajaura Makka	2008	CVRC	CSKHPKV, Bajaura	Orange flint grain; medium maturity; resistant to TLB and MLB; yield 47 q/ha.
Composite	Vivek Sankul Makka 35	2009	CVRC	ICAR-VPKAS, Almora	Yellow semi-flint grain; maturity 85-90 days; yield 40-45 q /ha.
Composite	Pratap Kanchan 2	2009	CVRC	MPUA and T, Udaipur	Yellow semi-flint grain; extra-early maturity; moderately resistant to TLB, MLB, BLSB, ESR and PFSR; yield 56 q /ha.
Single-cross hybrid	HSC 1	2010	CVRC	CCSHAU, Karnal	Light yellow semi-dent grain; medium maturity; resistant to MLB and CR; green ear yield 120 q /ha.
Single-cross hybrid	Vivek Maize Hybrid 39	2012	CVRC	ICAR-VPKAS, Almora	Yellow semi-flint flat grain; maturity 85-90 days; tolerant to TLB; 55-60 q /ha.
Single-cross hybrid	Vivek Maize Hybrid 45	2013	CVRC	ICAR-VPKAS, Almora	Yellow semi-flint flat grain; maturity 85-90 days; tolerant to TLB and MLB; yield 50-55 q /ha..
Single-cross hybrid	Vivek Maize Hybrid 47	2014	CVRC	ICAR-VPKAS, Almora	Yellow semi-flint; maturity 90-95 days; moderately resistant TLB; MLB; CR and resistant to BSDM and PFSR; yield 50-55 q /ha.
Single-cross hybrid	Vivek Maize Hybrid 53	2014	CVRC	ICAR-VPKAS, Almora	Yellow semi-flint grain; maturity 85-90 days; moderately resistant to TLB; MLB; CR; BSDM and PFSR; yield 50-60 q /ha.
Single-cross hybrid	Pant Sankar Makka 2	2015	SVRC	GBPUAT, Pantnagar	Early; medium plant height; flint yellow kernel and fairly field tolerance to insect pest except sporadic incidence of SB.
Single-cross hybrid	Pant Sankar Makka 4	2015	SVRC	GBPUAT, Pantnagar	Early; medium plant height; flint yellow kernel and fairly field tolerance to insect pest and diseases.
Single-cross hybrid	CMVL Sweetcorn 1	2016	CVRC	ICAR-VPKAS, Almora	Sturdy plant with medium height; yellow dent flat grains; maturity 90-95 days; yield 100-110 q/ha (dehusked green cob); TSS 16%

VL Amber pop corn 1　Pant Sankar Makka 4
CMVL sweet corn 1　Vivek QPM 9

Maize Seed Production in the State

In the state of Uttarakhand, apart from ICAR-VPKAS, Almora and GBPUA and T, Pantnagar, maize seed production is also undertaken by Uttarakhand Seed and Tarai Development Corporation (UAS and TDC). During 2001-2015, breeder seed indent of 4.37 q and 12.16 q was received by ICAR-VPKAS, Almora for composite varieties and parental inbreds of hybrids, respectively. A total of 9.09 q breeder seed of composites and 29.52 q breeder seed of parental inbreds was produced against the indent. Apart from this, 28.30 q seed of the composites was produced (Table 4). A total of 69.55 q seed of different hybrids was also produced by ICAR-VPKAS, Almora (Table 5).

Table 4. Maize seed production at ICAR-VPKAS Almora during 2001-2015.

Name of cultivar / inbred	BS indent received (q)	Quantity of BS produced (q)	Quantity of TL seed produced (q)
Composites			
VL Makka 16	0.55	2.50	1.24
VL Makka 88	1.09	0.00	3.38
VL Amber popcorn	0.60	3.10	6.88
Vivek Makka 11	0.33	9.05	6.54
VL Babycorn 1	0.00	0.90	1.75
Vivek Sankul Makka 31	1.80	23.24	5.55
Vivek Sankul Makka 35	0.00	0.30	2.79
Vivek Sankul Makka 37	-	-	0.17
Total	**4.37**	**39.09**	**28.30**

Parental lines			
CM 128	1.11	0.39	-
CM 129	1.12	0.23	-
CM 141	0.31	0.63	-
CM 212	3.48	6.71	-
CM 145	0.62	0.73	-
CM 502	0.14	0.38	-
CM 153	1.30	1.82	-
CM 152	0.12	0.50	-
VQL 1	1.10	13.00	-
VQL 2	1.07	3.40	-
V 341	0.05	0.05	-
V 351	0.10	0.10	-
V 372	1.64	1.55	-
V 373	0.00	0.03	-
Total	**12.16**	**29.52**	-

Table 5. Production of maize hybrid seed at ICAR-VPKAS Almora during 2001 2015.

Name of hybrid	Quantity of TL seed produced (q)
Him 129	29.95
Vivek Maize Hybrid 9	3.25
Vivek Maize Hybrid 15	1.55
Vivek Maize Hybrid 17	1.25
Vivek Maize Hybrid 25	2.50
Vivek QPM 9	24.25
Vivek Maize Hybrid 45	6.80
Total	**69.55**

During 2000-2016, breeder seed indent of 80.65q and 39.78q was received by GBPUA and T, Pantnagar for composites and parental inbreds , respectively. Against the indent, a total of 197.78q breeder seed of composites and 79.72 q breeder seed of parental inbreds was produced (Table 6).

Table 6. Maize Breeder Seed Production at GBPUAT, Pantnagar during 2000-2016.

Name of cultivar / inbred	BS indent received (q)	Quantity produced (q)
Composites		
Kanchan	21.91	37.74
Surya	14.38	40.50
Gaurav	16.72	35.37
Sweta	4.0	18.53
Amar	13.94	26.30
Navin	1.40	10.55
Pragati	5.84	22.49
Pant Sankul Makka 3	2.46	6.30
Total	**80.65**	**197.78**
Parental lines		
CM 300	9.25	15.89
CM 400	12.98	32.43
CM 600	17.55	31.4
Total	**39.78**	**79.72**

In addition to this, a total of 8222.54 q Certified Seed and 150.64 q Foundation Seed of various composite varieties has been produced by UAS and TDC during 2000-2016 (Table 7).

Table 7. Maize seed production by Uttaranchal Seed and Tarai Development Corporation during 2000-2016.

Year	Crop / Variety	Class of Seed		Total
		C/S	F/S	
2008	Kanchan	109.55	-	109.55
	Sweta	39.41	-	39.41
	Vivek Sankul Makka 11	-	0.07	0.07
2009	Kanchan	119.98	-	119.98
	Vivek Sankul Makka 31	10.01	42.98	52.99
	PEHM 2	1319.41	-	1319.41
	Local	5.74	-	5.74
2010	Vivek Sankul Makka 31	5.60	4.41	10.01
	PEHM 2	918.42	-	918.42
	Local	43.26	-	43.26
2011	Kanchan	-	72.66	72.66
	PEHM 2	796.16	-	796.16
2012	Vivek Sankul Makka 31	19.32	-	19.32
	PEHM 2	1344.58	-	1344.58
2013	Vivek Sankul Makka 31	28.00	7.21	35.21
	PEHM 2 / PEHM 5	1665.15	-	1665.15
2014	Vivek Sankul Makka 31	2.52	0.00	2.52
	Kanchan	-	4.41	4.41
	PEHM 2 / PEHM 5	1282.60	-	1282.60
2015	Vivek Sankul Makka 31	109.70	14.40	124.10
	Local	365.40	-	365.40
2016	Vivek Sankul Makka 31	37.73	4.50	42.23
	Grand Total	**8222.54**	**150.64**	**8373.18**

Among the varieties developed and released for the state, Vivek QPM 9 has been the most prominent in the recent times. Vivek QPM 9 is the QPM (Quality Protein Maize) version of popular hybrid Vivek Maize Hybrid 9 developed using Marker-assisted Selection. The hybrid is released for cultivation in the states of Jammu & Kashmir, Himachal Pradesh, Uttarakhand, NE states, Maharashtra, Karnataka, (former) Andhra Pradesh and Tamilnadu. It possesses high tryptophan (0.83%) and lysine (4.19%) and is nutritionally twice as superior as normal corn. Ever since its release in 2008, the hybrid is estimated to have covered 50 th ha area across the country (on assumption of complete utilization of breeder seed for hybrid seed production through the standard seed production chain). The hybrid has been commercialized with six private sector seed companies that produce and market the hybrid in different states of the country.

Farmer Participatory Hybrid Seed Production

For ensuring sustained supply of seed of VL maize hybrid varieties at affordable prices to farmers, ICAR-VPKAS Almora has initiated Farmer Participatory Hybrid Seed Production at Vikasnagar in district Dehradun. In view of the demand for Vivek Maize Hybrid 45 in the Jaunsar tribal region of district Dehradun, National Seed Corporation has initiated seed production of the hybrid at their Jawalgera Farm (Raichur , Karnataka). Efforts have also been made to link hybrid maize seed producers to the seed agencies to ensure procurement of the produced seed.

Changes in cropping pattern over the years

Maize crop is generally used for both food and fodder purpose in Uttarakhand hills. Earlier maize was generally grown as an intercrop or mixed crop with radish, rajmash, lablab bean and french bean, besides others. However, the cultivated area under maize is progressively declining in Uttarakhand hills due to wild animal menace. And as a result, maize is being replaced by other crops which are relatively less vulnerable to wild animal damage even if they are less remunerative than to maize.

In the plains of Uttarakhand, maize-wheat has been a prominent cropping system. However, in due course of time cropping patterns have changed mainly because of development of industries, market and market linked road infrastructure and profitability per unit area and time. Changed cropping patterns include deviation in sowing of maize to adjust the succeeding vegetable crops. In addition to traditional maize-wheat cropping system, maize (May sowing)-tomato-potato, maize (May sowing)-tomato-onion have also been adopted.

Production technologies recommended and package of practices for maize cultivation in hills and plains

Owing to differences in agro-ecology, the cultivars as well as maize production technology vary considerably for the hills and the plains of the state. For example, while only extra-early and early cultivars fit into the prevalent cropping systems in the hills; early, medium as well as late maturity cultivars fit into different cropping systems in the plains. Moreover, in hills maize is cultivated under

rainfed conditions, whereas in the plains maize cultivation is mostly irrigated. The package of practices for maize cultivation in the hills and the plains are given in Tables 8 and 9, respectively.

Table 8. Package of practices for Maize cultivation in Uttarakhand hills.

Practices	Early and extra early maturity maize	Sweet corn	Baby corn	Popcorn
Sowing time	Third week of June with pre-monsoon rain and before onset of monsoon			
Seed rate	20 kg	15 kg	40 kg	15 kg
Planting density	50 cm × 20 cm	50 cm×20 cm	50 cm×10 cm	50cm×20 cm
Weed management	Pre-emergence application of Atrazine @ of 1.0 kg a.i/ha in 600 litre water or Alachlor @ 2 kg a.i/ha			
Earthing-up	Earthing-up must be provided during knee high stage			
Water management	Must be provided life saving irrigation as per requirement			
Nutrient management	200-60-80 kg/ha N-P_2O_5-K_2ONitrogen is to be applied in three equal splits at basal, knee high and tasseling stage. Full dose of phosphorus and potash is to be applied as basal			

Table 9. Package of practices for Maize cultivation in Uttarakhand plains.

Operation	Normal maize	Sweet corn	Baby corn	Pop corn
Sowing time	*Kharif*: Mid June to July first week*Rabi*: 15 October to 15 NovemberSpring: Third to last week of February			
Seed rate/ha	18-20 kg	10-12 kg	40-45 kg	12-14 kg
Planting density	75 cm × 20 cm 60 x 25 cm	75 cm×20 cm 60 x 25 cm	45cm×20 cm	75cm×20 cm 60 x 25 cm
Weed management	Pre-emergence application of Atrazine @ of 1.0 kg a.i/ha in 600 litre water or Pendimethalin 30 EC@ 1.0 kg ai/ ha (i.e.3.33 L/ha)			
Earthing-up	Earthing-up must be provided during knee high stage			

Water management	Irrigation as and when required during *Kharif* however regular irrigation during winter and spring		
Nutrient management	Early: 100 -120 kg N + 60kg P_2O_5 + 40kg K_2O/ha Medium and late: 150 – 180 kg N+60kg P_2O_5+40kg K_2O/ha.Nitrogen is applied in five splits at 20% basal, 25% at 4- leaf stage, 30% at 8- leaf stage, 20% at tassel emergence and remaining 5% at early grain filling stage. Full dose of phosphorus and potash is to be applied as basal. Zinc sulphate @ 20kg/ha, if required	Baby corn 180 kg N, 60 kg P2O5 and 40 kg K_2O/ha. Zinc sulphate @ 20kg/ha, if required	Early: 100 - 120 kg N + 60kg P_2O_5 + 40kgK_2O/ha. Medium and late: 150 – 180 kg N+ 60kg P_2O_5 + 40kg K_2O/ ha. Nitrogen is applied in five splits at 20% basal, 25% at 4- leaf stage, 30% at 8- leaf stage, 20% at tassel emergence and remaining 5% at early grain filling stage. Full dose of phosphorus and potash is to be applied as basal. Zinc sulphate @ 20kg/ha, if required

Changes in disease and pest scenario over the years

In the hills of Uttarakhand maize crop is affected mainly by turcicum leaf blight (TLB), banded leaf and sheath blight (BLSB), maydis leaf blight (MLB) and common rust (CR). The hill ecosystem offers highly conducive weather conditions (moderate temperature and high humidity) for TLB development and for that reason ICAR-VPKAS Almora is known as a hot-spot for TLB screening. TLB incidence has been observed to increase over the years and under severe conditions yield losses up to 30-40 per cent may occur. The incidence of maydis leaf blight, in general, remains lower than TLB. However, constant rise in temperature during the cropping season is causing increase in the incidence of maydis compared to the past years. The incidence of common rust is sporadic and occurs only in years with higher number of rainy days. The disease remains largely restricted to the susceptible entries, in which the incidence may reach as high as 30-40%. Such a situation was observed in *Kharif* 2016 at ICAR-VPKAS.

However, the disease that has assumed major significance in the last ten years is banded leaf and sheath blight disease. The incidence of BLSB is rising at an alarming rate primarily due to availability of abundant pathogen inoculum in the soil, which is building further up constantly. The situation is aggravated by the fact that there are no known sources of resistance to BLSB. BLSB, therefore, is being seen as a major threat to maize cultivation in the hill regions of the state. To counter this threat, rigorous screening of maize germplasm has been initiated at ICAR-VPKAS for identification of tolerant genotypes with the aim to dissecting QTLs governing the trait followed by their incorporation into popular cultivars through aid of molecular tools. BLSB and Erwinia stalk rot are common diseases of maize in the plains. However, in recent years, incidence of Curvularia leaf spot and Carbonum leaf spot has also attracted attention of scientists as well as farmers.

The insect pest scenario in maize in NW Himalayan hills is distinct from other maize growing areas in the country with the pink borer (*Sesamia inferens*) being the predominant pest causing significant damage. Apart from pink borer, sporadic incidences of blister beetles and aphids are also observed. The incidence of aphids in maize has been observed to be increasing during the recent years, and so is of their natural enemies (syrphids). One spray of an effective chemical like acetamiprid or imidacloprid is sometimes required to keep the pest under economic threshold level. Stem borer (*Chilo* sp.) has long been a serious pest of maize grown in lower and foothills of the state but its incidence in the mid-hills (at ICAR-VPKAS Experimental Farm, Hawalbagh) was observed for the first time during 2013 and is increasing year after year ever since. Cut worms (*Agrotis* sp.) affect the crop during seedling stage and their incidence has remained largely unchanged over the years. Infestation of helicoverpa (a polyphagous insect pest prevalent in hills during Feb-May) was found high in February planted maize. At ICAR- VPKAS, research on identification and use of pheromones of pink borer, *Sesamia inferens* is under progress. Stem borer and shoot-fly are major insect-pests of maize in the plains as well. The infestation is particularly heavy in May and early June planted crop. The infestation declines mid-June onwards and becomes negligible after onset of monsoon. Stem borer and shoot-fly damage winter crop also especially when the temperature higher than the usual. These insects also cause severe damage to summer/spring crop during early stages.

Management of pest and diseases

The diagnosis, identification and management of major diseases and insects are given in Table 9 and 10.

Table 9. Important diseases of maize: symptoms; identification and management.

Disease	Causal organism	Symptoms	Management
Turcicum leaf blight	*Exserohilum turcicum* Teleomorph: *Setosphaeria turcica* (syn. *Trichometasphaeria turcica)*	An early symptom is the easily recognized, slightly oval, water-soaked, small spots produced on the leaves. These grow into elongated, spindle-shaped necrotic lesions . They may appear first on lower leaves and increase in number as the plant develops, and can lead to complete burning of the foliage.	• Use of resistant varieties • Sowing before third week of June helps in avoiding more damage • Spray of mancozeb or zenab @ 2gm /litre of water
Banded leaf and sheath blight (BLSB)	*Rhizoctonia solani* Telomorph: (*Thanatephorus Cucumeris*)	Symptoms are characteristic concentric spots that cover large areas of infected leaves and husk. The main damage in the humid tropics is a brownish rotting of ears, which show conspicuous, light brown, cottony mycelium with small, round, black sclerotia	• Sowing should be done on ridges to avoid high moisture • Stripping of lower 2-3 leaf along with their sheath considerably lowers incidence • Seed treatment with Trichoderma sp. @ 4gm /kg seed • Foliar application of azoxystrobin 23 SC @ 0.1%
Maydis leaf blight	*Helimnthosporium maydis* (Telomorph: *Cochliobolus heterostrophus*)	Young lesions are small and diamond shaped, as they mature they elongate. Growth is limited by adjacent veins, so final lesion shape is rectangular	• Use of resistant varieties • Spray of mancozeb or zenab @ 2gm/litre of water

		and 2 to 3 cm long. Lesions may coalesce, producing a complete burning of large areas of the leaves .MLB is prevalent in hot, humid, maize-growing areas. The fungus requires slightly higher temperatures for infection than *E. turcicum*, however, both species are often found on the same plant.	
Common rust	*Puccinia sorghi*	It may be recognized by small, elongate, powdery pustules over both surfaces of the leaves . Pustules are dark brown in early stages of infection, later, the epidermis is ruptured and the lesions turn black as the plant matures.	• Use of early maturing varieties • Spray of mancozeb @ 2-4 gm/litre of water and repeated at 15-day intervals based on severity of disease
Erwinia stalk rot	*Erwinia chrysanthemi* pv. *zeae*	Symptoms are first observed in mid season when plants suddenly wilt and lodge. Most distinctive symptom of *Erwinia* infection is the foul smell of diseased plants.	• Incorporation of debris of crop residue to avoid flooding • Avoid high nitrogen application with optimum doses of phosphorous and potassium

Table 10. Major insect-pests of maize in Uttarakhand and their management.

Insects	Identification	Damage	Management
Blister beetles (*Mylabris phalerata*)	Robust black insect with reddish or orange bands	Feeds on the pollen in the tassel and also cut and feed the silk	• Hand picking (with gloved hand) and killing by immersion in water • Spray of Deltamethrin 1 Ml or Imidacloprid 0.5 ml/ l.
Aphids (*Rhopalosiphum maidis*)	Small ash-green insects in colonies especially in growing tips	Nymphs and adults suck sap causing stunted growth of plant and curled leaves.	• Syrphid grubs and coccinellids prey upon the insect. • In severe cases; spray Imidacloprid 0.25 Ml/L
Pink borer; *Sesamia inferens*	Larva pinkish brown in colour and seen inside tunneled stem especially in the early stage of crop. The adult moth is straw coloured with white hind wings.	Larva bores inside the stem causing drying of main shoot	• Weeding and clean cultivation reduces incidence as it is a polyphagous pest. • Spray Chlorpyriphos 2 Ml or Chlorantranili-prole 0.3 ml/l.

Cutworms (*Agrotis* sp.)	Moths lay eggs in the soil near crop plants. Larva dark black in colour and attacks the small plants.	Plants cut off at the ground level. Feeds at night and usually not found during day	• Moths get trapped in light traps. Collect and kill the hiding larva in the soil near fresh attack. • Batain (*Melia azedarach*) seed powder repels the larva. • Drench with Chlorpyriphos 3 ml/l.
Stem borer (*Chilo partellus*)	Yellowish brown larva with brownish black head. Adult is straw coloured and medium in size.	Withering of main shoot causing 'dead heart'. Small larva feeds on tender folded leaves causing typical 'shot hole' symptom. The insect bores holes in the stem.	• Release egg parasitoid *Trichogramma chilonis* @250000/ha coinciding egg laying period. • Larval parasitoid *Cotesia flavipes*@5000/ha is also reported effective • Spray of Indoxacarb 0.5 Ml /L or Chlorantranili-prole 0.3 ml/l.
Shoot-fly (*Atherigona orientalis*)	Small grayish coloured flies. Eggs are white and flat. Grub period is 8-10 days and pupates in soil.	Maggot enters through the leaf sheath and feeds on the growing point causing 'dead heart'	• Seed treatment with Chlorpyriphos 4 ml/ kg of seed. • Fish meal trap @ 12/ ha at the seedling stage is very effective. • Spray of Chlorpyriphos or Malathion @ 2 ml/l.
Weevil, *Sitophilus* sp.	Storage pest; grub and adults feed on maize grains.	Adults make holes on whole grains and feed on them and make into powder. Grubs feed by residing inside the grains.	• Tolerant varieties should be used. • Storing maize under modified atmospheric conditions with 15% CO2 is reported effective. • Adding sweet flag powder, neem seed kernel powder @1 and 5% during storage. • Storage bags can be treated with 5% Malathion solution and dried before storage.

Extension: FLDs; yield gaps and exhibitions/farmer fairs

Extension of improved maize varieties, agronomical practices and crop protection measures in weed control through front line demonstration has been integral part of the AICRP activities for the popularization of improved maize technologies. In addition, strip trials of pre-released hybrids on farmer's fields are also being conducted to obtain feedback about the pipe-line hybrids.

Analysis of yield gap in maize

In trials comprising maize hybrids developed by ICAR-VPKAS and the local maize cultivar Dhanpau Local conducted at ICAR-VPKAS, Almora, best performance over the local cultivar was exhibited by VMH 47 followed by VMH 45. In the demonstrations conducted in the Dhanpau-Lakhwad village cluster also, VMH 45 exhibited 46.6 per cent superiority over the local cultivar. The local cultivar also yielded 79.4 per cent higher under improved cultivation practices in the trial conducted at ICAR-VPKAS, Almora. The results indicated that a wide yield gap existed between the local practices followed by the tribal farmers and the improved technology, which can be potentially be filled by introduction of maize hybrids and improved cultivation technology.

Technology demonstrations

Demonstrations of ICAR-VPKAS maize hybrids were organized in Dhanpau-Lakhwad and Kwanu tribal cluster in the Jaunsar tribal region of district Dehradun. At crop maturity, Field Days were organized to demonstrate the superiority of hybrid varieties over the local cultivar.

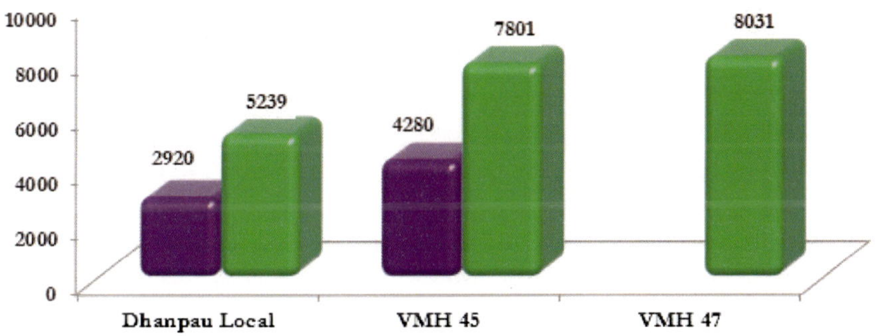

Fig. 5. Performance of maize cultivars (kg/ha grain yield).

Maize Hybrid Seed Production demonstration

Hybrid seed production demonstration of Vivek Maize Hybrid 45 was organized at Yamuna Khadar village (Vikasnagar). Yamuna Khadar is situated in the plains area adjoining Dhanpau-Lakhwad cluster.

Single cross maize hybrids demonstration

The objective behind organizing the demonstration, besides acquainting farmers with maize hybrid seed production methodology, was to attract village youth for taking up maize hybrid seed production on a commercial scale, thereby, linking the entire activity with entrepreneurship development.

Rainfall in Uttrakhand

The rainfall data clearly indicates that South-west monsoon during *Kharif* season in general and during the month of August in particular is continuously decreasing for the last two decades. This period coincides with the reproductive phase of maize crop in Uttarakhand hills. The *Rabi* season in Uttarakhand hills is mid October-May, and during *Rabi* season, the month of January, February, March and May receive good rainfall compared to other months. The rainfall during the month of May is continuously increasing and during the months of November and December is continuously decreasing.

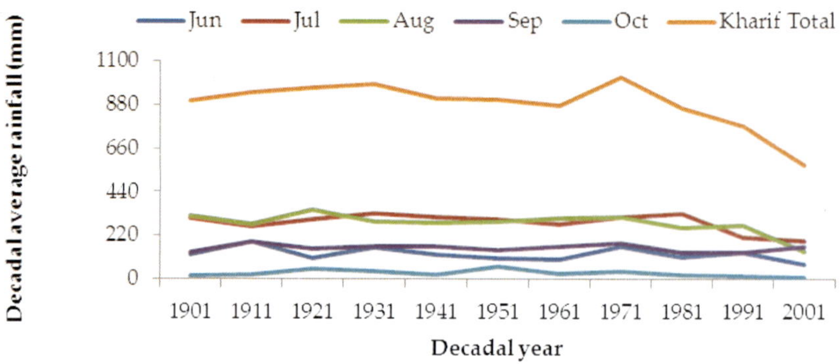

Fig. 6. Month-wise decadal average *Kharif* season rainfall (Only two years' data is included in the decade of 2001) of Uttarakhand (Source: Fig 6 and 7. http://www.indiawaterportal.org/met_data/ ; http://www.euttaranchal.com/uttarakhand/districts-in-uttarakhand.

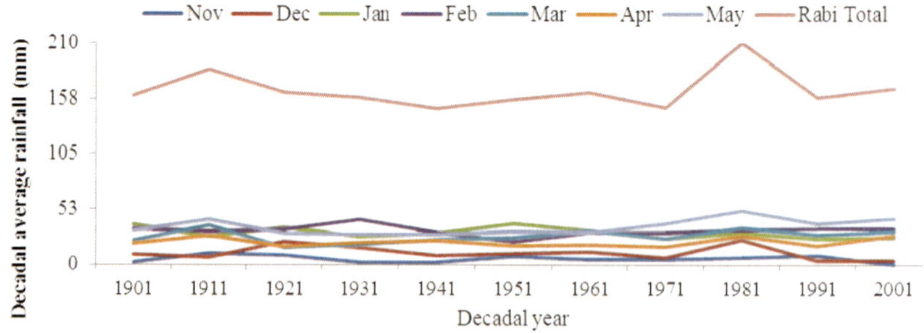

Fig. 7. Month-wise decadal average *Rabi* season rainfall (Only two years' data is included in the decade of 2001) of Uttarakhand.

Key R&D challenges and their probable solutions in maize for the state

Development of high yielding normal and specialty corn hybrids with competitive edge over private sector hybrids: About 25 per cent of the total maize area in the state is estimated to be under hybrid varieties (mainly single- and three-way crosses) and this area is mostly under private sector hybrids for two reasons: lack of availability seed of public sector hybrids and sub-par performance of public sector hybrids compared to those of private sector. While ensuring availability of seed is an issue that involves government seed agencies and changes at the policy level, developing hybrids that match up to or are better than the private sector hybrids calls for thorough reorganization of the ongoing public sector maize breeding programmes, which must essentially include

Development of new heterotic pools/reinvigoration of existing heterotic pools: Although new maize varieties are being developed and released for the state of Uttarakhand with fairly regular frequency, systematic utilization of existing heterotic pools is largely lacking, which is bound to be reflected in the performance of upcoming hybrids in the form of diminishing increments. In order to check this decline, the existing heterotic pools need to be systematically invigorated with infusion of robust heterotic stocks to ensure continuous supply of elite inbreds with defined heterotic pattern.

Rapid generation of inbred lines: Rationalizing breeding programmes is an imperative in the present times of resources constraint. Convention breeding methods to develop inbreds and transfer traits consumes enormous time and resources. Adoption of methods to cut short the time involved in the above procedures can lead to significant savings for building up/strengthening non-conventional breeding infrastructure. Use of doubled haploid technology in combination with molecular tools is an effective approach for rapid generation of inbreds with desired combination of traits.

Emerging biotic and abiotic stress: The rising incidence of previously insignificant diseases and pests and emergence of their newer forms pose a serious threat to maize cultivation in the state. Unavailability of sources of resistance/tolerance to some of them increases the complexity involved in tackling the threat. The situation is aggravated by changes in climatic pattern that adversely affect crop performance. A multi-pronged approach is need to deal with the collective threat of biotic and abiotic stresses which would include (i) Expeditious utilization of available sources of resistance (ii) rigorous screening of available germplasm at identified hot-spots in cases where sources of resistance are not available (3) exposure of breeding material to diverse climatic regimes to broaden their adaptability (through shuttle breeding and other means).

Value-addition/ biofortification for nutritional security and enhancing returns: Value-added/ biofortifed varieties of maize are important not only from nutritional security viewpoint but can also benefit industry enormously. QPM for example, can reduce the dependence of poultry and cattle feed industry on synthetic lysine. Likewise, provitamin A and Fe/Zn enriched maize can reduce expenditure on external sourcing of these supplements. Procurement of value-added varieties by the industry at premium price will serve to boost maize growers leading to overall growth in maize cultivation in the state.

Economic viability of hybrid seed production: Sub-par performance of public sector hybrids is just one of the reasons for their conspicuous absence of from the national scene, the chief problem lies in the failure of the parental inbreds to perform well in large scale hybrid seed production programmes while, parental inbreds of private sector hybrids are subject to rigorous testing with regard to their commercial seed production ability, this aspect remains largely ignored in case of public hybrids. Public hybrids often fail either on the count of male-female synchronization or fall short of the threshold economic yield. The imperative to validate the seed production ability of new hybrids before their recommendation for cultivation, therefore, needs to be built into the varietal release system at the state as well as national level.

Summary

Maize has been one of the important cereals in Uttarakhand. Diminishing returns from the crop owing chiefly to slow diffusion of improved agro-technologies, lack of organised market and maize promoting industries, inadequate policy support and damage by wild animals have been the major reasons for declining interest in maize cultivation. Establishment of starch, maize based food processing industries and poultry industry in the state in the recent years has led to creation of an in-house market for maize and has encouraged farmers to grow maize not only as a food commodity but also as an industrial raw material. The rapidly growing tourism industry and increasing demand of premium-priced specialty corn is expected to add further to expansion of maize in the state. Awareness generation and capacity building for adoption of scientific cultivation of the crop along with facilitating access to essential inputs will play an important role in transforming maize from sustenance farming to substantial farming in Uttarakhand. Ensuring availability of quality seed of maize hybrids at affordable prices to the farmers through active involvement of government seed agencies will be the key to sustained increase in productivity and production. Adequate policy support in terms of assured disposal of produce and tackling wildlife menace is expected to contribute significantly towards reclaiming lost area. Formulation of a comprehensive strategy integrating these approaches backed by strong policy support followed by concerted implementation holds promising prospects for enhancement of area, production and productivity of maize to the gap between demand and production of maize in Uttarakhand.

Future research and development strategy

Low productivity of maize in Uttarakhand is a realised fact. Rainfed cultivation during the *Kharif* season, lack of quality seeds and yield supporting inputs are amongst the important factors responsible for low productivity of maize in the state. Rainfed specific germplasm development and diversification, enrichment of existing heterotic pools and creation of new heterotic pools would be key strategies to ensure continuous availability of superior inbreds lines for hybrid development and would entail inter-institutional collaborations for exploitation of allelic diversity in development of heterotic hybrids. Area-specific maize production practices along with dynamic and profitable cropping system need to developed to increase resources utilization along with enhancing farm profitability. Strengthening of hybrid seed production especially in Uttarakhand hills is essentially required to ensure timely availability quality seed at affordable prices. To ensure nutritional security, development of biofortified hybrids with enhanced levels of provitamin A and Fe-Zn levels, in addition to lysine and tryptophan, will be a major future research area. Owing to the premium fetched by high starch and high oil maize hybrids in the industry, research requires to be initiated for development of specialised germplasm. A combination of doubled haploid technology and molecular breeding tools would be employed for expeditious accomplishment of the goals. To counter new races of insects and pathogens and the changed dynamics of major and minor biotic factors, strengthening of research along with continuous monitoring of emerging pathogens and insect pest races are among the important thrust areas and would involve development of integrated or eco-friendly management strategies along with utilization of wild alleles for strengthening and diversifying resistance sources. Similarly, to deal with existing as well emerging abiotic stresses due to climate change, the principal approach would involve exposure of breeding materials to diverse stresses for enhancing adaptation to a wider range of abiotic stresses. Maximized utilization of water and nutrient resources along with monitoring and enrichment of soil would also be an integral component of future research initiatives.

Acknowledgment

The authors express their gratitude to all the past maize workers at ICAR-VPKAS, Almora whose contributions find mention in the chapter. The authors also thank Department of Agriculture, Uttarakhand for providing APY data and UAS&TDC, Pantnagar for providing information related to FS/CS seed production in the state.

References

Anonymous. (2016). Crop Cutting Report. Directorate of Agriculture, Agriculture Department. Dehradun, Uttarakhand.

Annual Report 2013-14. (2014). Vivekananda Parvatiya Krishi Anusandhan Sansthan (ICAR), Almora– 263601, Uttarakhand.

Annual Report 2014-15. (2015). ICAR-Vivekananda Parvatiya Krishi Anusandhan Sansthan, Almora – 263601, Uttarakhand.

Annual Report 2015-16. (2016). ICAR-Vivekananda Parvatiya Krishi Anusandhan Sansthan, Almora – 263601, Uttarakhand.

Babu, R., Mani, V.P., Pandey, A.K., Pant, S.K., Singh, Rajesh., Kundu, S. and Gupta, H.S. (2004). Maize Research at VPKAS-An Overview. *Technical Bulletin 21 (1/2004)*. Vivekanand Parvatiya Krishi Anusandhan Sansthan, Almora– 263601 (Uttarakhand).

Bisht, A.S., Bhatnagar, A., Shukla, A. and Singh, V. (2015). Quantitative and qualitative implications of plant density and nitrogen management in quality protein maize (*Zea mays* L.). *Madras Agric J.* 102(4-6): 151-154.

Lala, K., Pal, M.S. and Bhatnagar A. (2013). Response of planting geometry and nitrogen on quality production of baby corn (*Zea mays* L.). *New Agriculturist*. 24(2): 141–146.

Directorate of Economics and Statistics. Government of Uttarakhand.

Mani, V.P., Singh, R., and Gupta, N.N. (2001). Development of open pollinated and hybrid varieties of maize for the hills of India, Pp. 31-37. In: Rajbhandari, N.P., Ransom, J.K., Adhikari, K., Palmer, A.F.E., (Eds.) Sustainable maize production system for Nepal. Proceedings of maize symposium. Kathmandu (Nepal), 3-5 December, 2001. Pp. 305.

Panyuli, A., Pal, M.S., Bhatnagar, A and Bisht, A.S. (2014). Effect of planting methods and irrigation schedules on growth, yield and nutrient content of sweet corn (Zea mays saccharata) in Tarai region of Uttarakhand. *International J Basic Applied Agric Res.* 12 (3): 332-338.

2

Maize Research and Development in Himachal Pradesh

S.K. Guleria, D.R.Thakur, S.D. Sharma, Rakesh Devlash
Gopal Katna, V.K. Rathee and Uttam Chandel

CSK, Himachal Pradesh Krishi Vishvavidyalya Hill Agricultural Research and Extension Centre, Bajaura, (Kullu) 175125 (H.P) India
**Corresponding Author's Email: skg0612@rediffmail.com*

Maize *(Zea mays L.)* is one of the most versatile crops having the highest genetic yield potential among cereals and wider adaptability under varied agro-climatic conditions rightly earning it the name of 'the queen of cereals'. Maize is cultivated in all agricultural areas around the world and is an economically significant crop for all human population. It is grown from 58^0 L north to 40^0 L south and extends from the sea level up to an altitude of above 3,800 m; covering regions with rainfall varying from 250 to 10,000 mm (Hallauer and Miranda, 1988). Globally, maize occupies an area of about 184.0 million ha with production of 1065 million tons and productivity of 55.2 q/ha (FAOSTAT, 2014). It is the third most important cereal crop and a major source of energy, protein and other nutrients for human and livestock in the world (Jompuk *et al.*, 2011). India is the second most important maize growing country in Asia, and now ranks as the world's fifth largest consumer of maize. It is an important cereal crop in India covering an area of 9.42 million ha with a production of 24.35 million tonnes and productivity 25.83 q/ha (Anonymous, 2015). Maize is cultivated in the country throughout the year mostly during rainy or *Kharif* season (June to September, 80%) followed by winter or *Rabi* (November to April,18%) and *spring* (February to May, 2%) for various purposes including grain, fodder, green cobs, sweet corn, baby corn and popcorn.

In Himachal Pradesh, maize is mainly grown in *Kharif* season under rained conditions. Predominately, it is crop of tropical and sub-tropical areas but can be cultivated successfully under temperate climatic conditions. One such example is its cultivation in temperate regions of Himachal Pradesh. Presently, it is cultivated in an area of 300 thousand hectares with production of 752.7 thousand tonnes and productivity of 25.09 q/ha. The trends from years 1966 to 2014-15 indicate that area under maize remained almost stagnate but there has been substantial increase in the production from 445.3 thousand tonnes to 752.7 thousand tonnes. This has been possible due to better input delivery system, adoption of improved production technologies, high yielding hybrids and spread of maize cultivation in new high productivity areas. Maize is grown in all the districts of the state except Lahaul Spiti and Kinnaur. The trends from 1972 to 2014 indicate that Kangra had maximum area under maize cultivation followed by Mandi, Una, Chamba and Bilaspur districts, whereas in case of production Mandi ranked on top with a production of 104.6 thousand tonnes followed by Kangra (89.4 thousand tonnes), Chamba (63.4 thousand tonnes) and Sirmour (58.2 thousand tonnes). Sirmour district of the state registered the highest productivity (23.47 q/ha).

Table 1. Area, Production and Productivity of Maize in different districts of H.P.

District		1972-75	1976-80	1981-85	1986-90	1991-95	1996-00	2001-05	2006-10	2011-14	Mean
Bilaspur	**A**	29	23	24	26	26	26	26	28	26	26
	P	39	34	37	39	45	46	50	53	52	44
	Y	2039	1556	1501	1516	1728	1767	1933	1900	2034	1775
Chamba	**A**	35	26	27	27	27	27	28	28	28	28
	P	51	51	45	62	77	70	66	65	83	63
	Y	2208	2048	1673	2268	2883	2588	2341	2317	2959	2365
Hamirpur	**A**	36	28	30	34	33	36	32	32	36	33
	P	42	46	47	55	57	54	59	57	60	53
	Y	1767	1748	1538	1625	1717	1695	1854	1801	1645	1710
Kangra	**A**	63	50	55	57	58	58	58	58	55	57
	P	72	78	91	87	96	94	90	101	97	89
	Y	1791	1623	1642	1519	1667	1626	1536	1732	1772	1656
Kullu	**A**	17	14	15	15	17	18	17	17	16	16
	P	26	27	33	35	34	43	41	42	60	38
	Y	2371	2103	2174	2234	2020	2461	2411	2531	3661	2441
Mandi	**A**	49	42	41	45	50	48	48	48	49	47
	P	62	73	78	91	121	129	131	129	128	105
	Y	1941	1831	1877	2013	2426	2696	2755	2706	2881	2347
Shimla	**A**	27	21	22	22	20	17	14	11	9	18
	P	38	36	39	48	41	39	34	23	21	35
	Y	2063	1786	1753	2212	2043	2229	2395	2035	2424	2104
Sirmaur	**A**	33	25	26	26	26	25	23	22	20	25
	P	46	49	57	59	63	67	60	64	59	58
	Y	2116	2044	2203	2291	2475	2686	2588	2918	2983	2478
Solan	**A**	32	24	24	26	25	24	23	24	24	25
	P	41	40	44	50	52	52	44	53	64	49
	Y	1949	1733	1817	1923	2049	2167	1898	2253	2797	2065
Una	**A**	29	30	29	29	30	30	29	31	32	30
	P	35	43	39	45	53	55	57	65	70	51
	Y	1691	1459	1344	1528	1735	1864	1951	2104	2195	1763

Area (A): (000 ha), Production (P) :(000 tonnes) and Yield (Y): (kg/ha)

Table 2. Area, production and yield of maize in H.P.

	Area (000 ha)	Production (000 t)	Yield (kg/ha)	% increase over base year 1970 Area	Production	Yield
1966-70	252.7	445.3	1759.9			
1971-75	254.7	427.8	1675.3	0.79	-3.93	-4.81
1976-80	282.2	477.7	1689.9	11.67	7.28	-3.98
1981-85	295.1	508.0	1717.6	16.78	40.804	-2.40
1986-90	308.9	555.2	1791.6	22.24	24.68	1.80
1991-95	311.0	637.8	2050.6	23.07	43.23	16.52
1996-2000	304.4	649.3	2134.9	20.46	45.82	21.31
2001-05	303.8	652.0	2143.2	20.22	46.42	21.78
2006-10	297.7	689.7	2314.9	17.81	54.88	31.54
2011	294.2	715.4	2432.2	16.42	60.66	38.20
2012	294.3	657.2	2233	16.46	47.59	26.88
2013	292.1	679	2325	15.59	52.48	32.11
2014	300	752.7	2509	18.72	69.03	42.56

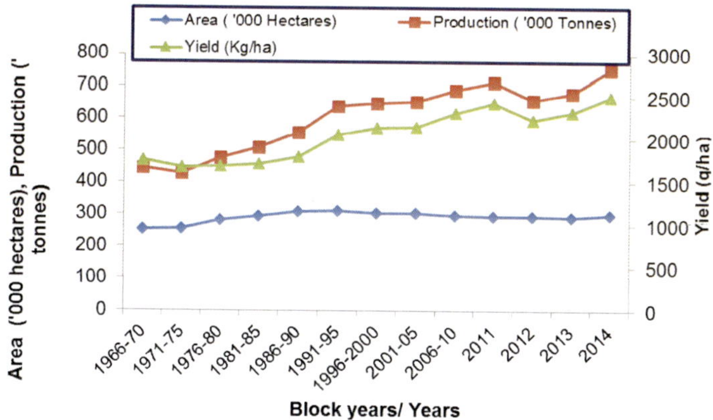

Fig. 1. Area, Production and Productivity in H.P.

Taking 1966-70 as the base year and dividing this period of into 9 blocks of five years each it is evident that area, production and productivity increased by 18.72, 69.03 and 42.56 per cent, respectively in the state. Area under maize continued to increase till 1994-95, but afterward it showed decreasing trend. This is probably due to the fact that area which was brought under irrigation during this period went under cash crops like vegetables, floriculture and horticulture. However, production as well as productivity continued to increase till date.

In Himachal Pradesh maize is consumed directly as food mainly in winter in various forms – chapatti of maize flour with green leaf preparation of brassica (*Sarson ka sag*) and curried pulses (Mash and Rajmash). Every part of the maize plant has economic value viz. grain leaves, stalk, tassel and cob which can all be used to prepare a large variety of food and non-food products. Maize production and consumption has been rising consistently, the consumption pattern also has been changed over the years. However, there are no precise data available in the utilization pattern in the state. Although, on the basis of some estimates, maize is utilized for human consumption (28%), animal feed (9%), kinds of payment (3%), seed (4%) and the remaining (56%) is surplus and sold to the traders of neighbouring states.

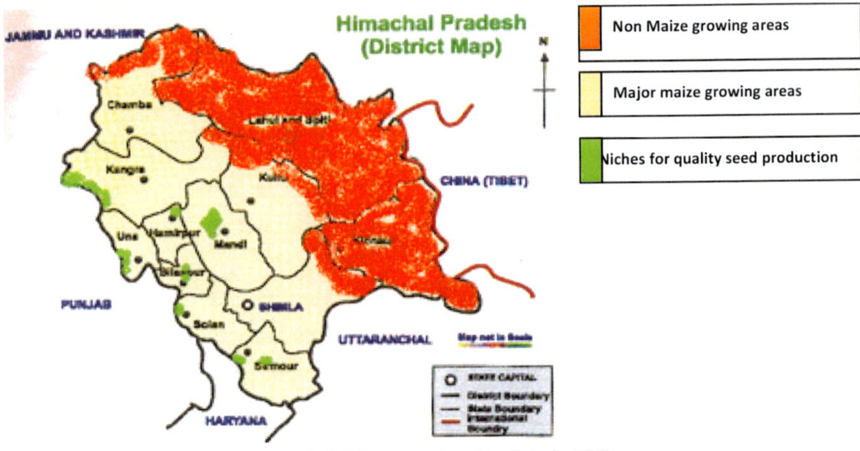

Fig. 2. Maize growing districts in H.P.

Different districts under Agro-climatic zones for maize cultivation in the state.

Zone	Area	Districts	Requirements of Zone for maize variety	Area under Hybrid/ composites and locals
Sub-tropical	Lower-hills and valley area	Sirmour, Una, Hamirpur, Bilaspura and lower parts of Kangra, Chamba and Solan districts	Medium maturing hybrids/ composites	Maximum area under hybrids
Sub-Temperate	Mid hills	Parts of Solan, Sirmour, Shimla, Mandi, Kangra and Chamba districts	Medium and early maturing hybrids/ composites	60% area under hybrid and remaining area under composites/ locals
Temperate (wet)	Higher hills	Parts of Shimla, Mandi, Kullu and Chamba districts	Early and Extra early maturing hybrids/ composites	Maximum area under locals/ composites and very less area (25%) under hybrids
Temperate (Dry)	Higher hills	Kinnaur and Lahaul&Spiti districts, Pangi and Bara Bhangal area of Chamba and Kangra districts	Extra early maturing	No area under maize in Kinnaur and Lahaul Spiti districts and in other areas only local maize varieties

The marketing of food grains in the state is not strictly regulated and there is no notified operating food market in the state. The farmers are selling their produce immediately after harvest in the local market due to paucity of storage structures and also to avoid storage loss. The buyers also reach to door steps of farmers so there is no marketing problem of maize in the state. The maize starch and chemical units of Punjab (Sukhjit Industries) and Haryana (Bharat Starch Industry) also procure maize directly from the farmers through their agents and traders. Himachal Pradesh supplies nearly 4.0-5.0 lakh MT of maize to other states every year after meeting its own requirement of food and feed.

AICRP centre's genesis and mandate

Earlier, maize improvement programme was handled by State Economic Botanist under the Millet Improvement work and in the late years as an *ad hoc* research project financed by Indian Council of Agriculture Research. Mainly work on improvement of maize crop in the state was initiated in 1957 at Kangra under All India Co-ordinated Maize Improvement Project and it functioned as a part of maize section in the department of Agriculture, Punjab. The research work was shifted to Bajaura (Kullu) in 1962 because the climatic conditions particularly high rainfall at Kangra were not conducive to large scale of pollination work in maize breeding programme. The technical control of Bajaura station remained with the Maize Breeder Punjab Agricultural University Ludhiana till 1970. Thereafter, it formed a part of the Agricultural Complex, Himachal Pradesh University, Shimla. Another sub-centre of the Coordinated Maize Improvement Project functioned at Solan in 1962 in the Department of Agriculture, Himachal Pradesh and it became a part of the Himachal Pradesh University in July, 1971. This centre was shifted from Solan to Salooni in Chamba district, which later on was shifted to Kangra. ICAR also sanctioned a national centre for evaluating maize genotypes against bacterial stalk rot disease of maize at Dhaulakuan. Various centres under AICRP are working with different mandates in various districts of the state.

AICRP Centre	Mandate
Hill Agricultural Research and Extension Centre Bajaura, Kullu	• Development of medium and early maturing hybrids and composites for different types of maize • Studies of prevalent diseases • Development of agro techniques for newly released cultivars and INM in maize based cropping system • Integrated management of major insect and pests
Shivalik Agricultural Research and Extension Centre, Kangra	• Development of medium maturing hybrids and composites for high rainfall areas
Hill Agricultural Research and extension centre, Dhaulakuan.	• Screening of maize genotypes against bacterial stalk rot disease.
Krishi Vigyan Kendera, Berthin, Bilaspur and Hill Agricultural Research and extension centre Dhaulakuan, Sirmour.	• Volunteer center (AICRP) for evaluation of maize genotypes

As a result of breeding work done under this project, number of composite/ hybrids were developed/identified and recommended for cultivation in the state. The most popular among these were normal maize hybrids viz. HIM 123, Ganga 101, Ganga-5, Ganga Safed-2, Sartaj, VL42, HIM 129 and composites viz. Early Composite, Parvati, Naveen, Girija and Bajaura Makka and speciality corn namely HQPM 1 (QPM), Bajaura Sweet corn, Bajaura Popcorn and VL 78 (Baby corn).

Impact of crop varieties developed

As far as the role of high yielding maize varieties in raising the maize production in the state is concerned, till the year 1970 maize hybrids viz. Ganga-101, Ganga-3, HIM123, Ganga-5 and composite Vijay were the varieties released through All India Coordinated Maize Improvement Project which benefitted the farmers and enhanced their income. Seed of these hybrids/composites was being produced by National Seed Corporation on indent basis and supplied to the farmers through Dept. of Agriculture. These varieties were high yielding, responsive to higher doses of fertilizer and resistant to the foliar diseases. These varieties played a significant role in raising the production and productivity of maize in the state. During the decade (1970-1980), the production and productivity recorded an increase of 8.73 and 6.72 percent, respectively. In the mean time H.P Agricultural University with its centres at Bajaura, Dhaulakuan and Solan also started work on development of maize varieties and released Early Composite, Paravti and Naveen during the period 1980 to 2000. A maize hybrid Sartaj was also identified for release in 1997 and its seed was produced and distributed to the farmers. Maize variety Early Composite because of its early maturity, high yield and resistance to foliar diseases became popular with the farmers of mid and high-hill regions. After that during period 2000 to 2008, research mandate remained only on composite breeding and Bajaura centre has developed and released two composites viz. Girija (medium maturing) in 2001 through SVRC and another composite Bajaura Makka (Early Maturing) in 2008 through CVRC. At present, three composites are in seed production chain and every year the centre is producing the seed because variety Girija is a preference of farmers of mid hill zone and Bajaura Makka and Early Composite of high altitude areas. This shows that improved varieties, production and protection technologies have always played a significant role in increasing the maize production and productivity in the state in the past. Despite of all efforts of the state as well as university, around more than 40% area in the state is still under open-pollinated varieties and farmers are using their produce as seed for sowing in the next crop season. Therefore efforts have been intensified to develop improved cultivars for different agro-climatic zones especially high altitude areas for which we have suitable cultivars. During the period starting from 1990 many national and multinational private sector companies also introduced their maize hybrids in the state. On the basis of testing and recommendations, some hybrids which became popular with the farmers were; PSCL 3436, PSCL 3438, PSCL 4640, KH101, KH510 and KH581. This period (1990 to 2000) recorded all time high rise in maize production (45.82 per cent) and productivity (21.31 per cent). This boom in maize production was because of the fact that affluent farmers of the valley areas adopted high yielding hybrids whereas poor farmers from remote and high altitude areas adopted composites and the overall result was higher maize production in the state. A large number of private sector hybrids are under cultivation in the state. A state level programme of private sector hybrid testing and recommendations for their cultivation has been established by the university. The hybrids are being evaluated at 7-8 locations representing different agro-climatic situations throughout the state. On the basis of testing over locations hybrids viz. KH 510, KH 2005, KH 2001, PSCL 4642, PMZ 4, 9572-A, HM-95, HM 105, C 816, IA 8101, Sakha 105, PG 2465, PG 2475, Polo, JKMH 2475, JKMH 1512, X 640, X 453, X 121, X 717, X 789, S 621, DK 7074, Apoorva, Bisco 855, 900M Gold, 110-08-01, Bisco 1141, Euro 1201, Hishell, HP333, 30R77, PAC 740, VMH 4102, P3420, 30V92, PSC 3322, Bisco Victory, B52, KF 105, GA95, PG 2474, P3441, SAMBA 3033 and DMH 6042, DKC 9125, P3303, P3337, XB 52-A, PG2424, PG2403, B54, K25, DKC9106, Rassi 3591, Biscoshiva, Deklab 9164, Deklab 9179, Deklab 8174, Deklab 8164, Deklab 8173, LG34.05 and P3542 were found promising for cultivation in the state.

After 2006-07, hybrid breeding programme started with a mandate of medium maturing and early maturing single cross hybrids for normal maize and quality protein maize of the state. The local germplasm collected from different maize growing areas and introgressed with the germplasm procured from different sources (Directorate of Maize Research, CIMMYT, VPKAS, Almora and SAUS) for deriving maize inbred lines. The heterotic pools for medium maturing viz. Bajaura Pool 95 and Bajaura Pool 98 and Hill early yellow pool have been established. Recently, two new heterotic pools i.e. Early Pool (A) & Early Pool (B) were constituted from the diverse early maturing lines categorised in two groups. A large number of inbred lines of Early and Medium for normal and QPM developed under this programme are being utilized for developing single cross hybrid. The inbred line development programme is tedious and time consuming which takes a time of more than ten years for developing a single inbred. For that AICRP has provided off season facility at winter maize nursery, Hyderabad which is free of cost to the northern states where only one crop season is available for advancing the generations and seed multiplication of those material which are being tested in the AICRP trials. With the result single cross hybrid Palam Sankar Makka 1 (Early maturing) has been developed and released through Central Variety Release Committee (CVRC) for the states of

Gujarat, Chattisgarh and Madhya Pradesh (Zone V) and Medium maturing hybrid, Palam Sankar Makka 2 (CVRC) for Himachal Pradesh, J&K and Uttrakhand.

Key production contraints

i) Adverse Climatic Conditions

In hills, due to prevailing low temperature, particularly at initial stages of crop growth and towards maturity and less sunshine hours prolong the crop duration. This allows cultivation of only short duration cultivars in maize, which have lower production potential as compared to medium/long duration cultivars. Temperate climate of Himachal Pradesh does not favour its cultivation during *Rabi* season.

ii) Predominance of Land Races

Traditional cultivars dominate the scenario in high hills, though these cultivars are highly adapted at specific locations but possess low yield potential. Susceptibility to diseases and insect pests further limit their productivity.

iii) Dependence on Rainfall

By and large, crop cultivation in higher elevation is rainfed. The production of maize suffers due to lack of rains at critical growth stages.

iv) Poor Soil Fertility

The Himalayas are comparatively much younger mountains and composed of sedimentary and metamorphic rocks and are very much susceptible to erosion due to action of wind and water. Due to topsoil erosion, gradually less productive soil is available with each advancing year for crop production.

v) Wild animals

Wild animals are emerging as a serious threat to maize cultivation in the state. Increased monkey menace and wild boars cause tremendous damage and in many areas farmers have been forced to abandon the maize cultivation or switch over to some other crops like paddy, vegetables etc., especially in irrigated areas.

vi) Lack of seed corporation

There is no state seed corporation in the state to take the maize seed production of public hybrids developed and released. There is a problem in producing the seed of hybrid suitable for hilly areas in other states such as Bihar, Andhra Pradesh and Karnataka because of susceptibility of inbreds to diseases and insects which are more in plains as compared to hills.

vii) Non-availability of vital inputs

The availability of vital inputs for increasing production and productivity, such as seed, fertilizer and pesticides, is not adequate. Organized system inputs is not yet established in higher hills. These inputs are available only at block headquarters and are not easily accessible to many of the farmers or are not always available. Many farmers in the state use the produce as seed for sowing in the next crop because of non-availability of seed of high yielding varieties.

viii) Low adoption of production technology

Low adoption of improved production technology by the farmers is major challenge in the state. Presently, only less than 40 per cent area of maize is under high yielding hybrids. Poor agronomic management of crop and weed infestation in the *Kharif* season in maize –based cropping system causes sub-optimal realization of yield potential of cultivars. The proper attention for the development of production technology in high altitude areas have not been received due to remoteness, inaccessibility and harsh climatic conditions. The technology developed at mid and lower altitude is not always successful in higher hills due to entirely different agro-ecological situation there.

ix) Post harvest handling

After harvesting the maize crop, cobs and subsequent grain drying and storage is also becoming a big issue that discourage the farmers to grow the maize on large scale. Maize crop is grown in *Kharif* in the state wherein the early onset of winter after maize harvest necessitates extra efforts for proper drying and storage of the produce. Maize is very much vulnerable to the post harvest problems, which account for more than 20% losses. These losses intensify due to poor drying and storage leading to weevil damages and formation of mycotoxins.

x) Non-regulated market and support price

The most challenging issue relates to the farm income of the maize growers as there is no system in place for regulated marketing to sell their produce. There is no support price announced for this crop like fruit crops, wheat and paddy and produce is not procured by the government agencies. Under such a situation private traders exploit the framers to sell their produce at minimum price.

AICRP technologies

Popular land races

Development of maize hybrids and high yielding composites /synthetic have replaced the low yielding maize land races at faster rate in major maize growing areas of the state. However, little impact of these high yielding varieties in non-accessible areas i.e high altitude areas of the state has been witnessed. The high yielding varieties are developed for favourable environments and add a non significant gain in the marginal land. The high cost of seed and non availability of suitable maturing hybrid further aggravate the situation for these areas. Some land races are existing and still popular among the farmers. A number of locals / land races have been documented and preserved in the National Bureau of Plant Genetic Resources but presently few are very much popular among the farmers because of one or another reason. They have niche specific adaptations being early maturing and those maturing before the onset of low temperature and fit in the cropping system. These land races are usually grown in the marginal land under rainfed condition with specific adaptability and thus thrive best in these environments. These are grown as sole crop and intercropped with pulse crops such as rajmash, mash, soybean to increase the cropping intensity and fertility of the land. The land races are serving as stable food and fodder for high altitude areas because of their good grain quality with thin and succulent stem which is good for fodder. The most popular maize land races of Himachal Pradesh along with their features are:

Name of land races	**Grain**	**Salient features**	**Remarks**
Hachhi kukdi (Salooni local)	White	Semi flint, tall, high cob placement, dark green leaves	Popular in Chamba area
Ratti	Yellow	Flint to semi flint, early maturing, tall , high cob placement	Popular in Chamba area
Chitku	White	Popcorn type , early maturing and medium height	Popular in Chamba area
Bharmour local	Yellow	Flint to semi flint, early maturing, high cob placement and susceptible to TLB and MLB	Popular in Bharmour area of Chamba district
Dwani	Yellow	Flint to semi flint, medium maturing, high cob placement and susceptible to TLB and MLB	Banjar and Sainj area of Kullu
Sathu	Yellow	Flint to semi flint, very early maturing, medium cob placement and susceptible to TLB and MLB	Banjar and Sainj area of Kullu
Ghyagi local	Yellow	Flint to semi flint, very early maturing, medium cob placement and susceptible to TLB and MLB	Banjar area of Kullu
Katrain Local	Yellow	Flint to semi flint, early maturing, medium cob placement and susceptible to TLB and MLB	Shangan & Katrain areas of Kullu
Jawalapur local	Yellow	Flint, long and thin cob, early maturing	Jawalapur area of Mandi
Popcorn local	White	Popcorn type, small cob, medium height	Jawalapur area of Mandi

Dehara local	Yellow	Semi flint, medium height and medium maturing	Dehara Gopipur area of Kangra
Kangra local	White	Tall, semi flint grain, high cob placement	Nurpur area of Kangra
Changgar Local	Yellow	Semi flint, medium height and medium maturing	Changgar area of Kangra
Hamirpur Local	Yellow	Semi flint, tall and medium maturing	Hamirpur
Bilaspur Local	Yellow	Semi flint, tall and medium maturing	Berthin area of Bilaspur
Sirmour Local	Yellow	Semi flint, tall and medium maturing	Sirmour area
Lahaul Local	Yellow	Flint, medium tall, weak stem , low yielding and cold tolerant	Lahaul area

These maize land races are losing their popularity and are gradually going out of farmers' domain and becoming extinct due to low yielding potential, lodging, susceptibility to major insects and diseases such as *Turcicum* leaf blight , *Maydis* leaf blight and common rust.

Cultivar released

The maize Improvement work was started in 1962, with the functioning of All India Coordinated Project in the state. As a result, numbers of composites/ hybrids have been developed and released/recommended. The details are given below:

Table 3. Cultivars released in the state.

Type	Cultivar name	Year	Released by	Type of cultivar/Special features
Yellow	Ganga Hybrid Makka 101	1962	-	Double cross hybrid, plant stay green at harvest.
Yellow	Hybrid Ganga 103	1964	-	Double top-cross hybrid, moderately susceptible to leaf blight and rust but tolerant to stem borer.
Yellow	Him 123	1964	-	Double cross hybrid, tall plants; recommended for mid and high hilly areas.
Yellow	Ganga Hybrid Makka 5	1968	-	Double top cross hybrid, grains bold, yellow and semi flint.
Yellow	Hybrid VL54	1962	-	Resistant to leaf blight but susceptible to stalk rot, brown stripe and powdery mildew.
White	Ganga Safed 3	1964	-	Double top cross hybrid with white grain, resistant to lodging, bacterial stalk rot and pythium stalk rot but susceptible to BSDM.
Yellow	VL 42	1978	-	Suitable for high hills. ,possess good cooking quality
Yellow	Early Composite	1980	SVRC	Orange flint grains, early maturing (98-100), suitable for higher regions of the hills.
Yellow	Parvati	1987	CVRC	Medium maturity tolerant to bacterial stalk rot.

Yellow	Naveen	1992	SVRC	Early maturing, and suitable for lower region of H.P.
Yellow	Girija (Comp.)	2001	SVRC	Medium maturing (105-108 days), high yielding, suitable for growing in lower and mid hill regions of H.P.
Yellow	VL 78 (Comp.) (Baby corn)	2005	SVRC	Early, high yielding and prolific bearing habit, silking in 50 days and gives 4-5 pickings, having light yellow colour of baby corn.
Yellow	Bajaura Makka	2008	CVRC	Early maturing (90-95 days) and high yielding and responsive to moderate to high doses. There is less reduction in yield at lower doses of fertilizers. Moderately resistant to lodging and diseases (TLB & MLB).
Yellow	Bajaura popcorn	2009	SVRC	This is a special type of maize having the ability to pop twenty times in volume upon heating. It matures in 90-95 days. Tolerant to TLB and MLB.
Yellow	Bajaura sweet corn	2009	SVRC	This variety has a tendency to bear two cobs with tight husk cover; grains are less shrivelled and yellow golden in colour. It matures in 100-105 days.
Yellow	HQPM 1	2009	SVRC	It is a single cross high yielding quality protein maize hybrid. The hybrid is resistant to *maydis* leaf blight, tolerant to t*urcicum* leaf blight. It matures in 105-110 days and is suitable for low and mid hills zones of the state.
Yellow	Sartaj (Hybrid)	1997	SVRC	Medium maturity resistant to lodging and leaf blight, recommended for sub-mountain and lower mid hills.
Yellow	Palam Sankar Makka 1	2016	CVRC	Early Maturing single cross hybrid suitable for irrigated conditions. Tolerance against major diseases viz; TLB MLB and ESR. Released for Rajasthan, Gujarat, Chattisgarh and Madhya Pradesh (Central West Zone)
Yellow	Palam Sankar Makka 2	2016	CVRC	Medium single cross hybrid suitable for irrigated conditions in *Kharif* season. Moderate resistance against major diseases viz; TLB, MLB, BSR and PFSR. Released for Himachal Pradesh, Jammu and Kashmir, Utarakhand (Northern Hill Zone).

Maize Seed Production in the state from the year 1995 onwards

Year	Name of cultivar(s)	Quantity (kg)
1995-96	Early composite	100
	VL 78(Baby corn)	40
	Sweet corn	14
	Popcorn	82
	CM124 X CM125	41
	CM122 X CM123	25
	CM 205	10
1996-97	Early composite	190
	CM124 X CM125	28
	CM122 X CM123	17
1997	Early composite	105
	CM124 X CM125	27
	CM122 X CM123	17
	CM 205	10.0
	CM112	5.0
	CM113	10.0
	CM122	8.0
1998	Early composite	115
	CM 205	10.0
	CM112	10.0
1999	Early composite	162
	Girija	110
	CM 205	7.0
	CM112	3.0
	CM113	12.0
2000	Early composite	105
	Girija	90
	CM 205	7.0
	CM113	12.0
2001	Early composite	154
	Girija	157
2002	Early composite	130
	Girija	150
2003	Early composite	186
	Girija	130
2004	Early composite	170
	Girija	215
2005	Early composite	195
	Girija	225
	VL78	40
2006	Early composite	240
	Girija	690
2007	Early composite	375
	Girija	930
	VL78	70
2008	Girija	450
	Bajaura Makka	258

	Bajaura Popcorn	35
	Early Composite	360
	Bajaura Sweet Corn	30
2009	Girija	260
	Bajaura Makka	250
	Bajaura Popcorn	45
	Early Composite	300
	Bajaura Sweet Corn	45
	VL 78	65
	HQPM	93
2010	Girija	400
	Bajaura Makka	350
	Bajaura Popcorn	45
	Early Composite	250
	Bajaura Popcorn	45
	Bajaura sweet corn	45
	HQPM 1	700
2011	Girija	292
	Bajaura Makka	177
	Early Composite	206
	Bajaura Popcorn	80
	Bajaura sweet corn	55
	VL78	30
	HQPM 1	6500
	HKI193-1	363
	HKI 163	490
2012	Girija	362
	Bajaura Makka	282
	Early Composite	205
	HQPM 1	1363
2013-14	Girija	300
	Early Composite	180
	Bajaura Makka	220
	HQPM 1	200
2014-15	Girija	347
	Bajaura Makka	164
	Bajaura Popcorn	97
	HQPM 1	36
2015-16	Girija	220
	Bajaura Makka	180
	Early Composite	140
	Bajaura Popcorn	60
	HQPM 1	69
2016-17	Girija	130
	Bajaura Makka	130
	Early Composite	190
	Bajaura Popcorn	42
	HQPM 1	35
	Palam Sankar Makka 2	15
	BAJIM 08-26(Female Parent of Palam Sankar Makka 2)	25
	BAJIM 08-27 (Male Parent of Palam Sankar Makka 2)	25

Maize based cropping system followed in the state are:

Rainfed situations :Maize – wheat; Maize+ pulses/wheat +oilseed/pulses; Maize +pulses-peas/cauliflower/cabbage/potato

Irrigated situations: Maize-toria-wheat (Lower and mid hills); Maize-peas/ cauliflower/cabbage/potato

Name of Package	Package Developed			
Fertilizer management	**Hybrid and composite**	**Nutrients (kg/ha)**		
		N	**P_2O_5**	**K_2O**
	i) High rainfall	120	60	40
	ii) Low rainfall	90	45	30
	Locals			
	i) High rainfall	80	40	30
	ii) Low rainfall	60	30	20
	Apply one third of N and whole of P and K at the sowing time. The remaining 2/3 of N should be applied in two equal splits one at knee-high stage of the crop (40-45 days after sowing) and another at pre-tasselling stage. In low hill sub- mountain areas, apply 1/8 N at the time of sowing, ¾ N at knee high stage and 1/8N at tasselling stage. The second and third dose of N should be applied. Under rainfed conditions, in case N fertilizers are not available at sowing, 1/3 N can be applied 2-4 weeks after sowing with hand weeding and remaining 2/3 N in two equal splits at knee high and pre-tasseling stages.			
Weed control in maize	To control both grassy and broad leaved weeds, apply Atrataf or Masstaf or Rasayanazine 50 WP in 750-800 L water per hectare with high volume sprayer within 2 days of sowing. In case if herbicide could not be applied within 2 days of sowing, it could be applied within 10 days of sowing i.e. 2-3 leaf stage of the weeds. Spray of herbicide shall be preferred over broad cast application. However, if the spray pump is not available, it could be applied after mixing with 150 kg sand per hectare but uniformity in application and sufficient moisture has to be ensured. While mixing the herbicide and its broadcast, the farmer must wear rubber gloves or polythene bags on the hands.			
Weed control in Intercropping	In maize and blackgram/horsegram intercropping system weeding and hoeing should be done within 5 weeks of sowing.			
Management of *Cypress rotundus L.*	Application of glyphosate 0.75 L/ha + 0.5% $(NH_4)_2SO_4$ on actively growing Motha (*Cyperus rotundus*) plants as foliar spray after harvest of wheat and atleast one or two weeks before the sowing of maize.			
Fertilizer management in intercropping system	In maize+blackgram intercropping system, no additional fertilizer dose is required to be added but, in maize+horsegram system 50 per cent of the recommended (7.5 kg N+ 22.5 kg P_2O_5/ha) dose should be applied to horsegram.			
Contingent crop planning for late *Kharif* crops.	In mid-hills sub-humid zone, if the major crop of maize could not be sown due to delayed onset of monsoon or fails due to adverse climatic conditions, the sowing of blackgram/ horsegram could be done up to first week of August.			

Package of Practices for Baby corn cultivation

Varieties:

Hybrid : VL 42, MEH 114, MEH 133 **Composite :** Early Composite, VL 78

Tassel removal just after its emergence results in higher number of cobs per plant, reduction in discarded baby corn cobs and barrenness and gives 15-18 % higher yield.Method of sowing: The crop must be sown in lines behind the plough at a spacing of 40 cm between rows and the spacing between plant to plant should be kept 20 cm (1,25,000 plants/ha) for Composite variety (average plant height of 200 cm) and 17.5 cm (1,43,000 plants/ha) in hybrid variety (average plant height of 165 cm).

Manuring

Nutrient (kg/ha)			Fertilizers kg/bigha (800 m^2)					
N	P_2O_5	K_2O	UREA	SSP	MOP	UREA	SSP	MOP
150	60	40	326	375	66.5	26	30	5.3

In addition to above fertilizers, incorporate farmyard manure @ 10 t/ha at the time of field preparation before sowing. Apply 1/3 of N and whole P and K at sowing time as band application. Remaining 2/3 of N should be applied in 2 equal splits one at knee high stage 25 days after sowing) and another at pre-tasseling stage (40 days after sowing).

Harvesting: Young baby cobs should be carefully handpicked within 2-3 days of silk emergence from the leaf sheath, so that the upper stem portion and lower leaves do not break. The picking of baby corn should be done every third day and 7-8 pickings are required in existing identified varieties.

i) Disease Management

Disease	Recommendation for management
Turcicum Leaf Blight (TLB)	1. Early sowing, preferably before 10th June, results in less disease. 2. Apply recommended doses of nitrogenous fertilizers. 3. Spray crop with zineb or mancozeb @ 0.25% or propiconazol @ 0.1% on appearance of disease especially in case specialty corn or seed crop. 4. Use resistant variety/hybrids. Most of the hybrids are resistant to disease. 5. Practice crop rotation, sanitation and destroy crop debris.
Maydis Leaf Blight (MLB)	1. Early sowing, preferably before 10th June, results in less disease. 2. Apply recommended doses of nitrogenous fertilizers. 3. Spray crop with zineb or mancozeb @ 0.25% or propiconazol @ 0.1% on appearance of disease especially in case specialty corn or seed crop. 4. Use resistant variety/hybrids. Most of the hybrids are resistant to disease. 5. Practice crop rotation, sanitation and destroy crop debris.
Banded Leaf and Sheath Blight (BLSB)	1. Grow resistant variety with erect growth habit and high leaf attachment from soil. 2. Strictly follow the row to row and plant to plant spacing.

	3. Stripping of lower leaves. 4. Spray crop with mancozeb (0.25%),propiconazol (0.1%), Thiophanate methyl (0.1%), validamycin (0.27%) with the appearance of disease and repeat it after 10 days depending upon crop stage and disease severity.
Brown Stripe Downy Mildew (BSDM)	1. Seed treatment with metalaxyl @ 4 gm/kg of seed. 2. Eradicate collateral and wild hosts growing near the field. 3. Grow resistant/ tolerant variety. 4. Spray the crop with mancozeb or mancozeb + metalaxyl @0.25% with the appearance of disease and repeat it at fort nightly interval.
Bacterial Stalk Rot	1. Apply judicious doses of nitrogenous and potassic fertilizers. 2. Field should be well drained and avoid water logging. 3. Destroy diseased plant debris. 4. Use resistant/ tolerant varieties/ hybrids. 5. Apply bleaching powder thrice @ 16.5 kg/ha. First application in furrows at sowing time, second at earthing up near base of the plant and third one week before tasseling. Do not apply bleaching powder mixed with urea.

ii) Insect Pest Management

Insect	Management
Cutworm, white grubs and black beetles	1. Use well decomposed FYM 2. High seed rate should be used to compensate the mortality due to severe damage by soil pests. 3. During sowing, mix 2 litres of chlorpyriphos 20EC with 25 kg sand/ha in the soil.
Stem borer	1. Remove the weeds, other undesirable plants and grasses by ploughing the fields. 2. Use high seed rate and remove the stem borer infested plants having small holes. 3. The granules of carbofuran (Furadan)3G or phorate (Thimet) 10G @ 2g/m row can be applied either in the soil before sowing or sprinkled over the central whorl especially of those plants showing shot holes. 4. At harvesting, cut the plants from the level of soil surface and collect and destroy the stubbles.
Hairy caterpillars and grasshoppers	1. Collect and destroy the caterpillars which are growing up gregariously. 2. Dust methyl parathion (2% dust) @ 25-30 kg/ha.
Blister beetles	1. Whenever attack of blister beetles reaches at economic injury level, methyl parathion (0.05%) (625 ml Metacid 50EC) or deltamethrin (0.0028%) (625 ml Decis 2.8EC) or cypermethrin (0.0075%) (470 ml Ripcord 10EC) in 625 litres of water per ha can be sprayed at tasseling stage

IPM of maize insect pests

For the management of soil dwellers, deep ploughing of fields and exposing the soil inhabiting larvae and pupae of different insect pests few days before sowing to bird is a cheap cultural method of control. Well decomposed FYM should be used to avoid the multiplication of white grubs and termites. High seed rate should be used to compensate the mortality due to severe damage by soil pests.

In addition to other practices, methyl parathion (2% dust) or malathion (5% dust) or carbofuran 3G@ 25-30 kg/ha should be mixed in soil at the time of field preparation or sowing (Panwar, 1998). Thakur and Vaidya (2000) found aqueous root extract (0.5%) of Alberry (*Rumexnepalensis*) and chlorpyriphos as an effective treatment against black cutworm, *A.ipsilon.* For the management of foliage feeders, clean cultivation by removing weeds, ploughing and removing alternate hosts must be adopted. Bunds and fields must be cleared off from grasses/weeds and then dusted with methyl parathion (2% dust) or malathion (5% dust) @ 25-30 kg/ since these are breeding places for grass hoppers. Larvae of hairy caterpillars and fruit borers when appear on foliage can be controlled by dusting methyl parathion (2% dust) @ 25-30 kg/ha. For the management of snails and slugs, weeds and other grasses near to crop fields should be removed because these are the breeding places for these pests. Pellets of snail kill bait (metaldehyde 2.5%) can be broadcasted or sprinkled in the field @ 37.5 to 62.5 kg/ha.

It is obligatory to protect the crop against the attack of maize borer, a key pest of maize, otherwise heavy investment in the form of fertilizers and seed may be totally lost. The borer attack is discernible soon after germination. When the plants are 10 days old, they should be sprayed with 1000 ml quinalphos (Krush 25%) in 500L water/ha. In order to avoid washing off of the insecticides by rains, the granules of carbofuran (Furadan)3G or phorate (Thimet) 10G @ 2g/m row can be applied either in the soil before sowing or sprinkled over the central whorl especially of those plants showing shot holes. Whenever attack of blister beetles reaches at economic injury level, methyl parathion (0.05%) (625 ml Metacid 50EC) or deltamethrin (0.0028%) (625 ml Decis 2.8EC) or cypermethrin (0.0075%) (470 ml Ripcord 10EC) in 625 litres of water per ha can be sprayed at tasseling stage (Kumar *et al.*, 2001). Presently an integrated pest management (IPM) approach involving bio-control agents and non chemical methods are advocated to manage maize insect pests to reduce the load of pesticides.

Changes in disease scenario over the years:

Plant diseases are of paramount importance to humans because they damage plants and plant products on which humans depends for food, clothing, furniture, the environment and in many cases the housing. The kinds and amounts of losses caused by plant diseases vary with the plant or plant products, the pathogen, the locality, the environment, the control measures practiced and combination of these factors. About 112 diseases of maize have been reported so far from different parts of the world. Of these, 65 are known to occur in India (Saxena, 2002). The major diseases in different agro-climatic regions of Himachal Pradesh are: seed rots and seedling blight, leaf spots and blights, downy mildews, stalk rots, banded leaf & sheath blight, and smuts & rots. The total loss in economic products of the crop due to diseases has been estimated to be the tune of 13.2% (Payak and Sharma, 1985). Most of hybrids and composites released from public and private sector have fairly high level of resistance to foliar diseases. Sources of resistance to major diseases have also been identified and being utilised in the breeding programme for development of maize varieties. Girija nad Bajaura Makka have marked resistance to foliar diseases, However, Hybrid like Palam Sankar Makka-2 have also shown substantial resistance to TLB, MLB and other diseases.

The spectrum of diseases has changed due to climate change and introduction of high yielding genotypes/ hybrids in the state and some diseases which were considered minor are now more conspicuous. Since the incidence of foliar blights appears to have waned, *Rhizoctonia* (banded leaf and sheath blight) and stalk rots are now becoming the dominant disease groups. Brown spots, brown stripe downy mildew and *Erwinia* stalk rot are endemic in nature. Seed rot and seedling blight, Common Rust (*Puccinia sorghi),* Corn Smut (*Ustilago maydis)* and Curvularia Leaf spot (*Curvularia lunata*) are diseases of minor importance in the state.

Changes in Pest scenario over the years

The multi-pest complex of maize crop poses serious limitations in its intensification in different agro climatic regions of the state. In one way or another, maize is subjected to insect attack from the time it is sown until it is consumed as a food or feed. Other crops particularly with small grains, forage, grasses and legumes frequently are the sources of insects that attack maize and must be reckoned as part of the maize-pest problem. Maize in the state is attacked by a wide array on pests which cause it a variable economic damage under field and storage conditions (Kumar *et al.,* 2001). The newly introduced high yielding varieties/ hybrids and heavy application of fertilizers provide luxuriant growth to the maize crop which attract a number of insect and non insect species in comparison to local cultivars. A comprehensive list of insects and non insect species found damaging maize in Himachal Pradesh has been given in Table 1. Recently some non traditional pests like larvae of tomato fruit borer (*Helicoverpa armigera*) have become pests of maize by feeding on its cob, flower and foliage Also new insects like metallic green beetles have also become serious pests of maize by feeding on its pollen and adversely affecting pollination process . Similarly Rice weevils (*Sitophilus oryzae*) has also become a serious problems particularly in early sown seed production plots.

Table 4. Important pest of maize in H.P.

Pest type	Common name	Scientific name	Pest status
1.Soil dwellers	a) Cutworms	*Agrotis spp*	Regular
	b) Black beetle	*Heteronychus lioderes* Retd.	Sporadic
	c) White grubs	i) *Holotrichia* spp ii) *Melolontha* spp iii) *Anomala* spp iv) *Brahmina* spp v) *Maladera* spp	Sporadic
	d) Termites	*Odontotermes obesus* (Rambur)	Sporadic
2.Foliage feeders	a) Grass hoppers	i) *Chrotogonus trachypterous* Blanchard and ii) *Hieroglyphus nigrorepletus* (Bol,)	Regular
	b) Hairy caterpillars	Spilosoma obliqua Walker	
	c) Fruit borer	*Helicoverpa armigera* Hubner	Sporadic
	d) Snails	i) *Bensonia monticola* Hutton ii) *Macrochlamys* sp	Sporadic Regular
	e) Slug	*Anadenus altivagus* Theobald	Regular
3. Stem borers	a) Maize stem borer	*Chilo partellus* (Swinhoe)	Regular
	b) Pink borer	*Sesamia inferens* (Walker)	Sporadic
	c) Shoot fly	*Antherigona naqvii* Steyskal	Sporadic
4. Flower feeders	a) Blister beetles	i) *Mylabris pustulata* Thunb. ii) *M. macilenta* Marshall iii) *M. phalerata* Pall.	Sporadic
	b) Maize stem borer	*Chilo partellus* (Swinhoe)	Regular
	c) Defoliating beetle	*Chiloloba acuta*Weild	Regular
	c) Fruit borer	*Helicoverpa armigera* Hubner	Sporadic
	d) Maize aphid	*Rhopalosiphum maidis* Fitch	Regular
5. Cob damaging pests	a) Fruit borer	*Helicoverpa armigera* Hubner	Sporadic
	b) Maize stem borer	*Chilo partellus* (Swinhoe)	Sporadic
	c) Blister beetles	*Mylabris* spp.	Sporadic
	d) Parakeets, Crows, Indian Myna	---	Regular
	e) Jackals, porcupines, wild pigs	---	Regular
	f) Rats		Regular

Table 5. Front Line demonstration conducted (1999-2014)

Year	FLDs	FLD yield (q/ha)	State yield (q/ha)	Change (%)
1999	30	49.57	22.72	118.1
2000	60	38.26	22.93	66.8
2001	60	49.56	25.50	94.4
2002	100	34.62	16.12	114.7
2003	100	47.62	24.44	94.8
2004	175	34.71	22.72	52.8
2005	150	33.45	18.39	81.9
2006	157	36.96	23.26	58.9
2007	150	33.45	28.73	16.4
2008	256	37.44	22.73	64.7
2009	300	34.24	18.39	86.2
2010	200	34.94	22.63	54.3
2011	600	35.65	24.32	46.6
2012	35	34.35	22.33	53.8
2013	75	45.43	23.25	95.4
2014	20	45.06	25.09	79.6
Total/Av.	**2468**	**39.08**	**22.72**	**72.0**

The concept of Frontline demonstration (FLD) was evolved by Indian Council of Agricultural Research (ICAR) with the inception of technology mission on oil seed crop (ISOPOM) during mid eighties. It is a unique approach to provide a direct interface between researcher and farmers as the scientists are directly involved in planning, execution and monitoring of the demonstrations for the technologies developed by them and get direct feedback from the farmers' field about the crops like cereals and pulses production in general and technology being demonstrated in particular. This enables the scientists to improvise upon the research programme accordingly.

A large number of frontline demonstrations (2468) were organized over the years (1999-2014) covering major maize growing areas of the state. The farmers were provided basal inputs like basal dose of fertilizer, seed of improved composite/hybrid and weedicide at the time of sowing. The mean yield of maize FLDs varied from 33.45 to 49.57 q/ha with overall mean yields 39.08 q/ha as compared with corresponding state average yield of 22.72 q/ha. Impact of the FLDS was observed as per cent increase over state average which comes to tune of 72 per cent.

Rainfall Pattern

During *Kharif* season, maximum rainfall is received during the months of July and August with uniform distribution over the years which favors proper grain filling and resulting in good yield of the maize crop.

Table 6. Month-wise rainfall pattern in the state

Years	Rainfall (mm)				
	June	July	Aug.	Sept.	Oct.
1984-88	55.6	177.7	102.1	87.0	43.3
1989-93	41.2	147.1	113.9	54.3	7.2
1994-98	79.2	121.2	145.9	107.8	56.8
1999-03	91.8	151.4	116.8	54.7	2.2
2004-08	51.4	151.9	141.5	85.8	40.6
2009-13	95.6	160.3	170.4	109.7	10.8
2014	39.5	135.0	117.9	50.4	16.6
2015	93.8	189.7	91.3	45.2	9.5
2016	38.3	83.1	207.6	23.9	2.3
Total	**586.54**	**1317.5**	**1207.44**	**618.92**	**189.38**

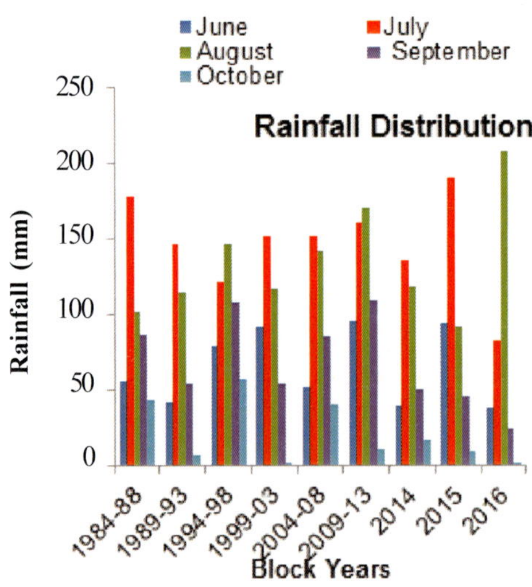

Fig. 3. Rainfall distribution pattern in H.P.

Future research and development strategy to enhance maize production

1. Genetic enhancement of maize germplasm

Maize productivity can be enhanced by exploiting genetic gains in near future by breaking negative correlation between yield and early maturity, continuous introgression of more and more diverse temperate germplasm, improving harvest index and nutrient use efficiency. Recycling of existing elite lines without disturbing their heterotic groups can further increase the yield in short term. Broadening of germplasm base for different traits like erect leaf type for high plant density, good stand ability, superior nutritional quality, diseases, and insect pests' resistance is the need of hour. The systematic work on maize germplasm for their heterotic grouping in different maturity group will help to derive the lines to develop productive hybrids.

2. Thrust on adoption of single cross hybrid

There is a tremendous scope to further increase the productivity in the state and make maize cultivation more remunerative to farmers. At present, area of maize under hybrids is still very less in the state which can be further increased by adoption of single hybrid in maximum areas along with complete package of practices which will help to increase the productivity.

3. Dual purpose maize hybrids/ varieties

The importance of livestock will continue to increase due to livelihood security especially in hilly area where there is a scarcity of fodder during winter for animals. The focused efforts to develop dual purpose maize varieties which can meet the requirement of food, feed and stover will be more promising for further increasing the value and importance of maize cultivation by farmers. The germplasm and released hybrids of "stay green nature" would enhance stover quality. Besides it, development of new sweet corn and baby corn hybrid would also help to provide good green fodder.

4. Innovative agronomic research

Innovative agronomic research can be undertaken focussing on tillage practices, weed management, soil fertility and water management, to identify germplasm having good performance under low input situations.

5. More emphasis on developing speciality corn suitable for industrial purpose

With the increase of urbanization, change in food habit and the improved economic status, the speciality corn like sweet corn popcorn and baby corn have gained significant importance in peri- urban areas of the state. Further, their cultivation can help to provide the quality nutritional food and green fodder and also generate employment to the unemployed youth for their income enhancement. Also starch industry requires genotypes with higher starch and food industry with high lysine maize, corn flakes, corn flour, pop corn and other products. Therefore, concerted efforts need to be intensified to develop such special corns.

6. Specialty and health corn food

Maize occupies an important place among cereals in Himachal Pradesh and forms staple diet of sizeable population in the state. The demand of speciality and health food is on the rise. QPM has a potential in achieving nutritional security and thereby help in reducing poverty with improving health. Availability of QPM with higher contents of vitamin-A, Fe, Zn and methionine is likely to be reality in future making maize as multi-nutrient grain. Breeding programme needs a great focus to develop products for target consumers.

7. Accelerating breeding

Development of inbred lines in maize and their maintenance requires a lots of efforts, labour, resources and time. The doubled haploid (DH) technology enables rapid development of completely homozygous lines and efforts offers significant opportunity for fast track development and release of cultivars (Prasanna, 2012). Doubled haploid technology using inducer lines can reduce the time span by 3-4 generations to get advanced generation of uniform inbred lines. With the development of number of inbreeds through DH technology new combinations of relevant adaptive traits including biotic and abiotic stress tolerance will be provided. The work on DH has also started with the haploid inducer stock procured from CIMMYT in 2014 by using their protocol .The early maturing lines have been developed and are being used in the breeding programme to develop the single cross hybrid.

8. Composite breeding

The marginal farmers in the state grow the maize crop under rainfed condition. They cannot afford to purchase the costly hybrid seed every year from the market and even it is not easily available to the farmers in the far flung areas. Farmers generally use their produce of open pollinated high yielding maize varieties as seed for sowing of next crop. Therefore, composite breeding needs to be strengthened for sustainable production for these areas.

Summary

Maize occupies an important place among cereals in Himachal Pradesh and forms staple diet of sizeable population in the state. The agro-climatic conditions of this hilly state are quite different from those prevailing in plains. Low temperature during sowing and harvesting extends crop maturity and heavy rainfall on steep slopes results in high nutrient loss. The variation in soil type and climate conditions makes the job of a breeder very challenging. The breeding programme of maize is underway to develop high yielding, disease resistant hybrids and composites for different types of maize viz; normal, QPM, sweet corn, popcorn and baby corn for cultivation in all the maize growing areas of the state. A large number of high yielding hybrids and composites have been developed and released for general cultivation in the state. These varieties have contributed significantly to the increased productivity. The productivity of maize has increased from 1520 kg/ha in 1966-67 to 2509 kg/ha in 2015-16 and there has also been a substantial increase in the total production from 445.3 thousand tonnes to 752.7 thousand tonnes. As a result of breeding work done under All India coordinated

Maize Improvement project, a number of composites/ hybrids were developed/ identified and recommended for cultivation in the state. The most popular among these are normal maize hybrids viz. HIM 123, Ganga 101, Ganga-5, Ganga Safed-2, Sartaj, VL42, HIM 129, Palam Sankar Makka 1 and Palam Sankar Makka 2 and composites viz. Early Composite, Parvati, Naveen, Girija and Bajaura Makka and speciality corn namely HQPM1 (QPM), Bajaura Sweet corn, Bajaura Popcorn and VL 78 (Baby corn). Besides this, new initiatives have been taken to accelerate breeding programme using doubled haploid techniques for developing the maize productive inbred lines and conversion of normal maize inbred lines to its QPM version with high pro vitamin A, Zn and Fe content. Suitable production and protection technique have been developed/ recommended for the cultivation of maize in the state under this project and maize growers have been benefitted by adopting these techniques.

The maize crop with myriad phases of production and utilization under different environmental conditions, encounters many interrelated problems in which diseases and insect pests are commonly involved. The multi-pest complex of maize crop poses serious limitations in its intensification in different agro climatic regions of the state. In one way or another, maize is subjected to disease and insect attack from the time it is sown until it is consumed as a food or feed. The major diseases (foliar blights, stalk rots, downy mildews etc.) in different agro-climatic regions of Himachal Pradesh cause loss in economic products of the crop which has been estimated to the tune of 13.2%. Also more than 40 insect pests have been reported on maize crop in the state but only a few viz. cutworms, stem borers, white grubs, rice weevils, aphids, chafer and blister beetle are serious, cause considerable losses to this crop, and thus warrant control measures. The high yielding, disease and pest resistant varieties with appropriate maturity when grown with proper package of practices will greatly help to increase the maize production and productivity in the state.

References

Anonymous. (2012). Package of Practices *Kharif* Crops 2012. CSK Himachal Pradesh Krishi Vishvavidyalaya, Palampur.

Anonymous. (2014). Area, production and productivity maize in Himachal Pradesh. Directorate of Agriculture, Govt. of Himachal Pradesh, Shimla.

Anonymous. (2015). Annual Progress Report *Kharif* Maize 2015. All India Coordinated Research Project on maize. Indian Institute of Maize Research, New Delhi, India.

FAOSTAT, ProdSTAT. (2014). Module on crops containing detailed agricultural production data. www.faostat.fao.org.

Hallauer, A.R. and Miranda, J.B. (1988). Quantitative genetics in maize breeding. 2nd ed. Iowa State University, Press, Ames . Pp. 64-71.

Jompuk, P., Wongyai, W., Jampatong, C. and Apisitvanich, S. (2011). Detection of quality protein maize (QPM) using simple sequence repeat (SSR) markers and analysis of tryptophan content in endosperm. *Kasetsart J Natural Sci.* 40: 768-774.

Kumar, J., Kashyap, N.P. and Sharma, S.D. (2001). Pests of maize and their management in Himachal Pradesh-a review. *Agricultural Reviews.* 22(1): 47-51.

Panwar, V.P.S. (1998). Insect pests and their management in maize. *India Farm.* 98(4): 97-99.

Payak, M.M. and Sharma, R.C. (1985). Maize diseases and their approach to their management. *Tropical Pest Manage.* 31: 302-310.

Prasanna, B.M., Chaikam, V. and Mahuku, G. (eds) (2012). Doubled Haploid Technology in Maize Breeding: *Theory and Practice.* Mexico, D.F.: CIMMYT.

Saxena, S.C. (2002). Bio-Intensive Integrated Disease Management of Banded Leaf and Sheath Blight of Maize. *Proceedings of the 8th Asian Regional Maize Workshop, Bangkok, Thailand: August 5-8, 2002* Pp 380-390.

Thakur, S.S. and Vaidya, D.N. (2000). Management of cutworm, *Agrotis ypsilon* (Rottenburg) in maize. *Insect Environ.* 6(2): 91.

3

Maize Research and Development in Jammu and Kashmir

Bashir A. Alie, Zahoor A. Dar, Ajaz A. Lone, Gulzaffar, F.A. Nehvi M.I. Makdoomi, S.A. Dar, R.S. Sudan#, Abu Manzar, M.A. Ahangar H.R. Naik and S.Z. Hussain*

AICRP on Maize- Srinagar Centre (SKUAST-K),

[#]AICRP on Maize- Udhampur Centre (SKUAST-J)

Corresponding Author's Email: baelahi@gmail.com

Agriculture is the predominant sector in the economy of Jammu and Kashmir. Directly and indirectly, it supports about 80 per cent of the population besides contributing nearly 60 per cent of the state revenue. The total geographical area of the state is nearly 22.2 million hectare and population of 12.5 million, which is approximately 1.04% of the national total. The J&K is divided into three agro-climatic zones: Cold arid desert areas of Ladakh, temperate Kashmir Valley and the humid sub-tropical region of Jammu. Each has its own specific geo-climatic condition which determines the cropping pattern and productivity profile.

In Jammu province, a small portion of the land lies in the plains along the borders of Punjab while the rest of the area is hilly, dominates both in maize and wheat production. About 67 per cent of the area is under maize and wheat production with the production of 21.25q/ha maize and 15.36 q/ha wheat. This region contributes 79.56 per cent and 95.69 per cent of total production of these two cereals respectively. Even though the yield is not high, the region makes appreciable contribution to the cereals production.

The second agro-climatic zone Kashmir is also known as 'cultivator's paradise'. The region practically depends on irrigation, which is easily available. A large area of level land has alluvial soil. Extensive elevated plateaus of the alluvial or lacustrine material (locally called *Karewas*) also exist in the Kashmir valley. These *Karewas* are productive only in the face of sufficient rainfall or adequate irrigation facilities. Rice is the chief crop of this zone, followed by maize, oat, barley and wheat.

Ladakh zone is endowed with bare rocky mountains and bare gravel slopes. Villages are located near pockets of land with level ground and irrigation facilities, where cultivation is viable. In this region, barley is the major cereal crop followed by summer wheat. Millets and wheat rank second in importance and are grown in the warmer belt of the region.

Agricultural exports from Jammu and Kashmir include rice, wheat, maize, vegetables, apples, cherries, peaches, pears and saffron. From very long past till date, agriculture of J&K has been maize centric and will remain so for very long, as very large proportion of soil remains rainfed stipulation (85%) leaving very little scope of growing any important crop other than maize. With 34% share of gross cropped area maize is the most predominant crop which alone contributes 35% to the total food grain production. Thus, the goal of achieving self-sufficiency in food grain and also the sustainable growth of agriculture sector depend largely on the growth in maize production; Table 1 illustrates the current area, productivity and production statistics of major crops in Jammu and Kashmir.

Table 1. Crop Production Statistics of J&K (2015-16).

Parameters	Rice	Wheat	Maize	Pulses	Oilseeds
Area*	261.7	290	315.8	26.7	64.8
Production^	818.1	462.4	633.2	14.2	51.1
Productivity#	3126	1595	2004	530	789

* Area ('000 ha); ^Production ('000 tonnes); #Productivity (kg/ha)

Maize is the major crop of Jammu and Kashmir in terms of acreage under any crop, nearly one-third of the total cropped area is devoted to its cultivation. It is gown in the state during *Kharif* season and about 85 % of the copped area is rainfed. In Jammu & Kashmir maize is grown over an area of 315.8 thousand hectares with production of 633.2 thousand tones and productivity of 20.04 qtls per hectare (Anonymous, 2015). Besides, maize is the only alternative to rice in the circumstances of low rainfall which are of recurrent occurrence. For the last one decade the demand for maize grains has also been felt from every nook and corner of the state. Rough maize grain flour (Satoo) and fine grain flour for bread purposes is usually taken as breakfast food and in after noon working tea. It is the staple food of Gujjars and Bakar-wals, living in the Kandi and hilly areas. Moreover, the grains form an important cattle food, being fed to farm cattle and horses. The different parts of the plant and the grain are put to a number of industrial uses. The silk threads of maize are used as a filter; husks for making of mattresses. Maize provides huge quantities of fodder to the cattle in the state.

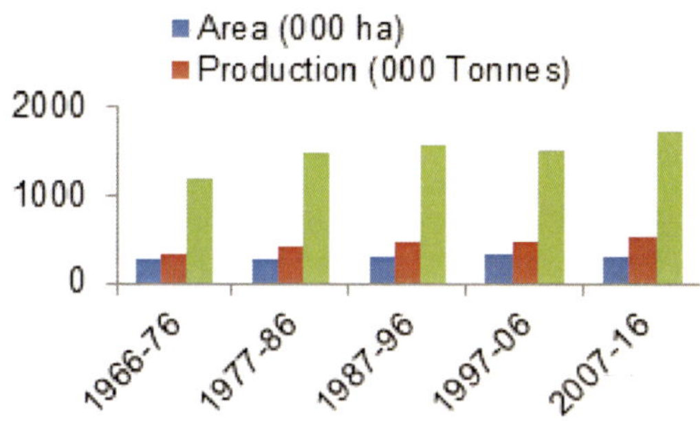

Fig. 1. Trends in Area, Production and Productivity of Maize in J&K state.

Maize production in J&K is increasing year by year (Figure 1). The increase in maize production in terms of area, output and yield indicate that maize is becoming one of the prominent agro produces in State. Maize is cultivated in almost all districts in State. The leading maize producing districts in State are Kupwara, Baramulla, Budgam and Anantnag. The crop is grown in two seasons – *Kharif* and *Rabi* . More than 90 per cent is grown during *Kharif* season. For the last few years there is a slight decline in the production due to less precipitation during flowering time in the state. In terms of production and yield, Kupwara stands first among the maize producing districts. The state has also witnessed maximum increase in maize area except Leh and Kargil districts.

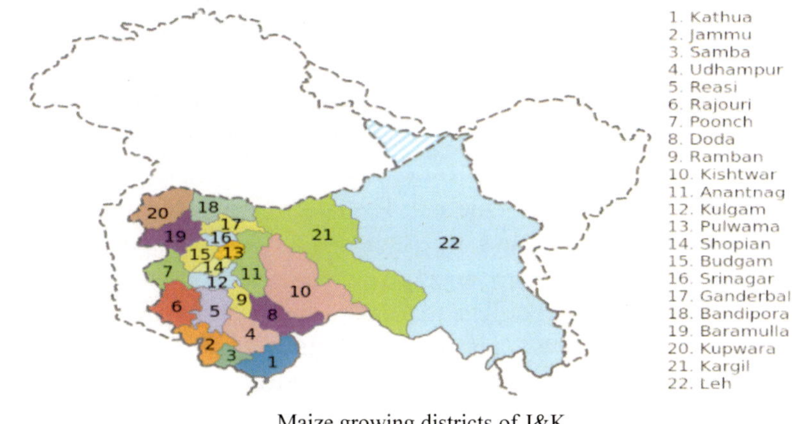

Maize growing districts of J&K

Cropping Pattern

Owing to variations in climate, soil and nature of irrigation, agricultural operations and the system of cultivation naturally vary from region to region. In the Jammu province, there are usually two crops a year, namely, *Rabi* in winter and *Kharif* in summer. The winter crops, consisting chiefly of wheat and barley, are sown between mid September and mid January, depending upon the moisture in the fields. These are harvested in May-June in the low-lying areas and in July-August at higher altitudes. The summer crops like rice, maize and millet are sown from mid July, according to the geographical location of the place and character of the soil. They are harvested between mid-August and mid November. As regards the rotation of crops, maize is often followed by wheat or sometimes by toria or barley and mustard, or by some fodder crop. The fodder crops are sometimes sown with cotton, especially on the irrigated lands. Sugarcane fields are frequently left fallow or a fodder crop is succeeded by two fallows and wheat, or by one fallow and cotton, or sugarcane. Cotton is generally preceded and followed by a fallow. Rice is generally grown on the same field year after year in the spring, the land being left fallow or some fodder crop being grown. Wheat is also sometimes grown on rich- manured fields but its output is generally poor. The rotation of crops is, however, often upset by scanty rainfall. In Kashmir province, land usually produces one crop a year; therefore it is known as *Ekfasli*. There are, of course, exceptions where farmers produce more than one crop in a year. Ploughing for rice, maize and other autumn crops in the Kashmir province commences in the middle of March. In April and May, seeds of these crops are sown. In June and July, oat, barley and wheat, sown in the previous autumn, are harvested.

Maize, rice and other *Kharif* crops are harvested in September and October. In November and December, ploughing for oat, wheat and barley is undertaken. During the October-November, rice and maize as well as other autumn crops are threshed/shelled. In Ladakh, like Kashmir, no customary rotation of crops is followed. However, wheat is not grown on the same land for more than two or three consecutive years, as this process is believed to deteriorate the soil. Wheat is always followed by some pulse. If the soil were much improvised, Pea is sown for a year as it strengthens the soil. The fallow is allowed to restore the exhausted strength of the soil. In some villages, land called *Dofasli*, gives two crops a year. Trumba, China or kangni give preference to gram. The time of sowing in the frontier districts differs from area to area. Generally, it commences early in the spring. In the low-lying areas, where the *Kharif* crop maize follows wheat, the wheat crop is sown anytime from 15 November to 15 January when the soil is not frosty. Maize is sown in July and August. In J&K, among the food grains, the main crops are rice (29.06 per cent), maize (31.90 per cent) and wheat (23.31 per cent) accounting for 84 per cent of the total cropped area whileas the balance 16 per cent is shared by inferior cereals and pulses. The commercial crops of significance grown in the state are apple and saffron. The consequence of such a cropping pattern is that the bulk of the cultivators have little to spare for buying other necessities of life.

Keeping in view the importance of maize cultivation in the State, a Maize Research Station at Srinagar was started in the year 1958 and was further strengthened with the sanctioning of the Coordinated Maize Improvement Project by ICAR in 1959. This Station called K. D. Farm (Srinagar) functioned as the main centre of Zone-1 which represented the Himalayan Hill region of J & K, U.P, Sikkim, H. P, West Bengal and North Eastern States. K. D Research station lies between 34.6^0 N and 74.5^0 East at an altitude of 1650 meters. It is about 29 Km away from SKUAST-K main campus at Shalimar and adjacent to Old Airport Srinagar.

Maize Research Station has made significant contribution towards achieving a breakthrough in the food production targets of the State. Sher-e Kashmir University of Agriculture Sciences and Technology of Kashmir has a mandate for development of composites and hybrids suitable for low to high altitude conditions of Kashmir valley and development of production and protection technologies for yield gains. Adaption of latest technologies by the farmers improved the production and productivity of maize in Kashmir by three fold.

Production issues for maize in J&K

Major constraints for maize cultivation primarily involves predominance of cultivated land races, non-availability of vital inputs, in accessibility, scattered and small land holding and cultivation of maize over a wide range of environmental conditions ranging from approximately 1650 m to above 2,600 m amsl, mostly under rainfed environments. Low productivity levels observed at farmers field is a big challenge for researchers and extension workers to ensure sustainability of maize. Lack of location specific varieties, senile fields, inadequate plant population, moisture stress, (rainfed cultivation) nutrient depletion, poor germination are the major issues at present that have resulted in the decline in production and productivity. The low and highly variable productivity levels are an impediment to investment by resource-poor farmers. Climatic risks like drought and sometimes heavy rainfall and edaphic constraints like poor water and nutrient retention capacity, low soil organic matter (SOM) make maize farming highly vulnerable.

Varietal Development

Exploitation of maize heterosis through the development of modern high yielding hybrids and synthetics has gradually replaced the low yielding maize populations at a faster rate in maize growing regions of the world. However little impact was realized in Kashmir in the form of area enhancement under these modern varieties. The most important reason behind the fact was that hybrids are developed for environments that are more favorable and add a non-significant gain in the marginal environment. The expensive seed cost further exacerbate the situation. In some cases diminishing returns were realized because of their poor adaptability under cold temperate conditions of Kashmir. Some land race populations of maize are still popular with the farmers of Kashmir. The main reason of their popularity even in the circumstances of availability of high yielding varieties bred by public and private sector are adaptability, greater resistance to natural factors, early maturity, good grain quality, great food and fodder value and ability to thrive best even under low input conditions. However, these are losing their popularity and are gradually going out of farmer's domain and becoming extinct because of low yielding potential, low resilience to some biotic stresses, and lower sensitivity to inputs.

An overview of Maize Land Races of Jammu & Kashmir

During the last more than three decades of its existence, SKUAST-K has developed altogether 15 maize varieties adapted to diverse maize growing situations meeting varied needs of the farmers. Its AICRP center has established itself as a front runner among the maize AICRP centers of the country with its significant accomplishments in temperate maize technologies. The maize technologies developed by the University are now ruling the maize fields of J&K and even few are popular in the neighboring states/national level as well. Srinagar centre since its inception has given top priority to early maturing, cold tolerant and disease resistant germplasm in addition to production and productivity augmentation. In this context Maize germplasm is being collected on routine from both indigenous and exotic sources for its evolution at the centre. SKUAST-K is in the process of significant contribution towards achieving a breakthrough in the food production targets of the state. The increase of 150 per cent in production of maize as compared to marginal increase of nearly 28 percent in cultivable area under maize over the last two decades speaks of the contribution made by the maize Scientists in the state. Composite breeding programme has been so far a major thrust of research with the aim of development of Extra Early /Early/ Medium maturing composites with high yield potential and tolerance to biotic and abiotic stresses

A large number of high yielding maize varieties mostly composites *viz.* C6, Super-1 (Mansar), C8, C14, C15, Shalimar KG Maize-1, & Shalimar KG Maize-2 have been developed by Srinagar Centre and released for commercial cultivation in J & K State. These composites have been developed through hybridization of indigenous and exotic materials. During the last five years Srinagar centre has developed two high yielding Composites viz Shalimar Maize Composite-6 and Shalimar Maize Composite-7, two hybrids viz Shalimar Maize Hybrid-1, Shalimar Maize Hybrid-2 and two specialty corn composites viz Shalimar Popcorn-1 (National release) and Shalimar Maize Composite-5 (QPM) along with their production technologies suitable for cultivation at high altitude (1800-2250 m amsl) and low medium altitude areas of J&K (1400-1800 m amsl) respectively, Table 2 illustrates the Cultivars released for state by AICRP centre Srinagar.

Table 2. Details of Composites/Hybrids Released by AICMIP Srinagar Centre.

Composite/Type	Year of Release/by	Salient features
C_6	1976SVRC	It has vigorous medium tall plants with tendency to bear two ears /plant. It has a yield potential of about 55-60 q/ha. , matures in 155-160 days in the Valley and 125-130 days in the mid elevations. It is recommended up to an elevation of about 2000 meters.
C_{15} (Rehmat)	1976SVRC	It has short plants with low ear placement; ears are long orange yellow semi-dent to dent grains. It has yield potential of 50-55 q/ha. It matures in 135-140 days in the Valley and is recommended for higher elevations up to 2400 meters.
Super-1 (Mansar)	1978SVRC	Yield potential 50-60 q/ha. Matures in about 140-145 days in the valley. 105-110 days in the int6ermediate zone and 90-95 days in the lower hil8ls and plains. Fairly resistant to foliar diseases.
C_8	1996SVRC	Yield potential 55-60 q/ha. Matures in 150-155 days in Kashmir Valley 110-115 days in the intermediate Zone and 95 days in the lower foot hills of J&K.
C_{14}	1996SVRC	It is an early maturing composites and ear replace C_6. The composite matures in 135-145 days in the Valley and can be grown up to 2250 meter elevation. Yield potential: 55-60 q/ha.
KG-1	2005SVRC	Medium tall plants with white yellowish grains. Extra early maturity composites, matures in 120-125 days. Recommended to an altitude of 2250 mamsl. Yield potential: 45-55 q/ha.
KG-2	2005SVRC	It is an extra early maturity composite. The composite mature in 125 days can be grown at an altitude of 2250 mamsl. Yield potential: 35-40 q/ha.
Shalimar Maize Composite-3	2009SVRC	An early maturing composite matures in 135-140 days (1800-2000 m amsl) and 150-155 days (2000-2250 m amsl). Suitable for early sown conditions due to season limitations the average yield potential is 39.5 q/ha with 10.6% protein.
Shalimar Maize Composite-4	2009SVRC	A medium maturity composite and matures in 140-145 days when grown at an altitude of 1800 m amsl

		and in 100-105 days when grown at 1300 m amsl. Yield potential is 58-63 q/ha with good protein content of 10.63 per cent.
Shalimar Maize Hybrid-1	2009SVRC	Early maturing hybrid matures in 125-130 days (1800-2000 m amsl) and 140-145 days (2000-2250m amsl). Yield potential is 50-55 q/ha with moderate resistance to TLB and resistant to stem borer.
Shalimar Maize Hybrid -2	2013SVRC	Recommended for Mid altitudes of Jammu & Kashmir between 1650-1800 m amsl (irrigated/rainfed). Te hybrid has a yield potential of 70.0 quintals/ha, it is yellow in colour, early in maturity.
Shalimar Maize Composite -5	2013SVRC	Recommended for High altitudes areas of Kashmir between 1800-2250m amsl. (rain fed). The variety has a yield potential of 50.0 quintals/ha, it is white in colour, early in maturity.
Shalimar Maize Composite-6	2013SVRC	It has yield potential of 55-60 q/ha. It matures in 135-140 days in the Valley and is recommended for higher elevations up to 2200 meters. It is yellowish orange bold grain composite.
Shalimar Maize Composite-7	2013SVRC	Recommended for Mid altitudes of Jammu & Kashmir between 1650-1800 m amsl (irrigated/rainfed). The composite has a yield potential of 65.0 quintals/ha , it is yellow in colour, early in maturity.
Shalimar Pop Corn-1	2016CVRC	Recommended for Zone-1, 2, 3 and 5 across India, it's a yellow flint maize with yield potential of 5 t/ha and a popping of 80%. It's early in maturity and disease resistant.
Shalimar Sweet Corn-1	2017SVRC	High yielding and extra early maturity sweet corn composite with distinct sweet corn traits, tender cobs and inbuilt tolerance to biotic and abiotic stresses prevalent in the mid altitude areas of Kashmir
Shalimar QPMH-1	2017SVRC	The single cross QPM hybrid is suitable for areas up to 1850 m in elevation in Kashmir Valley. It's an early maturing, orange yellow flint hybrid with a tryptophan content of 0.81 % of the total protein and a yield potential of 60 q/ha. It is resistant to TLB and Common Rust Disease.

Maize seed production in the state

Desired quantity of Breeder seed of released varieties of SKUAST-K was produced and supplied to the indenting agencies (Table 3).

About 9775 Kgs of Breeder Seed of released varieties was made available to the Nodal Officer, SKUAST (K) for its onward distribution to the indenting agencies.

Table 3. Summary of breeder seed production by Srinagar centre (2010-2016).

Variety	Seed produced (Kg)						
	2010	2011	2012	2013	2014	2015	2016
C 6	150	250	725	450	500	400	—
C 15	160	550	300	450	400	600	—
C 8	90	100	—	50	350	250	—
C14	50	—	—	—	—	—	—
SMC-4	450	150	—	—	180	—	500
SMC-3	100	—	—	—	50	100	600
SMC-5	80	—	—	—	100	80	300
Shalimar KG-2	35	—	—	—	75	350	150
Super-1	30	20	—	—	—	—	—
SMC-6	—	—	—	—	—	—	60
SMC-7							400
SMH-2	—	—	—	—	—	—	117 (FS)
Male parent of SMH-2							50
Female parent of SMH-2							90

Changes in cropping pattern

The cropping pattern of a region reveals the proportion of area of land under different crops at a point of time, the rotation of crops and the area under double cropping. The cropping pattern changes in space and time. In fact, no cropping pattern can be good and ideal for all times to come. The cropping systems of a region are decided mostly by a number of soil and climatic parameters which determine overall agro-ecological setting for nourishment and appropriateness of a crop or a set of crops for cultivation. Nevertheless, at the farmer's level, potential productivity and monetary benefits act as guiding principles while opting for a particular crop/cropping system. These decisions with respect to choice of crops and cropping systems are further narrowed down under the influence of several other forces related to infrastructure facilities, socio-economic factors and technological developments, all operating interactively at micro-level. The prevalent cropping system of any locality is, therefore, the cumulative result of the past and present decisions by individuals, communities or governments and their agencies. The cropping pattern plays a vital role in determining the level of agricultural production and reflects the agricultural economy of an area/region. A change or shift in cropping pattern implies a change in the proportion of area under different crops which depends, to a large extent, on the facilities available to raise crops in the given agro-climatic setting. Moreover, the natural, social, economic and historical factors, which determine the cropping pattern of a region, the cropping, pattern also changes in consonance with the government policies and technological innovations especially in agriculture. It is, however, pertinent to mention that in most of the areas it is the availability of water more than any other input, which determines the nature of agricultural production. Cultivators in regions and localities without sources of water- supply to supplement rainfall are constrained in their choices of cropping patterns and have to suffer enforced adjustments when the rains are irregular, either in their timing or in their quantity. Hence the state of agriculture of a region depends vitally upon the security and flexibility provided by irrigation facilities. Characterized by mountainous and undulating terrain, the state of Jammu and Kashmir has micro-level variations in the agro-climatic conditions. Consequently, the cropping patterns and crop combinations differ significantly at the macro and micro levels. Not surprisingly, therefore, the cropping pattern of the state had been dominated by paddy– the staple food of the majority of the people- in those areas which had adequate irrigation facilities and by other food crops like maize in the rain fed areas. The peculiar physical character and climate has from times past been a serious handicap to intensive farming and diversification of crops in the valley. As the land remained under snow for 2-3 months a year, therefore, rice was the main *Kharif* crop of the valley. Moreover, the harvesting of the crop in September-October left very little time for the sowing of another crop in the same land and, therefore, *Rabi* crops were sown only in such lands as had not been cultivated during the *Kharif* season. The farmer of Kashmir had, therefore, to subsist on one crop economy, either *Kharif* or *Rabi*. Of all the *Rabi* crops only such crops were sown which germinated before the snowfall and started growing after the melting of the snow in the months of March and April. Hence, it goes without saying that the *Rabi* crops which took 4-5 months to mature in other parts of India took 5-6 months for the same in the valley of Kashmir. The important *Rabi* and *Kharif* crops grown in the valley are: oats, wheat, rapeseed, barley, pea, rice, maize, saffron and moong.

The greater shift of area towards maize wheat production was because of the technological breakthrough, which brought about spectacular enhancement in productivity of these two crops. Rape and Mustard emerged as the main oil seed crops, mustard being found remunerative on the paddy lands of Kashmir on which it was cultivated as winter crop. The cultivation of fodder assumed a respectable position in the cropping pattern of the state during the report period.

Maize package of practices not only differs for various cropping systems in different regions of the state but also requires some adjustment to meet the specific needs of the individual farmer so as to help him to increase his productivity and profit. For realizing high yield in maize, it is essential to follow the package of practices at the appropriate time. Any lapse in the execution of any one component of the package or a delay in operation is likely to adversely affect the yield level.

Changes in disease and pest scenario

Among the biotic stresses estimated annual yield and quality loss in maize due to the different diseases is 6-11 per cent. In Kashmir valley, the impact of diseases on maize production and productivity has increased over time. In current maize-cropping environment, a few major diseases have caused significant yield losses in maize under temperate agro-climatic conditions of Kashmir. Also in the present climate change scenario, maize crop is facing the tough competition of new diseases, which were otherwise not touching the economical threshold level.

Maize cultivation in Kashmir is subjected to the diseases mainly caused by phytopathogenic fungi however, the disease risk of maize crop in Kashmir due to bacteria and viruses cannot be ruled out. The current maize-cropping environment favours the emergence and specialization of new pathogen species by providing a new ecological niche. Introduction of new diseases is also quite possible because more than twenty diseases of maize can be transmitted on or inside maize grains.

Of the various diseases commonly observed on maize under Kashmir agro-climatic conditions the more destructive ones are Turcicum Leaf Blight also called as Northern leaf blight of maize caused by fungus *Exserohilum turcicum* (Pass.) Leonard and Suggs (Syn. *Helminthosporium turcicum* Pass.), teleomorph; *Setosphaeria turcica* Luttrell (Syn. *Trichometasphaerium turcica* Luttrell), common rust (*Puccinia sorghi*), Common smut (*Ustilago maydis*) and Maydis leaf blight (Teleomorph: *Cochliobolus heterostrophus*, Anamorph: *Bipolaris maydis*, syn. *Helminthosporium maydis*). Besides some minor diseases such as banded leaf and sheath blight [*Rhizoctonia solani* (Kuhn); Teleomorph: *Thanatephorus cucumeris* (Frank)],Brown Spot (*Physoderma maydis*), Gray Leaf Spot (*Cercosporazeae-maydis*), Eye spot (*Aureobasidium zeae* previously known as *Kabatiella zeae*) and Downy mildews also contribute in yield loss of maize under temperate agro climatic conditions of Kashmir. A group of pathogens are responsible for downy mildews of maize. The important species causing downy mildew in maize in the temperate region are crazy top downy mildew (*Sclerophthoram acrospora*), Philippine downy mildew [*Peronosclerospora philippinensis* (Weston) Shaw] and sorghum downy mildew [*Peronosclerospora sorghi* (Weston & Uppal) Shaw].

Maize crop is also severely attacked by seed rots and seedling blight, a disease complex caused by several seed borne and soil borne fungi including species of *Pythium*, *Fusarium*, *Acremonium*, *Aspergillus*, *Penicillium* and *Sclerotium.* The diseases is favoured by cold and wet environmental conditions and reduce plant stands in temperate and high-altitude areas where soil temperatures are low at planting time.

Table 4: Package of Practices for various types of Maize

Parameters	Normal Maize	Baby corn	Sweet corn	Pop Corn	QPM	Fodder Maize
Land Selection	Maize can be grown successfully in variety of soils. However specialty corn fits well in peri urban agriculture. It can be very profitably grown in areas surrounding large cities and towns. It can successfully be grown in well-drained soils with a pH of 5.5-7.0.					
Land Preparation	If land is fallow one deep ploughing before winter is recommended for moisture conservation. Before sowing 2-3 ploughings across the slope along with FYM in the soil may be given followed by proper leveling for better establishment of the crop					
Time of Sowing	For higher belts 1st of April to 30th of April For lower belts 1st of April to 31th of May	1st of April to 15th of June	1st of April to 5th of June	1st of April to end of May	1st of April to end of May	1st of April to 15th of June
Varieties	KG-1, KG-2, SMC-3, SMC-4 SMC-6, SMC-7, SMH-1, SMH-2	HM-4, VL-78 and Prakash.	Madhuri, Priya, Win Orange Sweet Corn , Sugar-75 and Mishti	Amber Pop corn, VL Pop corn, and Shalimar Pop corn-1	SMC-5, HQPM-1, VivekQPM-9, HQPM-5	African Tall, J1006, Pratap Makachari-6
Seed Rate (kg/ha)	Composites Line sowing-25 Broadcasting-30 Hybrids Line sowing-20 Broadcasting-25	Composites Line sowing-35 Broadcasting-40 Hybrids Line sowing-30 Broadcasting-35	Composites Line sowing-14 Broadcasting-18 Hybrids Line sowing-10 Broadcasting-12	Composites Line sowing-16 Broadcasting-20 Hybrids Line sowing-14 Broadcasting-18	Composites Line sowing-25 Broadcasting-30 Hybrids Line sowing-20 Broadcasting-25	Composites Line sowing-60 Broadcasting-80 Hybrids Line sowing-60 Broadcasting-70
Method of Sowing	Line sowing at a spacing of 60cm or 70 cm x 20 cm. Sow the seed 3-5 cm deep.	Line sowing at a spacing of 50cm or 55 cm x 20 cm. Sow the seed 2-4 cm deep.	Line sowing at a spacing of 70cm or 75 cm x 20 cm. Sow the seed 2-4 cm deep.	Line sowing at a spacing of 60cm or 65 cm x 20 cm. Sow the seed 3-5 cm deep.	Line sowing at a spacing of 65cm or 70 cm x 20 cm. Sow the seed 3-5 cm deep.	Line sowing at a spacing of 25cm x 10 cm. Sow the seed 3-5 cm deep.
Nutrient Management (FYM t/ha:N:P:K:Zn kg/ha	Irrigated hybrids 15:150:75:40:20 Irrigated composites 15:120:60:30:15	Irrigated hybrids 15:175:75:40:20 Irrigated composites	Irrigated hybrids 15:150:75:40:20 Irrigated composites 15:120:60:30:15	Irrigated hybrids 15:150:75:40:20 Irrigated composites	Irrigated hybrids 15:150:75:40:20 Irrigated composites	Irrigated hybrids 15:200:75:40:20 Irrigated composites 15:170:60:30: 0

	Unirrigated hybrids 15:90:50:30:15 Irrigated hybrids 15:75:40:20:10	15:150:60:30:15 Unirrigated hybrids 15:120:50:30:15 Irrigated hybrids 15:90:40:20:10	Unirrigated hybrids 15:90:50:30:15 Irrigated hybrids 15:75:40:20:10	15:120:60:30:15 Unirrigated hybrids 15:90:50:30:15 Irrigated hybrids 15:75:40:20:10	15:120:60:30:15 Unirrigated hybrids 15:90:50:30:15 Irrigated hybrids 15:75:40:20:10	Unirrigated hybrids 15:150:50:30:15 Irrigated hybrids 15:120:40:20:10
Weed management	Application of atrazine@ 1 kg a.i./ha diluted in 600-800 litres of water is recommended as pre-emergence herbicide followed by running a weeder in between the rows at 30-40 DAS that will help weed control and better root aeration apart from soil water conservation. Subsequently, a tractor mounted ridger is run to prevent plant lodging and minimize volatilization losses of urea					
Irrigation	Most of the maize area is rainfed. If required provide 2-3 irrigations at the most critical periods i.e. at knee high, silking and grain filling stages					
Intercropping	To fetch additional income corn can be intercropped with other highly remunerative crops like pulses, vegetables etc					
Pest and Disease Management	A preventive spray of Endosulfan 35EC @2ml/1 of water is given to 20-25 days old maize plants to take care of stalk borers *Chilo partellus*. Disease problems tend to be sporadic. Troublesome diseases include leaf blight and post-flowering stalk rots if it appears, Use crop rotation and avoid sequential planting in adjacent fields to minimize disease. A spray with Bavistin @1 g/litre takes care of foliar diseases.					
Harvesting of Cobs	Cobs are harvested at 16% grain moisture when leaves turn completely brown and silk dries	Picking should be done on alternate days within 1-3 days of silk emergence from the leaf sheath depending upon the variety	Cobs are harvested when kernels are fully developed and exude a milky liquid when punctured. Delaying the harvest will progressively reduce the sugar content in the kernels.	Harvesting should be done accurately and carefully at 16-18% moisture so that kernel damage is prevented.	Harvesting is done at 16% grain moisture when leaves turn completely brown and silk dries	Harvested at 50% flowering

Post-Harvest Handling	Do not heap ears, Shell cobs and store in the gunny bags.	Do not pile ears, keep harvested ears in cool and shaded areas for use	Do not puff up ears, keep harvested ears in cool and shaded areas for use	Do not heap ears, Shell cobs and store in the gunny bags.	Do not heap ears, Shell cobs and store in the gunny bags.	----
Yield	80-85 q/ha	12-20 q/ha	80,000-95,000 cobs/ha	45-60 q/ha	55-75 q/ha	-----
Fodder/ Green Fodder yield	150-250 q/ha fodder yield including cob husk	350-400 q/ha green fodder yield including, detasseled tassel, cob husk	300-350 q/ha green fodder yield including husk	130-200 q/ha fodder yield including cob husk	150-225 q/ha fodder yield including cob husk	450-550 q/ha green fodder yield

Yield loss in maize caused by diseases can be minimized by proper management strategies. An integrated approach that uses combination of different methods to control disease is cost effective and sustainable way to minimize the yield losses. Varietal resistance is primary element of integrated disease management and is most acceptable method for farmers particularly under high altitude conditions of Kashmir. Maize resistance breeding in Kashmir by SKUAST-Kashmir has gained large achievements in managing the losses due to many important diseases particularly Turcicum leaf blight , Common rust and Common smut(Ahangar *et al*., 2015; Ahangar *et al*., 2016). In the present climate change scenario some other diseases in temperate maize production ecosystem could collectively cause chronic yield losses in maize. To meet such a challenge research on host plant resistance is focused to develop resistant varieties with broad resilience to a range of diseases by using donor sources like VL-102, NAI-209, NAI-175, NAI-138, KDL-211, VL-1034, V-370, CML-451, CML-474, CML-472, CML-360, CML-470, CML-165 and PS-39.

Cultural practices such as optimum plant density, planting date, crop rotation, healthy seed, balanced fertilization, weed and water management as per the recommendations are very essential tools for maize disease management. Different production technologies as well as protection technologies have been standardized by SKUAST- Kashmir for successful maize production under temperate agro-climatic conditions of Kashmir.

Stem borer (*Chilo partellus*)

Stem borer is the most serious pest of maize in Kashmir valley. The adults emerge at the end of April or in early May. After mating, the females lay eggs generally on the underside of the leaves of host plants. The freshly hatched caterpillars feed on tender leaves at seedling stage causing symptoms of pinhole, windowing and finally drying of central whorl and dead heart, resulting in wilting and drying of infested plants. In older plants it cuts central leaves and also feed from margins of the leaves. Presence of perforated terminal leaves is an indication of infestation. Maximum activity of stem borer is recorded between 18th June (10 DAG) to 29th June (30 DAG) and activity continue till August (45DAG).The pest remains active from May to September.

Management

- Increase seed rate (25% of the normal).
- Crop rotation with non host crops reduces the pest incidence.
- Remove and burn all the left over stubbles and trash from the field after harvesting of the crop which harbour borer and act as source of infestation for the next crop.

- Roughing of infested plants and removal of dead heart.
- Growing of resistant varieties.
- Deep ploughing and proper spacing
- Whorl application of carbofuran 3G @ 3-5 granules/ whorl 15 days after germination.
- Application of *Trichogramma chilonis* @ 1.5 lac/ha at 15 days after germination followed by second application after two weeks.

Cut worm (*Agrotis ipsilon*)

A serious pest of maize in Kashmir valley. The caterpillars feed on young plants by cutting them off a little below or above the surface of soil. It destroys many more plants than it feeds upon, leading to reduction in plant stand and yield. Although infestation of cutworms typically is not uniform within a field, the damage to standing maize crop therefore appears in patches only. The pest causes damage to the crop from 2nd week of April to 3rd week of June and its incidence is recorded up to July in maize crops at higher altitudes.

Management

- Crop rotation with non host crops reduces the non host crops.
- Mechanical collection and killing of larvae sheltering under provided small heaps of grasses kept in rows/bunds.
- Seed treatment of maize either with imidacloprid 200 SL (Gaucho) @ 4 ml/ kg of seed or carbosulfan 25 ST (Marshall) @ 30 gm/ kg of seed or deltamethrin 2.8 EC (Decis) @ 36ml/ kg of seed are very effective in reducing the crop damage.
- Higher seed rate of 30-40 kg/ hectare is recommended in compensating the plant damage caused by the cutworm in maize. This should be followed by thinning to maintain required plant population in case cutworm damage is not severe.
- Early sowing of maize (2nd to 3rd week of April) reduces the damage.
- Ploughing and pre-sowing weeding in the fields 15 days before sowing of maize reduces the cutworm damage.
- Under heavy infestation, application of carbofuran 3G @ 32.5 KG/ ha or Quinalphos10 dust @25 kg/ ha or carbaryl 10% dust @ 20-25kg/ ha before sowing of seeds during last ploughing of field is most effective in managing cutworm.
- Deep summer ploughing helps to expose pupae/ larvae to scrotching heat and bird predation.

- Installation of light traps @ 5 per hectare and pheromone traps @ 20 per hectare for monitoring the pest.
- Use pit fall traps so that the larvae are drowned, collected and destroyed.
- Use of resistant varieties.

Future Scenario: Borer Complex Stem borer (*Chilo* spp.), Pod borer (*Helicoverpa* spp.) and Armyworm (*Mythimna* spp.). Among borer complex these two *viz.*, Pod borer (*Helicoverpa* spp.) and Armyworm *Mythimna* spp.) are have infested on maize , graminaceous crops and weeds. These are sporadic pest. These pests mostly attack the crops at the whorl forming stage. Young larvae feed on plant whorl, while as, grown up caterpillars feed on other leaves leaving mid rib and skeletonise them. These pests remain active from April to June. The prevailing weather conditions *viz.*, dry spell, cloudy, followed by intermittent rainfall, wind and rainstorm helps in migration of moths to long distance; abundance of alternate host can lead to serious damage in future.

Post Harvest Management

Corn flour has become an attractive ingredient in the extrusion industry due to its best expansion and puffing (Gujral *et al*., 2001, Anonymous, 2012).

Proximate composition of corn

Corn variety	Moisture (%)	Ash (%)	Fat (%)	Protein (%)	Total sugars (%)	Starch (%)
C8	12.20	1.60	3.40	8.20	1.20	69.20
C4	11.6	1.60	3.38	8.97	1.60	68.82

Food extrusion is one of the latest multidimensional food processing techniques. Great possibilities are offered in food processing field by the use of extrusion technology to modify physicochemical properties of food components. Extrusion cooking technology has almost limitless applications in the processing of cereal-based foods and other materials, and is associated with partial or complete gelatinization of the starch, complex formation, transformations and interactions involving biopolymers. The technique may be used to precook, instantize and agglomerate food components (Cheftal, 1990). Extrusion has for years provided the means of producing new and creative foods.

Starch is the dominant polymer in corn and plays important role in developing extruded products by giving good expansion, binding and mouth feel to the product. The quality of the final product depends on the processing conditions used during extrusion and this includes the composition of the raw materials, feed moisture, barrel temperature, screw speed and screw configuration. Moreover ingredients and formulation play an important role in developing the texture of the extruded product and ultimately the acceptability of the product to the consumer. Most snack foods made from cereal grains (oats, rice, corn) are usually low in nutritional components such as vitamins and minerals. Therefore the snacks that are rich in fibre, vitamins and minerals and low in fat, are more popular and high in demand. Incorporation of fruits and vegetables in the extruded snack shall give a new perspective and alternative for healthy snack food production. Division of Post Harvest Technology has explored the possibility of corn (C4& C8) for the development of nutritious ready to eat foods by extrusion technology.

C8 variety was studied with water chestnut and C4 was blended with apple for development of nutritious extruded snacks.

Optimization of the process for development of products was done by Response surface methodology. The optimum conditions for development of corn based water chestnut incorporated extrudates were 40%, 14%, 300 rpm and 170^{0}C for water chestnut incorporation, feed moisture, screw speed and barrel temperature respectively. The optimum conditions for development of corn based apple incorporated extrudates were 10%, 15%, 450 rpm and 170^{0}C for apple incorporation, feed moisture, screw speed and barrel temperature respectively. The optimized products were evaluated for following nutritional attributes.

Proximate Analysis of optimized products.

Product	Moisture	Protein	Fat	Ash	Total sugars	Starch
40 % water chestnut: 60% corn	6.64	3.8	2.04	1.54	5.43	58.20
10% apple:90% corn	4.36	6.4	1.53	1.71	3.23	56.8

Impact of production technologies through FLDs at farmers field

By increasing the productivity of maize in the valley we can remove food, fodder & feed deficit to a greater extent therefore primary concern of the researchers aims at transfer of technologies to the maize farmers that ensure high productivity. In this pursuit SKUAST-K Shalimar has conducted 4726 Front Line Demonstrations in all the 10 districts of Kashmir valley from *Kharif* 2005 to 2016 with the following objectives:

- To popularize cultivation of high yielding maize composites and hybrids suitable for temperate conditions of Kashmir
- To promote and popularize scientific methods of maize cultivation recommended for this region
- To educate progressive maize farmers to save their own seed with special reference to composites and hybrids

In order to have better impact of the generated technologies, Front line demonstrations in maize were conducted at various farmers fields in District Budgam, Baramullah, Kupwara, Bandipora, Anantnag, Pulwama, Srinagar, Shopian and Ganderbal, (Table-5) to demonstrate the production potential benefits of latest production technologies viz-à-viz traditional farming practices.

The purpose of these FLD's was to understand the yield gaps between FLD's and farmers practices and to find out the reasons for low yield and specific constraints with the maize farmers. The impact of production technologies was generated on the basis of output data and inputs used per hectare were collected from the frontline demonstration.

Table 5. Year-wise distribution of FLDs and farmers covered for popularization maize production technologies in Kashmir.

Year	Area (ha)	Farmers covered	Districts covered
2004	77.20	193	06
2005	520.78	1447	06
2006	132.80	368	06
2007	250.00	715	06
2008	69.20	198	06
2009	143.60	376	06
2010	91.60	297	10
2011	135.20	486	10
2012	138.00	456	10
2013	134	321	10
2014	171	478	10
2015	12	21	10
2016	10	19	10
Total	**1720**	**5421**	

The data in (Table 6 and Fig 2) showed that the yield variations were quite large during the year 2004-2016. In total 4726 front line demonstrations were conducted during the period on improved practices (IP) v/s Farmers Practices (FP).

Table 6. Year-wise average yield variation between improved practices and farmers practices.

Year	FLDs (no.)	FLD yield q/ha)	Farmers practice yield (q/ha)	Increase over FP (%)
2004	193	39.30	29.15	34.82
2005	1302	39.49	27.50	43.60
2006	332	36.39	24.70	47.33
2007	625	41.32	23.40	76.58
2008	173	42.48	28.90	46.99
2009	359	37.71	30.10	25.28
2010	229	63.04	27.30	130.92
2011	358	44.55	25.75	73.01
2012	345	43.56	26.65	63.45
2013	335	48.43	24.25	99.71
2014	435	44.35	28.25	56.99
2015	21	48.70	24.65	97.56
2016	19	46.20	23.65	95.34
Total/Av.	**4726**	**44.27**	**26.48**	**68.58**

In IP mainly recommended dose fertilizer (90 kg N,50 kg P2O5&30 kg K_2O/ha) under rainfed situation and (120 kg N,60 kg P2O5&40 kg K_2O/ ha) in irrigated condition along with seed treatment, optimum plant population load, proper weed management were practiced.

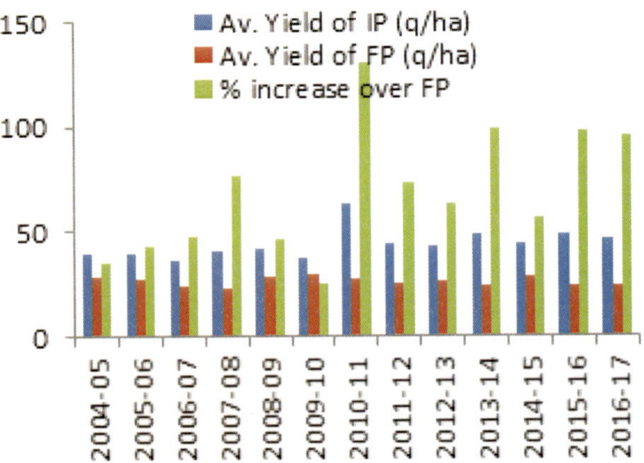

Fig. 2. Year-wise average yield variation between improved practices and farmers practices.

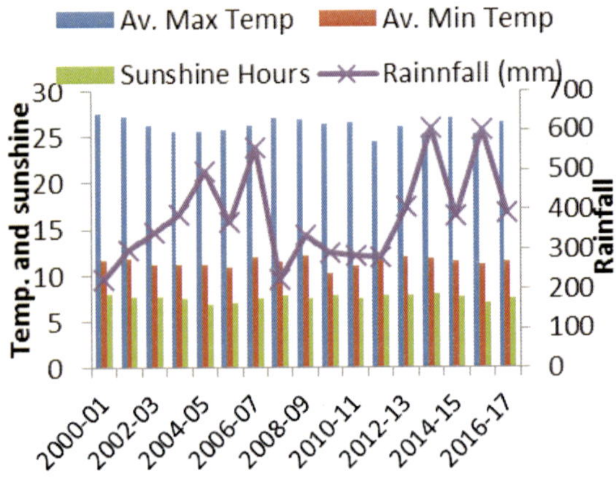

Fig. 3. Pattern of weather elements in Kashmir valley.

The average yield of maize with farmers practice in 4726 demonstrations remained 26.48 qtls/ha whereas with the inclusion of location specific improved practices the yield levels were enhanced to 44.27 q/ha with a significant yield increment of 68.58 & 164.67 over the farmers practices and state average yield respectively.

Figure 3 is a graphic depiction of behavior of different weather elements in Kashmir valley since the turn of the millennium in the maize growing season.

New Initiatives

A) Identification of drought resilience in the germplasm resource available

Elite lines identified include KDM-361A, CM-129, KDM-372, KDM-331, KDM-1051, KDM-402, KDM-463, KDM-717, KDM-912A, KDM-932A, KDM-343A, KDM-961, KDM-918A, KDM-1156 and KDM-1236. These exhibited earliness in maturity and superiority in morphological, physiological, yield, quality, seedling and root traits indicating the inbuilt potential of these elite lines. These elite lines exhibited maximum per cent decrease over population mean for days to 50% tasseling (13.29%), days to 50% silking (15.07%), days to maturity (8.76%), ASI (42.64%) and stomatal count (14.0%) in stress regime. Elite lines exhibited maximum per cent increase for plant height (12.8%) and ear height (14.01%), leaf relative water content (17.63%), canopy temperature before flowering (8.06%), canopy temperature before maturity (12.4%), chlorophyll content before flowering (29.3%), chlorophyll content before maturity (49%), ears plant^{-1} (60%), grain yield plot^{-1} (42%), kernels row^{-1} (21.3%), 100 grain weight (19.1%), protein content (22.2%), germination % (28.4%), number of seminal roots (74.24%), number of crown roots (75.20%), primary root length (85.10%), fresh root weight (90.09%) and dry root weight (89.28%) in stress regime as compared to well watered regime.

Molecular characterization of elite lines showing signs of moisture stress tolerance was carried out using SSR markers. Major allele frequency ranged from 0.53 (Phi-061) to 0.13 (umc-1424) with a mean of 0.29 respectively. The 32 SSR makers detected a total of 239 alleles in 15 maize lines. The number of alleles per locus ranged from 11.00 (umc-1766) to 4.00 (umc-1664) with a mean of 7.47. Heterozygosity ranged from 0.60-0.00 with a mean of 0.02 and the polymorphic information content (PIC) values ranged from 0.886 (umc-1766) to 0.608 (Phi-051) with an average of 0.782 respectively. DARwin derived cluster analysis grouped 15 elite maize lines in three major clusters.

B) Participatory Varietal Evaluation

A client oriented comprehensive initiative was undertaken to address location specific issues pertaining to maize cultivation in various maize growing pockets of intermediate altitude zones of valley. In this regard four trials were constituted comprising of University bred varieties, advanced pipeline genotypes and extra early maturity genotypes received from VPKAS, Almora. The validation trials were laid in three replications involving test entries and local checks and were laid in collaboration with KVK's in the Districts of Pulwama (Harpora), Kupwara (Warnow), Budgam (Sithaharan) and Bandipora (Sumlar), and one set of trial was also laid at the Station for metric trait estimation and validity of findings.

The basic aim of the trials was to access the performance of cultivars in actual farm conditions and to accumulate feedback from the farming masses so that a redefined and comprehensive strategy can be devised for the betterment of maize productivity. The information generated from the trials was collected in the form of questionnaire pertaining to diverse facets of maize cultivation, utilization and farming pattern of the individual farmers. The cumulative average results of the trials are based on individual experience of the farmers.

C) Multi-locus profile and SSR Barcode of five single cross maize hybrids

Multi-locus profiling of five single cross hybrids was carried out using 45 genome wide SSR markers. Ten most polymorphic markers were identified based on their PIC values. These included the Markers Umc-283, Umc-1664, Umc-1147, Umc-2245, Umc-2372, Umc-1594, Phi -061, Umc-1002, Umc-1424 and Phi - 452693. These ten Markers were used to generate barcode or a DNA profile for five single cross hybrids. SSR Marker Umc-2383 amplified 135 bp allele in hybrids H1, H4 and 150 bp allele in H2, H3, and H5. Marker Umc-1664 was found to be polymorphic across all the five hybrids amplifying a fragment size of 140 and 120 bp in each of the five hybrids. Umc-1147 showed polymorphism between constitutive parents of hybrids H1, H2 and H4 and were monomorphic in H3 and H5. H1 with respect to this marker had an allele size of 100 and 85 bp .The marker amplified 120 bp and 85 bp in H2. The fragment size of 120 and 85 bp was scored in hybrid H4. The marker Umc-2245 amplied 100 bp fragment in H1, H2, H4 and H5. Umc -2375 was again polymorphic between corresponding parents of all the single cross hybrids. Umc-1594 showed fragment size of 140 bp and 100 bp in H1 and unique allele of 140 bp in rest four hybrids. Ph-061 amplified 2 bands each in hybrids H1, H4 and H5 with a size of 80 and 60 bp. Hybrid H1 and H3 were characterized having two alleles each against marker Umc-1002. Umc-1424 amplified two fragments each in H2 and H3. Phi 452693 amplified 125 bp across all the five hybrids. Taken together each hybrid revealed distinct allele profile across a set of ten markers. Hybrids H1, and H5 recorded 16 and 13 alleles against 10 SSR markers, while H2, H3, H4 amplified 14 alleles each.

Research and Development Priorities for Maize in J&K State

- Development of diverse productive inbred lines resistant to biotic and abiotic stresses from temperate x tropical hybridization programmes.
- Majority of Maize crop is grown rainfed (>75%) and on poor fertility status soils with little or no amelioration done by farmers over a long period of time except a very few pockets in Jammu Province, so the priority should be to develop products which have drought and low fertility resilience along with high water and nutrient use efficiency.

- As the state is predominantly hilly and possesses climatic variations across geographical boundaries varieties/hybrids have to be developed for regional needs. Funding needs to be provided to public sector institutions in the state to breed for stressed ecologies as compared to irrigated ones.
- The Seed Multiplication Ratio and the Seed Replacement rate are pathetically low and seed village scheme promotion for improving both needs to be a focal point of policy makers. In this case community based seed groups may be organized at block level. Likewise facilities for drying, packaging and safe low cost storage need to be provided at the block level.
- Improvement in input supply system needs to be focused at grass root level.
- Mobile soil testing facilities in PPP mode on pay and use basis should be made available.
- Development of facilities for demonstrating benefits of improved varieties/ hybrids and other production and post-harvest practices through KVKs.
- Popularizing specialty corn in peri urban areas by organizing exhibitions, food festivals with specialty corn products so as to tap the vast tourist market
- Promoting QPM under nutrition mission-mode approach to augment nutrient intake of school going children especially in tribal areas of the state.
- To meet the rising demand for feed and other purposes supporting services like roads and warehousing need to be improved
- Maize farmers are very poor and need subsidized seeds of improved varieties as a priority.

References

Anonymous. (2012). Food processing sector report, ministry of food processing Industries,. Government of India. http/www.mfpi.nic.in.

Anonymous. (2015). Digest of Statistics. Government of Jammu & Kashmir. Pp. 151-152.

Ahangar, M.A., Najeeb, S., Sheikh, F.A., Rather, A.G. and Teeli, N.A. (2015). Enhancing Food and Livelihood Security of Highland Temperate Rainfed Ecologies Using Improved Maize Cultivars. *International J Bio-resource Stress Manage.* 6(5): 572-578.

Ahangar, M.A., Bhat, Z.A., Sheikh, F.A., Dar, Z.A. Ajaz, A.L., Hooda, K.S. and Reyaz, M. (2016). Pathogenic variability in *Exserohilum turcicum* and identification of resistant sources to turcicum leaf blight of maize (*Zea mays* L.). *J Applied Natural Sci.* 8(3): 1523–1529.

Cheftal, J.C. (1990). *Extrusion cooking, operating principles, research trends and food principles*. London : Elsevier science publishers. Pp. 1-60.

Gujral, H.D., Singh, N. and Singh, B. (2001). Extrusion behavior of grits from flinth and sweet corn. *Food Chemistry*. 74: 303-308.

4

Maize Research and Development in Punjab

J.S. Chawla, Tosh Garg, Gurjit Kaur Gill, Mahesh Kumar Jawala Jindal and Harleen Kaur*

Department of Plant Breeding and Genetics, Punjab Agricultural University, Ludhiana
**Corresponding Author's Email: jschawla-pbg@pau.edu*

Maize is one of the most important cereal crops of the world. Being a C_4 plant, it is physiologically more efficient than rice, wheat and other important cereal crops which belong to C_3 category. In Punjab, traditionally, maize is grown during *Kharif* season but from last few years the area under spring maize is increasing because its cultivation is highly profitable and commercially viable after harvesting of potato crop. No doubt, Punjab has been the leading state in contributing to the national food security but simultaneously paddy-wheat rotation resulted in depletion of underground water, deterioration in soil health and environmental pollution. These problems need to be addressed for the sustainability of these natural resources and at the same time meeting the food grain production for the ever-increasing human population. Under such circumstances, maize crop is a viable alternative to replace some area occupied by rice and can help in breaking rice–wheat dominant cropping pattern in the state. Promising maize hybrids possessing high grain yield potential and offering comparable economic returns to the farmers are available with Punjab Agricultural University as well as with the private sector.

In Punjab, maize is commonly used in the form of *chapatis*, boiled or roasted green cob, breakfast food like cornflakes, popped grains etc. Maize *chapatis* are used as a delicacy with *saag*. The high carotene content of yellow grained maize is considered very useful in imparting yellow colour to egg yolk. Therefore, a substantial quantity of maize grain is utilized as an important ingredient of animal/ poultry feed industry. The use of maize grains for such purposes is going to increase as poultry and dairy industries are expanding. Maize plant is extensively used as green as well as dry fodder. The digestibility of maize fodder is better than sorghum, bajra and other non- legume forage crops. Maize plant does not have any anti quality factor hence can be fed to cattle at any stage. Maize can be utilized as an industrial raw material for number of industries *viz.*, starch, ethanol, rubber, paper, cosmetics and pharmaceuticals etc. Quality Protein Maize (QPM) can be promoted for nutritional security of poor masses whereas pop corn, baby corn and sweet corn can help in increasing the farm income due to high market value. In Punjab, major industries which are utilizing maize are The Sukhjit Starch & Chemical Limited, Kapurthala, and M/s Field Fresh Food Pvt. Ltd, Ladhowal, Ludhiana.

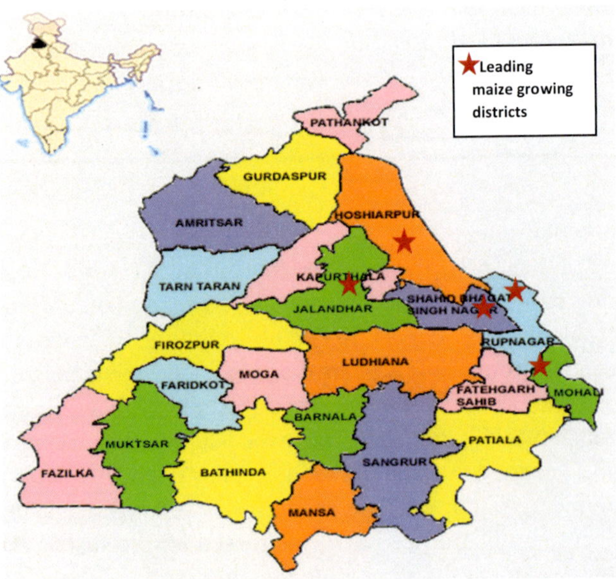

In 1950, the area under maize in Punjab was around 2,52,000 ha with an average productivity of 565 kg/ha which increased to 3,26,000 ha with almost double the productivity (1135 kg/ha) in 1960 (Fig.1) During 1975-76, the area under maize

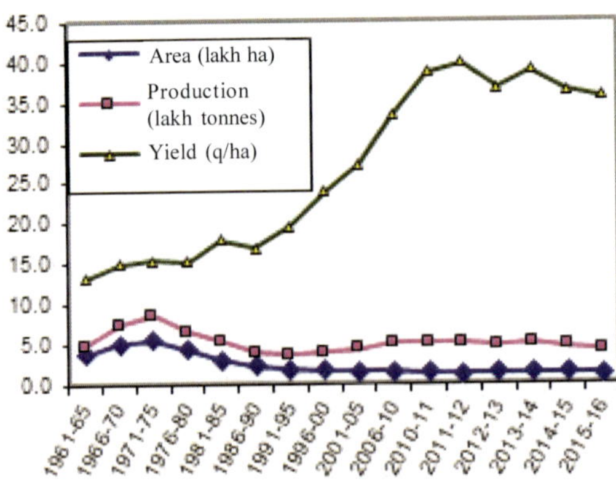

Fig. 1. Area, Production and yield of maize in Punjab

reached to maximum in Punjab. Afterwards, there was a sharp decline in acreage under maize and since then the area decreased up to 2010. Since 2010, the area varied from 1,15,000-1,35,000 ha in different years. The decrease may be due to several factors, such as expansion of irrigation facilities, introduction of high yielding dwarf varieties of rice and assured pricing of rice.

History of maize improvement in Punjab

Maize improvement work in Punjab started long back in 1945 at Layalpur now in Pakistan under Maize Breeding Scheme jointly financed by Punjab Government and Imperial Council of Agricultural Research (presently ICAR). After independence maize research work started at Jalandhar in the cereal section. Later, in 1957 an independent maize section was established. In 1958, the programme was shifted to Nariangarh, Distt. Patiala, afterwards, shifted to its present location at Ludhiana in 1962 with the creation of Punjab Agricultural University (PAU). The scheme for the production of hybrid maize started in 1955 at Chandigarh under the Indo-US agreement was also transferred to Ludhiana in 1963. In 1970, with the creation of Himachal Pradesh Agricultural University, a substation at Bajaura, Kullu established in 1957 shifted to this newly formed university (Khehra and Dhillon, 1982). In 1956, Punjab Hybrid No. 1 the first hybrid in India was released using parental lines derived from local varieties. At national level, systematic and planned research work on maize improvement was taken up with the formulation of All India Coordinated Maize Improvement Programme which was initiated in 1957. Initially several composite varieties such as Vijay, Rattan, Makki Safed-1, Ageti 76 and Partap were developed by PAU and released for cultivation in the state of Punjab. All these varieties except Makki Safed 1 were also released at national level. These varieties were high yielding than the locals and possessed features like early maturity, with disease and lodging resistance. Rattan, a composite variety was developed from a cross of Opaque-2 × J 1. This became the first QPM composite variety released for the state in 1972. These varieties contributed significantly in increasing the productivity of maize in the state from 1.1 t/ ha in 1960-61 to 1.6 t/ha in 1980-81.

In addition to these, early maturing high yielding hybrids Ganga 3, Ganga 5 and Him 123 were also released in Punjab during this period. The second wave of maize improvement in the state was introduced with the development of varietal and double cross hybrids by PAU such as Sangam and Sartaj. Sartaj was the high yielding hybrid in the country at that time that was released for cultivation throughout the country. With the advent of double cross hybrids, maize productivity increased from 1.6 to 1.8 t/ha in Punjab from 1987 to 1995. Double cross hybrids soon gave way to single cross hybrids. In 1995, PAU took the lead to develop first high yielding single cross hybrid Paras in India. On arrival of maize single cross hybrid, productivity level raised from 1.8 to 3.8 t/ha in Punjab from 1995 to 2015. Research efforts were directed to develop hybrids and composites of different maturities, possessing high yield, multiple disease resistance, tolerance to abiotic and biotic stresses and management responsiveness to cater the specific requirements of different agro climates, crop sequences and seasons.

Since inception of PAU, 42 high yielding hybrids and composites have been released for cultivation (Table 1). These varieties have contributed significantly to the increased productivity in the State. New breeding methods *viz.*, 'Selfed-plant Mass selection', 'Modified Selfed (S_1) family selection' and 'Alternate selection of S_1 and half–sib families' for the improvement of a population were also developed along with 'Hico Breeding' (Dhillon and Khehra, 1989) for the development of hybrids and composites from two populations coupled with statistical parameters of multiple disease resistance to identity desirable germplasm. During the last twenty years, several promising single cross hybrids like PMH 1, PMH 2, PMH 3, PMH 4, PMH 5, PMH 6, PMH 7, PMH 8, PMH 9, PMH 10, PAU 352, Parkash, JH 3459, Sheetal and Buland developed by this institute for varied agro-ecological situations.

Crop diversification in Punjab

Maize can play an important role in crop diversification policy of the state. It is used in poultry and animal feed, human food and for the manufacture of starch, glucose and corn flakes. Baby corn is eaten as salad and used for cooking vegetables and preparing pickles, *pakoras*, soups, etc. Main objective of the scheme is to enhance maize production through field demonstrations, adoption of improved seed and dissemination of best production technology. The following objectives are expected to achieve through implementation of crop diversification programme in original Green Revolution states including Punjab.

1. To demonstrate and promote the improve production technologies of alternate crops for diversion of paddy cultivation
2. To restore the soil fertility through cultivation of leguminous crops that generates heavy biomass and consumes lesser nutrient intake crops

The central government made provision of Rs. 500 crore to start the programme of crop diversification out of which Rs. 224.50 crores were allocated to Punjab. Diversification in agriculture is possible and essential to save the crumbling agriculture economy and environment of Punjab. However, the process and strategies of making it happen are not as easy as said. People who are actual players in the field have a definite mindset and conditioned behaviour. Conducive conditions are another aspect. During *Kharif* 2014, large scale demonstration at the farmers fields were conducted by the Department of Agriculture, Punjab

State in collaboration with Punjab Agricultural University on an area of 80,000 hectares to demonstrate the production technology and potential of maize hybrids to convince the farmers for assured higher returns.

Key maize production problems in state

The most challenging issue relates to the declining farm income as there is no system in place for regulated marketing for maize. Though minimum support price (MSP) is announced every year for maize crop but it is not procured by the government agencies unlike rice and wheat. Under such a situation private traders exploit the farmers to sell their produce at low price. Further, maize being a management sensitive crop, its productivity is affected adversely by a number of environmental and biological factors such as heavy rains , high velocity winds, drought, wild animals (wild boar, monkeys, wild cows etc.) and birds etc. Post harvest handling of the maize cobs, subsequent grain drying and storage is also a big issue that discourages the farmers to opt for its cultivation on a large scale. No doubt combine harvesters and other machines like dryers, dehusker-cum-shellers are now available for maize harvesting and shelling but there use is to a limited extent.

Table 1 : Cultivars released by the PAU centre

Cultivar	Year	Released by	Distinguishable characters	Remarks
Punjab Hybrid-1	1956	SVRC	Varietal hybrid	Grain type
Ganga Hybrid Makka 101	1961	SVRC	Double cross hybrid, ears long, well filled with tight husk, grains attractive, bold, and orange yellow, plant stay green at harvest. Resistant to downy mildew and other diseases.	Grain type
Ganga Safaid Hybrid Makka 2	1964	SVRC	Double top cross hybrid, ear placement medium with good husk cover; grains round, medium and white; moderately susceptible to leaf blight and rust but tolerant to stem borer.	Grain type
Ganga Hybrid Makka 3	1964	CVRC	Double cross hybrid, early maturing, plants vigorous with good husk cover; grains bright orange yellow, medium, bold and flint; resistant to downy mildew.	Grain type
Him 123	1964	CVRC	Double cross hybrid, tall plants; medium ear placement; grains yellow, bold and semi flint; recommended for hilly areas.	Grain type
Vijay	1967	CVRC/SVRC	Dual purpose composite; leaves broad; ears bold thick with light husk	Grain type

			cover; average to low ear placement; tassel open, generally white silks; grains moderately bold, semi-dent to flint and yellow to orange.	
Ganga Hybrid Makka 5	1968	SVRC	A medium stature double top cross hybrid with medium maturity; grains bold, yellow and semi flint.	Grain type
Rattan	1972	CVRC	A medium maturity composite; leaves broad; medium ear placement; grains light yellow, semi flint, soft and dull.	Grain type
Makki Safed 1	1974	CVRC	Early maturing, erect plant type; grains dull white.	Grain type
Ageti 76	1976	CVRC/SVRC	Early maturing; grains bold, attractive, medium round and orange yellow, usually flint; wider adaptability.	Grain type
Partap	1980	CVRC/SVRC	Medium tall composite with low ear placement; stem strong, medium thick; ears long, thick with tight husk cover; grains flint to semi-flint, medium and orange yellow.	Grain type
Sangam	1981	SVRC	Medium tall, vigorous hybrid with strong stem; average ear placement; grains white, bold, semi-flint to dent.	Grain type
Navjot	1982	CVRC/SVRC	Early maturing composite, compact plant type; grains attractive, medium sized, yellow orange and mostly flint type.	Grain type
Partap 1	1983	SVRC	Cold tolerant, high yielding winter maize composite, resembles Partap for morphological traits.	Grain type
Parbhat	1987	CVRC/SVRC	Medium tall plants; stem medium thick, lodging resistant; medium ear placement; ears medium long and thick with well developed husk cover; grains bold, yellow orange and semi-flint to flint.	Grain type
Sartaj	1987	CVRC/SVAC	Double cross hybrid; strong stem, medium plant height, low ear placement; leaves dark green and narrow, relatively erect; tassel light; ears long and medium thick; grains medium bold, yellow orange and flint.	Grain type
Kiran	1988	CVRC/SVRC	Average ear placement; leaves medium broad; early maturity; grains usually orange flint.	Grain type
J 1006	1989	SVRC	Tall, vigorous, broad leaves; ear placement medium, ears are long, thick and with white grains.	Fodder type

Megha	1990	CVRC/SVRC	Early maturity, medium tall plants with medium ear placement; drought tolerant; grains yellow orange and flint.	Grain type
Kesri	1992	SVRC	Early maturity, plant height and ear placement medium; ears thick at the base and tapering towards the tip; good tip filling and husk cover; grains deep orange flint. Also recommended for baby corn.	Grain type
Punjab Sathi 1	1994	SVRC	Heat tolerant, extra early maturity; plants short with medium ear placement; leaves narrow, short, semi-erect to erect; tasselslight and open; silks generally light green; ears short, thin with light husk cover; grains small, orange and flint.	Grain type
Paras	1995	SVRC	Single cross hybrid, tall, medium stout stem and ear placement; medium sized semi erect leaves; tassel medium; anther and silks green; ear thin and long; grain yellow bold flint with a few semi dents.	Grain type
Pearl Popcorn	1995	SVRC	Medium plant height and ear placement; ears thin long and cylindrical; grains small, round and flint.	Specialty corn (Popcorn)
Parkash	1997	CVRC/SVRC	Short duration and drought tolerant hybrid; plants medium tall, medium ear placement; leaves medium sized and semi erect; tassel medium sized and open; ear long and tip slightly blank; grains yellow orange and flint. Also recommended for baby corn.	Grain type
Bio 9637*	1999	SVRC	Tall plants, medium ear placement; leaves broad and medium long; tassel open and medium sized; ears medium long and thick; grains yellow and semi-dent to flint; cobs white; medium maturing hybrid developed by a private company.	Grain type
Sheetal	2001	SVRC	Medium tall winter hybrid, medium ear placement; leaves medium size; ears uniform, medium long, medium breadth; ear tip slightly blank; tassel medium and open; anther colour cream; grains orange, yellow flint, few grains with caps.	Grain type

JH 3459	2000	CVRC/SVRC	Plants medium tall, medium ear placement; tassel open with spreading branches and medium; ears uniform, medium long with shinning orange flint grains; plants have stay green characteristics.	Grain type
Buland	2002	CVRC/SVRC	Medium tall plants; ear placement medium; tassel open and medium in size; glumes green with anthocyanin pigment at the base; ears thick; grains bold, yellow orange and flint; Cold tolerant.	Grain type
F 9572 A*	2003	SVRC	Tall plants, medium ear placement, plants have stay green trait at harvest; stem thin; leaves narrow; tassel medium; ears thin, long with good tip filling; grains attractive, orange and flint; resistant to lodging.	Grain type
PMH 1	2005	CVRC/SVRC	Tall plants; stem with purple colouration, zig-zag and sturdy; leaves medium broad; tassel open and medium; ears medium long with yellow orange and flint grains.	Grain type
PMH 2	2005	SVRC	Short duration hybrid, medium plant height; medium ear placement; leaves medium sized and dark green; tassel medium and semi open; silks green; ears medium long; grains yellow orange, flint with yellow caps, resistant to lodging.	Grain type
PAU 352	2007	CVRC	Single cross, short duration hybrid for North-western plains of India; medium tall plants, drought tolerant; tassel medium sized, semi compact; silks green; orange flint kernels with yellow caps; cobs white.	Grain type
Punjab Sweet Corn 1	2008	SVRC	Tall plants, medium ear placement; tassel open with creamish anthers; silks usually creamish at the time of emergence; husk cover well developed; grains orange at maturity.	Specialty corn (Sweet corn)
PMH 3	2008	CVRC	Tall plants, medium ear placement; ears conical and long; tassels open; grains orange and flint; cobs white; recommended for North-West plains of India.	Grain type
PMH 4	2010	CVRC	Medium plants, medium ear placement; ears conical and medium; tassels open; silk green: grains yellow orange flint with caps; recommended for North-West plains of India.	Grain type

PMH 5	2010	CVRC	Medium plants, medium ear placement; ears conical and medium; tassels open; silk green: grains orange red flint with slight capping; recommended for Zone V of India.	Grain type
PMH 6	2013	CVRC	Medium duration, thick ears; orange yellow flint grains with caps; recommended for irrigated conditions during *Kharif* season for Zone III and V.	Grain type
PMH 7	2013	SVRC	Short duration; attractive shinning deep orange flint grains; moderately tolerant to high temperature stress; suitable for irrigated conditions during *spring* season in Punjab	Grain type
PMH 8	2014	SVRC	Orange flint grains with yellow caps; moderately tolerant to high temperature stress; suitable for irrigated conditions during *spring* season in Punjab.	Grain type
PMH 9	2014	SVRC	Suitable for irrigated conditions during *winter* season in Punjab	Grain type
PMH 10	2015	SVRC	Suitable for irrigated conditions during *spring* season in Punjab.	Grain type
DKC 9108*	2015	SVRC	Suitable for irrigated conditions during *spring* season in Punjab.	Grain type

*Hybrids from private sector

Climatic conditions during crop season

During *Kharif* season, July and August are main rainy months in Punjab except 2011 and 2013 when rainfall was more in June month as compared to July and August. Whereas May, June and July month are the hottest months in Punjab, in these months average mean temperature remain near or more than 30°C (Fig. 2).

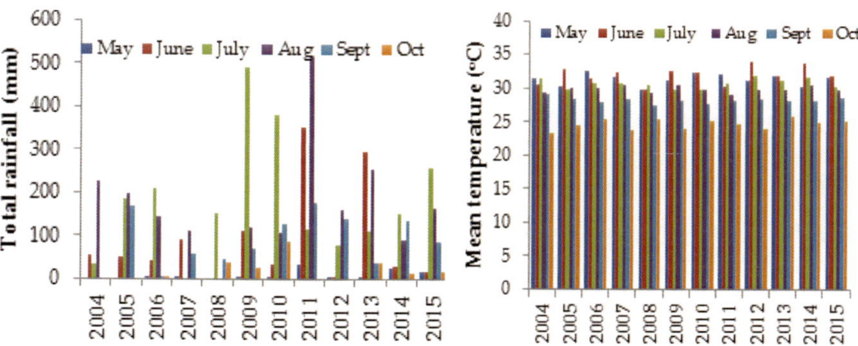

Fig. 2. Rainfall and temperature during *Kharif* season in Punjab.

Heterotic pools developed for hybrid Breeding Programme

Three pairs of heterotic pools (Late maturity, early maturity and winter season) have been established. In late maturity programme, the emphasis is on high yield, exploitation of better management, resistance to lodging and disease; whereas, in early maturity programme, the objectives are high yield and tolerance to drought and other stresses so as to cater the needs of the farmers in rainfed areas and those cultivating maize under medium and low fertility in irrigated areas to fit in multiple cropping system and also areas with limited irrigation conditions.

Heterotic pools in late maturity

The MS and Tux pools were established using the parents of commercial varietal hybrid Sangam as base populations (after two cycles of reciprocal recurrent selection). The parents of Sangam were MS1-DR and Tuxpeno Planta Baja C-7 Lothian (Tux PBL). MS1-DR was an improved version (for resistance to brown stripe downy mildew) of a local collection from Punjab. Tux PBL was a locally adapted version of an exotic introduction (Tuxpeno Planta Baja C-7) from CIMMYT, Mexico. A large number of germplasm including introductions from different regions of the world and locally developed materials were crossed with testers and those locally developed showing promise (heterosis and performance *per se)* were introgressed into these depending upon their heterotic patterns. Four cycles of reciprocal recurrent selection accompanied by introgression of new germplasm were followed to complete in this programme.

Heterotic pools in early maturity

Two heterotic pools, indigenous (Ind) and semi-exotic (SE) were developed in early maturity programme. The early maturity indigenous collections and other germplasm were crossed with two testers, namely JS 2 and CM 124 x CM 125. Composite JS 2 developed from selected locals of *Kandi* (rainfed) areas of Punjab whereas CM 124 x CM 125 is a single cross parent of hybrid Sartaj and has shown high heterosis with locals.

Heterotic pools for winter season

A pair of heterotic pool was synthesized for winter season. The sibbed generations of (CM 202 x Mo 17) and (B 73 x B 79) were used as testers. The large numbers of materials including U.S. Corn Belt were crossed with these testers. The materials were assigned to the two pools as per the heterotic pattern expressed in crosses.

Germplasm improvement for drought, heat and excess water stress

Drought Stress

A pair of heterotic pool specifically for the development of early maturing drought tolerant hybrids for cultivation in rainfed areas of the state has been synthesized. The inbred lines from the pools meant for rainfed areas were evaluated under rainfed situations. The promising lines were further evaluated in top cross performance using opposite pool as tester. The outstanding lines possessing good general combining ability (GCA) are further used in the development of hybrids. As a result of this breeding programme, a number of hybrids have been evaluated and identified. A drought tolerant hybrid Parkash was released for cultivation under rainfed situations in 1997 in Punjab and at national level in 1999. Hybrid JH 3459 performed better under rainfed conditions (20 and 46 % superiority over Parkash and Megha, respectively) and released in Punjab and North Western Plain Zone in 2000 and 2001, repectively. Hybrid PMH 2 with better drought and lodging resistance has been released for rainfed situation in Punjab in 2005.

Heat tolerance

Spring maize cultivation is picking up very fast under Punjab conditions in the fields vacated after potato cultivation in the month of January and February. High productivity is witnessed due to assured irrigation and fertile fields vacated by potato crop. The crop duration varies from 110-120 days depending upon the choice of variety and date of sowing of the crop. The crop experiences quite low temperature during its germination and high during its growth and flowering period. Need was felt to develop the materials with tolerance to cold at sowing and high temperature tolerance at the time of pollination to ensure good seed setting. The screening work was initiated in the year 2000 to develop agronomically superior and productive inbred lines for use in the development of hybrids for spring season. Consequently, hybrids PMH 7, PMH 8 and PMH 10 were released specifically for cultivation during spring season in the year 2013, 2014 and 2015, respectively.

Cold tolerance

High yield is being obtained during *Rabi* season. However, there are problems like damage (yellowing and drying of leaves) due to cold and frost, long duration of the crop and the susceptibility to rust. Research efforts have been intensified to screen the inbred lines for reaction to cold and rust. Field screening method for evaluating inbreds has been developed. It includes sowing the material early in the season and applying light irrigation when there is incidence of cold/frost.

Spraying of the rust inoculums for creating artificial disease epidemic, is done to identify the resistant lines. The promising lines are being used in the development of cold tolerant high yielding hybrids. Cold tolerant hybrids *viz.*, Sheetal and Buland were released for commercial cultivation in the state during winter season in the year 2001 and 2002, respectively. Later, PMH 9 a better hybrid than Buland was released for Punjab during 2014.

Excess water stress tolerance

Changing climate and erratic rainfall pattern has created the necessity to reorient our breeding objectives. The heavy downpour spells over shorter periods have demanded to screen the germplasm for excess water stress conditions. The work has been initiated to assess the suitability of the inbred lines to withstand the adverse effect of flooding water over a period of time without compromising its yield potential.

Composite Breeding

The production and supply of quality hybrid seeds to farmers has always remained a major challenge. In areas with limited irrigation facilities or under rainfed conditions and limited resources, farmers demand a suitable high yielding variety so that they do not depend on the fresh hybrid seeds every year. A strong composite breeding program was in place before the introductions of single cross hybrids in 1995. Punjab Agricultural University developed and released a number of land mark maize composites *viz.* Vijay, Navjot, Ageti 76, Parbhat, Kiran and many others to cater to the needs of farmers in different parts of Punjab. The composite Vijay was also adopted for cultivation in neighbouring countries like Nepal over a large area. The fodder purpose maize composite J 1006 developed and released during 1989 by this centre is still popular among dairy farmers. The research work on composite breeding is being further strengthened to develop superior composite varieties of grain as well as fodder to meet the demand of the farmers in remote areas.

Specialty maize and value addition

Baby corn

The work on the identification of genotype suitable for baby corn cultivation was initiated during 1993. All the released cultivars and germplasm in advanced stage were evaluated. Hybrid Parkash gave 17.5 q/ha baby corn yield with 23.2 % superiority over composite Kesri (14.2 q/ha). Parkash produced more number of ears/plant (2.65) of greater length (9.3 cm) than other genotypes which were of good quality yellow colour and regular row arrangement. Hybrid **Parkash** and Composite **Kesri** (early maturing genotypes) were found most suitable for baby corn cultivation and were recommended in the state for this purpose in 1998.

Sweet corn

A number of introductions of Sweet Corn were obtained and evaluated. The performance was not encouraging and most of them were susceptible to diseases. The backcross programme was taken up using Parbhat, an outstanding disease resistant composite as recurrent parent and five sweet corn introductions as donors. The resultant 10 populations were evaluated for yield and sugar content. Sweet corn composite JC (Sw) 1 performed well in station as well as in all India coordinated trials and was released in the state for cultivation in 2008 under the name Punjab Sweet corn 1.

Germplasm development for high plant density

Increasing productivity per unit area is a major challenge to boost the income of small and marginal farmers. Accommodating higher number of plants per acre /hectare can be a viable option to harvest good yield. Therefore, to address the issue the research program have been reoriented to develop agronomically superior and productive inbred lines with erect leaves to develop maize cultivars that produce optimally when grown under high plant population environment and having maximum photosynthetic efficiency with efficient conversion of photosynthate to grain, short interval between pollen shed and silk emergence, prolificacy, small tassel size, photoperiod insensitivity, long grain-filling period and slow leaf senescence. The experiments are being conducted to evaluate hybrids at different nutrients levels along with high plant population.

New Initiatives

Accelerated breeding

Maize is a cross-pollinated crop and development of inbred lines in maize and their maintenance requires a lot of attention, labour, resources etc. Doubled haploid (DH) technology using inducer lines can cut short the time span by 3-4 generations to get advanced generation uniform inbred lines. The development of lots of inbreds through DH technology will provide new combination of genes within a very short duration to the breeders to use in single cross hybrids development. It is important to select the breeding materials in relation to various abiotic/ biotic stresses, in addition to grain yield and other desirable agronomic traits. At PAU, haploid inducer stock was procured from CIMMYT in 2013. The centre has standardized the DH technology and developed doubled haploid lines for further evaluation.

Development of quality baby corn using CMS source

Baby corn means harvesting of maize cobs within 2-3 days of silk emergence before pollination. In Punjab, a complete package of practices (sowing time, seed rate, spacing, picking stage & packing) for baby corn has been developed and recommended for the cultivation of baby corn.

But the major problem of baby corn is to maintain the quality of the baby corn. To ensure the quality of baby corn detasseling is needed to avoid pollination which makes its cultivation highly labour intensive thus leads to decrease in farm income. In order to maintain baby corn quality and reduce costs, *cytoplasmic male sterility (cms)* based baby corn is the need of hour. Keeping this in view, cytoplasmic *male sterile* system used for production of *baby corn* hybrids. The objective of this research is to develop non-detasseled baby *corn*. Indigenous male sterile source was used as a non recurrent parent to cross with two highly prolific inbred lines LM 15 and CM 139 as recurrent parent (BC_4). Male sterile version of LM 15 is highly stable and crossed with other lines to develop hybrids for baby corn whereas; CM 139 (male sterile version) still needed further refinement due to occurrence of pollen shedders. New male sterile baby corn hybrids based on *cms* version of LM 15 have been synthesized and are at par with the ordinal hybrid in baby corn production.

QPM through Marker Assisted Selection (MAS)

Normal maize is deficient in two essential amino acids *viz.* lysine and tryptophan which can be well managed in the diet through the development of Quality Protein Maize (QPM) to overcome the prevailing malnutrition in poor masses. With this objective the effort have been made to convert normal parental inbred lines of commercial hybrids into QPM through MAS. Converted QPM version of elite normal inbred lines (LM 11, LM 12, and LM 13 & LM 14) was used to develop new QPM single cross hybrids (QPM version of hybrids Buland and PMH 1). These hybrids were evaluated in spring and *Kharif* seasons along with national released QPM hybrids (HQPM 1 and HQPM 7) as checks. The QPM version of commercial hybrid Buland and PMH 1 recovered with comparable yield and are being further evaluated for commercial cultivation.

All these cultivars and their parental lines were mainly used for seed production at Ludhiana centre from the year 2000 onwards given in Table 2.

ICAR- CIMMYT Collaborative Program

CIMMYT, Mexico has always been helpful in strengthening the Indian Maize Improvement Program. Trait specific germplasm materials are continuously received from CIMMYT and are being utilized in maize improvement programmes throughout the country.

Table 2. Indent and production of maize seed from 2000 onwards at Ludhiana centre.

Year	Cultivar/ inbred	Indent (kg)	Produced (kg)	Year	Cultivar/ inbred	Indent (kg)	Produced (kg)
2000	CM 122	70	100	2003	CM 122	2	10
	CM 123	35	40		CM 123	2	10
	CM 124	37	100		CM 124	2	10
	CM 125	26	40		CM 125	2	10
	Megha	90	300		CM 139	21	50
	Navjot	333	160		CM 140	11	40
	Ageti 76	20	25		LM 5	16	700
	Kesri	20	320		LM 6	8	600
	Kiran	10	20		LM 9	16	80
	Vijay	170	350		LM 10	6	100
	Partap	8	20		Parbhat	16	250
	Partap 1	8	400		Megha	16	260
	Parbhat	150	300		Navjot	250	250
	Pb. Sathi 1	8	50		Ageti 76	3	30
2001	CM 125	22	80		J1006	100	150
	CM 139	12	1350		Vijay	80	150
	CM 140	6	10	2004	CM 139	14	188
	LM 7	12	-		CM 140	8	125
	LM 8	6	-		CM 143	5	150
	Vijay	180	-		CM 144	3	130
	Megha	30	200		Kesri	10	655
	Navjot	515	400		Navjot	150	470
2002	CM 122	2	10		Ageti 76	5	100
	CM 123	2	10		Vijay	105	400
	CM 124	2	10	2005	CM 139	400	440
	CM 125	2	10		CM 140	200	200
	Navjot	435	450		CM 143	8	980
	Ageti 76	10	10		CM 144	4	84
	Kiran	200	280		Kesri	16	128
	Vijay	100	100		Navjot	213	377
2005	Ageti 76	12	180		Vijay	57	150
	Vijay	137	200		Kiran	500	520
2006	CM 143	8	40	2009	Navjot	600	700
	CM 144	4	10		Ageti 76	7	15
	Navjot	454	600	2010	Navjot	50	150
	Ageti 76	22	110		Ageti 76	7	10
	Vijay	62	225	2011	Navjot	100	200
2007	CM 143	8	30		Vijay	70	290
	CM 144	4	80	2012	Navjot	100	700
	Kesri	8	60		LM 18	13	15
2008	CM 143	3	20	2014	CM 139	4	40
	CM 144	3	15		CM 140	1	50
	Kesri	8	200	2015	No indent	-	-
	Navjot	910	1000	2016	No indent	-	-
	Ageti 76	27	150				

Change in cropping system over the years

Recommendations regarding complementary production technology have been made from time to time for newly developed maize varieties/hybrids by conducting experiments on different agronomic aspects *viz.* sowing time, sowing methods, spacings, and fertilizer requirements etc. for *Kharif*, *Rabi* and spring seasons. Agronomic data have been generated by evaluating large number of genotypes under different agronomic factors to identify some promising genotypes being significantly better than checks.

The agriculture in Punjab is highly intensified with assured irrigation. The principal crop rotation is rice-wheat; however, maize is being cultivated throughout out the year and due to diverse uses, it holds an important position in Punjab. It is traditionally grown in *Kharif* season. The crop rotations involving maize are Maize-Wheat/Barley/Potato/*Berseem*, Maize-*Senji*-Sugarcane-Cotton, Maize-Wheat-*Moong*, Maize-Wheat-Green manure, Maize-Potato/Toria-Sunflower, Maize-Potato-Wheat/Sunflower, Maize-Early Pea-Sunflower, Maize-Wheat-Cowpea (fodder), Maize-Raya/*Gobhi* Sarson, Maize-Potato-Summer *Moong*, Maize-Potato-Mentha, Maize (August)-Mentha, Maize-*Gobhi* Sarson-Summer *Moong*. However, a major change in cropping pattern has been observed with season like high temperature, irrigation requirement, insect pests etc. To overcome these challenges single cross hybrids specially developed for spring season like PMH 7, PMH 8 and PMH 10 recommended. The bed planting of maize and drip irrigation are recommended for judicious use of water and to save the crop from high temperature stress.

Improved agronomic practices recommended by PAU, Ludhiana (Table 3) has addressed the major productivity constraints of maize cultivation. *Kharif* maize being a rainy season crop gets infested with variety of weeds and subjected to heavy weed competition, which often inflicts huge losses ranging from 28 to 100 per cent (Patel *et al.*, 2006). Uncontrolled weed growth reduced the grain yield of maize up to 52% as reported by Walia *et al.* (2005). The farmers principally adopt the herbicides recommended by PAU for the control of weeds. The pre-emergence herbicides i.e. Atrazine and Alachlor are supplemented by the recommendation of new group of post emergence weedicides. For the control of dila/motha (*Cyperus rotundus*), spray 2, 4-D amine salt @ 58%, 20-25 days after sowing by dissolving in 150 litre of water. Laudis OD is a new herbicide for the selective control of broad leave weeds and annual grasses in maize. The active ingredient of Laudis OD is tembotrione, a member of the triketone class of herbicides is recommended by PAU for Punjab. Weeds can also be controlled through cultural and mechanical methods. Growing one or two rows of cowpea in between maize rows, due to high foliage of cowpea, it competes with weed flora for space, light and suppress the weed growth. Harvest the cowpea crop at 35 to 45 days after sowing before it starts twining the maize plants. Mechanically, two hand hoeing with *khurpa* or *kassola* should be given 3 and 5 weeks after sowing the crop. Second hoeing / interculture can also be given with bullock drawn *triphalli* or tractor driven cultivators.

The water logging is another major constraint in *Kharif* season. To avoid the flooding, earthing up of crop is recommended. In case of flooding, two sprays of 6 kg urea/acre (3% solution) at weekly interval or broadcasting of additional nitrogen @ 12 to 24 kg (25-50 kg urea) per acre is recommended.

Table 3. Package of practices for released composites/ hybrids in *Kharif, Rabi* and spring seasons.

Practice/ Item	*Kharif*	*Rabi*	Spring
Field preparation	Well prepared seed bed with proper irrigation and drainage system	Well prepared seed bed with proper irrigation and drainage system	Well prepared seed bed with proper irrigation and drainage system
Seed Rate	20 Kg/ ha	25 Kg/ ha	25 Kg/ ha
Seed Treatment	Bavistin or Derosil or Agrozim 50 WP (carbendazim) @ 3 g/kg of seed.	Bavistin or Derosil or Agrozim 50 WP (carbendazim) @ 3 g/kg of seed	Gaucho (imidacloprid) 600 FS@ 6.0 ml/kg + Bavistin or Derosil or Agrozim 50 WP (carbendazim) @ 3 g/kg of seed.
Sowing time	End May - end June	25th Oct-30th Nov	20th Jan-15th Feb
Sowing Method and spacing	Sowing can be done 3-5 cm deep in lines with a maize planter or seed cum fertilizer drill with a spacings of 60 x 20 cm	Sowing can be done on 6-7 cm high ridges by dibbling on the southern side of East-West ridges and maintaining 60 x 20 cm spacings.	Sowing can be done on 6-7 cm high ridges by dibbling on the southern side of East-West ridges with spacings 60 x 20 cm. or on beds 67.5 x 18 cm
Weed Control	Spray Atrataf 50 WP (atrazine) @ 800 g/acre on medium to heavy soils and 500 g/acre in light soils using 200 litres of water within 2 days of sowing.	Spray Atrataf 50 WP (atrazine) @ 500 g/acre using 200 litres of water within 2 days of sowing.	Spray Atrataf 50 WP (atrazine) @ 800 g/acre on medium to heavy soils and 500 g/acre in light soils using 200 litres of water within 2 days of sowing.
Nutrients' requirement	N: 110 kg/acre of urea P: DAP 55 kg/acre or superphosphate 150 kg/acre or nitro phosphate 125 kg/acre. K: On the basis of soil test only. Apply one third of nitrogen and entire amount of phosphorus and potassium	N: 155 kg/acre of urea P: DAP 55 kg/acre or superphosphate 150 kg/acre. K: On the basis of soil test only. Apply one third of nitrogen and entire amount of phosphorus and	N: 110 kg/acre of urea P: DAP 55 kg/acre or superphosphate 150 kg/acre or nitro phosphate 125 kg/acre. K: On the basis of soil test only. Apply one third of nitrogen and entire amount of phosphorus and

	at the time of sowing, one third nitrogen at knee high stage and remaining at pre tasseling stage. Note: If 55 kg of DAP is used then apply only 90 kg of urea. Apply 60 kg of urea if 125 kg of nitro phosphate is used.	potassium at the time of sowing, one third nitrogen in mid Jan and remaining at pre tasseling stage. Note: If 55 kg of DAP is used then apply only 135kg of urea.	potassium at the time of sowing, one third nitrogen at knee high stage and remaining at pre tasseling stage. Note: If 55 kg of DAP is used then apply only 90 kg of urea. Apply 60 kg of urea if 125 kg of nitro phosphate is used.
Irrigation	4-6 irrigations Critical stages for irrigation: pre-tasseling, silking, grain filling	Apply first irrigation 3-4 weeks after germination and subsequent irrigations after 4-5 weeks interval upto mid March thereafter 1-2 weeks interval.	Depending upon rains apply first irrigation at 25-30 days after sowing, then after 15 days interval up to 10th April and thereafter at 7 days interval. Irrigation should be ensured during grain filling stage to evade high temperature stress.
Plant protection	Maize borer is serious pest from June to September. Spray Coragen 18.5 SC (chlorantraniliprole) @ 30ml/acre or Desis 2.8EC (deltamethrin) 80 ml/acre in 60 liter of water. Maize borer can also be controlled using *Trichogramma* cards in 10-17 days old crop.	No serious insect pest attack is there	Maize shoot fly: Seed treatment with Gaucho or apply Furadan 3G (carbofuran) @ 5 kg/acre or Thimet 10G (phorate) @ 4 kg/acre in the furrows at the time of sowing During March to May to control Jassid, thrips, pyrilla and other leaf feeding insects spray Metasystox 25 EC (oxydemeton methyl) or Rogor 30 EC (dimethoate) @ 200 ml/acre in 50 litres of water.

Late sown *Kharif* (August) sown maize

Experiments were conducted to study the possibility of growing maize in late *Kharif* season (end of August) as maize sown during already recommended sowing time (end May- end June) requires greater amount of water because of hot dry weather conditions. Based on the results of experiments, recommendation was made during 2003 that maize can also be successfully grown by sowing in second fortnight of August. Short duration hybrids Parkash and JH 3459 were recommended to be sown in the fields needed to be vacated early otherwise hybrids Paras should be preferred to get high yield. Studies conducted further endorsed that the newly released hybrids PMH 1 and PMH 2 also perform good in August sowing. In this season, grain yield ranging from 6 to 8 t/ha can be obtained because of perfect plant stand, more ear girth and 1000 grain weight due to longer grain filling period as a result of prevailing low temperature conditions. The water requirement of August sown crop is comparatively less and insect pest and disease incidence is also less.

Insect pest scenario of maize in Punjab

The generation of new agricultural technology in India not only helped to increase production of maize but also has given rise to new cropping patterns. This has led to round the year cropping of maize. It provided continuous supply of food for insect pests avoiding any dearth period throughout their generations. In Punjab, maize cultivation is recommended in *Kharif, Rabi* and recently in spring season. Amongst the insect pests maize stem borer, *Chilo partellus* (Swinhoe) in *Kharif* and shoot fly species, *Atherigona naqvii* Steyskal in spring are of major importance in Punjab. An overview of the insect pest problems indicates that there has been a gradual change in the pest scenario, pest status and pests hitherto unknown in maize have started posing a serious threat to its cultivation. In Punjab conditions, during *Kharif* season the *C. partellus* is invariably present and produces dead hearts by boring in the young plants. The dead hearts incidence varied from 5.14 to 7.49% in protected and 13.00-17.00% in unprotected conditions during 2013 and 2014, respectively. The mean maize grain yield losses caused by *C. partellus* were found to be 10.99-13.57 and 12.65-14.69 per cent during 2013 and 2014, respectively (Kaur, 2016). The quantum of losses occurred due to *C. partellus* infestation; necessitate the protection of crop from this pest. Similarly, planting of maize in spring has led to establishment of polyphagous shoot flies species as the key pests on spring sown maize. These infests young seedlings, causing deformed leaves and ultimately results in dead heart formation. The dead hearts incidence due to shoot fly varied from 1.52 to 4.03 % in protected and 16.55 to 25.01 % in unprotected conditions during 2011 and 2012, respectively. The grain yield losses due to shoot fly increased with delay in sowing of spring maize as it caused 27.93 & 37.97, 38.21 & 38.66; and 38.99 & 45.04 per cent losses in end January, mid and end February sown crop, respectively during 2011 and 2012 (Jindal, 2013). The grain yield losses were more in late planted end February crop. The shoot fly incidence increased with rise of minimum air temperature in maize crop sown at different times; however, it was not correlated with relative humidity.

Apart from the incidence of shoot fly, the high incidence of small brown planthopper (*Laodelphax striatellus* Fallén) was observed for the 1st time in spring sown maize in the Punjab during spring 2011 at Ludhiana. The highest population counts of *L. striatellus* were observed in 3rd week of April (96.06 to 186.08/ sweep). The occurrence of small brown plant hopper, being a vector of number of viruses on cereal crops including maize could be a limiting factor in future (Jindal *et al.*, 2014a). Pink stem borer (*Sesamia inferens* Walker) is a well established pest of rice and wheat crops in northern India as well as winter maize grown in the peninsular India. But in Punjab conditions, it is coming up in *Kharif* maize as well (Singh *et al.*, 2014).

The sporadic incidence of cob borers especially *Helicoverpa armigera* (Hubner) in spring and grasshoppers infestation in *Kharif* is also observed in Punjab conditions at few places. In the current scenario of climate change with an IPCC prediction of 1.4±5.8°C increase in the global-average surface temperature and atmospheric carbon dioxide concentrations expected to rise between 540 and 970 ppm by 2100 (Houghton *et al.*, 2001) and hence the increasing adaptation of *S. inferens* and other insect pests to Punjab conditions is needs to be taken care of.

Integrated pest management (IPM)

The management package of insect pests of maize includes host plant resistance, cultural, biological and judicious use of chemical considering most susceptible stage of cop synchronizing with pest infestation. The single cross hybrids released for *Kharif* and spring cultivation in maize are moderately susceptible to insect pest infestation. The good seedling vigour and fast growth in vegetative stage help them to overcome the pest injury. The cultural practices like planting of spring maize from January 20 to February 15 to avoid synchronization of peak shoot fly infestation and susceptible stage of crop (Jindal, 2013); removal of borer infested plants at the time of weeding, destruction of crop stubbles to reduce carryover of the borer, use of judicious amount of fertilizers, sowing of crop on ridges are integral components of package for the management of key pests in Punjab.

The biological control of insect pests is the main stay of any IPM programme. The single release of egg parasitoid *Trichogramma chilonis* @ 40,000 parasitized eggs of *Corcyra* at 10-15 days old crop is successfully demonstrated at the farmer's field for the management of maize borer in *Kharif* (Aggarwal and Jindal, 2013). The practice is further strengthened by recommendation of two releases of *T. chilonis* on 10 and 17 days old crop. The demonstration of two releases on 124 acres during 2015-16 revealed 59.41 % reduction in deadhearts incidence and 13.75 % increase in yield over untreated control, which is significantly better than single release (In press). The judicious use of chemical insecticides is recommended for the management of insect pests of maize. For the management of maize borer in *Kharif* spray of synthetic pyrethroids on susceptible sage of the crop i.e. 2-3 week old crop was recommended in 90's. The new green chemistry molecule Coragen 18.5 SC (chlorantraniliprole) @ 30 ml per acre has been added up for the management of the pest (Jindal *et al.* 2014b). The shoot fly infestation in spring maize is managed by seed treatment with Gaucho 600 FS (imidacloprid) @ 6 ml/ kg seed (Jindal and Hari 2011). The seed treatment is cost effective and environment friendly also. The future management strategies for insect pests in maize need to address the challenges coming up with new copping systems vis-à-vis climate change.

Changes in maize disease scenario over the years in Punjab

Worldwide losses in maize due to fungal and bacterial diseases were estimated to be 9% in 2001-03 (Oerke, 2005). On global basis, out of 112 diseases recorded on maize, 35 have been reported in India. In India, losses due to maize diseases have been estimated to be 13.2% (Payak and Sharma, 1985). With the introduction of high yielding genotypes, extensive use of fertilisers and continuous cropping system, there has been a phenomenal shift in the occurrence of maize diseases.

Blights and Leaf Streak

Seed rot or seedling blights are not usually widespread but are endemically present in localized areas. This disease becomes prevalent with the use of old seed stored under high temperature and humidity and sown during winter. Maydis leaf blight (MLB) caused by *Drechslera maydis* was reported for the first time from Punjab by Mitra in 1931. Both Race O and T of the pathogen occur in India, however, the most prevalent one remains to be Race O. In absence of utilization of male sterility in Indian maize programme, Race T does not pose any threat to maize cultivation. Most of the hybrids recommended for cultivation in Punjab are resistant/ moderately resistant to MLB, however, this disease is economically important in breeding blocks, crossing blocks and seed production plots.

The incidence of banded leaf and sheath blight (BLSB) caused by *Rhizoctonia solani* has increased over the years owing to the intensive cropping pattern and availability of susceptible hosts in the field. It is more prevalent in humid weather with temperature around 28^{0}C (Singh and Shahi, 2012). Its soil borne nature and scarcity of resistant germplasm are the major factors contributing towards its spread. During recent years, incidence of bacterial leaf streak (*Acidovorax avenae* subsp. *avenae*) has been seen in Ludhiana and Hoshiarpur districts of Punjab during *Kharif* sown maize crop which can cause significant losses to this crop in future (Dhkal *et al.*, 2015).

Banded leaf and sheath blight

Bacterial leaf streak

Stalk and Ear Rots

Bacterial stalk rot of maize (*Dickeya zeae*) is becoming a serious concern for maize cultivation during *Kharif* season (Kumar *et al.,* 2016). The high incidence of the disease is favoured by high temperature (e"28°C) and high moisture, which commonly prevails in Punjab, 3 to 4 weeks after planting. Susceptible hybrids grown at high nitrogen levels in heavy soils with poor drainage favours the disease development. Among post flowering stalk rot complex, Fusarium stalk rot caused by *Fusarium moniliforme* and charcoal rot caused by *Macrophomina phaseolina* cause economic damage to maize crop. The incidence and severity recorded were comparatively high on private sector hybrids/locals as compared to public sector hybrids (Kaur and Mohan, 2012).

Recently, ear rot samples collected from five different districts of Punjab revealed the presence of *Aspergillus* spp., *Fusarium* spp. *and Penicillium* spp. Field observations suggest that their incidence is high in humid moist environments. The maximum frequency of *A. flavus* was found in Hoshiarpur district, closely followed by Jalandhar (Singh *et al*., 2016). Temperature shift within maize growing regions are highly likely to change the distribution and predominance of *Fusarium* spp. which will ultimately result in a change in mycotoxins scenario.

(a) **(b)**

Aspergillus ear rot (a) *A. flavus* (b) *A. niger*

Disease Management

The major approach of management of these diseases has been the utilization of host resistance. Hybrids released for Punjab state possess fairly good level of resistance to maydis leaf blight and post flowering stalks. The genetic variability for resistance to banded leaf and sheath blight and bacterial stalk rot has been found to be limited which puts limitation on a resistance breeding programme.

Seed treatment with Bavistin/ Derosal/ Agrozim @3 g/kg seed is a cheap, simple and effective method of protection against seed rot and seedling blights. In seed production plots, spray Indofil M-45 @ 200g/100 litres of water after about fortnight of sowing to control leaf blights. Give two more sprays at ten days interval. Recently, Kumar *et al.* (2016) concluded that severity of bacterial stalk rot was reduced by drenching stable bleaching powder @ 100ppm concentration in three maize cultivars-Dekalb Double, Punjab Sweet Corn-1 and PMH-1. The increase yield per cent as compared to control was 52.4% in Dekalb Double, 64% in Punjab Sweet Corn-1 and 57.9 % in PMH-1 maize cultivars. Chemical control of BLSB has also been taken over in AICRP maize programme in which several fungicides have been tested at Ludhiana centre. Among them, Azoxystrobin (Amistar) @ 0.05% proved highly effective with minimum disease severity (34.0 %) and maximum grain yield (76.5 q/ha). Validamycin, Trifloxystrobin + Tebuconazole (Nativo) and Difenoconazole (Score) were next in the order of their efficacy.

In addition, some cultural methods need to be followed for integrated disease management. Late sown crop (July-Aug) generally develops more infection of MLB than the crop sown earlier during May-June. Removal of lower leaves touching the soil is very effective technique in reducing the spread of BLSB. To minimise the yield loss of bacterial stalk rot, farmers are advised to keep their fields well drained and should not allow the water to stagnate. Ridge sowing should be preferred than flat sowing in disease prone areas of Punjab. Dense planting should be avoided.

Summary

In maize, 42 high yielding hybrids and composites have been developed and released for general cultivation by PAU. These varieties have contributed significantly to the increased productivity in the state in spite of the fact that the better managed and irrigated areas under maize shifted to rice cultivation since early eighties. The productivity of maize has increased from 1135 kg/ha in 1960-61 to 3570 kg/ha in 2015-16. Varieties Vijay and Ageti 76 have also played a significant role in enhancing the productivity of maize at the national level. At present, the major emphasis is on the development of single cross hybrids. Hybrids Paras was the first single cross hybrid released in the country in 1995. Among all the released varieties, it was the highest yielder. Since then the research work on inbred line development was strengthened at PAU and a number of single cross hybrids have been developed namely, PMH 1, PMH 2, PMH 3, PMH 4, PMH 5, PMH 6, PMH 7, PMH 8, PMH 9, PMH 10, PAU 352, Parkash, JH 3459, Sheetal and Buland. Special purpose varieties; Pearl Popcorn, Punjab Sweet Corn 1 and Hybrid Parkash for baby corn have also been recommended.

The other modern practices like use of leaf colour chart, use of consortium biofertilizer to increase the nutrient efficiency, diversification of the crop with the adoption of baby corn ensures maximum returns to the farmers. For higher price of farmer produce the provision of seed drier in grain market is an asset to popularize the maize crop in Punjab.

The maize stem borer, *Chilo partellus* (Swinhoe) in *Kharif* and shoot fly species, *Atherigona naqvii* Steyskal in spring causing 10.99-14.69 and 27.93 to 45.04 % yield loss, respectively are of major importance in Punjab . The insect pests like small brown planthopper, *Laodelphax striatellus* (Fallén) and cob borer *Helicoverpa armigera* (Hubner) in spring and pink stem borer, *Sesamia inferens* Walker in *Kharif* are also observed in Punjab conditions at few places. PAU has recommended various management practices of these insect pests which includes host plant resistance, cultural, biological and judicious use of chemical.

Diseases constitute a significant factor in lowering yield and productivity of maize. The spectrum of diseases is continuously changing as some diseases which were considered major (downy mildew) have waned and minor ones (*Rhizoctonia* and stalk rots) have become dominant in Punjab. For their effective management, specific integrated management modules by combining host plant resistance, cultural practices and biological control should be encouraged. There is a need to develop techniques and strategies to monitor virulence shifts in different pathogens through differential host lines as well as through molecular techniques.

Besides that new initiatives are taken to accelerate breeding using doubled haploid lines, conversion of normal maize productive inbred lines to its QPM version, development of quality Baby corn using CMS source, breeding for higher biomass and fodder purpose, and improvement of germplasm for heat stress and water use efficiency.

References

Aggarwal, N. and Jindal, J. (2013). Validation of biocontrol technology for suppression of *Chilo partellus* (Swinhoe) on *Kharif* maize in Punjab. *J. Bio Ctl.* 27: 278-284.

Anonymous (2016). *Package of Practices for Kharif Crops of Punjab.* Punjab Agricultural University, Ludhiana. pp 21.

Anonymous. (2006). *Package of Practices for Kharif Crops of Punjab.* Punjab Agricultural University, Ludhiana. pp 27.

Dhillon B.S. and Khehra A.S. (1989). Modified S1 selection in maize improvement. *Crop Sci.* 29: 226-228.

Dhkal, M., Hunjan, M.S. and Kaur, H. (2015). Molecular identification of the pathogen causing bacterial leaf streak disease of maize and standardization of inoculation technique. *Pl Dis Res.* 31**:** 79-85.

Houghton, J.T., Ding, Y., Griggs, D.J., Noguer, M., van der Linden, P.J., Xiaosum D., Maskell, K. and Johnson, C.A. (2001). *Climate Change 2001: The Scientific Basis.* Cambridge University Press, Cambridge.

Jindal, J. (2013). Incidence of insect pests and management of shoot fly, *Atherigona* spp. in spring sown maize. Ph.D. Dissertation, PAU, Ludhiana.

Jindal, J. and Hari, N.S. (2011). Efficacy of seed treatments against shoot fly, *Atherigona naqvii* Steyskal on spring-sown maize in Punjab. *J Insect Sci.* 24: 86-90.

Jindal, J., Brar, D.S. and Singh, G. (2014a). Occurrence and population dynamics of small brown planthopper, *Laodelphax striatellus* (Homoptera: Delphacidae) on spring sown maize in Punjab. *Ind J Ent.* 76: 69-73.

Jindal, J., Singh, G., Kumar, V. and Aggarwal, N. (2014b). Bioefficacy of chlorantraniliprole against *Chilo partellus* in *Kharif* maize. *J Insect Sci.* 27: 215-218.

Kaur, A. (2016). Estimation of losses and management of maize stem borer, *Chilo partellus* (Swinhoe) in *Kharif* maize. Ph.D. Dissertation, PAU, Ludhiana.

Kaur, H. and Mohan, C. (2012). Status of post flowering stalk rots of maize and associated fungi in Punjab. *Pl Dis Res.* 27: 165-70.

Khehra A.S. and Dhillon B.S. (1982). A handbook on maize for extension workers. Punjab Agricultural University, Ludhiana.

Kumar, A., Hunjan, M.S., Kaur, H., Kaur, R. and Singh, P.P. (2016). Evaluation of management of bacterial stalk rot of maize (*Dickeya zeae*) using some chemicals and bioagents. *J App Natural Sci.* 8: 1146-1151.

Oerke, E.C. (2005). Crop losses to pests. *J Agric Sci.* 144: 31 – 43.

Patel, V.J., Upadhyay, P.N., Patel, J.B. and Meisuriya, M.I. (2006). Effect of Herbicide mixtures on weeds in *Kharif* maize (*Zea mays* L.) under Middle Gujarat conditions. *Ind J Weed Sci.* 38 (1 and 2): 54-57.

Payak, M.M. and Sharma, R.C. (1985). Maize diseases and approaches to their management in India. *Trop Pest Manag.* 31: 302-10.

Singh, A. and Shahi, J.P. (2012). Banded leaf and sheath blight: an emerging disease of maize. *Maydica.* 57: 215-219.

Singh, G., Jindal, J. and Kumar, M. (2014) Incidence and extent of losses by cob borer complex on *Kharif* maize in Punjab. *Proc Int Conf: Changing Scenario of Pest Problems in Agri-horti Ecosystem and their Management*. pp 171. Entomological

Research Association, Maharana Pratap University of Agriculture and Technology, Udaipur.

Singh, H., Kaur, H. and Hunjan, M.S. (2016). Prevalence and morphological characterization of *Aspergillus* isolates of maize ear rot in Punjab. *J App Natr Sci.* 8: 1229-34.

Walia, U.S., Brar, L.S., Singh, B. (2005) Recommendations for weed control in field crops. *Research Bulletin.* Department of Agronomy. PAU, Ludhiana. pp. 5.

5

Maize Research and Development in Haryana

M.C. Kamboj, Narender Singh, Maha Singh Jaglan Rakesh Mehra, and Preeti Sharma*

CCS HAU, Regional Research Station, Karnal, Haryana, India

**Corresponding Author's Email: karnalmaize@gmail.com*

Maize is one of the most important food, feed, fodder and industrial crop in Haryana and is cultivated in all the three seasons (*Kharif, Rabi* and *spring*). It was cultivated on 6 thousand hectares in *Kharif* season with average productivity of 2.83t/ha during 2015-16. The maize is also increasing in spring season which was grown on ~ 7 thousands hectare with productivity of 4.5t/ ha. The specialty corn (baby corn and sweet corn) is also grown in >2 thousand hectare area of Haryana in national capital region. The demand of maize is increasing every year due to growth in poultry industry and expansion of maize based industries. There is a good scope of cultivation of specialty types of corn (baby corn, quality protein maize and sweet corn) in Haryana as it is very near to national capital, Delhi. The cultivation of specialty corn and its value addition by establishing

Map: Major maize growing districts in Haryana

industries in rural area may be helpful to the Haryana's farmers in crop diversification, generation of employment in rural area, solving the problem of green fodder and earning foreign currency.

Although, the area and production of maize in Haryana decreased over the years but productivity increased during 1966 to 2015 (Fig. 1). The QPM may be helpful in food and nutritional security, nutritious feed and promoting maize based entrepreneurship in the country (Jat *et al.*, 2009).

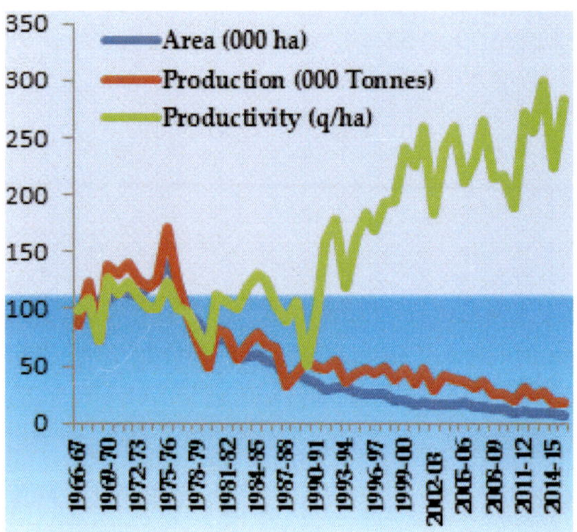

Fig. 1. Decadal seasonal trends in APY of maize in the Haryana from 1966 to 2016.

Major industries utilizing maize in the Haryana

There are two starch industries *viz.*, Bharat Starch Industries, Yamunanagar and Sada Sat Corn Product Pvt. Ltd. Garhi Singha, Kurukshetra. These industries utilize approximately 200 and 75 tonnes corn grain, respectively on daily basis. In addition to these, there are four baby corn industries *viz.,* Pratibha Foods, Village Aterna, Sonepat; Shimla Farm, Village Aterna, Sonepat; Gulab Fruit and Vegetable Growers and Marketing Co-operative Society Ltd. Food Park, HSIIDC Rai, Sonepat and Integrated Unit for Mushroom Development, Kundli, Sonepat in the state, each of these industries are processing approximately 150 -200 tonnes baby corn every year.

Major cropping systems and production ecologies

Recently, due to changing scenario of natural resource, maize based cropping system become essential in Haryana state. In *peri*-urban interface sweet corn, baby corn and maize based high value intercropping systems are also gaining importance due to market driven farming. Studies carried out under various soil and climatic conditions under All India Coordinated Research Project on Cropping Systems revealed that compared to existing cropping systems like rice-wheat, maize based cropping systems are better user of available resources and the water use efficiency of maize based cropping systems was about 100 to 200 % higher at different locations. The major cropping systems in the state are rice-wheat, cotton-wheat, pearlmillet-wheat, fallow-rapeseed and mustard and sugarcane. There is considerable decline in water table in rice-wheat areas in the north eastern zone of Haryana. The delay in planting and low yield of wheat in cotton based cropping system is some of the important issues related to these cropping systems and diversification. Inadequate availability of quality seeds of vegetables, flowers, spices, planting material for fruits, their higher cost and timely availability are some major constraints.

Different maize based cropping systems in Haryana are:

- Maize (*Kharif*)- wheat.
- Maize (*spring*) - paddy-potato.
- Maize (*Kharif*) - potato- sunflower.
- Maize (*Kharif*)- potato/mustard/maize (*spring*).
- Maize (*Kharif*)-wheat-urdbea/mungbean.

AICRP centres genesis and their mandate

Maize research was initiated in CCS Haryana Agricultural University at its Regional Research Station, Karnal during 1982-83 but AICRP on Maize centre was established during the year 2001. The mandate of maize research programme at Karnal centre firstly is generation of maize breeding material, screening of maize germplasm for various traits of economic importance and development of superior hybrids of normal maize and specialty corn with respect to yield, quality, disease and insect-pest resistance, secondly development of agro-technologies for enhancing the productivity of commercial and seed crop of maize based cropping system. The centre is working for three seasons in a year on normal maize, quality protein maize (QPM), baby corn and sweet corn on these mandates. Since inception of the maize program, this centre has developed a number of single cross hybrids for normal and specialty corn and also the technologies of their production and protection, seed production of single cross hybrids, intercropping in maize and product recipes of baby corn. Proven technologies are being demonstrated every year at farmer's field through Front Line Demonstration. The centre also imparts trainings to progressive farmers on maize and baby corn to promote the maize cultivation in the state, as well as at national levels.

Future approach in research and development will focus on development of superior single cross hybrids. Priority will be given on evolving hybrids not only with high yield potential but also for specialty corn like quality protein maize, baby corn, sweet corn, pop corn and for specific industrial application.

The mission for future research activities will be increasing the productivity and profitability of maize and maize based cropping system with socio-economic upliftment and conservation of natural resources leading to generation of wealth and employment in farming and industrial sector.

Coverage of cultivars

Almost whole of the cultivated area during spring season is under single cross hybrids while in *Kharif* season from 6000 ha cultivated area approximately ~30% area is under single cross hybrids and rest of the area is multi-parent hybrids. However, in specialty corn 100% area is under single cross hybrids.

Popular cultivars of maize in Haryana

No land race of maize is popular in Haryana. The maize hybrids *viz.*, DKC 9125, Hyshel, DKC 7074, P 3408 and PAC 740 are popular in *Kharif* while maize hybrids 31Y45, P 1844 and DKC 9108 are most popular in spring season. For baby corn cultivation, hybrid HM 4 and NK 5414 are most popular cultivars among the farmers. Among the QPM, the HQPM 1 is the most popular hybrid among the farmers while in sweet corn Sugar 75 is most popular cultivar.

Key maize production problems in state.

- Subsidy on electricity by state Govt. is the key factor responsible for promoting rice cultivation and exploitation of underground water in the state which decreased the area under maize cultivation in the state.
- Increased population of blue bulls is another factor for reduction of area under maize in the state.
- Poor marketing of maize is also a limiting factor for maize cultivation. There is good scope of specialty corn (QPM, baby con and sweet corn) but sufficient processing industries i.e. QPM based food and feed industries and baby corn and sweet corn processing industries are not there in the state.
- Seed of recommended hybrids is not available in sufficient quantity to the farmers while the seed of private companies is very costly.

AICRP technologies

Cultivars released from Karnal Centre

CCS HAU, Regional Research Station, Karnal has developed more than 450 inbred lines of normal maize and specialty corn (QPM, baby corn, sweet corn and popcorn) with different desirable traits and maturity groups. Approximately 3940 experimental hybrids developed by the centre have been tested at the station. The superior experimental hybrids have been submitted for testing in different coordinated trials. The single cross hybrids of maize developed and released from the centre are given in Table 1.

Table 1. Maize hybrids developed and released from the Karnal centre.

Cultivar	Year	Released by	Other /Characteristics
Yellow maize hybrids			
HHM 1	2000	SVRC	Early maturing yellow, dent grain, drought tolerant SCH for *Kharif* and *Rabi* season, Matures in 83-84 days in *Kharif season. A*verage grain yield is 53-55 q/ha and 60-65 q/ha during *Kharif* and *Rabi* season respectively.
HM 4	2005	CVRC	Dual purpose SCH released for grain and baby corn purpose, attractive colour of grain and baby corn, released for *Rabi* and *Kharif*, matures in 84-86 days in *Kharif* season. Average grain yield during *Kharif* is 55-60 q/ha and 67-72 q/ha during *Rabi* season. Its baby corn is harvested in appox. 60 days during *Kharif* season, baby corn average yield is 13-15 q/ha and released for central western zone of India.
HM 8	2007	CVRC	Attractive orange and flint grains, SCH released for peninsular zone resistant to TLB, SDM and PFSR, tolerant to cold. Matures in 85-86 days in *Kharif* season and grain yield 55-62 q/ha and 67-72 q/ha during winter respectively.
HM 9	2007	CVRC	Attractive orange and flint grains, SCH released for North Eastern Plain Zone resistant to MLB and common rust, tolerant to frost/cold, matures in 85-88 days and grain yield is 57-62 q/ha during *Kharif* and 70-75 q/ha during winter.
HM 10	2008	CVRC	Yellow semi dent grain, SCH for winter season, resistant to common rust, MLB, stem borer and cold, released for whole country , matures in 88-90 days in *Kharif* season, average grain yield during *Kharif* is 57-60 q/ha and 70-75 q/ha during *Rabi* season.
HM 11	2011	CVRC	Semi dent, yellow grain SCH recommended for *Kharif* season, responsive to high dose of nutrient, high level of resistance against MLB and common rust, matures in 88-90 days in *Kharif* season, average grain yield during *Kharif* is 57-60 q/ha and 67-70 q/ha during *Rabi* season.
HM 13	2014	CVRC	High starch, yellow grain SCH recommended for Northern Hilly Zone for cultivation in *Kharif* season, resistance against MLB, common rust, BLSB and TLB. Matures in 82-85 days in *Kharif* season. Average grain yield during *Kharif* is 55-60 q/ha and 65-68 q/ha during *Rabi* season.

White maize hybrids			
HHM 2	2000	SVRC	First white shining flint grain, medium maturing SCH released for *Rabi* and *Kharif* season for cultivation in the state, matures in 88-90 days in *Kharif* season, average grain yield during *Kharif* is 54-55 q/ha and 65-67 q/ha during *Rabi* season.
HM 5	2005	SVRC	Very productive white dent maize hybrid, suitable for intercropping, released for *Rabi* and *Kharif* season, tolerant to common rust, MLB and cold, matures in 88-90 days in *Kharif* season, average grain yield during *Kharif* is 60-65 q/ha and 70-75 q/ha during *Rabi* season.
HM 12	2012	CVRC	First white SCH recommended for North Eastern plain zone for cultivation in *Kharif* season, resistant to MLB and common rust, matures in 85-87 days in *Kharif* season, average grain yield during *Kharif* is 55-60 q/ha and 63-68 q/ha during *Rabi* season.
Quality protein maize hybrids			
HQPM 1	2006	CVRC	Yellow grain, SCH of QPM released for *Rabi* and *Kharif* season in the whole country, resistant to MLB and common rust, matures in 88-90 days during *Kharif* season, grain yield is 57-62 q/ha during *Kharif* and 65-70 q/ha during winter.
HQPM 4	2011	CVRC	QPM, SCH attractive light orange semi dent grain, released across the country, matures in 90-93 days during *Kharif* season, resistant to MLB and tolerant to stem borer, grain yield is 58-62 q/ha during *Kharif* and 65-72 q/ha during winter season.
HQPM 5	2007	CVRC	QPM SCH with attractive orange flint grain released for cultivation across the country, resistant to MLB, common rust, BLSB and PFSR, matures in 88-91 days, grain yield is 60-62 q/ha during *Kharif* and 67-72 q/ha during winter season.
HQPM 7	2008	CVRC	QPM SCH released forpeninsular zone, grain is light orange, resistant to MLB and stem borer, matures in 88-91 days, grain yield is 60-64 q/ha during *Kharif* and 67-72 q/ha during winter season.
Sweet corn hybrid			
HSC 1	2011	CVRC	First public sector bred single cross sweet corn hybrid, released for cultivation in Hilly zone (Jammu, Himachal Pradesh and Uttarakhand) for *Kharif* season, green cob average yield is 120 q/ha, ready to harvest in 80-82 days, resistant to MLB and common rust.

Public-public Partnership

This centre has played a leading role in public-public partnership by sharing maximum number of inbred lines to different AICRP centres. Different centres are using these inbred lines for the development of inter-institutional maize hybrids.

A large number of such hybrids have been tested in coordinated trials and promising hybrids are as follows:

a) Hybrids released

i) Malviya Hybrid Makka-2 developed by BHU, Varanasi has been released and notified during 2007 for cultivation in North Western Plain zone of country. The male line (HKI 1105) of this hybrid belongs to Karnal centre.

ii) QPM hybrid EHQ 16 (Pratap QPM 1) developed by MPUA and T, Udaipur has been released and notified for cultivation in Central Western Zone of country during 2013. The female line (HKI 193-1) of this hybrid belongs to this centre.

iii) Inter-institutional early maturing normal maize hybrid EHL 162508 developed by CSKHPKV, HAREC, Bajaura has been released and notified for cultivation in Northern Hilly Zone and central western zone of the country. The female parent of this hybrid (HKI 1040-7) is of Karnal centre.

b) Hybrids under testing

IARI New Delhi has converted three inbred lines of normal maize *viz.*, HKI 323, HKI 1105 and HKI 1128 developed by CCS Haryana Agricultural University into QPM version by marker assisted back crossing. These inbred lines are parental lines of HM 4, HM 8, HM 9, HM 10 and HM 11. The QPM version of these normal maize hybrids is under advance stage of testing at national level.

Maize seed production in the state

The seed production of maize is difficult in Haryana state as the environment is not stress free. The problem of cold in winter and banded leaf and sheath blight in *Kharif* is a serious problem for maize seed production in Haryana. The QPM hybrids (HQPM 1, HQPM 4, HQPM 5 and HQPM 7) and baby corn hybrid (HM 4) developed by CCS Haryana Agricultural University is very popular among farmers of whole country. This center is meeting out the breeder seed requirement of government agencies and MOU signed companies. The detail of breeder seed of parental lines of maize hybrids produced is as follows in Table 2.

Table 2. Breeder seed production (kg) of different maize inbred lines at Karnal.

Year	HKI 193-1	HKI 163	HKI 161	HKI 1105	HKI 1128	HKI 193-2	HKI 1344	HKI 1348-6-2	HKI 323	Total
2006-07	146	50	-	-	-	-	-	-	-	**196**
2007-08	60	50	-	28	-		4	4	15	**161**
2008-09	100	55	10	15	15	-	40	45	100	**395**
2009-10	122	75	97	98	126	21	-	16	30	**585**
2010-11	70	40	35	87	25	75	30	20	30	**437**

2011-12	82	155	42	42	-	-	-	-	25	**346**
2012-13	75	85	32	-	-	25	-	-	-	**217**
2013-14	525	665	652	50	107	40	-	-	22	**2061**
2014-15	-	486	112	50	42	185	-	-	15	**890**
2015-16	127	108	62	-	15	107	-	-	-	**419**
2016-17	-	150	70	10	-	-	-	-	11	**241**

Changes in cropping system over the years

Haryana is predominantly an agriculture economy with preponderance of wheat, rice, bajara, mustard, sugarcane and cotton. In the recent years, commercial orientation of the state agriculture is more associated with mustard, vegetables, fruits etc. and the area under pulses declined considerably. Among the cereals, the annual compound growth rates for fine cereals, *viz.* wheat and rice were 3.7% and 2.4 % while for coarse cereals, *viz.*, maize, bajra and jowar the growth rates were 1.4% and -0.76% during the period 1976-77 to 2011-12, respectively. These changes in cropping pattern have especially marked in the regions, which have witnessed the advent of green revolution. For example, the pulses production in Haryana has declined alarmingly from 952.0 thousand t in 1975-76 to 100 thousand t in 2015-16. Contrary, during last four decades, wheat production in the state increased about eleven fold; from 1059 thousand tons in 1966-67 to 11352 thousand t in 2015-16. Similarly, the rice production has increased from 223 thousand t to 4145 thousand t during the same period, which is more than eighteen-fold. The indices of cropping intensity in the state have increased from 1975-76 from 121.02 in 1975-76 to 181.47 in 2015-16. Sangwan (1985) reported that changes in the state cropping pattern resulted from increase in irrigation facilities to a large extent. Considering the diversity of soil, agro-climatic conditions and availability of canal irrigation and infrastructure services (e.g., roads and regulated markets) across the sub-regions, potentiality to cultivate varied types of crops exists in the state.

Ramphul (2012) reported that the cropping pattern and performance of different districts in growing different crops like maize in Ambala and Yamunanagar, cotton in Hisar, sugarcane in Yamunanagar and Ambala, mustard in Rewari and Mahendragarh, gram in Mahendragarh, bajra in Rewari were the highest. The specialization of wheat in Panipat, Hisar and Faridabad, rice in Kurukshetra, Kathal and Karnal, jowar in Rohtak and Faridabad is highest area under this crop. The highest specialization of bajra was observed in Mahendragarh, Rewari and Gurgaon.

Production technologies recommended

The CCS HAU, Regional Research Station, Karnal has given complete production technology for maize cultivation in Haryana.

Maize can be grown in all seasons *viz.*, *Kharif* (monsoon), post monsoon, *Rabi* (winter) and spring. During *Rabi* and spring seasons to achieve higher yield at farmer's field assured irrigation facilities are required. During *Kharif* season, it is desirable to complete the sowing operation 12-15 days before the onset of monsoon. However, in rainfed areas, the sowing time should be coincided with onset of monsoon. The optimum time of sowing, spacing and seed rate in different seasons is as follow by table:

Season	Planting time	Spacing	Seed rate
Kharif	Last week of June to 20 th of July	75 x 20 cm	18-20 kg/ha
Rabi	Last week of October to mid November	75 x 20 cm	20-22 kg/ha
Spring	1st week of February to 15th of March	60 x 15 cm	20-22 kg/ha

Method of sowing

Sowing of maize on southern sides of east-west ridges at depth of 5-6 cm has been found to give better germination and growth.

Ridge and furrow sowing of maize

The traditional practice of growing maize crop has several limitations such as inconvenient input management when sown by broadcasting, improper plant geometry and uneven plant stand resulting in insufficient utilization of space and plant competition leading to low productivity and input use efficiency. Ridge and furrow method of sowing can help to maintain proper plant competition and better input management leading to higher productivity and input use efficiency. It conserves soil moisture during prolonged dry spell in water scarcity areas for longer availability of moisture to sustain life and act as drainage channel in high rainfall/water logged area and provide better micro- climate for plant growth and root development. However, pre-requisite for ridge and furrow are:

- Laser leveling of the field.
- Prepare beds in East-West direction.
- Sowing on the sides of the ridges.
- Row to row spacing (60-75 cm) and plant to plant spacing (20-25 cm).

Fertilizers requirement and schedule

Application of nutrients @ 180-60-60-25 kg/ ha (N-P_2O_5-K_2O-$ZnSo_4$) have been found optimum for winter maize and spring season, whereas nutrients @ 150-60-60-25 kg/ ha (N-P_2O_5-K_2O-$ZnSo_4$) gave higher yield during *Kharif* season. Fertilizers schedule i.e. 1/3rd N + full P, K & Zn at sowing, 1/3rd N at knee high stage, 1/3rd at tasseling stage gives higher productivity and nutrient use efficiency.

Weed management

Weeds compete with crop for nutrient and cause yield loss up to 30–40% depending upon season of planting and crop rotation followed. Atrazine applied as pre-emergence (2-3 days after sowing) @ 1.0 -1.50 kg/ha followed by one hoeing at 25-30 DAS was found to give satisfactory control of all type of weeds in *Kharif* maize. Herbicide tembotrione 34.4% SC w/w (Laudis 42% SC w/v) @ 287.5 ml/acres + surfactant @ 1000 ml/ha in a spray volume of 200 liter water at 2-4 leaf stage of weeds (10-15 days after sowing) for control of complex weed flora in maize has been recommended.

Irrigation management

Irrigation should be applied when required depending upon rains and moisture holding capacity of the soil. Water should not overflow on the ridges. Irrigation applied in furrows up to 2/3rd height of the ridges/bed helps in improving water use efficiency. Inbreds responded to light and frequent irrigations. Irrigation schedule for different season are given in following table:

Rabi **season**	*Kharif* **season**
Young seedling	Young seedling
20-25 days interval up to 15th February	Knee high stage
Tasseling stage	Tasseling stage
Grain filling stage	Grain formation stage
Dough stage	Dough stage

Earthing up

Earthing up after second application of fertilizer has been found effective to avoid lodging of crop.

Suitable intercrops in maize for Haryana

Rabi **season**	*Kharif* **season**
Garlic, onion, cabbage, knol-khol, cauliflower, potato, fenugreek , coriander, spinach, gladiolus, lentil, pea	Mung bean, urdbean, cow pea

Changes in Disease scenario over the years

Among the factors adversely affecting productivity, incidence of diseases in pre-harvest stage is prominent. Considering the losses caused by diseases in Haryana from last ten years, five diseases *viz.*, maydis leaf blight (*Bipolaris maydis*), banded leaf and sheath blight (*Rhizoctonia solani*f.sp.*sasakii)*, common rust (*Puccinia sorghi*), bacterial stalk rot (*Erwini acrysanthemi* pv. Zeae) and pythium stalk rot (*Pythium aphanidermatum*) are of economic importance. However, incidence of Curvularia leaf spot is also increasing from last few years and posing new threat to maize crop in Haryana. Banded leaf and sheath blight (BLSB) and maydis leaf blight (MLB) diseases were found most important diseases in maize growing area during *Kharif* and common rust during *Rabi* season. Maydis leaf blight appears every year and incidence ranges from 20-85%. Banded leaf and sheath blight observed in severe form every year and losses up to 100% causes if continuous rain prevails during July-August. Common rust ranging from 5-60 % also noticed during February-March.

Epidemiology

Prevalence and severity of different diseases in research and farmers field trials were recorded during both the seasons. More than 80% prevalence and 42% severity of MLB was recorded in non-recommended varieties grown on farmer's fields. Incidence of BLSB was recorded in severe form during 2009 and 2010 due to continuous rains in the months of July and August. Banded leaf and sheath blight disease appears at pre-flowering stage on leaf and sheath in 40 to 45 days old plants. Out of three dates of sowing of maize, the highest banded leaf and sheath blight disease intensity was observed in 1st date of sowing (29 June) followed by 2nd date of sowing (13July) and least disease intensity in 3rd date of sowing (28 July). Maximum temperature range of 31.9-32.3^{0}C, and minimum temperature 24.6-25^{0}C with relative humidity morning (94-95%) and evening (69-83%), sunshine (2-7.7 hrs) per day and rainfall (34.6-55.8 mm) were most congenial for BLSB disease. However, temperature and relative humidity plays important role in BLSB progression. Regression equations on relationship between disease development and weather parameter have been developed for five maize hybrids/inbreds.

The *Rabi* crop gets infected with rust during the month of February and March. Moderate temperature ranging from 16-25^{0}C and high relative humidity favours rust development and spread.

Host resistance of diseases

Major thrust is for the development of hybrids/inbreds lines resistant to various diseases under artificial stress conditions. All the hybrids and inbred lines developed at the station and coordinated trials were screened artificially and natural disease inoculation condition. Every year about 80-100 inbred lines screened for the major diseases. Several hybrids and inbred lines were identified resistant to various diseases particularly MLB, BLSB and common rust. Hybrids resistance to MLB *viz.,* HM 5, HM 9, HM 10, HM 11, HM 12, HM 13, HQPM 1, HQPM 4, HQPM 5, HQPM 7 and HSC 1 released at national and state level. Hybrids resistant to common rust *viz.,* HHM 1, HM 2, HM 4, HM 5, HM 9, HM 10, HM 11, HM 12, HM 13, HQPM 1 and HQPM 4, HQPM 5 and HSC 1 also released at state / national level. For identification of stable sources of resistance to major diseases of maize, the following lines were found promising against major diseases of maize.

Pedigree	Resistant	Moderately resistant
HKI PC 4B	-	PFSR ,BSDM
HKI-PC-5-1	-	BSDM
HKI-PC-7	-	BSDM
HKI 1040-5	CLS	PFSR,, BSDM
HKI 1094-WG	-	MLB, PFSR
HKI C 78	BSDM	TLB, MLB, PFSR
HKI 141-2	BSDM	-
HKI C 323	BSDM	TLB, PFSR
HKI 1352-5-8-9	TLB, MLB,	-
HKI 34(1+2)-1	BSDM	PFSR
HKI 164-7-4 ER-3	BSDM	MLB, TLB, PFSR
HKI 164-4-(1-3)	BSDM	
HKI 191-1-2-5	BSDM	MLB
HKI 193-2-2-4	BSDM	PFSR
HKI 193-1	BSDM, PFSR	TLB
HKI-MBR-139-2	BSDM, TLB	PFSR

Disease management

Several integrated management strategies were developed for various diseases of maize. Stripping of the lower 2-3 leaf sheaths, applications of propiconazol (0.1%) and validamycin (0.2%) provide maximum banded leaf and sheath blight disease control and also increased seed yield significantly.

- Among plant extracts evaluated against *R. solani* f.sp. *asakii,* garlic clove and neem leaves extracts at 15% concentration were found most effective both under *in vitro* and considerably reduces BLSB and also increase seed yield significantly. Bio agents, *Trichoderma harzianum*, *T. viride, Pseudomonas fluorescens* and *Bacillus subtilis* also found effective both under *in vitro* and under pot conditions against BLSB .
- Spraying of crop with mancozeb (0.2%) with the appearance of the disease at ten days interval recommended for the control of MLB .Similarly, applications of propiconazol (0.1%) and leaf extracts (10 %) of medicinal plant Sarpgandha (*R. serpentine*), neem and garlic clove and bio agent (*T. harzianum*) were also found effective against MLB under *in vitro* and field conditions and also increase yield. Sanitation and destruction of crop debris will reduce the initial amount of inoculums. Rotation with non host species also found most effective against MLB.
- Seed treatment with thiram @ 4g/kg of seed before sowing found effective in controlling the seed rot and seedling blight disease. Application of 150g of Captan and 33g of stable bleaching powder in 100 liters of water and drench the soil near plant roots when crop is 5-7 weeks old is effective in checking

the Bacterial stalk rot *(Erwinia chrysanthemi* pv. *zeae)* and Pythium stalk rot *(Pythium aphanidermatum)* diseases . To manage both the above disease (Bacterial and Pythium stalk rot) destroy the diseased plant debris, keep the fields well drained, maintain required plant population and use improved varieties of maize is recommended.

- Foliar spray with mancozeb @ 0.2 to 0.25% at ten days interval starting with the appearance recommended for the control of common rust. Several fungicides were also found effective against common rust as foliar spray. However, Trifloxystrobin 25% + Tebuconazole 50% @ 0.05% were found best followed by Difenconazole @ 0.1% and Propiconazole @ 0.1%.

Changes in pest scenario over the years

Among the major factors of low productivity is the infestation by different insect–pests at various stages of crop. Different species of insect-pests infest maize crop during *Kharif, Rabi* and spring seasons. In India, 130 species of insect pests have been reported to damage this crop. Amongst these, the most serious pests are the maize stem borer, *Chilo partellus* (Swinhoe), the key pest throughout during rainy season, pink stem borer, *Sesamia inferens* (Walker), serious in peninsular India in post rainy season and two species of shoot fly, *Atherigona soccata* Rund and *Atherigona nuquii* Steyskal, serious in spring maize in Northern India, which cause economic yield losses.

Recently, *S. inferens* has emerged as a new pest and is likely to pose serious threat to the successful cultivation of maize in the north-western plains of India under largely adopted rice-wheat/maize cropping system. *S. inferens* has also been reported to infest maize during *Rabi* season in Haryana. Pink stem borer infestation in hybrid HQPM 1 ranged from 3.2 to 8.4% and 5.6 to 17.6% from November end to March at an average temperature of 10-25°C during 2014-15 and 2015-16, respectively.

Chaffer beetle, *Chiloloba acuta* (Weidemann) is a serious pest in maize seed production area. It was first time observed in August 2002 at Research Farm, CCS HAU Regional Research Station Karnal. It appears in first fortnight of August and feeds on pollen grain and hence causes serious problems in seed setting. It prefers to lay eggs on organic matter in soil and larvae/grubs remain in soil till the emergence in next rainy season. There is only one generation in a year.

Cob borer (*Helicoverpa armigera*) is also a regular occurrence in maize at silking/ tasseling stage during last decade in Haryana. Female moths lay eggs on the stem, leaves (both sides), tassels, silks and husks on the upper two thirds of plants. Caterpillars hatching prior to silking cause little damage to tassels but may cause damage when migrating to cobs. Per cent incidence and severity of infestation by cob borer was reported as 3.9 and 0.61%, respectively. Often the damage to the grain is not much yet the infestation marks reduces the market value of green cobs.

Pest management

a. Germplasm resistant to *Chilo partellus*

- Most promising inbred lines *viz.* HKI 46, HKI 47, HKI 139, HKI 170 (1+2), HKI 193-2, HKI 332, HKI 586, HKI 1128, HKI 1344, HKI 1348-6-2, HKI 1347-4 Lt (1+2+3), HKI 295, HKI 332, HKI 1015-6, HKI 1105, HKI 1352 and HKI 1378 tolerant to *Chilo partellus* were identified. Now these lines are being employed in breeding program to develop resistant hybrids.
- Fifteen hybrids released by station *viz.* HHM 1, HHM 2, HM 4, HM 5, HM 9, HM 10, HM 11, HM 12, HM13, HQPM 1, HQPM 4, HQPM 5, HQPM 7 and HSC 1 showed resistant to moderately resistant reaction against *Chilo partellus.*

b. Biological control

- In survey and surveillance, parasitoids like *Apantalis* sp. and braconid were found to parasitize on larvae and pupae of *Chilo partellus.* Natural parasitization of egg parasitoid, *Trichogramma chilonis* was observed to parasitize on *Chilo partellus* eggs. Predators (*Chrysopa* and lady bird beetle) have also been found feeding on aphids.
- Release of egg parasitoid, *Trichogramma chilonis* @ 125000 parasitized eggs/ha at 10-12 and 20-22 days after crop germination gave significant parasitization of *Chilo partellus* eggs and increased yield as compared to control and may be used in IPM strategy. While evaluating bio control agents of maize pest (egg and larval parasitoids), egg parasitization was reported zero while larval parasitization by *Cotesia* sp. was 7.69% at 40 DAG and severity of infestation was recorded 19.08%.

c. Seed treatment

- Shoot fly (*Atherigona* sp.) is a very serious pest of spring maize. It causes dead hearts in early stage of the plants. If the incidence is serious then re-sowing of the crop becomes imminent. Studies indicate that sowing of *spring* maize during first fortnight of February resulted in escape of crop damage by this pest. Seed treatment with Imidacloprid 600 FS or Fipronil 5SC @ 7ml/ kg of seed was also found effective against shoot fly. Seed treatment with Thiamethoxam @ 6 g/kg seed was found effective up to 3 months of sowing for control of termite.

d. Chemical control

Different insecticides have been recommended and included in package of practices for control of various insect pests of maize under ago-climatic conditions of Haryana.

- Carbaryl 50 WP (Sevin, Hexavin, Carbavin) @ 1000 g in 500-625 liter water/ ha recommended for the control of maize stem borer, (*Chilo partellus*). Methyl parathion 2 D @ 25 k/ha for control of grass hopper and army worm, monocrotophos 36 SL (Monocil, Nuvacron) @ 625ml, dichlorvas 76 EC (Nuvan) @ 500ml, quinalphos 25 EC (Ekalux) @ 1250ml for the control of hairy caterpillar, malathion 50 EC @ 1000ml for control of grey weevil and malathion 50 EC @ 625 ml/ha in 500-625 liter water per ha for the control of thrips and jassids have been recommended.
- Chlorantaniliprole 20 SC @ 0.4 ml/liter water was found most effective for the control of *Chilo partellus*. The chlorantaniliprole 20 SC @ 0.3 ml/ liter water, flubendiamide 480 SC @ 0.1and 0.2 ml/liter and deltamethrin 2.8 EC @ 0.8 ml/liter water also found effective for control of *Chilo partellus* .

e. Mechanical control

- Collection of beetles congregating on the tassels and put in kerosene oil or buried deep in soil has been found effective for management of Chaffer beetle, *Chiloloba acuta*.

f. Habitat modification

- Habitat manipulation has been found to be very potential pest management tactics in maize ecosystem. Based on per cent plant infestation by *Chilo partellus*, leaf injury rating and yield, maize intercropped with cowpea in the ratio of 2:1 row was found at par with recommended insecticide. Moreover, in intercrop there is additional gain of cowpea produce and nitrogen-fixation.

Post harvest management

Maize can store for a considerable period in unprocessed form without undergoing deterioration. Its shelf life greatly depends on the prevailing ambient temperature and relative humidity, and other factors like the inherent moisture, pests and diseases. Therefore, recommended post harvest handling and managing operations involve the manipulation of the above factors in order to obtain high quality maize grains. Quality control starts with harvesting. The optimum time of harvesting maize is when the stalks have dried and moisture of grain as about 17-20%.

Requirements during the Harvesting Process

- Dry maize on cement floor or use tarpaulin to reduce chance of contamination.
- At home, do not first heap the cobs in any room, kitchen or in the yard because this will expose them to all the dangers that cause post harvest losses. Transfer them to the drying place (like thecrib) immediately.
- Dry on concrete or canvas not on bare soil.

Activities after Harvesting

After harvesting, farmers should clean all the materials used in the process of harvesting and store them properly, away from sources of contamination and insect breeding places. The same materials may be needed during the proceeding operations e.g. to transport cobs from the crib for threshing or to transport grains to the store. If the materials are not cleaned properly, they can easily contaminate clean grains or become source of pest infestation since pest infestation starts from the field.

Drying

Wet grains attract insects and mould. Therefore, the grain must be dried as soon as possible after harvesting. Drying is the systematic reduction of crop moisture down to safe levels for storage, usually up to 12-15.5% moisture content.

- **Popularization of recommended technologies**: A total of 760 FLDs on maize have been conducted by this station to popularize the maize hybrids and other technology developed by this station. The grain yield gap was observed more than 100% during *Kharif* season while it was observed 56.9 to 89 % during spring season.
- **Public Private Partnership:** The popularity of the hybrids is clear from the fact that nineteen seed producing companies from different parts of the country have signed with the University for seed production of these hybrids.

Rainfall pattern: month-wise *Kharif* and *Rabi* separately for Haryana

Haryana state is considered the breadbasket of India along with the Punjab. The effects of drought and mitigation of those effects are therefore of considerable importance for the state. Haryana has a semiarid climate in the southwest and a Gangetic plain environment in the rest of the state. About 50% of the state has a moisture deficit. One of the reasons for adverse crop production in the state during June–September is the early withdrawal or late onset of monsoon rains, which contribute nearly 80% of the state's annual rainfall. The monsoon rain during June–September ranges from 284 mm to 521 mm in the drier western and southern plains and from 333 mm to 721 mm in the eastern districts of the state.

Protection of Plant material under PPV &FR Act, 2001"

a) Lines registered in NBPGR, New Delhi

Thirty six inbred lines have been registered with NBPGR New Delhi (Table 3).

Table 3. Inbred lines registered at NBPGR for various traits by Karnal centre.

Inbred line	Year	IC No.	IGNR	Characters
HKI-209	2003	IC 0405277	03055	Protogynous, drought tolerant, early maturing
HKI-335	2003	IC 0405279	03056	Prolific rabbit ear, tolerant to BLSB, SDM, PFSR and water logging, good GCA
HKI-1025	2004	IC 0405280	04067	Broad erect leaves, bold seeded and good GCA
HKI-295	2004	IC 0408327	04073	Medium, good pollen shedder, Resistant to MLB and rust, good GCA
HKI-1105	2004	IC 0405282	04069	Good GCA, bold seed
HKI-323-8	2004	IC 0405278	04066	Medium, MLB resistant, orange grain , good GCA
HKI-139	2004		04065	Attractive grain color, early maturing
HKI-586	2004	IC 0408328	04074	Early maturing, dark green leaves, Cold and MLB resistant, orange grain, good GCA
HKI-1040-7	2004	IC 0405281	04068	Medium , bold white grain, Resistant to MLB, bold seed, good GCA
HKI-1344	2004	IC 0408330	04075	Medium, bold white grain, Resistant to MLB, good GCA
HKI-1348-6-2	2004	IC 0405283	04070	White grain, good pollen shedder, Resistant to MLB and rust, good GCA, productive
HKI-1352-58-9	2004	IC 0405284	04071	Bold white grain , good pollen shedder, Resistant to MLB and rust , good GCA
HKI-1354	2004	IC 0405285	04072	Long cob, white grain, good pollen shedder, Resistant to MLB and rust, good GCA
HKI-1332	2004	IC 0408327	04120	Medium, dark green erect leaves, good GCA
HKI-288-2	2009	IC 0563956	8071	Late, yellow and flint grain, resistant to MLB
HKI-1126	2009	IC 0563958	8072	Late, yellow and flint grain, resistant to MLB
HKI-1040-4	2009	IC 0563959	8073	Medium maturity, orange and flint grains, MLB resistant
HKI-1015WG-8	2009	IC 0563961	8074	Medium maturity, orange and flint grains, MLB resistant
HKI-1347-4LT	2009	IC 0563964	8075	Late, white, flint grains, MLB resistant
HKI-164 D-4(O)	2009	IC 0563965	8076	QPM , late, yellow and semi dent grains, MLB resistant
HKI-164-7-6	2009	IC 0563966	8077	QPM, late, orange and semi dent grains, resistant to MLB

HKI-47	2009	IC 0563953	09057	Late maturing, bright orange, flint, good combiner
HKI-287L	2009	IC 0563954	09058	Late, yellow, flint, long cob, productive
HKI-327T	2009	IC 0563955	09059	Tall, late, yellow, flint
HKI-326	2009	IC 0563957	09060	Late, yellow, flint, productive
HKI-1040-5	2009	IC 0563960	09061	Late, yellow, flint, good combiner, highly productive
HKI-1341	2009	IC 0563962	09062	Late, white, flint, productive, resistant to rust
HKI-1342	2009	IC 0563963	09063	Late, white, flint, long cob, resistant to rust and MLB
HKI-170 (1+2)	2009	IC 0563967	09064	Late, yellow, flint
HKI 322	2010	IC 0584587	10081	Medium, white, flint, productive, strong plant, dark green leaves
139-2	2010	IC 0584588	10082	Medium, white, flint, good combiner, dark green leaves
5072-2- BT	2010	IC 0584589	10083	Medium , yellow , flint, high tryptophan, attractive grain colour , dark green leaves
PC -4B	2010	IC 0584599	10093	Medium, high popping, good pollinator
PCBT-3	2010	IC 0584600	10094	Medium, high popping, good pollinator
HKI -6	2010	IC 0584601	10095	Yellow, flint, high oil content
HKI -1 (T)	2010	IC 0584602	10096	Yellow, flint, high oil content

Note: GCA stands for Good combining ability

Fifteen hybrids have been released and recommended from this station. Out of these twelve hybrids *viz.* HHM 1, HHM 2, HM 4, HM 5, HM 8, HM 9, HM 10, HM 11, HQPM 1, HQPM 4, HQPM 5 and HQPM 7 were registered under PPV and FR act, 2001 and rest three hybrids are under the process of registration.

Public Private Partnership

CCS HAU has played an important role in commercialization of maize hybrids developed by the centre (Table 4).

Table 4. The list of MOU signed companies with CCSHAU, Karnal centre for maize hybrid seed production.

MOU signed Organization	Maize hybrids
M/S Kamboj Exports, Karnal, Haryana	HQPM 1, HQPM 5, HM 4 & HM 5
M/s PI Industries, Guru gram, Haryana	HQPM 1, HQPM 5, HM 4 & HM 10
Akash Seeds & Co., Ambikapur, Chhattisgarh	HQPM 1 & HQPM 5
Bhartiya Beej Nigam Ltd., Uttarakhand	HQPM 1& HQPM 5
Nuziveedu Seeds Pvt. Ltd., Ranga Reddy Distt, Andhra Pradesh	HM 8, HM 9 & HM 10
Arpan Seeds Pvt. Ltd., Udaipur, Rajasthan	HQPM 1, HQPM 7, HM 4, HM 5 & HM 1

Charoen Pokphand Seeds Pvt Ltd., Bangalore, Karnataka	HQPM 1 & HM 5
Balaji Seeds, Kurnool, Andhra Pradesh	HQPM 1
Sansar Agropol Pvt. Ltd., Bhubaneswar, Odisha	HQPM 1 & HQPM 5
Vibha Agrotech Ltd. Hyderabad, Andhra Pradesh	HM 9 & HM 11
Nath Bio Genes India Ltd., Aurangabad, Maharashtra	HM 9, HM 10 & HM 11
Bhartiya Beej Nigam Ltd., U. S. Nagar, Uttarakhand	HM 8, HM 10 & HM 11
Siri Seeds (India) Pvt. Ltd., Hyderabad, Andhra Pradesh	HQPM 1 & HQPM 5
Balaji Agri Biotech Pvt. Ltd., Bargarh, Odisha	HQPM 1 & HQPM 5
J.K. Agri Genetics Limited, Hyderabad, Andhra Pradesh	HQPM 1 & HQPM 5
Bioseed Research India, Hyderabad, Andhra Pradesh	HQPM 1
Sampoorna Seeds, Adoni, Kurnool District, Andhra Pradesh	HQPM 1 & HQPM
Sri Sai Seeds, Kurnool, Andhra Pradesh	HQPM 1 , HQPM 5, HQPM 5 & HQPM 7
Ganga Kaveri Seeds, Hyderabad, Andhra Pradesh	HM 9 & HM 10

Future research and development strategy

- Development of maize hybrids with grain yield of 10 t/ ha during *Kharif* season.
- Development of cultivars responsive to high density.
- Development of specialty corn (baby corn. Sweet corn and pop corn) hybrids for better income and solution to fodder problem.
- Development of nutritional rich value added products from QPM, baby corn, sweet corn and pop corn to double the income of farmers.
- To develop hybrids tolerance to extreme moisture conditions during *Kharif* season.
- Enhancing water use efficiency in spring maize by breeding, resource conservation, etc.
- Development of fodder maize varieties.
- Development of QPM hybrids for silage purpose.
- Improving performance of parental lines *per se* (Breeding, production and protection).
- Emphasis on management of emerging pests and diseases like banded leaf and sheath blight and bacterial leaf streak virus.
- Improvement in mechanization of maize cultivation.
- Strengthening of the seed production programme public breed hybrid.

Summary

CCS Haryana Agricultural University has developed very good hybrids of normal maize and specialty corn but now it is challenge for the center to develop normal maize hybrids which may give grain yield more than 10 t /ha and specialty corn comparable with the private companies. Efforts will be there to promote area under maize cultivation replacing paddy. Establishment of maize based industry particularly baby corn, sweet corn and QPM may be helpful in doubling the income of the farmers.

References

Sangwan, S.S. (1985). Dynamics of Cropping Pattern in Haryana: A Supply Response Analysis, *The Developing Economies*. XXIII (2): 173-186.

Jat, M.L., Dass, S., Yadav, V.K., Sekhar, J.C. and Singh, D. K. (2009). *DMR tech bull.* 2009/4.

Ramphul. (2012). Performance and Suitability of Growing Crops in Haryana: District-level Analysis: *Agricultural Situation in India.* pp. 27-32.

6

Maize Research and Development in Uttar Pradesh

*J.P. Shahi[1], Varsha Gayatonde[1], Anima Mahato[1], M.V. Singh[2]
R.K. Srivastava[2] and H.C. Singh[3]

[1]BHU, Varanasi, [2]NDUA &T, Bahraich, [3]CSAUA & T Kanpur
**Corresponding Author's Email: jpshahi1@gmail.com*

Uttar Pradesh, a fourth largest state of India geographically, where agriculture is a prime occupation and this is the traditional maize growing state, accounts 7.7% of area and 5.4 percent of total maize production of the country with an average yield of 1791 kg/ha (IIMR Annual Progress Report, *Kharif* Maize 2015). Maize is mainly taken as a *Kharif* crop but, gradually *Rabi* maize is gaining popularity with the availability of high yielding varieties and advanced production technologies. The rice, wheat, potato, sugarcane based cropping is more popular, and maize is given a moderate priority. The current level of growth in productivity definitely fall short to meet the demand of rapidly growing population even it may increase the poverty, unless vigorous measures are taken to accelerate yield, growth, employment generation, increasing the income of farming family and ultimately targeting for the sustainable climate resilient maize production.

Maize is mainly used as a staple food and quality feed for animals. In addition to that it also serves as a basic raw material for several industries like, starch, oil, protein, alcoholic beverages, food sweeteners, pharmaceuticals, cosmetic, textile, gum, package, paper etc. A major portion of the maize produce is directly consumed as a food in the form of boiled or roasted green ears, chapattis, porridges of various forms, popcorn, cornflakes etc. and as livestock feed *viz.*, cattle, poultry and piggery both in

the form of grain and fodder The central rachis of maize cob, to which the grains are attached remains as an agricultural waste after threshing, has many important agricultural and industrial uses. It is approximately 15 to 18% of the total ear weight and contains 35% cellulose, 40% pentose and 15% lignin (used as a raw material in cardboards). Its uses in agriculture includes as a litter for poultry and as a soil conditioner.

Trends of the maize cultivation in Uttar Pradesh

Maize, being an important *Kharif* crop of UP, is grown in 6.43 lakh ha area in the season however, *Rabi* maize occupy only small portion i.e. 0.69 lakh ha during 2014-15. A total area of 7.12 lakh ha is under maize cultivation contributing 5.38% to the Indian maize basket. Maize cultivation in UP has undergone several changes since 1950, when it was being grown in 8.34 lakh ha area with 6.5 lakh tonnes production and 7.81 q/ha productivity. During 1960s, green revolution entirely transformed the maize production scenario of the state. With adoption of double cross hybrids/ composites, a hike of 51.19, 65.53 and 9.21% was recorded in area, production and productivity, respectively. But their fortune in farmer's field to a great extent was affected by the problem encountered in their economic seed production which eventually affected maize productivity in next decade. However, a different scenario was observed during 1985-86 where in spite of decrease in 14.05% area, as compared to 1975-76, there was 67.76% hike in productivity. In 1995-96 also, although there was a decline of 21.86% in maize growing area in comparison to 1975-76 but per hectare production was increased by 80.44%.

From 1975 onwards, the area under maize cultivation has been continuously declining due to increased population, urbanization and other industrial and developmental activities. But this has not affected production potential of the state. With high yielding cultivars utilization of improved production technologies, production per unit area is tremendously increasing year after year. Although, there was a slight decline of 1.22% during 2005-06 (Table 1) in comparison to productivity of 1995-96 but the loss was soon recovered and huge gain in productivity was recorded later. In five decades, from 1950 to 2010, the maize growing area in UP decreased by 19.06% but production and productivity has shown a massive rise of 43.38 and 87.58%, respectively. The 2014-15, productivity increased by 129.32% over 1950. In past ten years, maize productivity has shown increase of 38.40% by virtue of improved varieties and technology development in maize research (Vision, 2050).

Table 1. Status of area, production and productivity of UP from 1950 to 2014-15.

Particulars	1950	1966-67	1975-76	1985-86	1995-96	2005-06	2009-10	2014-15
Area (lakh/ha)	8.34	12.61	13.95	11.99	10.90	8.55	6.75	7.12
Production (lakh t)	6.5	10.76	10.14	14.62	14.28	11.13	9.32	12.75
Yield (q/ha)	7.81	8.53	7.26	12.18	13.10	12.94	14.65	17.91

(Source: http://www.indiastat.com)

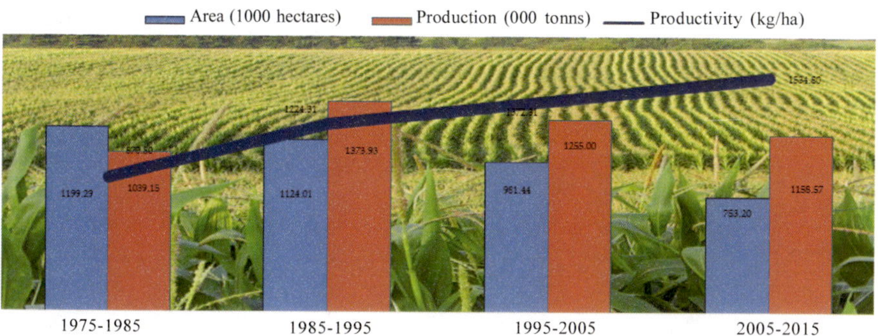

Fig. 1. Trends in maize area, production and productivity in the state during 1975-2015.

The graph representing area, production and productivity of Uttar Pradesh clearly indicates that the area under maize cultivation gradually decreases due to partitioning of the state as well as industrialization and urbanization since 1975 but it does not affect the production and productivity much. It shows a hike in productivity during 2000-02 and 2004-05 and, after a huge decline in 2005-06 but there is a gradual increase till 2015 (Fig. 1).

District-wise trends in maize cultivation

Uttar Pradesh, is divided into 75 districts and 18 divisions. District-wise area, production and productivity of maize (map 1 to 3 shows) that Mainpuri (49.7 thousand ha) occupies maximum area under maize cultivation followed by Bahraich (66.6 thousand ha). The top three district, Manipuri (49703t) followed by Jaunpur (45029t) and Gonda (43274t) shares the highest production. In case of productivity, Etah (31.77 q/ha), Firozabad (27.21 q/ha) and Etawah (26.67 q/ha) bags a high share (Map 1).

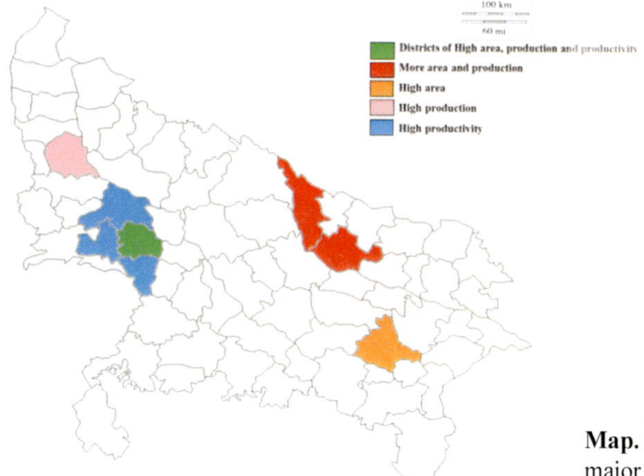

Map. 1. Uttar Pradesh map showing major maize growing districts.

Map 2. Maize area under cultivation in Uttar Pradesh (District-wise).

Map 3. Scenario of the maize productivity in Uttar Pradesh (District-wise).

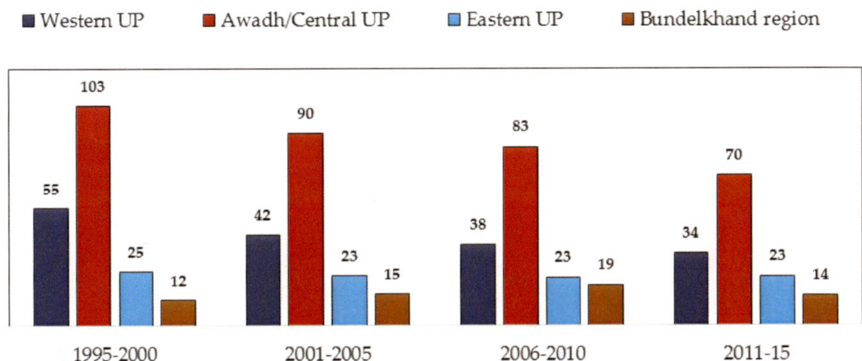

Fig. 2. Area (000 ha) under maize cultivation in four different regions during 1995-2015.

Among the four regions of UP, Central part is having the maximum and Bundelkhand is having minimum area under cultivation. Over the years, from 1995-2015 area under maize in Central UP decreased gradually.

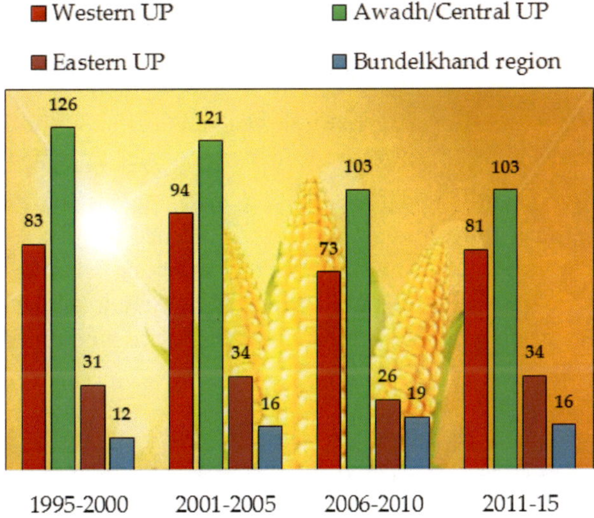

Fig. 3. Region wise maize production (000 t) during 1995-2015.

In Bundelkhand region from 2006-2015, sudden reduction in area occured due to severe drought. In terms of production Central UP is in the top but comparatively gradual reduction observed in this region. Eastern UP contributes less for the total state production but over the years increasing production is observed due to the effective transfer of technology. The Western region of UP has shown continuous increase from 2005 and the same trend has been observed in the central part from 2010 onwards (Fig. 2 to Fig. 4).

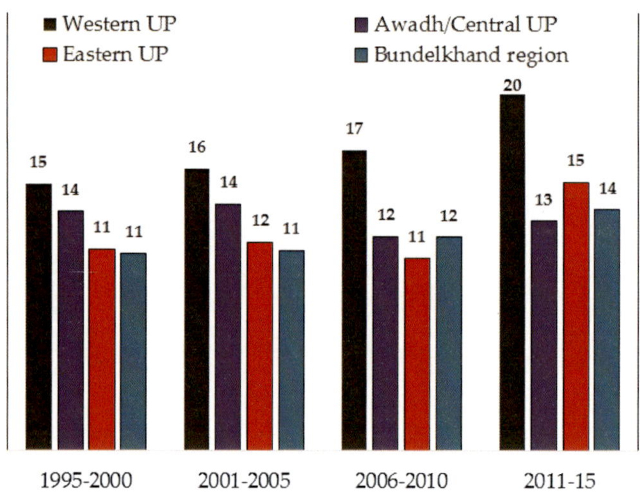

Fig. 4. Maize productivity (q/ha) status in four different regions during 1995-2015.

Major cropping systems and production ecologies

India is gifted with heterogeneous landforms and variety of climatic conditions based on which it has been classified into 15 agro-climatic zones and the state fall under zone II and III. A large part of the state falls under Middle Gangetic Plains whereas its Central and Western parts come under Upper Gangatic Plains.

Based on heterogeneity in edaphic and climatic factors, Uttar Pradesh can be divided in nine agro-climatic zones which facilitate better utilization of available resources and prevailing growing conditions for attaining maximum benefit from various cropping patterns. Several districts share more than one agro climatic zone. The area of districts sharing different zones are mentioned (Table 2) in percentage (%).

Table 2. Districts falling under different agro-climatic zones.

Agro-climatic zone	District
Tarai and Bhabar	Saharanpur (58%), Bijnor (79%), Moradabad (21%), Rampur (40%), Bareilly (19%), Pilibihit (79%), Khiri (39%), Bahraich (47%), Shravasti (71%)
Western Plain	Saharanpur (42%), Muzaffar Nagar, Meerut, Bagpat, Gaziabad, GautamBudha Nagar, BulandShahar
Mid Western Plain	Bareilly (81%), Badaun, Pilibihit (25%), Moradabad (79%), Jyotibaphule Nagar, Rampur (60%), Bijnor (21%)
South Western Dry Plain	Agra, Firozabad, Aligarh, Hathras, Mathura, Mainpuri, Etah
Central Plain	Shahjahanpur, Kanpur Nagar, Kanpur Dehat, Etawa, Auraiya, Farrukhabad, Kannauj, Lucknow, Unnao, Raibareilly, Sitapur, Hardoi, Khiri (61%), Fetehpur, Allahabad (58%), Kaushmbi
Bundelkhund	Jhansi, Lalitpur, Jalaun, Hamirpur, Mahoba, Banda, Chitrkoot

North Eastern Plain	Gorakhpur, Maharajganj, Deoria, Kushi Nagar, Basti, St. KabirNagar, Siddhartha Nagar, Gonda, Baharaich, Balrampur, Shravasti (29%)
Eastern Plain	Azamgarh, Mau, Balia, Pratapgarh, Faizabad, Ambedkar Nagar, Barabanki, Sultanpur, Varanasi, Chandauli, Jaunpur, Gazipur, St. Ravidas Nagar (86%)
Vindhyan	Allahabad (42%), St. Ravidas Nagar (14%), Mirzapur, Sonebadhra

(Source: http://planning.up.nic.in)

Cropping system vary based on climatic requirements of crops (Table 3 and 4). Western and South Western dry plains are the major maize growing zones.

Table 3. Maize based cropping systems followed in Upper and Middle Gangetic Plains.

Agro-climatic Zone	**Cropping systems**		
	Irrigated	**Rainfed**	**Spring**
Upper gangetic plains (Rainfall 1200-1470 mm)	Maize-early potato-wheat-mungbean, Maize-wheat Maize-wheat-mungbean/urdbean	Maize-wheat	
Middle gangetic plains and Eastern region (720-1200 mm)	Maize-wheat-mungbean Maize-potato-wheat/ sunflower/ onion/sugarcane ratoon Rice-potato-maize Maize -Blackgram/Greengram/ cowpea/ Groundnut/Pegionpea /Mesta	Maize-wheat Maize-barley Maize-safflower Maize – Lentil/Rajmash/ Backla/Pea/ Linseed/ Garlic/ Spinach/ Coriander	Maize-Sugarcane/ Cowpea

Source: http://agridaksh.iasri.res.in

Table 4. Maize based cropping system in different districts.

Districts	**Associated cropping system**
Kannauj	Maize-Potato-Mung /Sunflower,Maize-Wheat-Mung
Kanpur dehat	Maize-Mustard
Chandauli and Ballia	Maize-Lentil
Mainpuri	Maize-Wheat/ Mustard
Bahraich	Maize-Lentil/ Wheat/ Mustard
Gonda	Maize-Wheat, Arhar-Maize
Etah	Maize-Potato-Sunflower/ Groundnut/ OnionMaize-Wheat
Jaunpur	Maize-Chickpea/ Pea/ Mustard /Wheat
Varanasi	Maize-lentil/ Chickpea, Maize-wheat
Hardoi	Maize-Wheat/ Potato, Maize-Potato-Vegetable
Unnao	Maize-Toria-Wheat-Fallow
Bulandshahr	Maize-Potato-Sorghum (Fodder)
Auraiya	Maize-Toria-Wheat-Fallow

Source:http://agropedia.iitk.ac.in

AICRP centers, genesis and their mandate

The Ministry of Food and Agriculture, Government of India and Rockefeller Foundation in 1956, the All India Co-ordinated Research Project was started in 1957 with a view to utilize heterosis in maize improvement after its great success in U.S.A. and other countries.

In Uttar Pradesh, the AICRP on maize started in C.S. Azad University of Ag. and Tech. in 1972. This was the first central project in U.P. and was further extended through establishment of more centers at Institute of Agricultural Sciences, Banaras Hindu University, Varanasi in 1976 and Crop Research Station, NDUAT, Bahraich in 1976 with the mandates to carry out basic, strategic and applied research aimed at enhancement of production and productivity of maize. The various objectives as follows

- Development of suitable high yielding single cross hybrids for *Rabi* and *Kharif* seasons.
- Germplasm collection, evaluation, maintenance and its enhancement.
- To develop specialty corn cultivars and their associated technologies such as Quality protein maize, baby corn sweet corn, biofuel etc. for diverse uses.
- To conduct training, frontline demonstrations and on-farm research to maximize and accelerate adoption of research findings and innovative technologies.
- Development of suitable agronomic package of practices for commercial cultivation.

Key maize production problems and prospects in different regions

1. Western and Central UP

Key research problems includes, inadequate availability of quality seed, moisture stress, unbalanced fertilizer use, lack of early maturing varieties, problems of broadleaf and grassy weeds. Public and private seed sectors are weak and most farmers in these sub-regions are resource poor and cannot afford to buy improved seed, even if available. Under such a scenario, the agricultural research institutions and state agricultural universities based in the sub-regions may initiate seed multiplication programs and make available superior cultivars at reasonable prices, and hence pass on the benefits of research to farmers confronted with poverty and water scarcity.

2. Bundelkhand region

In this region, the area under maize cultivation is restricted to only two districts and maize based cropping system is quite poor.

3. Eastern Uttar Pradesh

Maize production is gradually shifting from the rainy to the winter season, when the crop is grown mainly under irrigation, so that yield levels are higher and unit costs are lower. Lack of quality seed, inappropriate crop establishment, and a lack of balanced use of nutrients during the winter season were the top three researchable issues in this region. A strong policy research analysis, assessing the reasons for non-availability of improved seed and developing appropriate strategies to overcome this constraint, would allow further expansion of the area under winter maize. Similarly, diagnostic surveys to understand the reasons for inappropriate crop establishment and lack of balanced use of nutrients would provide deeper insights for undertaking in-depth research programs. Other problems for winter and irrigated maize are post-harvest losses, management of Turcicum leaf blight, and post-flowering stalk rot Inter-cropping with maize and transplanting maize under late sown conditions are other high-priority research areas for which public and private sector research needed. The development of medium- and full-season cultivars for high rainfall regions and of extra-early or early cultivars for medium rainfall regions are of the highest priority. Weeds have a higher priority in high rainfall regions than in the medium rainfall areas. Development of drought-escape varieties along with appropriate management practices may be the best research strategy (Sharma, 2007).

General production problems

The common major problems are summarized below:

1. Small land holding and marginal land use for maize cultivation restrict the farmers to grow maize.
2. Use of poor quality seeds/old seeds, lack of seed and its availability at right time.
3. High cost of hybrid seed supplied by private sector.
4. Poor management practices like use of low nitrogen, other fertilizers/micro-nutrients, non-availability of assured irrigation facilities and less emphasis on winter maize.
5. Lack of mechanization of maize cultivation and maize based industries in the state.
6. Biotic (*Chilo partellus),* leaf blight and BLSB and abiotic stresses (drought, water logging, salinity, nutrient imbalance, high temperature, cold etc).

AICRP technologies

Popular land races

During sixties several local varieties in the name of their places like T-41 (first improved variety of maize developed by using population improvement technique with local collections of Jaunpuri, Rudrapur local, Farukhabad local, Pilibhit local, Tinpakhia (an extra early maturing race), and Jaunpur safed (white seeded type) etc. were under cultivation.

Systematic breeding program was initiated with a greater emphasis on composite breeding utilizing locals and other materials in the state during seventies. A number of varieties were developed through population improvement techniques e.g. Diara composite was released to fulfill the demand of farmers in diara area. Various composites *viz.* Kisan, Vijay, Vikram, Jawahar and Sona. Tarun, Navin, Kanchan, Surya, Pragati were released as early to medium maturing composites. Besides normal maize, four QPM composites i.e. Shakti, Ratna, Protina, Shakti-1 were also developed. After few years, hybrid breeding specially double and double top cross hybrids *viz.*, Ganga series were developed but couldn't popularize. Greater emphasis on hybrids especially single crosses including speciality corn hybrids started during 1990 onwards and very good progress in production and productivity was observed. Jaunpur local resistance to stalk rot and other diseases as well as tolerant to excess water and drought is an important landrace of U.P. Composite populations VC1, VC7 and VC2 were synthesized and improved from Jaunpuri local and other material at the Varanasi center. Azad Uttam, an early maturing composite, was developed by using early maturing germplasm and was identified and released by State Varietal Release Committee during 1991-92.

Popular varieties/hybrids

The centers were initially focused in improving the local germplasm through the development of composite varieties. Later on, with availability of improved inbred lines the work area has been shifted towards exploiting heterosis and developing high yielding hybrids. **Malviya Hybrid Makka-2** was the first high yielding hybrid variety of the state with 50-55 q/ha yield and 85 days maturity developed at BHU, Varanasi and released for cultivation in UP, Bihar, Jharkhand, Odisha, Assam, Chhattisgarh and West Bengal by CVRC in 2007. Another hybrid **Malviya Hybrid Makka 3** having yield potential 50-60 q/ha and tolerance to TLB and MLB was identified for notification by SVRC in 2010. But its successful release as a variety was failed due to parental inbred lines becoming susceptible to diseases.

In 2008, **Sharadmani,** a composite of maize for cultivation during winter season was notified and released by SVRC, Lucknow and **Azad Kamal** was the first composite developed at C. S. Azad University of Ag. and Tech. Kanpur.

Chandramani, an early maturing composite of the Kanpur center, identified and released by CVRC in the year 2008. In 2009-10 Annual Maize workshop, the first single cross hybrid developed at Kanpur center **Azad Hybrid Makka-**

1 (REH-2003), was identified. As the entire UP is concerned hybrid seed use for commercial purpose is only 21%. (FICCI, 2011-12). Besides a large number of composites recommended for the state, some popular hybrids from the public sector are namely, PMH 3, PMH-2, PMH-4, Deccan 105, HM-10, HM-12; HQPM 1 and HQPM-5 (QPM hybrids; HM-4 (babycorn), HSC-1 (sweet corn; Partap Chari 6 and African Tall (fodder maize) etc. There were many varieties/ hybrids released through AICRP testing for the state by private sector also. (Table.5).

Table 5. List of private sector hybrids.

Kernel color	Name of the cultivar
Yellow	Pinnacle, BIO 22027, KMH-218 Plus, NK-6240, NMH-803, ADVSW-1, GK3150, DKC-7074.
White	Bisco555
Orange	KMH 3712, KMH-3426, Dragon
Yellow orange	SMH-3904, NK 30, Kaveri 50, Bisco X 6573, Bisco X 6573, (LG 34.03), X35C537 , P3544), KMH2589, DKC9120, D2244, BIO 9782
Others	KH9541, MMH133, PRO 303-(3461), Kh5991, 4210, Bio 9637, PRO-316, NK61

*DMR, 2011

Seed production

Seed production of Azad Uttam (75q) and Azad Kamal (50q) composites and Malviya hybrid Makka-2 (3q) and HQPM-1 various other inbred varieties like Naneen, Shweta, CML-142, CML-150, CML-182, CML-142xCML-150 and Shaktiman-1 has produced as per the demand from various agencies and farmers. Besides these quantities the state seed corporation also produce the Certified seed of Azad Uttam on various farms of state.

Maize Utilization

From 1950 to 1970s, maize had been utilized as a minor crop only to feed the cattle/poultry or as a source of raw maize (bhutta). During the same period, the local landraces of low productivity were used. Earlier some of the popular composites and local varieties occupied major area but gradually some part covered with the popular hybrids in around 30-35% area. There are some small scale industries scattered across the state involved in manufacturing starch based byproducts and snack items, however, the documented records at industrial scale not found. Gradually the farmers are growing sweet corn and popcorn to fulfil the demands of local markets but the area is quite negligible. The rapid urbanization attracts and encourage the farmers of the peri-urban area to grow specialty and other type of corn due to the popularity of green maize, the various snacks prepared by using maize as a prime ingredient (Singh, 2006). The changing utility pattern of the maize consumption clearly shows that the utility trend is shifting from poormen's meal to a major ingredient of high profile restaurants food.

Production technologies and package of practices

Production technologies for maize developed by AICRP

Intercropping with maize During the *Rabi* season intercropping of maize with table pea (1:1), the cultivation of onion and garlic as inter crop with maize (1:2) and the inter- cropping of coriander (*Dhania* for leaf), spinach, beetroot and cabbage (1:1) are important to improve the income of growers without harming the yield of winter maize.

Maize (HQPM-1) + groundnut (1:1) produced 112.4 q/ ha, maize + carrot (1:1) produced 113.93 q/ha in *Rabi* season. In *Kharif* maize, inter-cropping of cucumber (3:1) and cowpea (2:1) were found more profitable .Maize + rice (1:4) system was recommended as economically viable under rainfed/mid low land areas of tarai region of Eastern U.P.

Integrated weed management

Hand weeding at 25 DAS followed by an application either 2, 4-D Na salt as post emergence or atrazine 0.50 kg/ha as post or pre emergence was recommended as most remunerative method .

The highest grain yield and net profit recorded under 60x20cm plant geometry water management.

- Maize cultivar Syn-2 recommended for cultivation under excess soil moisture condition.
- *Rabi* maize irrigated at critical stage *viz.,* 30thdays at knee high, tassel emergence, grain filling stage and fertilized with 150N and 10ton FYM/ha recommended for harvesting the higher grain yield and monetary gain.
- A substitution of 25% through FYM of the recommended Nitrogen dose in the *Rabi* maize.

Nutrient management

- Application of 240:60:60kg NPK/ha where, nitrogen was applied 10% at basal; 20% at four leaf stage +30% at 8 leaf stage + 30% at tasseling stage +10% at grain filling stage produced highest grain yield (83.96 q/ha) and net profit .
- Application of 225:60:40 kg NPK/ha with 50q FYM produced highest grain yield (74.97 q/ha) and net profit.

Package of practices

Maize is the second most popular *Kharif* crop after rice. The monsoon dependent cultivation and lack of awareness regarding advanced production technologies like, high yielding varieties, time and amount of fertilization *etc* (Jat *et al.*, 2006) greatly limits the production and productivity of maize in the state. Hence, maize production can be boost up by adopting improved agricultural practices as mentioned below:

Cultivars

Recommended released maize varieties under cultivation should be grown cultivars.

Maize can be cultivated in variety of soils ranging from clay loam to sandy loam. Indo-Gangetic alluvial soils recommended for maize cultivation.

Date of sowing

To avoid the rain damage early seed at 10-15 days before the onset of monsoon in the *Kharif* season recommended for *Rabi*, sowing period 15th October to 10th November is recommended. For *zaid* sowing at first or second week of March is the best. Late sowing in summer may hinder the better establishment of young seedlings.

Sowing should be done by forming furrows and ridges at 60cm apart. It provides sufficient irrigation channel. Seeds should be placed at 22-25cm spacing.

Nutrient management

NPK fertilizers should be applied as per soil test recommendations as far as possible. If soil test recommendations are not available, NPK should be given as (kg/ha)

Season	N	P	K
Kharif	100-120	60	40-60
Rabi	120-150	60-75	40-60

One or two sprays of 2% urea (spray grade with low biorate content) and application of micronutrient mixture can be given in 3-5 weeks old crop to restore the vigor which may have received setback because of delayed weedings and inadequate drainage.

Irrigation management

Kharif: Maize is mainly grown as rainfed crop. Based on rainfall during the season irrigations can be given at critical stages, if required.

Rabi **and *Zaid***: Maize crop is sensitive to both moisture stress and excessive moisture; hence irrigation should be regulated as per requirement. Optimum moisture availability should be ensured during the most critical growth phase (Flowering phase: 45 to 65 days after sowing).

Weed management

Pre-emergence application of atrazine @ 1.0 kg/ha on 3-5 DAS followed by one hand weeding on 30-35 DAS or pre-emergence application of atrazine @ 1.0 kg/ha on 3-5 DAS followed by 2, 4-D @ 0.5 kg/ha is recommended for controlling major weeds in maize field. When pulse crop is grown as intercrop then pendimethaline should be used @ 0.75 kg/ha as pre-emergence at 3-5 DAS.

Seed treatment

To protect the maize crop from seed and major soil borne diseases and insect-pests, seed treatment with Thiram @ 2.5 g/kg seed is recommended.

Seed rate

About 20kg of seed would be needed to sow one hectare maize crop. Seeds should be sown about 5cm deep to ensure good seedling growth and vigor.

Plant population

- A population of 65-70 thousand plants/ ha at harvest is necessary for realizing high grain yield.
- It will be necessary to attain about 10% higher stand at germination.
- For attaining the desired level of plant density, it is desirable to use a row-to-row and plant-to-plant spacing of 75cm x 18cm or 60cm x 22cm.

Hoeing, hand weeding and earthing-up

Hoeing and hand-weeding at 30th days after sowing, which earthing -up is to be done to provides additional anchorage to the plants after of first split N application.

Harvesting

The maize crop sown for grain is harvested when the grains are nearly dry and at the physiological maturity more than 20% moisture. Ears to be removed from the standing crop but harvested ears should be dried in the sun before shelling. In the case of the late-sown crop, farmers prefer to harvest the whole plants and pile them, and the ears are removed are removed later. Maize stalks are used as cattle feed or fuel. In fact, no part of the maize plant, even the cobs from which the grains have been removed, is left unused.

Maize grown for fodder should be harvested at the milking to early dough stage as the early harvested crop is likely to yield less and have a lower protein content. For silage, however, the late dough is preferred. Both power and hand operated low priced maize shellers are available indigenously. These shellers are considerably more efficient than hand shelling or beating with sticks, the common practice.

Farmers using hybrid maize should not save their own seed for their next crop, as the advanced generation hybrid seeds are likely to give yield reduction of 25-30%. However, farmer can save seed from composites and open pollinated varieties, when grown in isolation. At least seeds from 500 to 100 ears of the best yielding and normally spaced plants resistant to prevalent diseases and pests should be bulked. Ears should be dried, shelled and treated with an insecticide to avoid the further damage by storage pests.

Yield

Considerable variation in grain yield is observed. The yield levels depend upon the variety, the amount of the fertilizer used, and the rainfall pattern etc. Under irrigated conditions and recommended cultural practices, an average yield of 4 t/ha in the Indo-Gangetic Plains is common under low fertility and rainfed conditions with poor yielding varieties, a grain yield of about one to 2 t /ha is obtained. However, 15-20 tonnes cobs/ha can be harvested if crop is for green ear purpose.

Post-harvest management

After drying the cobs to a moisture level of 12-14% and threshing grain/seeds should be stored in gowdons in gunny bags or high gauge polythene covers. Harvesting for green ears (bhutta) is done when the kernals are at late milky stage and soft. Without removing the outer husk, cobs should be transferred to the targeted market. Whole cob can be stored at low temperature for a longer time, whereas, for immediate marketing without cold storage cobs can be stored for maximum 5 days. The cobs for ear purpose are packed in thin (low gauge) polythene covers to avoid the further damage by respiration. The baby cobs (immature small cobs popularly known as babycorn) are processed, canned in the tins as it has a high export demand.

Major disease and insect

All the major diseases of maize found in the plains are reported in the state. Epidemic for several diseases appeared in past in different parts of the state.

Bacterial stalk rot (*Erwinia chrysanthemi* p. *zeae*) appeared in the foothills (Tarai) area of U.P. in 1959 on Hybrid Ganga 5. In 1965, an epidemic for brown stripe downy mildew (*Sclerophthora rayssiae* var. *zeae)* was developed in the tarai region of UP which completely devastated maize production in the state. There is a major shift in disease pattern during the past years as the major diseases like TLB, MLB, BSDM and ESR are gradually becoming diseases of lesser economic importance now a days due to availability and use of resistance sources in the newly developed hybrids and varieties. On the other hand the disease of minor importance became increasingly severe and assumed epidemic scale in coming years. Among diseases rust, brown spot, seed and seedling blight are the common across the state.

In early sixties the diseases like BLSB, late wilt, polysora rust and pre-harvest cob rots were considered as a disease of minor importance only, but due to changing weather condition and pattern of disease spectrum these diseases were considered as major diseases.

Major pest in western part of the state is *Chilo partellus*. In Bahraich and Hardoi, termite, stem borer, caterpillar, cut worm and leaf roller are the prevailing pests while, Bulandshahr district is mainly affected by termite, stem borer, caterpillar and cut worm.

Abiotic stresses in Uttar Pradesh

North Eastern Plain (Bahraich): Water stress, water-logging, late planting

Central Plain (Hardoi): Water stress, water-logging, late planting

Western Plain (Bulandshar): Water stress, late planting.

Short information on the ongoing research on maize in U.P.

There are 680 inbred lines maintained (Table 6) and utilized for various research programmes. Two maize hybrids UMH-9, UMH-8 (Early Maturity) and three composites (UMC-4, UMC-3, UMC-10) were identified by State Research Council of Agriculture Department, U.P. Government at Krishi Bhawan, Lucknow. Currently there are many composites and hybrids are under AER, IET and state varietal trial (Table 7 and 8). Hybrids/composites under adaptive trial (under multi-location testing) are UMH-8, UMH-9, UMC-3, UMC-4, UMC-10, UMC-1 UMC-12, UMC-11 identified from Kanpur centre. Shaktiman is a hybrids registered under PPV and FR in the state.

Table 6. Inbreds maintained and utilized for various purpose from Varanasi centre.

Germplasm received from (IIMR)		Number
Normal inbred	HKI	71
	CML	52
	Other	24
Sweet corn		29
High oil		10
QPM		32
Pop corn		17
New inbred line (winter nursery)		71
CIMMYT		90
Other		284
Total		680

Table 7. List of the experimental hybrids developed and tested in varietal trials

Kharif	IET	AET
2007	VEH-3017, VEHQPM-3027	V-37, VQPM-306
2008	VEHQPM-3018, VEH-07-2, VEH-07-6	VEH-3017, V-37, VEHQPM-3027
2009	VEH-09-1, VEH-09-2, VEHQ-3019	VEHQPM-3027, VEHQPM-3018
2010	VEH-10-1, VEHQ-3028	VEH-09-2
2011	VEH-11-1, VEHQ-3019	VEH-09-2
2012	VEH-12-1, VEHQ-3020	—
2013	1-VEH-13-1	1-VEH-11-1, 2-VEHQ-11-1
2014	1-VEH 14-1, 2-VEH 14-2VEHQ-14-1	1-VEHQ-11-1
2015	VEHQ-15-1, VEH-15-1	-
2016		VEH-16-1, VEHQ-16-1

Table 8. Number of experimental hybrids tested in state varietal trials

Season	Hybrids
Rabi-2013-14	1-VEH-12-1, 2-VEH-12-2, 3-VEH 13-1
Zaid-2014	1-VEH 11-1, 2-VEH-9-1
Kharif-2014	1-VEH-14-2, 2-VEH-12-1
Rabi 2014-15	1-VEH 12-1, 2-VEH 12-2, 3-VEH 13-1, 4-VEHQ 14-1
Zaid-2015	1-VEH 11-1, 2-VEH 9-1
Rabi-2015-16	VEH-14-2, VEH-12-1, VEH-12-2, VEHQ-3020
2016	VEH-12-1, VEH-12-2, VEHQ-3020, VEH-14-2

With the 45 crosses attempted for analyzing the better quality and yield traits of baby corn, HUZM- 221 X CM-119, HUZM-217 X HKI 323, HUZM-217 X CM-119, HUZM-217 -1-1 X CM-119 and HUZM-69 X HKI-323 showed promising results (Sarla Yadav *et al.*, 2012). Dulice Amanillo x HUZM 536, Dulice Amanillo x HKI 323, Win sweet corn x HUZM 536, DMSC 19 x CM 119 and DMSC 36 x HUZM 185 were the best combinations found when sweet corn X field corn were crossed. In an experiment conducted to evaluate the oil content where better combinations i.e. HUZM-53 X HUZM-265, HUZM-478 X HUZM-265, HUZM -478 X High oil, HUZM-265 X High oil, High oil X 53, High oil X HUMQPM-06 performed well (Langade *et al*., 2013). A study has been conducted to analyse the effect of Ozone on QPM and non-QPM maize. The non-QPM inbreds are more prone to ozone damage especially during the reproductive stage. When maize was subjected to different concentration of Ozone (upto 40 ppb), it decreased the photosynthesis (Singh *et al*., 2014).

Some of the current research programs are, breeding for improving the plant architectural and physiological traits, enhancing the yield and quality parameters in sweet corn, identifying the inbreds and parental combinations tolerant to cold, quality and yield enhancement of popcorn and assessing the suitable maintenance breeding methods for maintaining the inbreds and their performance over the different cross combinations. Different inbreds and their cross combination were also evaluated at different plant density to test the stress response and other physiological characters.

Extension activities

Twelve training programs were organized for the Agricultural Officers of U.P. Govt. Dept. of Agriculture and farmers to promote maize at farmer's field from 2006 to 2016. Kisan melas and field days have been conducted regularly.

A part of seeds produced in hybrid seed production program (Shaktiman, 7074, HQPM-1, 9544 etc.) are utilized in assessing their performance in farmers field in front line demonstration (FLDs). Various FLDs at farmer's field were executed in Jaunpur, Mirzapur, Varanasi, Kanpur Dehat, Bahraich, Gonda and few other districts regularly demonstrating improved varieties and package of practices. The performance of hybrids was upto 44% higher over the average of the popular varieties of the state, even appreciated by the farmers and tribal farmers of different districts and this helped farmers to get additional benefits (as a green cob giving higher income than matured cobs) when they sold it in green cob form. The total FLDs (Table 9) over the years range from 35-283 in various centres. Maize seeds multiplied and provided to the farmers under FLD trials are harvested as green ear stage and sold for 1500 Rs/q that gives around Rs 30-35 thousand profit to the farmers within 70 days. This thrives best in the rice –wheat, rice- potato grown area where maize is grown in summer with minimum irrigation.

Table 9. A brief view of FLDs conducted through Varanasi Centre during different years.

Year	FLD's Conducted (ha)	Mean yield (q/ha)
**Kharif*		
2005	50	52.1
2006	64	48.2
2007	57	43.4
2008	96	37.3
2009	103	61.2
2010	90	50.7
2011	90	52.2
2012	38	52.8
2013	282	51.48
2014	11	62.2
2015	27	42.8
2016	30	33
Rabi		
2004-05	20	61.4
2007-08	28	40.8
2008-09	57	57.7
2009-10	52	58.1
2010-11	51	58.5
2011-12	52	58.5
2013-14	37	57.1

*Estimated grain yield; As farmers harvested the green cob.

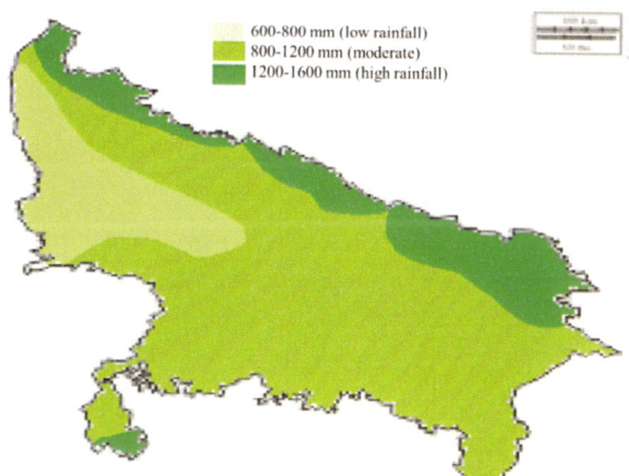

Map 4. Average annual rainfall pattern of UP. (Data source: India Meteorological Department, Pune).

Rainfall pattern in the state

The state is having some of the drought prone regions as well as heavy rainfall areas. The average rainfall distribution across the state is given in Map 4. The hills of western U.P. receives highest rainfall of 1600mm. Sitapur, Lakhimpur

Khiri, Sant Kabir Nagar, J. P. Nagar and Mau are the high rainfall district whereas Raebareli, Auraiya, Etawah, Kanpur are the drought prone regions. From 1950-2010, the changing rainfall trend is -4.42 annually (- 0.22 during winter, 0.02 during summer, -3.52 during the monsoon months and -0.33 in the post monsoon period). This indicates there is a major fluctuation of rainfall distribution over the years. Maximum productivity observed under 600-800 mm rainfall area in western UP. Lowest productivity is recorded in the moderate rainfall area which need to be focused in for further yield enhancement, however high rainfall areas maize is not recommended for cultivation.

Key R and D challenges and their probable solutions in maize

The current challenges are replacing the low yielding varieties and composites with the popular SCHs, developing and popularizing short duration hybrids. The research should emphasize on the trait, region and target specific character improvement in a variety to encourage its stability and further improvement in a desirable direction. E.g. a popular QPM line can be further subjected to bio-fortification with key elements and strengthening it with various stress response. This should be linked with extension work to popularize it among the farmers; the seed multiplication and enabling the farmers to get the seeds at a fair price. This target oriented research definitely pave a way to increase the profitability, adoptability and certainly increment in the overall production. If exclusively research is concerned, the high throughput phenotyping strategies should be focused for the better crop genotypic strata and minimum environmental interactions, as the crop may face the future problems like emergence of new pest and diseases, ozone injury and other climate issues.

Future research and development issues

- Develop high yielding hybrids in early and medium maturity group including specially corn for *Kharif*, *Rabi* and *spring* season.
- To focus on the further improvement in nutrient management, weed control, irrigation, intercropping and other cultural operations.
- Priority should be given on trait specific crop improvement like cold, drought, insect, disease and other stresses.
- Bio-fortifying the varieties which are adopted to the specific regions and already under cultivation to attain the food and nutrition security.
- Availability of seed through participatory seed production program to meet the huge demand of quality seeds at reasonable price at right time.
- Integrated post-harvest management to improve food safety and post-harvest losses.

- The regular training programmes required for generating awareness to reduce production cost and governmental policies made in favour of the beneficiaries.
- Emphasis on value addition technologies for enhancing profitability and employment generation. There is an ample scope to manufacture corn germ oil, ethanol, minor food items, baby food, dahlia, Sooji, sweet dishes etc.
- Farm mechanisation in maize cultivation especially suitable for small holdings.

Among the infrastructure and policy priority issues, efficient marketing, quality and timely seeds availability need to be emphasized.

Though UP is having tremendous manpower and currently there is lot of scope for initiating small scale industries in the rural areas with low production cost. The only need is Government should increase the minimum support price, popularize and emphasize the maize cultivation.

References

Jat, M.L., Sharma, S.K., Rai, H.K., Srivastava, A. and Gupta, R.K. (2006). Effect of Tillage on Performance of Winter Maize in Northern India. Maize Association of Australia 6th Triennial Conference, Griffith, NSW, Australia February 21-23, 2006.

Langade, D.M., Shahi, J.P., Srivastava, K., Kumar, P. and Sharma, A. (2013). Heterosis breeding for maturity and yield traits in diallel crosses of maize. *J Appl Biosci*, 39(2) 91-96.

Sharma, R. (2007). Weed management techniques for *Kharif* maize in north India. *Ind Farm.* 59(5): 23-25.

Singh, A.A., Agrawal, S.B., Shahi, J. P. and Agrawal M. (2014). Investigating the response of tropical maize (*Zea mays* L.) cultivars against elevated levels of O_3 at two developmental stages. *Ecotoxicology*. DOI 10.1007/s10646-014-1287-6.

Singh, Prem. (2006). Alternate cropping systems in peri-urban areas. In: *PDCSR Annual Report*, 2005-06. Pp. 11.

Vision. (2050). Directorate of maize research, ICAR, New Delhi.

Yadav, S., Kumar, A., Shahi, J.P. and Singh, P.K. (2012). Baby corn: A value Added Crop of Maize. *Indian Farmers Digest* 45(9): 23-24.

7

Maize Research and Development in Chhattisgarh

S.K. Sinha and A.K. Sinha*

Raj Mohini Devi, College of Agriculture & Research Station, Indira Gandhi Krishi Vishwavidyalaya, Ambikapur

**Corresponding Author's Email: santoksinha@yahoo.co.in*

Chhattisgarh, the 26th state of the Indian Union came into existence on November 1st, 2000. The state is geographically situated in the central part of India, between the latitudes of 17° 46' N - 24° 5' N and the longitudes of 80° 15' E - 84° 20' E. Its proximate position with the Tropic of Cancer has a major influence on its climate. The total geographical area is around 137.90 lakh ha of which cultivable land area is 46.77 lakh ha and forest land area is 63.53 lakh ha with more than 2.55 crore population. In Chhattisgarh region about 22% of net cropped area was under irrigation. About 80 percent of the population in the state is engaged in agriculture and 43 percent of the entire arable land is under cultivation. Paddy is the principal crop and the central plains of Chhattisgarh are known as rice bowl of central India. Other major crops are coarse grains, wheat, maize, groundnut, pulses and oilseeds.

Importance of Maize in Chhattisgarh

Earlier maize was mainly cultivated in Surguja and Bastar division of Chhattisgarh but now it is emerging as main cash crop in entire Chhattisgarh state. There had been an increase of 20 per cent in maize crop acreage in Chhattisgarh during the past 10 years. Now the farmers of the area are cultivating maize in lines, applying atrazin weedicide, full dose of chemical fertilizers, etc. They are willing to buy the hybrid seed for cultivation from the market and taking maize as green cob and also utilizing as green fodder. Maize threshers have been purchased by some farmers in the region which is an indicative of large maize production.

For industrial utilization of the maize Chhattisgarh state has "Raja Ram Maize Factory" in Rajnandgaon district. They are producing and supplying Maize starches, liquid glucose, dextrose mono hydrate, dextrin, maltose liquid, malto dextrin, maize oil, maize gluten, maize grit, maize husk etc. But most of the maize production is used as poultry feeds in the state. Traders purchase it directly from the farmers' fields in the villages. It is also exported to the neighboring states either as grain or green cob. Around 10-15% is retained by the farmers for own consumption and seed purpose.

Maize cultivated by farmers in Bastar is already being exported to Malaysia, Hong Kong and Vietnam. In the *Kharif* season more than six lakh quintals of maize worth Rs'85 crore was exported from Bastar.

The maize area of the State has increased from 93.4 thousand ha in 2000-01 to 111.1 thousand ha in 2013-14, while the production has gone up from 125.7 thousand tons to 229.1 thousand tons (Fig. 1). There is 1.5 times increase in productivity in the state. The yearly growth rate of maize area, production and productivity of the State since 2000-01 indicated that there is a steady increase over the years.

The continued growth in maize production and productivity over years is due to adoption of high yielding cultivars (composites/ hybrids) and scientific crop management practices. A significant produced maize is from high yielding varieties, which accounts ~ 85% of the total maize area. Private sector companies like Pioneer India Monsanto, Syngenta Nuziveedu, Advanta, Bio-seeds, Mahyco, Rasi Seed etc. play important role in supply of hybrid seeds to the farmers. Maize crop can be extended from upland to midland situation in the state as in drought and prolonged dry spell condition this crop holds better potential than rice.

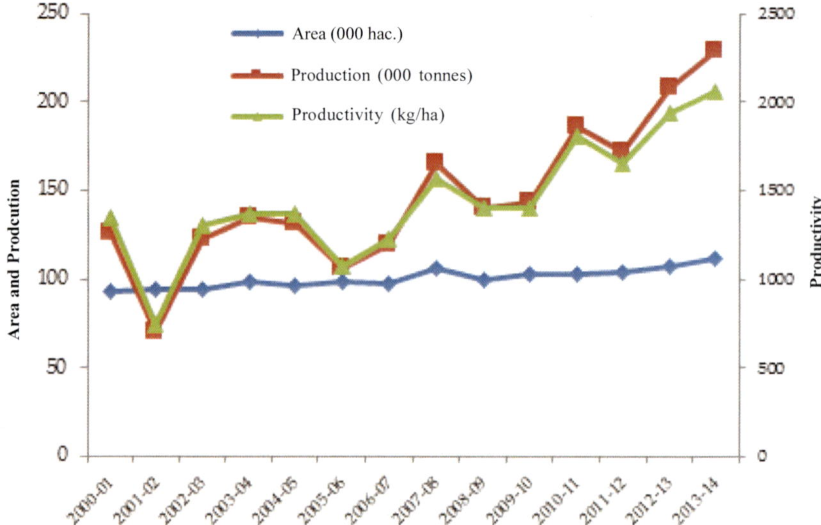

Fig. 1. Year-wise area, production and productivity of maize in Chhattisgarh State.

Major maize growing districts of Chhattisgarh

Maize is being cultivated almost in all the districts of Chhattisgarh (Table 1). However, districts of Northern hills and Bastar Plateau are important with a view of area and production. That are pre-dominantly tribal regions. Maize yield significantly improved in Balrampur, Surajpur, Surguja, Kanker, Bastar and Jashpur districts in the recent past. Though the *Rabi* maize accounts 5% area but it is gaining popularity in Balrampur, Surajpur, Korba and Kanker. Balrampur is the leading maize growing district of the state. While, Dhamtari has the highest productivity followed by Bijapur and Sukma.

Table 1. Crop production statistics of Maize for the crop year 2013-14 in CG.

District	Area(ha)	Production(t)	Yield (t/ha)
Balod	258	415	1.61
Baloda bazar	86	150	1.74
Balrampur	19661	39902	2.03
Bastar	9503	22351	2.35
Bemetara	24	21	0.88
Bijapur	580	1487	2.56
Bilaspur	3016	5654	1.87
Dantewada	3323	8515	2.56
Dhamtari	2074	6090	2.94
Durg	184	202	1.10
Gariyaband	10519	15934	1.51
Janjgir-champa	364	819	2.25
Jashpur	5563	12224	2.20
Kabirdham	2589	5071	1.96
Kanker	11511	25705	2.23
Kondagaon	13586	31831	2.34
Korba	4630	7233	1.56
Korea	8148	14470	1.78
Mahasamund	109	191	1.75
Mungeli	183	343	1.87
Narayanpur	1177	2759	2.34
Raigarh	990	2105	2.13
Raipur	218	411	1.89
Rajnandgaon	1338	2096	1.57
Sukma	2640	6765	2.56
Surajpur	11661	22187	1.90
Surguja	9438	19156	2.03

Agro-climatically, Chhattisgarh is divided into 3 distinct zones with immense potential for agricultural development. Out of these three zones climate of the two zones **Bastar Plateau** and **Northern hills** are very much suitable for maize crop production.

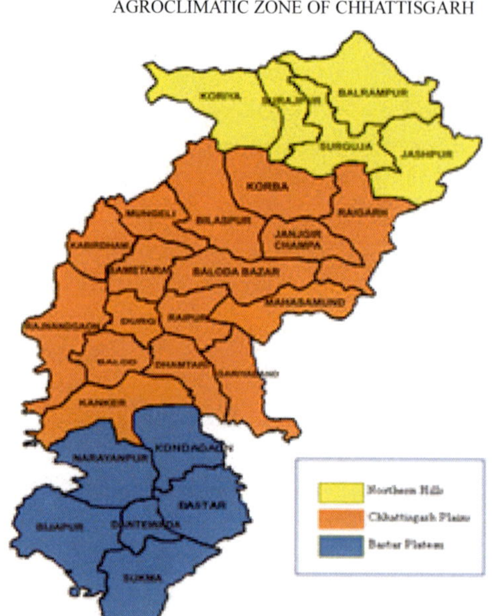

1. **Chhattisgarh Plains**: The plains cover districts of Raipur, Mahasamund, Dhamtari, Durg, Rajnandgaon, Kabirdham, Bilaspur, Korba, Janjgir-Champa, Balod, Baloda Bajar, Bemetara, Gariyaband, Mungeli and a part of Kanker district (Narharpur and Kanker blocks) along with a part of Raigarh district.
2. **Bastar Plateau**: Bastar plateau region comprises of Baster, Bijapur, Dantewada, Kondagaon, Narayanpur, Sukma and the remaining part of Kanker district.
3. **Northern Hills**: It covers districts of Sarguja, Surajpur, Balrampur, Korea, Jashpur and Dharamjaigarh Tehsil of Raigarh district.

In Chhattisgarh, maize is grown on 111.1 thousands hectare land with the productivity of 2062 kg/ha in *Kharif* 2013-14. It is a *Kharif* season crop and is the second most important crop after paddy in area and production. Though the productivity of maize in the state is quite low in comparison to All India and states like Tamil Nadu, Andhra Pradesh, Karnataka, West Bengal, Punjab, Odisha, Bihar etc. but there is an immense possibility to increase productivity. The state has got very good potential for maize but the productivity is very low due to cultivation of open pollinated varieties (OPVs) and improper input management practices. The real potential can be realized by the adoption of hybrid maize with full package of practices. As the major chunk of maize acreage of the state is rainfed, there is a need to popularize single cross hybrids (SCH) which have better adaptability under stress environments.

Climate

The climate of Chhattisgarh is tropical. It is hot and humid because of its proximity to the Tropic of cancer and its dependence on the monsoons for rains. Summer in Chhattisgarh temperatures can reach 45 °C (113 °F). The monsoon season is from late June to October and is a welcome respite from the heat. Chhattisgarh receives an average of 1,292 mm (50.9 in) of rain. Winter is from November to January. Though weather varies from region to region, it's warm in most of the places. Like any other part of India, Chhattisgarh enjoys three seasons, summers, winters and monsoons. During summers (April-June), the temperature sometimes goes up to 45°C. Late in the month of June, monsoons (July-September) arrives in the state as a respite from the scorching heat. Chhattisgarh receives pretty decent amount of rainfall with an average of 1292 mm. Since it falls under the rice agro-climatic zone, rainfall proves to be the main source of irrigation. A significant variation in the annual rainfall adversely affects the harvest. The elevated regions in the north and south observe moderate climate round the year. In October, cool breeze envelops the entire state as if heralding the arrival of winters. The winter season (November-February) doesn't necessarily mean wearing loads of woolens in Chhattisgarh. At this time, the temperature even drops down to 10°C. The climate favours the maize crop production. The normal weather parameters of Northern hills are given in Table 2.

Table 2. Monthly normal mean of values of weather parameter in Northern hills of Chhattisgarh.

Month	T. Max (°C)	T. Min (°C)	Rainfall (mm)	RH1	RH2	Wind Speed (kmph)	Sunshine (hrs)	Evap. (mm)	Rainy Days
January	23.5	8.5	21.0	87	41	2.3	8.2	3.0	1.8
February	26.8	11.5	19.2	81	37	3.1	8.7	4.2	2.0
March	31.5	15.8	18.3	67	30	3.7	8.8	5.5	1.7
April	36.3	20.7	12.8	52	23	4.6	9.1	7.9	1.4
May	39.0	24.6	14.9	49	25	5.4	8.9	9.4	1.7
June	34.8	24.5	205.7	71	51	6.3	6.0	7.2	10.5
July	29.7	23.2	384.9	90	76	5.6	3.7	3.9	18.0
August	29.1	22.5	366.9	93	77	3.8	3.6	3.5	16.9
September	29.7	21.6	233.6	92	71	3.0	5.1	3.7	12.4
October	29.5	17.6	54.5	87	53	2.4	7.1	4.1	4.2
November	27.2	12.1	15.1	86	41	2.0	8.0	3.8	1.1
December	24.2	8.3	11.3	87	37	1.9	7.7	3.3	0.9
Annual	**30.1**	**17.6**	**1358.3**	**78**	**47**	**3.7**	**7.1**	**5**	**73.0**

AICRP centre/s genesis and their mandate

Since time immemorial Indian farmers were growing land races/local cultivars of maize having low productivity which was a matter of concern at national level. In Chhattisgarh AICRP Maize Research programme started in *Kharif* 1999 (ICAR F. No. 18-2/98-FFC; Dt. 05.12.99) at Indira Gandhi Krishi Vishwavidyalaya, RMD college of Agriculture Research Station, Ambikapur-497001. (C.G.) with the following mandate:

- Collection, evaluation, utilization and documentation of the germplasm.
- Development of single cross hybrids.
- Development of early hybrids.

- Development of specialty corn hybrids.
- Standardization of agro techniques.
- Dissemination of maize production technology.

State government effort for maize in Chhattisgarh

The Agriculture Department is involved in the rapid growth in output and popularity of the maize crop among the farmers in Chhattisgarh. Mini-kits are supplied to the farmers and technical advice is provided Various demonstrations on improved maize cultivation conducted under ISOPOM, BGREI, RKVY and with new approach like Cluster demonstration etc.

In view of enormous prospect of promoting food processing units using Maize as raw material, the Chhattisgarh government recently signed Memorandum of Understanding (MoU) with three companies mainly- BMD Starch Pvt Ltd., Kakkad Udyog Ltd and Indian Agro and Food Industries. The aforesaid companies would establish their units in Bastar, Dhamtari and Rajnandgaon districts, with their captive power units.

Reasons for low production/ productivity in Chhattisgarh

The productivity of maize in Chhattisgarh is quite low as compared to the National average. States like Tamil Nadu, Andhra Pradesh, West Bengal, Punjab, Haryana, Karnatak and Orissa are far ahead in terms of maize grain productivity. As the major chunk of maize acreage of the state is rainfed, there is a need to popularize single cross hybrids which have better adaptability under stress environments.

The main reasons for low production/ productivity of the state are:

- Use of traditional local cultivars.
- Cultivation of open pollinated varieties.
- Improper input management practices.
- Maize hybrids have been adopted but could not exploit their potential:
    - Unavailability of suitable hybrids of rainfed condition such as drought due to prolonged dry spell.
    - More concentration on rice as it is the principal crop of *Kharif* in CG.
- Other reasons
    - Non-availability of quality seed in time.
    - High cost of hybrid seed.
    - Inadequate supportive irrigation.
    - Lack of facilities for post-harvest primary processing.
    - Lack of assured marketing.
    - Lack of facilities for value addition of the produces.

Maize production opportunities in Chhattisgarh

Maize production in Chhattisgarh can be increased by seeking and implementing opportunities through area expansion yield improvement, and others etc.

Area expansion

- Substitution of upland non-productive crops with maize during *Kharif.*
- Substitution of paddy (Upland & Mid land) with maize during *Kharif.*
- Increase in area under maize during *Rabi and summer.*

Yield Improvement

- Minimizing / reducing the yield gap at farmer's level.
- Availability and use of potent maize hybrids for biotic and abiotic stress environments.

Other measures

- Implementation of district-wise, zone-wise plan through cluster-approach for yield enhancement.
- Capacity building of the farmers through technology transfer for adoption of scientific crop management practices hybrid maize seed production.
- Promotion of farm mechanization in maize cultivation.
- Promotion of inter-cropping in maize system.
- Market linkage for better price to the farmers.
- Increase in irrigation facilities.
- Promotion of high quality protein maize production.
- Promotion of sweet corn and baby corn cultivation in peri-urban areas.
- Crop insurance facility.

Future Research Strategy

- Development and dissemination of potent hybrids resistant to biotic and abiotic stresses more particularly to drought, water-logging etc. for cultivation in state.
- Standardization of maize agro techniques for different agro climatic zone of the state.
- Standardization of maize based intercropping for different agro climatic zone of the state.

AICRP technologies

The exploration of maize germplasm can be worth, in transfer of the resistance to biotic and abiotic stress and also adoptability of the genotype for the region of the state there are large number of local landraces, most of them with the tribal community, which are keeping just for own consumption. These include flint, yellow to orange grained tall maize germplasms. Collection of these have been made and at present the centre has 48 local maize germplasm.

Popular hybrids

The state is dominated in use of maize hybrids; most of them are from private sector including **NK 30, Hishell, 900M Gold, Pro 4212**, Pro 311, DHM 117 etc. Earlier local landraces or cultivars were taken into large area for cultivation but presently even tribal farmers are also using single cross hybrids. About 85% area is under hybrids (mostly SCH) and the rest are under local landraces or germplasm.

Cultivars released for state by AICRP

So far none of the hybrid/variety has been released from this centre however, based on the performance of hybrids in various varietal evaluation trials maize, the following cultivars recommended for cultivation in the Chhattisgarh.

Full Season Maturity (95-110 days)	Hybrids	PMH 3, PMH 1, Bio 9681, 900M Gold, Hi-shell, NK 30, Seedtech 2324, NMH 731, DKC 9117
	Composites	Prabhat
Medium Season Maturity (85-95 days)	Hybrids	HM 10, DHM 117, HM 9, CMH 08-282, Bio-9637
	Composites	Navjot
Early & Extra Early Maturity (<85 days)	Hybrids	JH 3459, Prakash, Vivek 17, Pusa Hybrid 1, Vivek 21, Vivek 9, Pro 4212
	Composites	Jawahar Makka 216
Quality Protein Maize	Hybrids	HQPM 1, HQPM 5, Vivek QPM 9 Shaktiman 2, Shaktiman 4
Sweet Corn	Hybrids	HSC 1, Bisco Madhu.
Baby Corn	Hybrids	HM 4
Pop corn	Hybrids	VL Popcorn, Bajura Popcorn

Major cropping systems and production ecologies

Due to large production of rice, Chhattisgarh is known as the "rice bowl". Apart from paddy, cereals like maize, kodo-kutki and other small millets, pulses like arhar and kulthi and oilseeds like groundnut, soybean, niger and sunflower are also grown. The main *Rabi* crops of Chhattisgarh are jowar, gram, urd, moong and mothbean. About 80% of its population residing in villages depend mainly on agriculture and agro-based industry and 88% of the total cultivated area is un-irrigated and is under mono-cropping of which 72% of net sown area is under cereals (rice, niger, maize and ragi), predominantly paddy followed by pulses (gram, arhar and urd). All these crops have productivity lower than the national average.

Region	Cropping system
Northern Hilly region of Chhattisgarh	Rice/ Maize-Fallow
	Kodo Kutki-Fallow
	Niger-Fallow
	Fallow-Niger/ Horsegram
	Rice/ Maize-Groundnut
	Rice-Fallow
Plain region of Chhattisgarh:	Rice/ Maize/ Pigeon pea -Chickpea/ Lentil/Peas
	Rice/ Maize-sarson/
	Mustard/ Linseed
	Rice/ Maize-Potato/Wheat
	Maize/ Soybean- Vegetable
Bastar Plateau Chhattisgarh	Rice/ Maize-Fallow
	Kodo Kutki-Fallow
	Niger-Fallow
	Fallow-Niger/ Horsegram
	Maize/ Ground nut- Fallow

Chhattisgarh state has been divided into three Agro-climatic zones *viz.* Chhattisgarh plains, Bastar Plateau and Northern Hill covering 51, 28 and 21% of the geographical area, respectively. The location of the state is such that it is close to the Bay of Bengal, which is instrumental in bringing monsoon in the Northern part of the country. Due to high average precipitation in the state and undulating topography in north and south hilly region having porous soil characteristic results in high runoff and leaching of available rainwater which leads to loss of soil nutrients as well the soils are also left acidic with deficiency in micro-nutrients. Thus, soils do not retain adequate moisture even for *Kharif* crop. *Rabi* crop becomes impossible without tapping ground and surface water, which is scarce.

Characteristics of the Agro-ecological situation

The soils of Chhattisgarh state have been classified into two broad groups as per revenue class and productivity i.e. land on upper slope i.e. Dand and Goda Chawar (Upland soils), and the land following called Gadar Chawar and Bahra (Lowland soils). Upland soils are usually red and acidic (pH 5.0-5.5), lighter in texture and water holding capacity is poor (25-30%). Upland soils are low in organic matter, nitrogen, available phosphorus, sulphur, calcium and magnesium. The medium land soils are yellowish and slightly acidic (pH 5.5-6.0) with considerable water holding capacity. These soils are poor in organic carbon, available nitrogen, calcium and magnesium. The lowland soils are grayish and neutral or slightly alkaline (pH-7.0-7.3). These soils are heavy textured with high water holding capacity and medium in organic carbon and available nitrogen. Due to the presence of high amount of iron oxides, the soils becomes very hard when dry. The soils get easily saturated during rains but at the same time releases moisture very fast for which soil moisture depletion occurs quite frequently and the upland crops have to face moisture stress and physiological dryness.

Changes in cropping system over the years

All the three zones of Chhattisgarh receive rainfall mainly through south west monsoon during *Kharif* and very few through north east monsoon during *Rabi.* Rice is a staple food and is grown in about 3.8 million hectare and out of which 80% area is rainfed which suffers drought either at early or later stage of crop. There are several constraints of continuous rice cultivation over year of which insect-pest and weeds often pose a serious problem as well. Despite good amount of rainfall received in Chhattisgarh, most of the fields are monocropped usually with upland rice with very poor productivity (d"one tonne). The area under cereal crop has been dropping over time and on the other hand the rural focus on fruits and vegetables is growing rapidly. Under the prevailing and changing agro-climatic conditions of northern hills of Chhattisgarh which experiences drought due to wide variation in yearly rainfall, the cultivation of maize may improve farm productivity and uplift the economic status of farmers.

Upland cropping systems

Previous system	Current system
Rice-Fallow	Rice/ Maize- Fallow
Kodo- Kutki-Fallow	Maize + Pigeonpea (1:2) - Fallow
Pigeon pea/Blackgram-Fallow	Pigeonpea + Rice (1:3)- Fallow
Fallow-Niger/Horsegram	Pigeonpea + Maize (1:1)- Fallow
	Maize + Blackgram (1:2)- Fallow
	Maize/ Niger/ Horse gram- vegetable

Midland cropping systems (rainfed/ irrigated)

Previous system	Current system
Rice-Chickpea/ Lentil/ Peas	Rice/ Maize–Chickpea/ vegetable/peas
Rice - Mustard/ Linseed	Maize/ Moong/ Urd– Sarson/ Potato- Summer vegetable
Rice/ Maize-Potato/ Wheat	Ground nut- vegetable- vegetable
	Pigeonpea – Sarson- Summer Maize
	Maize + Blackgram (1:2)
	Maize/ Niger/ Horse gram- vegetable
	Til/ Soybean- Veg - Veg

Lowland cropping systems

Previous system	Current system
Rice-Fallow	Rice – Wheat (surface seeded on wet soil) Rice – Sarson-Summer vegetable *Kharif* Rice – Winter rice- Summer rice Rice – Linseed/ Lathyrus

Production technologies recommended and package of practices for maize in Chhattisgarh are as follows:

Suitable soil

Maize can be cultivated successfully on variety of soils ranging from well drained sandy loam to clay loam. Upland and midland soils of Chhattisgarh are most suitable for cultivation of maize. Soil pH of 5.5 to 7.0 supports good crop growth. Hilly regions of Chhattisgarh soils are acidic in nature, at pH lower than 6.0 application of 2.5 quintal of lime/ha in furrows should be done before time of sowing.

Land Preparation

2-3 ploughings followed planking and Fields should be free from weeds, clods and stubbles.

Selection of cultivars

Full Season Maturity (95-105 days): Bio 9681, 900M Gold, Hi-shell, NK 30, Seedtech 2324, PMH 3, PMH 1, NMH 731, DKC 9117 suited for timely sowing in adequate rainfall areas.

Medium Maturity (85-95 days): Bio 9637, HM 10, DHM 117, HM 9, CMH 08-282, KMH 3712 Suitable for rain fed areas and under delayed sowing. Early Maturity (80-85 days): Pro 4212, JH 3459, Prakash, Vivek 17,Pusa Hybrid 1, Vivek 21, , Pro 4212, Suited for intercropping in arhar and sugarcane.

Extra Early Maturity (75-80 days): Vivek 21, Vivek 9 Suitable for growing as short duration catch crop or an intercrop.

Baby corn: BVM-2 and HM-4

Sweet corn: HSC 1, Bisco Madhu, Madhuri and Priya; Popcorn: VL Amber Popcorn, Bajura Popcorn

Time of Sowing

Season	Optimum time of Sowing
Kharif	25 June to 31 July (sowing time coincide with the monsoon)
Rabi	15 to 30 November
Summer	1 to 28 Feburary

Sowing

In mostly cases, sowing is done with the help of plough. Sowing of seed and fertilizer application is done behind plough.

Seed Rate and Plant geometry

Purpose	Maturity	Seed rate kg/ha)	Plant geometry
Grain and green cob (*Kharif*)	Full	20-25	75 x 20
	Medium	25	60 x 20
	Early	18-20	50 x 20
Sweet corn	15	50 x 20	
Baby corn	20	50 x 20	
Pop corn	12	60 x 20	
Fodder	50	30 x 10	

Seed Treatment

To protect seed from soil borne diseases seed should be treated as follows:

Fungicide/ Pesticide	Application rate (g/kg seed)	Disease/ Insect-pest
Bavistin and Captan	2.0	BLSB, MLB Phythium stalk rot

Manures and fertilizer

Apply 4- 5 t/ha of farmyard manure at the time final land preparation at 10-15 before sowing and chemical fertilizer as follows:

Hybrids Maturity	Nutrient to be applied (kg/ha)		
	N	P_2O_5	K_2O
Early	100-125	50-60	30-50
Medium	125-150	60-80	40-60
Late	150-200	80-100	60-80

Full dose of phosphorus and potassium along with 1/3rd of nitrogen is applied as basal The remaining nitrogen is applied in two equal splits at knee high and tasseling stage.

Water management

Kharif maize is grown as rainfed crop. In case of less rainfall irrigation should be applied at critical stages but in case of heavy rainfall drainage should be done to remove excess water. Irrigation should be given if moisture stress at critical stages i.e. young seedling, knee high stage, tasseling, silking, grain filling stage happen.

Weed management

The critical period of crop-weed competition is 30-35 days after sowing, Application of atrazine 50 wp @ 2.0 kg/ha as pre-emergence is effective for the control of weeds 2 hand weeding at 25 and 40 days after sowing may also take care of weeds.

Harvesting

Maize grown for green fodder, the crop should be harvested at flowering stage. Maize grown for green cob, the cob should be harvested 20-25 days after pollination which is indicated by full sized cob with light green husk, dried silks, soft white kernels turning yellow that produce milky liquid when punctured. Harvesting of baby corn is done at 1-2 days after silk emergence in morning or evening. Harvesting of sweet corn is done 21-22 days after silk emergence. Harvesting of cobs is done when husk around the cobs turns brown. The moisture content in grain at harvesting is up to 25-30%.

Postharvest operations

Harvested cobs should be dehusked and sun dried for about 4-5 days till the grain moisture decreased to 12-15%. Shelling can be done manually or by power operated shellers. Shelled grain is dried for 2-3 days till moisture falls to 8-10% and then stored in gunny/ polythene bags and kept on wooden pallets in dry ventilated store.

Pest and disease management

In Chhattisgarh, stem borer is the major pest in *Kharif* and *Rabi* damaging the crop severely. Major diseases of maize in this state are maydis leaf blight and banded leaf and sheath blight.

Stem Borer

Symptoms

- Central shoot withers and leading to "dead heart".
- Larvae mines the midrib enter the stem and feeds on the internal tissues.
- Bore holes visible on the stem near the nodes.
- Young larva crawls and feeds on tender folded leaves causing typical "shot hole" symptom.
- Affected parts of stem may show internally tunneling caterpillars.

Mix any of the following granular insecticides with sand to make up a total quantity of 50 kg and apply in the leaf whorls on the 20th day of sowing:

- Phorate 10% CG@10 kg/ha
- Carbaryl 1 4% G@ 20 kg/ha.

- For stem borer, release egg parasitoid *Trichogramma chilonis* @ 2,50,000/ha coinciding to egg laying period. Three releases at weekly interval are desirable. Third release is to be accompanied with larval parasitoid *Cotesia flavipes* @ 5000/ha.
- If granular insecticides are not used, spray any one of the following :

Dimethoate 30% EC 660 ml/ha. Carbaryl 50 WP @ 1 kg/ha or on the 20th day of sowing in 500 l of water.

Termite

Symptoms

Termite can cause substantial damage to the maize crop. They establish colonies much deep into the soil.

Control

Frequent irrigation before land preparation and during the crop growth reduces its infestation.

Application of Fepronil granules @ 20 kg/ha followed by light irrigation controls termites to a reasonable extent.

Maydis leaf blight

Young lesion is small and diamond shape. As they, mature they elongate. Growth is limited by adjacent veins so, final lesions shape is rectangular and 2–3 cm long, lesions may coalesce producing a large area. The fungus require slightly higher temperature for infection. Seedling from the infected kernels may wilt and die within 3 – 4 weeks of sowing. If the environment is warm with humidity than this is, unfavourable condition for the occurrence of this disease management.

- Field Sanitation: Ploughing down of crop debris may reduce early infection.
- Growing of resistant variety.
- Spray of diethane M – 45, Captan Zineb @ 2 gm/ lit; 2 to 4 application at 7-10 days interval.
- Crop rotation for 2 – 3 years.
- Use of healthy and certified seed of resistant cultivars.

Banded leaf and sheath blight

Symptoms are first produced on lower leaves, the affected plants produces large gray or brown coloured areas alternating with dark brown bands on infected leaves and sheath. Developing ear is completely damaged and dries of prematurely with cracking of the husk leaves. Brown rotting of ears may develop which shows conspicuous light brown cottony mould with small round sclerotia.

Control

- Deep summer ploughing.
- Field sanitation.
- Seed treatment with *Pseudomonas fluroscense* @16 gm/kg of seed.
- Rhizolex 50WP @ 1-2 g/litre when the crop is 30-40 days old.

Yield gap in maize

Farmers of the state started growing improved varieties/hybrids and now maize crop is emerging as main cash crops in the state as it gives good market price selling of either as green cob or grain.

Demonstrations conducted on hybrid varieties (Pro 4640, 900M gold, Pro 4212, DHM 117, HQPM 1, NK 30, Hishell) of maize from 2001 to 2013 at mostly tribal farmer's field of benefitting 4258 farmers of Chhattisgarh. An additional yield advantage of 50-200% practices observed in these demonstrations. In similar way TSP demonstration has also been conducted during 2013-16 on 92 ha land that benefitted 96 tribal farmers.

8

Maize Research and Development in Bihar

Mritunjay Kumar and Ajay Kumar*

Dr. Rajendra Prasad Central Agricultural University, Pusa, Samastipur – 848 125 (Tirhut College of Agriculture, Dholi, Bihar – 843 105)

**Corresponding Author's Email: drmritunjay.rau@gmail.com*

Bihar is an important maize growing state in India after Karnataka, Andhra Pradesh, Telangana, Rajasthan and Maharashtra, where it is cultivated during all season. However, the yield of maize is much higher during *Rabi* and summer seasons than the *Kharif*. Due to climate change and other factors, the yield of maize is likely to further decline during the *Kharif* season. However, the cultivation of maize in Bihar during the *Rabi* season will be much more profitable as compared to wheat due to rise in temperature. There is need to develop maize varieties which can with stand biotic and abiotic due to changing climate factors.

Decadal trends of maize area, production and stresses yield in the Bihar from 1950–2015.

Year	Area (lakh ha)	Production (lakh tonnes)	Yield (kg/ha)
1950-60	6.1	3.31	538
1961-70	8.98	9.67	1072
1971-80	8.96	8.24	910
1981-90	7.36	9.17	1265
1991-00	7.31	13.79	1954
2001-10	6.25	14.99	2155
2011-15	6.90	19.63	2822

The recent trends of utilization pattern of maize in last decades that about 47% is used as poultry feed, 14% as cattle feed, 20% for direct human consumption, 12% for making starches and 7% for food processing brewery etc. It is estimated that nearly one fourth of the stock keeping units in the modern grocery store contains maize in one and other farm. These range from tooth paste, detergent, paper, dyes, soaps to artificial sweeteners, fructose its etc. maize also finds application in food containers, plastic food packaging, baby powder, diapers, medicines, vitamins tablets, textiles products, candies, breakfast cereals, cooking oils, snacks, pop corn, etc. Internationally its has been processed to produce bio ethanol in a big way for blending with auto fuels. Maize baby corn, sweet corn and green cobs have popularity in urban areas. In many areas maize is considered to be a cash crop as produce could be sold quickly in the market. It has tremendous export potential, considering utilization trend, varieties with better quality should be developed as per need of the market.

Improved local varieties eg. KT 41, the Kalimpong, Sathi and Tinpakhiya were maintained through mass selection by maize breeders and were distributed among farmers. All India Coordinated Maize improvement project (AICRP) was started in the year 1957. Dr. E.W. Sprague of the Rockefeller Foundation was the project coordinator and N.L. Dhawan of Indian Agriculture Research Institute, New Delhi was the associate project coordinator. As a result of cooperative investigations at various research centers of the maize project, the first set of four maize hybrids namely Ganga 1, Ganga 101, Ranjeet and Decan were released in the year 1961. These hybrids were of yellow colour, Ganga 1 was of early in maturity where as Ganga 101 was full season hybrid. For meeting the demand of white grain maize, two hybrids namely Ganga Safed 1 and Hi-starch were developed by the maize project, and released during 1963. Those hybrids were treated in Bihar at multi-location trials during the *Kharif* and *Rabi* seasons. It was soon observed that production of maize hybrids were higher during the *Rabi* season as compared to the *Kharif* season. Furthermore, white grain hybrids namely Ganga Safed 2 during *Kharif* and Hi-starch during the *Rabi* season became more popular in Bihar than yellow grain hybrids, even though certain yellow grain hybrids yielded more as the hybrids Ganga safed 2 was having male parent of indigenous origin. Seed production of these hybrids were taken up by the national Seed Corporation and other seed agencies. Ganga safed 2 during *Kharif* and HI starch during the *Rabi* season had a great acceptance by the farmers of Bihar.

Maize breeding strategies followed in Bihar are as follows:

Period	Varietal Breeding Strategy
1950-59	Local open Pollinated Varieties.
1960-67	Double crosses, double top crosses
1967-71	Composites
1971-89	Composites, double crosses, double top crosses, varietal crosses
1990-99	Double crosses, double top crosses, Three way crosses, Composite
2000-2007	Single crosses, double crosses, Three way crosses, Composites
2008-onwards	Single crosses and synthetics.

Just to make ease in timely availability of quality seeds among maize farmers of India breeding strategy shifted towards development of composites varieties in 1967. However, development of double crosses and double top crosses was continued. Composites like Vijay, Jawahar and Kisan were given to the farmers.

In Bihar, composites *viz.,* Laxmi, Suwan, Hemant, Devaki, and Diara were developed. These composites gained popularity among farmers of Bihar through large scale on farm trials and demonstrations. They were found to be equally good yielder as compared to prevailing hybrids at that time. Lakshmi, Hemant and Divaki were found to be suitable for *Rabi* season where as suwan was good for *Kharif* season. Those composites helped in increasing and stabilizing maize production in Bihar. Large number of private seed production companies came up in Samastipur and Muzaffarpur districts and started producing certified seeds. They distributed quality seeds of these composites at a very affordable price to the farmers. During these periods double top cross, Rajendra Hybrid Makka-1 and a varietal hybrid, Rajendra Hybrid Makka-2 was released but could not compete with these composite varieties, as private seed companies started marketing of three-way and single cross hybrids. In order to provide high quality protein maize, at Dholi, five Shaktiman QPM, three-way and single cross maize hybrids were released during 2001-13. Emphasizing importance, seeds of these QPM hybrids were first time distributed by the Nobel Laureate Dr N.E. Borlaug and then central Agriculture Minister, Shri Nitiesh Kumar at the Pilkhi Village to large number of farmers. In recent years several single crosses and other type of hybrids, synthetics and composites have been developed and given to farmers by the public and private seed agencies.

The crop of single crosses is highly productive and quite uniform that attract most of the farmers. It is estimated that approximately 30% maize area is covered under hybrids indicating low acceptance and reason of low productivity of maize in Bihar. Due to uniformity, the genotype under stresses of high and low temperatures, drought, storm and other abiotic factors single crosses suffer most. Furthermore, they require more fertilizers, irrigation and other inputs that increases cost of production that most often small and marginal farmers are unable to afford. Though, scientists are trying to minimize production cost through adoption of zero tillage technology but its popularity is also less due to lack of proper knowledge. Effect of a biotic stresses can be minimized if we have broad based open pollinated varieties. Therefore, a comprehensive breeding strategy may be followed having suitable heterotic populations are cyclically upgraded against prevailing abiotic and biotic stresses and productive inbred lines and families are spilled out for the development of suitable hybrids and open pollinated varieties. From this each cycle of selection, hybrids or open pollinated varieties should be the choice of farmers and both should be developed with equal emphasis.

The maize research in Bihar with co-ordinated scheme started in 1957 at Pusa (Samastipur) with the establishment of Agriculture College at Dholi in 1960, it was shifted to Dholi and the centre is working with the following mandates:

- To develop early, medium and full season maturity group composites, normal and QPM hybrids.

- To find out suitable agronomic practices for the developed composites and hybrids of different maturity groups.
- To screen materials against foliar diseases, downy mildew, rust and stalk rot.
- Screening of maize genotypes against stem borer and climbing cutworm.
- To produce breeder seeds of recommended composites and parental lines of hybrids.

Trends in APY of maize in Bihar district wise during 2014-15 .

District	Area (ha)	Production (ton)	Yield (kg/ha)
Patna	6779	21910	3232
Nalanda	5896	17999	3053
Bhojpur	3454	8713	2523
Buxar	1359	2026	1491
Rohtas	51	130	2549
Bhabhua (Kaimur)	375	450	1200
Gaya	7431	16682	2245
Jahanabad	451	1530	3392
Arwal	641	2152	3357
Nawada	1929	5032	2609
Aurangabad	349	710	2034
Saran	27816	80758	2903
Siwan	18155	44496	2451
Gopalganj	16690	45195	2708
Muzaffarpur	48768	191885	3935
E. Champaran	52231	102582	1964
W. Champaran	9825	28322	2883
Sitamarhe	5494	18598	3385
Sheohar	1824	6445	3533
Vaishali	32325	131451	4067
Darbhanga	14758	43242	2930
Madhubani	618	1686	2728
Samastipur	56886	156000	2742
Begusarai	47993	121659	2535
Munger	4644	13142	2830
Sheikhpura	663	1055	1591
Lakhisarai	5106	11620	2276
Jamui	2798	6350	2269
Khagaria	60157	214834	3571
Bhagalpur	46062	155808	3383
Banka	11033	35932	3257
Saharsa	28240	128110	4536
Supaul	12408	50821	4096
Madhepura	43420	229466	5285
Purnia	35604	77570	2179
Kishanganj	3033	9465	3121
Araria	44308	189297	4272
Katihar	46944	305622	6510
Total	**706518**	**2478745**	**3508**

Hybrid and varieties developed by AICRP on Maize, TCA, Dholi, Bihar

A. Composites

Variety	Pedigree	Year	For	Special features			Potential yield (q/ha)	Suitable for situation/ District	Time of sowing
				Seed	Maturity (days)	Biotic /abiotic stress			
Lakshmi	American Dent × Tuxpeno	1983	Bihar	White, bold, semi-dent, with white cap for milling	*Rabi*: 165-170	Tolerant to leaf blight and rust, Ability to Resist cold under late sowing in *rabi*	*Rabi*: 65	Bihar: All zones and Districts. *Rabi* season only	1 to 20 November
Suwan	Exotic germplasm	1985	Bihar and Jharkhand	Yellow, flint	*Kharif*: 85-90 *Rabi*: 160-165	Free from foliar diseased and stalk rot prevalent during *kharif* season	*Kharif*: 40 *Rabi*: 45	Bihar : All zones and Districts *Kharif* and *Rabi*	*Kharif* 25 May to 15 June *Rabi* : 1 to 20 November
Hemant	Dholi 7744 (AE)× Tuxpeno	1986	Bihar and Jharkhand	White, semi-dent flat and large in size	*Rabi*: 160-165	Disease resistant	*Rabi* : 60	Bihar : All zones and districts in *Rabi* Season only	1 to 20 November *Rabi*
Dewki	American early Dent ×Tuxpeno	1996	North eastern plains zone (NEPZ)	White, bold semi-dent and flat good for chapatti purpose	*Kharif*: 95-100 *Rabi*: 160-165	Resistant to common rust, TLB, *Fusarim* wilt, stem borer and climbing cut worn	*Kharif*: 45 *Rabi*: 65	Bihar : All zones and districts *Kharif* and *Rabi*	Kharif 25 May to 15 June *Rabi* : 1 to 20 November

B. Normal maize hybrids.

Hybrid	Pedigree	Year	For	Special features Grain	Maturity (days)	Biotic-Abiotic stress	Potential yield (q/ha)	Suitable for district/ season	Time of sowing
Rajendra Hybrid Makka -1	(CM 400 × CM300) × Pant 7421	1994	North eastern plains zone (NEPZ)	White, milky, semi-dent	*Kharif*: 95-100 *Rabi* 155-160	Resistant to TLB, Late wilt @ common Rust. R to stem borer and climbing cutworm	*Kharif*: 50 *Rabi* : 65	All zones and districts of Bihar in *kharif* and *rabi*	Kharif : 25 May to 15 June ; *Rabi* : 1 to 20 Nov.
Rajendra Hybrid Makka -2	(M13× Jogia local)	1996	NEPZ	White, bold, semi-dent often with xeina effect	*Kharif* : 100-105 *Rabi* 160-165	Highly Resistant to rust, *Fusarium* wilt and stalk rot and R to leaf blight	*Kharif*: 50 *Rabi* : 70	All zones and districts of Bihar in *kharif* and *rabi*	*Kharif* : 25 May to 15 June *rabi* : 1 to 20 Nov.
Rajendra Hybrid Makka -3	Dholi in bred 32× Dholi in bred in 40	2009	NEPZ	Yellow, Flint	*Kharif*: 80-85	Free from foliar diseased and stalk rot	*Kharif*: 45	All zones and districts of Bihar in *kharif* and *rabi*	*kharif* : 25 May to 15 June
Rajendra Hybrid Makka Deep Jwala	CM 400×CM30 0× M5	2010	Bihar	White, Bold	*Rabi*: 160 - 165	Resistant to TLB. Late wilt and common Rust. R to stem borer and climbing cutworm	*Rabi* :110	All zones and districts of Bihar in *rabi*	*Rabi*: 1 to 20 November

C. Quality protein maize hybrids.

Hybrid	Pedigree	Year	For	Special features Seed	Maturity (days)	Biotic- A biotic stress	Potential yield (q/ha)	Suitable for district/ season	Time of sowing
Shaktiman -1	(CML-142× CML 150) × CML 186	2001	North eastern plains zone (NEPZ)	White, bold, flint, tryptophan (1.01%), and lysine (4.03%) in protein (9.03%)	*Kharif* :90-95 *Rabi* : 150-155	Resistant to lodging, rust, TLB and MLB, tolerant to stem borer	*Kharif* : 60 *Rabi*: 80	All zones and district of Bihar in *kharif* and *rabi*	*Kharif* : 25 May to 10 June; *Rabi* : 1 to 20 November
Shaktiman-2	CML 176× CML186)	2004	NEPZ	White, flint, tryptophan (1.04%), and lysine (3.998%) in protein (9.3%)	*Kharif* : 90-95 *Rabi* : 150-160	Tolerant to rust and TLB	*Kharif* : 65 *Rabi* : 85	-do-	-do-
Shaktiman-3	(CML 161 × CML 163)	2006	NEPZ	Orange, yellow, flint, tryptophan (0.73%), and lysine (2.96%) in protein (9.63%)	*Kharif* : 95-100 *Rabi* : 155-160	Resistant to MLB, Downey mildew and rust	*Kharif* : 70 *Rabi* : 90	-do-	-do-
Shaktiman-4	(CML 161× CML 169)	2006	NEPZ	Orange, yellow, semi-flint kernels tryptophan (0.93%) and lysine (3.78%) in protein (9.88%)	*Kharif* : 95-100 *Rabi* : 155-160	Resistant to MLB and other diseases and pest	*Kharif* : 70 *Rabi* : 95	-do-	-do-
Shaktiman-5	CML -161× CML-165	2013	NEPZ	Orange, yellow, flint, tryptophan (0.63%) and lysine (3.86%) in protein (9.43%)	*Kharif* : 95-100 *Rabi* : 155-160	Resistant to MLB, Downey mildew and rust	*Kharif* : 70 *Rabi* : 100-110	-do-	-do-

Key production constraints in Bihar for maize

Kharif season

Poor drainage system and water management

- Difficulty in weed management.
- More diseases and pests attack.
- Non-availability of seeds of adapted improved varieties.
- Inadequate use of production input.
- Lack of awareness and adoption of production practices improved.

Rabi season

- Long, dry, cold growing season with terminal heat problem.
- Inadequate use of poor quality seed, fertilizers, pesticides and other production inputs.
- Lack of varieties tolerant to cold, drought, low nitrogen input and suitable for various inter cropping systems.
- Frequent weed, diseases and pest problem.
- Lack of awareness for diversification of maize cultivation.
- Non-remunerative prices of the produce.

Summer season

- Inadequate moisture in soil at the time of sowing.
- High temperature at flowering and grain filling period.
- Needs frequent irrigation as flood system followed mostly.
- Problem in weed and pest management.
- Lack of awareness about high yielding early maturity hybrids.
- Non remunerative prices.
- Difficulty in storage during rainy season which increased post harvest losses.

Maize production technologies recommended for Bihar

Field preparation

Soil should be loose up to 20-25cm depth by deep ploughing or discing or two or three ploughings. In case tractor is used one ploughing followed by couple of disking is generally adequate.

Time of seeding

Maize can be grown in all season *viz., Kharif* (monsoon), post monsoon, *Rabi* (winter) and spring. During *Rabi* and spring seasons to achieve higher yield at farmers field assured irrigation facilities are required. During *Kharif* season it is desirable to complete the sowing operation 12-15 days before the onset monsoon. However, in rainfed areas, the sowing time should be coincided with onset of monsoon. The optimum time of sowing ate given below.

Kharif - 15th June to 15th July

Rabi - November

Distance (cm^2)

Short duration variety -60 × 20 = 83.33 thousand plants/ha
Long duration variety -75 × 25 = 55 thousand plants/ha
Rabi season -75 × 20 = 66 thousand plants/ha

Seed rate

Hybrid - 18 to 20 kg /ha
Composites - 20 kg/ha

Methods of planting

Sowing of maize seeds across the slope to be done at 3-5 cm depth which depends on the moisture status and the type of soil.

1. Planting on the side of a ridge: this method adopted in high rainfall areas.
2. Planting in narrow furrow: this method is adopted in low rain fall areas.
3. Planting in flat bed with no earthing up : in normal conditions.
4. Planting in flat bed with earthing up after 40-50 days of planting: in areas of heavy storms during rainy days.

Nutrient management

The quantity of nutrients required by a crop depends on soil, climate, cultivar yield level, soil moisture and other managements factors.

Manures and fertilizer both play important role in the maize cultivation. Addition of 10 to 15 tonnes of well-rotted organic matter in the form of farm yard manure compost before sowing. The application of organic matter to the soil ensures good tilth and improved water holding capacity. The quantity of fertilizer to be applied depends on soil testing However, a general recommendation of fertilizer is given below:

Quantity of fertilizer (kg/ha) required for maize production in Bihar

Nutrient	At sowing	2nd split	3rd split
Kharif			
Nitrogen	50	40	30
Phosphorus	60	-	-
Potash	50	-	-
Rabi			
Nitrogen	60	50	40
Phosphorus	75	-	-
Potash	50	-	-
Spring			
Nitrogen	50	50	-
Phosphorus	40	-	-
Potash	30	-	-

In *Rabi* season, 2nd split of fertilizer at 50 to 60 days after sowing while in *Kharif* and short duration varieties 2nd split should be applied at 30 to 40 days after sowing. In the interval of 3 years zinc sulfate should be applied @ 25kg/ha at the time of field preparation.

Rotation and mixed cropping

The recommended crop rotations are given below:

Maize –potato	1 year
Maize –wheat	1 year
Maize –toria-wheat	1 year
Maize –potato-wheat	1 year
Maize –barseem	1 year
Maize –toria-sugarcane	2 years
Maize –wheat-sugarcane	2 years
Maize–wheat–sorghum- sugarcane	3 years

Harvesting

In absence of irrigation facilities and failure of rains, maize may be profitably harvested at any stage of its growth. At pre flowering stage, it may be used as fodder, just at silk emergence stage as baby corn, at early-dough to late dough stages for green-ear, and the stover may be used to feed cattle. Grain crop should be harvested when it reaches physiological maturity containing 25-30% moisture, and ears should be removed before cutting the stalks. Immediately after harvesting, stover can be used as cattle feed.

Post harvest management

Drying of cobs: The drying of the cob should not be done either on the kuccha or pucca flour, rather it should be dried on tarpoline sheets to avoid seed injury and during night cobs should be kept covered. To maintain the purity, dissimilar, diseases and pest infested cobs should be removed before shelling. The cobs should be dried up to 13-14% moisture content before shelling.

Shelling : Shelling can be done manually or by power operated maize sheller.

Seed processing : All under size, broken, and damaged seeds should be removed for maintaining the quality of seeds.

Stores and Marketing: Seed drying should be done till the moisture content of the seed reduced to 8% and it should be kept in aerated jute bags. Seed should be stored at cold and dry place preferably in cold storage.

Major diseases of maize and their control

Maydis leaf blight (*Helminthosporium maydis*)

In maydis leaf blight, individual spots are grayish, tan in colour, up to one and a half inches in length, oval-shaped with straight zonations. Young lesions are small and diamond shaped. As they mature, they elongate. Growth is limited by adjacent veins, so final lesion shape is rectangular and 2 to 3 cm long. Lesions may coalesce, producing a complete burning of large areas of the leaves. The symptoms described to the "O" strain of the fungus. In the early 1970s the "T" strain caused severe damage to maize cultivars in which the Texas source of male sterility had been incorporated. Lesions produced by the T strain are oval and larger than those produced by the O strain. A major difference is that the T strain affects husks and leaf sheaths, while the O strain normally does not.

Disease cycle

The fungus over-winters as mycelium and spores in maize debris in the field and on kernels. Conidia are carried by wind or splashing water to growing plants where primary infection occurs. Sporulation on the lesions produces additional primary or secondary inoculum. The disease cycle can be completed in about 60 to 72 hours (race T) under ideal conditions. Infected kernel may be potential means of overwintering and spread of *H. maydis*. However, infected grains are not toxic to livestock. Maydis leaf blight (on southern maize leaf blight) is prevalent in hot, humid, maize- growing areas where the temperature ranges from 20-30^{0}C during cropping period.

Management

- Ploughing down of crops debris may reduce early infection.
- Rotation with non-host species will help reduction of inoculum.
- Use tolerant varieties *viz.*, Suwan, Ganga Safed-2, Deccan, Ganga-4, Ganga-5, Kisan, Jawahar, PRO345, JH10655, MCH117and JHM1701.
- Seed treatment with Captan or Thiram @ 2.5 g/kg seed before sowing.

- Spray of Dithane M-45 or Zineb @ 2 g/l water as soon as first symptoms of disease become apparent on leaves or when the crop is at knee high stage. Two to four applications needed depending on disease intensity.

Turcicum leaf blight (*Exserohilum turcicum*)

An early symptom can be easily recognized, slightly oval, water-soaked, small spots produced on the leaves. These grow into elongate, long, elliptical, spindle shaped grayish green or tan lesions ranging from 2.5 to 15 cm in length. They may appear first on lower leaves and increase in number as the plant develops and can lead to complete burning of the foliage. They will later have dark, reddish-brown borders and occur on leaves, stalks, leaf sheaths, husks, and shanks. Cob rot occurs as well as ear drop, and severe infection causes a prematurely death and gray appearance that resembles to frost or drought injury. The fungus overwinters in corn debris and on seed. Wind and splashing water spread the spores rapidly in the field under ideal conditions, cycling in about 72 hours.

Disease cycle

Crop debris is the usual source of primary inoculum. As the ears are not affected the possibility of the disease being carried on the seed is remote. Once the infection takes place, it spreads through wind currents. The damage is severe in area with heavy dew and rainfall. Spores germinate and penetrate the leaves within a few hours in the presence of free water and when the temperature range is 25-30°C. Losses are high if the disease occurs in the early stages of the crop in a severe form, blighted leaves killed prematurely and lose their nutritive value even for fodder. The disease is prevalent in areas where cooler condition prevails and maize is planted in high lands, winter planting in the plains as the cool/moderate humid conditions (18-27°C) favours disease developments. When infection occurs prior to and at silking stage and conditions are optimum, it may cause significant economic damage.

Management

- Sanitation by plough down infected crop debris.
- Following proper cropping sequence.
- Apply recommended dose of different nutrients.
- Grow tolerant varieties like, Vivek 21, Vivek 23, Vivek 25, Pratap, Kanchan, Rajendra Hybrid Makka-1, R.H.M, M-2, Deoki, Lakshmi, PRO345, JH10655 and MCH117.
- Seed treatment with Captan or Thiram @ 2.5 g/kg seed before sowing.

- Spray the crop with Mancozeb (Indofil M-45) @ 0.2% or 2 kg fungicide/ 1000 liters water / ha if disease is noticed on the crop and repeat spraying at 15 days depending upon severity of disease.

Banded leaf and sheath blight (*Rhizoctonia solani*)

The symptoms of the disease appears at the pre-flowering stage in 40-50 day old plants but infection can also occur on younger plants. The disease is manifested on leaves, sheaths, stalks and ears; infection on tassels on tassels has not been observed. Symptoms develop on leave and sheath are characteristic concentric spots that cover large areas of infected leaves and husks, which show conspicuous light brown cottony mould with small, round, black sclerotia. In later case, severe blight occurs which is accompanied by death of the apical growing point.

Disease cycle

The primary source of inoculums are sclerotia in the soil and grass hosts that grow in the vicinity of maize crop. Secondary spread of disease is by contact of infected leaves with parts of adjoining healthy plants. The optimum temperature for *in vitro* growth of the pathogen is 30^{0}C and the highest level of disease is induced when RH is in the range of 90-100%. At RH 70% or lower, the disease development is negligible or absent. These conditions of moisture and temperature exist in North Indian plains, the month of July and August, a time when the crop is in vulnerable growth stage. The disease is prevalent in hot humid foothill region in Himalayas in plains.

Management

- Deep summer ploughing.
- Sowing should be done on ridges to avoid high moisture.
- Grow tolerant varieties like Pratap Kanchan 2, Pratap Makka 3, Pratap Makka 5, Shaktiman 1, Shaktiman 3, Deoki, Suwan, Ganga-11, Ganga Safed-2, Sweta and D941.
- Stripping of 2 lower leaves along with leaf sheath.
- Provide proper drainage in the field.
- Use recommended dose of different nutrients.
- Treat seed with Captan @ 2.5 g/kg seed before sowing.
- Seed treatment of peat based formulation @16g/kg of *Pseudomonas fluorescence* or as soil application @ 7g/l of water.
- Spray the lower part of the standing crop by Validamycine @ 2.7 ml/l of water.

Common rust (*Puccinia sorghi*)

Common rust is most conspicuous when plants approach tasseling. It may be recognized by small, elongate, powdery pustules over both surfaces of the leaves. Pustules are dark brown in early stages of infection; later, the epidermis is ruptured and the lesions turn black as the plant matures. These spores eventually turn black. Rust is favored by cool temperatures and high humidity. Older tissue are generally resistant to the disease.

Disease cycle

The life cycle of *P. sorghi* involves two hosts (maize and *Oxalis* species) and five spore stages (teliospores, basidiospores, spermatia, aeciospores and urediniospores). In tropical or subtropical regions, urediospores can over winter and serve as the primary source of inoculum in subsequent seasons. Urediospores are disseminated by wind over vast distances (hundreds of kilometers) and frequently spread from tropical/subtropical regions to temperate regions in spring and summer when maize is cultivated. The sexual stage of the life cycle occurs predominantly in tropical and subtropical regions. Generally, urediospores germinate within 6 hours. Infection structures penetrate the host via stomata on leaf surfaces, stems, and ear sheaths. Chlorotic spots as a result of infection can arise in as little as 24 hours in favorable climatic conditions. In favorable weather conditions the disease cycle can be completed in 5 days. The disease is common in subtropical, temperate and highland environment of moderate temperature (16-25°C) and high relative humidity.

Management

- Do not plant seed corn in a field where corn was the previous crop, unless absolutely necessary.
- Following proper cropping sequence.
- Cultivate early maturing varieties.
- Grow hybrids like Deccan, Ganga-5, Hybrid Makka-103 and DHM-1 which tolerant to this disease.
- Spray the crop with Zineb or Indiofil M 45 @ 2-3 g/l of water.

Brown stripe downy mildew (*Sclerophthora rayssae*)

Affected leaves become narrow with chlorotic spots (3.7 mm wide) with well defined margins extended in parallel fashion between the veins. The stripes later turn redish to purple. If the development of the chlorotic stripes occurs prior to flowering, seed development fails and plants die prematurely. Collapse of the tissues has never been observed. Malformation or deformation is also not on record.

Disease Cycle

S. rayssiae over season as in infected debris in the soil. On germination, the oospore produces a sporangiophore bearing a sporangium that liberates the zoospores that constitute the primary inoculum. In the presence of enough moisture or high temperatures, the sporangium may produce a germ tube that can also over season in the grass host as mycelium from which sprongia are produced. This is another source of inoculum. Secondary spread is by sporangia produced after primary infection is established.

The disease is most prevalent in warm, humid regions and common in the Himalayan areas of northern India. The disease is limited to location below 1500 masl.

Management

- Planting before rainy season begins can minimize the occurrence of disease.
- Use tolerant varieties such as Prabhat, Kohinoor, PAC 9401 etc.
- Follow recommended cultural practices with well drained field.
- Rogue and destroy infected plants.
- Seed treatment with systemic fungicide such as metalaxyl @ 2.5 g/kg seed.
- Brown stripe downy mildew can be controlled by three to four foliar sprays of Ridomil 25 WP @ 2.5 g/l . Give first spray as soon as disease appears in the field. The sprays should be given at 10 to 15 days intervals depending upon the severity of disease.

Insect

Maize satlk borer, *Chilo partellus* is a major pest during *Kharif* season throughout the country, while *Sesamia inferens* occurs during *Rabi* season in some of the areas of peninsular India.

Sometimes in a few pockets, maize crop may be attacked by some polyphagous pests like Pyrilla, armyworm, cutworm etc. for these; it is desirable to follow the locally recommended control measures.

Future research and development strategies for maize in the state are

- Apex organization should provide guide lines for proper production, storage, marketing, value addition, industrialization and export promotion of maize and its products.

- To promote industrial utilization of maize, development and cultivation of industry specific suitable varieties should be emphasized.
- Development and promotion of maize diversification by organic cultivation, processing and export of baby corn, sweet corn, pop corn, green cobs etc.
- Bridging the gap between potential and realized yield by targeted technology intervention in maize.
- Standardization of production techniques at agro- ecoregional basis.
- Conduct of large scale trainings and demonstrations, etc. for modern production, processing and value addition technologies.
- Local production, processing, storage and marketing of hybrid maize seed by establishing seed village.
- Replacing the common maize varieties with QPM varieties especially for use in food or feed purposes.
- Encourage private seed companies to provide seed of suitable varieties.
- Better co-ordination between public-private sectors in production, processing, marketing and exporting maize and maize products.

9

Maize Research and Development in Jharkhand

M. Chakraborty, C.S. Singh, V.K. Tudu and M.K. Chakravarty*
Birsa Agricultural University, Ranchi, Jharkhand
**Corresponding Author's Email: manigopa291061@yahoo.com*

Maize or corn (*Zea mays* L.) is second most important crop after rice in Jharkhand. Though considered as a *Kharif* season crop, it can be successfully grown in other two crop seasons *viz.*, *Rabi* and *zaid* also as the climatic condition in Jharkhand is conducive for growth of the plants. *Rabi* maize has comparative advantage over *Kharif* maize due to low incidence of diseases and insect pests slow growth of weeds and crop does not have to suffer on account of heavy rainfall hence, higher yield potentiality could be realized from *Rabi* maize. In Jharkhand, the *Kharif* crops including the maize are grown as rainfed as the 80 to 85% of the annual rainfall takes place during June to September. The share of maize under irrigated condition is restricted to only 15 to 18% of the total cropped area. Maize is not only an important source of nutrient to the human beings but also constitutes a basic component of animal and poultry feed. Here, farmers grow maize mainly for green cobs as cash crop whereas the grains are to be used for human consumption as well as for animals and poultry feed. The cultivation of the specialty corns like baby corn, sweet corn and popcorn is still in its infancy in Jharkhand and hence, large scale cultivation of specialty corn in the state will take some more time. Establishment of food processing unit in the state is at the nascent stage. However, the Government of India is implementing a Mega Food Park in Getalsud Industrial Area of Ranchi district in Jharkhand. The establishment of this mega food park is expected to give a significant boost to the food processing sector in Jharkhand.

The state has a total geographical area of 79714 sq km of which about 30% area is covered with forests. The state extends between 22° North to 25.5° North latitudes and 83° East and 87.75° East longitudes. The cultivable area is estimated around 41.80 lakh hectares out of which the net sown area is 18.08 lakh ha only. The land surface being undulated and is subjected to sheet and gully erosion, causing loss of soil and plant nutrients and about 23 lakh hectares are subjected to severe erosion every year. The state forms a part of Agro-climate zone-VII of the country known as Eastern Plateau and Hills Region. Based on rainfall, temperature, terrain and soil characteristics, this state has been divided into three Agro-Climatic sub zones, viz; Sub Zone IV (Central and North Eastern Plateau sub zone) Sub Zone V (Western Plateau sub zone) and Sub Zone VI (South Eastern Plateau Sub Zone). The annual rainfall in the plateau and sub-plateau region is 1400 mm on an average, of which 82.1% is received during the period of June to September and the rest 17.9% is received in remaining months. Jharkhand has one of the highest levels of poverty in India at 40.3% as against the all India average of 27.5%. There is sharp contrast between rural (49%) and urban poverty (23%) in the state. Almost 80% of its population depends on agriculture and allied activities for their livelihood contributing about 15% to states gross domestic product. About 82% of the households have holdings of less than two hectares land (average 1.18 ha).

In the agricultural sector there is scope for increasing production and productivity of food crops through vertical and horizontal expansion by bringing more area under irrigation and through optimum utilization of inputs like seeds, fertilizers, pesticides, agricultural tools and implements. Irrigation stands out as the most critical requirement for the development of agriculture in the state. Despite the fact that the state has a good rainfall, the surface water availability to agriculture is not sufficient due to inadequate water storage facilities. Ground water recharge rate is low; as a result, the water table in the plateau is going down. The water holding capacity of soil in most part of the state is low which a major constraint for successful agriculture.

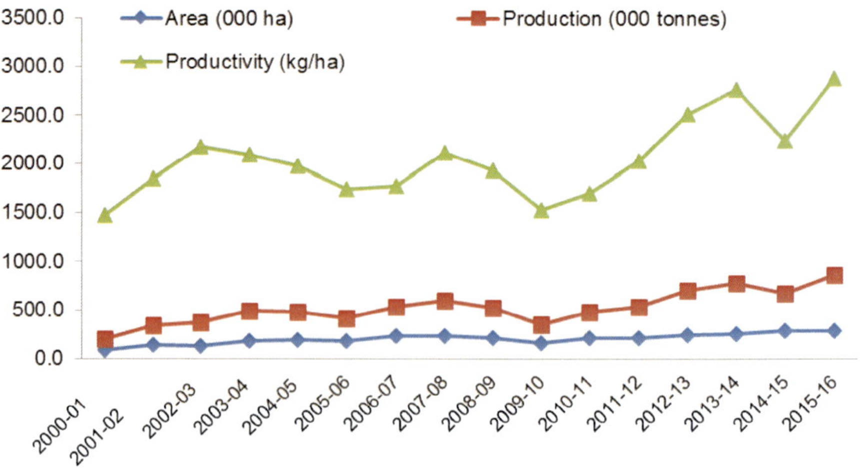

Fig.1. Trends in area, production and yield (APY) of maize in Jharkhand.

Trends in APY of maize in the Jharkhand

The area as well as the production of maize has been increasing continuously over last decades. Though the acreage has increased consistently, production pattern has been erratic owing to the variations in the yield as maize is mostly grown in *Kharif* as a rainfed crop in Jharkhand. There are many factors that have contributed towards variation in yield like, weather during crop season, pest and disease attack, technological advances and development of new hybrids and varieties, *etc*.

Major maize growing districts of Jharkhand

Maize is mostly cultivated as rainfed crop in upland soils of Jharkhand. It is being grown in all the 24 districts of Jharkhand. However, the major maize growing districts of Jharkhand are Garhwa, Palamu, Latehar, Chatra, Hazaribag, Giridih, Dumka and Godda.

Major cropping systems and production ecologies of maize in Jharkhand

In Jharkhand, 80 % of its population residing in 32620 villages depends mainly on agriculture and allied activities for their livelihood. Much of the farming is at subsistence level with virtually no surplus as the agriculture is susceptible to vagaries of monsoon which can be gauged from the fact that as much as 92%

of the total cultivated area is un-irrigated and is under mono-cropping. 72% of the net sown area is under cereals (Rice, wheat, maize and finger millet), predominantly rice followed by pulses (gram, pigeonpea and urd) with 13% and horticultural crops by 10%. The rest 5% area is covered under different oilseeds crops.

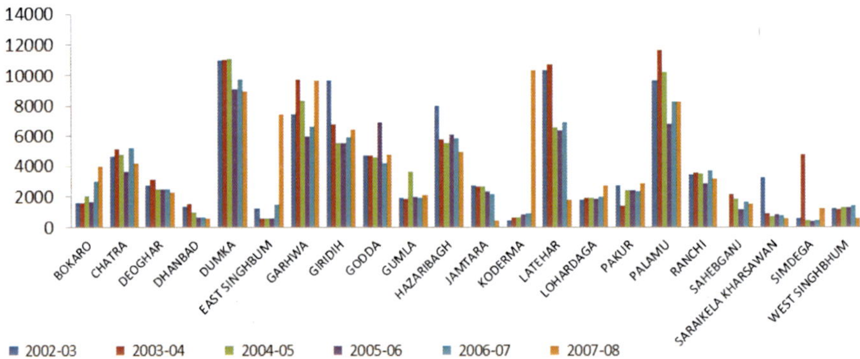

2a. Maize area (ha) in different districts of Jharkhand

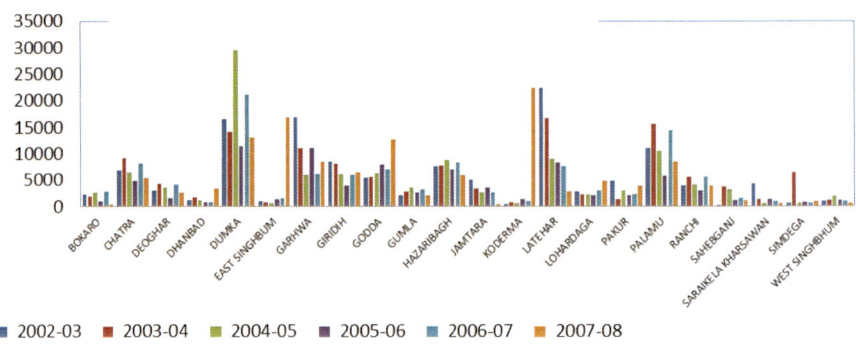

2b. Maize production (tonnes) of Jharkhand districts

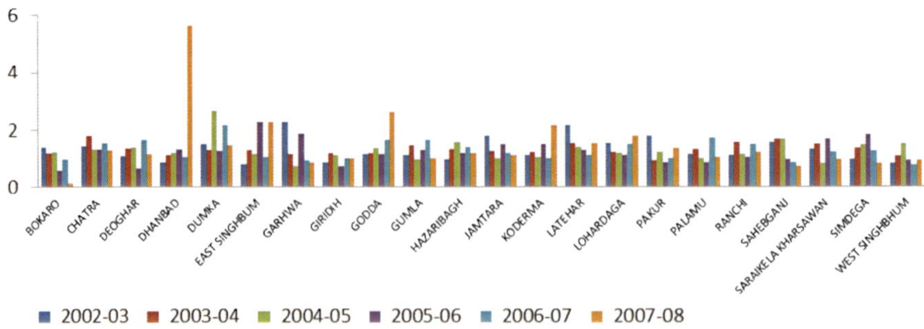

2b. Maize productivity (t/ha) of Jharkhand districts

Fig. 2. Trends in Maize APY of Different districts of Jharkhand.

Intercropping as a regular practice is not so common in the state. However, the intercropping systems found to be predominant among the tribal community includes legume-based mixed and intercropping practices like pigeon pea + rice, pigeonpea + blackgram, pigeonpea + fingermillet, pigeonpea + maize, cowpea + rice, gram + wheat, gram + linseed and gram + mustard. The crop rotations followed by the farmers are rice-gram, rice-lentil, rice-wheat-mungbean, rice-pea, groundnut-wheat, maize-wheat and different vegetables and potato based cropping systems and paira cropping in rice lowlands. Lathyrus seeds are broadcasted in standing rice when soils are saturated; the crop gives a good yield without irrigation. A close look in the above systems indicates that most of the tribal farmers include legumes in their planting which contributes to biological nitrogen fixation in soil and thus enriches the soil fertility.

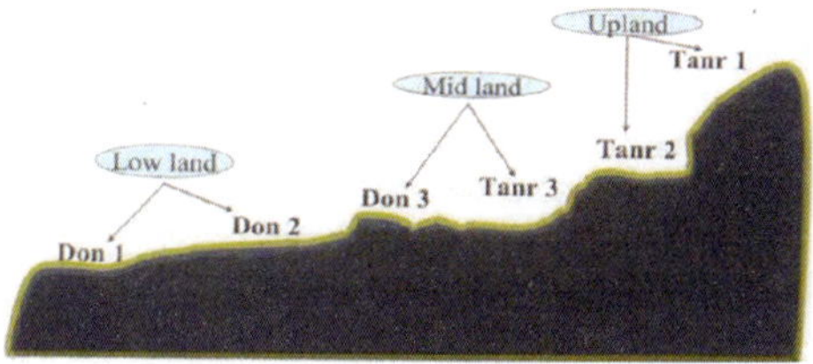

Land situation for technology generation

Due to high average precipitation in the state and undulating topography and porous soil characteristic, it results in high runoff and leaching of available rainwater which leads to loss of scantily available soil nutrients. The soils are also left acidic with deficiency in micro-nutrients. Thus, soils do not retain adequate moisture even for *Kharif* crop. *Rabi* crop becomes impossible without tapping ground and surface water, which is scarce.The region mainly comprises soils developed on granite gneiss (32.6%) and granite schists (14.2%). There is practically no problem of soil salinity or flooding. Soil acidity problem (pH< 5.5) is acute in 4 lakh hectare of cultivated area.

Topo-sequence in Jharkhand

The soils of Jharkhand state have been classified into two broad groups as per revenue class and productivity *i.e.* land on upper slope, *Tanrl* and (Upland soils) and the land following called *Don* land (Lowland soils). For technology generation, these are grouped again into three categories *i.e.* Upland (*Tanr I &Tanr II*), Medium land (*Tanr III & Don III*) and Lowland (*Don I & Don II*). Upland soils are usually red and acidic (pH 5.5-5.9), lighter in texture and water holding capacity is poor (30-35%). Upland soils are low in organic matter, nitrogen, available phosphorus, sulphur, calcium and magnesium. The medium land soils are yellowish and slightly acidic (pH 6.0-6.5) with considerable water holding capacity. These soils are poor in organic carbon, available nitrogen, calcium and magnesium. The lowland soils are greyish and neutral or slightly alkaline (pH-7.0-7.3). These soils are heavy textured with high water holding capacity and medium in organic carbon and available nitrogen. Due to the presence of high amount of iron oxides the soil becomes very hard when it get dry. The soil get easily saturated during rain but loss of moisture is also fast which causes frequent depletion of soil moisture and the upland crops have to face moisture stress and physiological dryness.

Characteristics	Upland	Medium land	Low land
Colour	Red or brown red	Yellowish	Greyish
Texture	Light textured (Sandy loam)	Medium textured (Sandy clay loam)	Heavy textured (Clay loam)
Drainage	Well Drained	Moderately drained	Poorly drained
Soil reaction	Low pH	Moderately acidic	Neutral pH
Soil fertility	Poor in organic carbon, Ca, Mg, N, P & S	Poor in organic carbon, N, Ca, Mg	Medium in N & Organic carbon

Efficient cropping systems for different land situation

Upland rainfed ecosystem having water retention capacity 180-200 mm/ m soil depth.

Existing	Diversification
Rice/ Maize-Fallow	Pigeonpea + Groundunt (1:2)
Finger millet/ Gundli-Fallow	Pigeonpea + Blackgram (1:2)
Pigeonpea Blackgram-Fallow	Pigeonpea + Rice (1:3)
Fallow-Niger/ Horsegram	Pigeonpea + Maize (1:1)
	Maize + Blackgram (1:2)
	Maize + Soybean (1:2)
	Blackgram-Safflower/ Linseed/ Lentil/ Chickpea/ Niger/ Horsegram
	Maize-Safflower/ Linseed/ Lentil/ Chickpea
	Soybean -Safflower/ Linseed/ Lentil/ Chickpea

Medium land: It is very potential land situation with limited to adequate irrigation facilities for sequential cropping.

Existing	Diversification	
Rice/ Maize-Chickpea/ Lentil/ Peas	Maize + Blackgram – wheat	Rice – Potato – Green Gram/ Blackgram
Rice/ Maize-Barley/ Mustard/ Linseed	Maize – Early potato late wheat	Rice – Early Potato – Late Wheat
Rice/ Maize-Potato/ Wheat	Maize - early potato – Late peas (green pods)	Rice – Toria – Late Wheat
	Maize – Wheat – Green gram	Rice – Early Potato – Late Peas
	Maize – Potato – Green gram	Rice – Early Potato – Onion
	Maize – Potato – Onion	Rice – Berseem
	Maize – Toria – Wheat	Deenanath Grass – Berseem – Maize + Cowpea
	Maize – Early Potato/ Toria- Late wheat - Green gram	Rice – Early Potato/ Toria – Wheat
	Maize – Wheat + Lentil (4:2)	Rice – Early Potato/ Toria – Wheat – Greengram
	Maize – Wheat + Mustard (8:2)	Rice – Wheat + Lentil (4:2)
	Maize – Lentil + Mustard (5:1)	Rice – wheat + Mustard (8:2)
	Maize – Potato + Wheat (1:1)	Rice – Potato + Wheat (1:1)
	Rice – Wheat – Green Gram/Black gram	Rice – Lentil + Mustard (5:1)

Lowland: Excessive moisture present in the soil after harvesting of low land rice does not permit tillage operation till January – February. Hence, land remains fallow. Under this condition, surface seeding of wheat–utilized residual moisture and increases cropping intensity.

Existing	Diversification
Rice-Fallow	Rice – Wheat (surface seeded on wet soil) Rice – Summer vegetable Rice – Summer rice Rice – Berseem

AICRP centre/s genesis and their mandate

All India Co-ordinateed Maize Improvement Project (AICRP) on maize has started functioning at Birsa Agricultural University, Ranchi from September, 2005 with the following mandate.

- To test the performance of maize hybrids/varieties under co-ordinated network.
- To conduct assigned and/or national location specific trials on agronomy.
- To develop varieties or hybrids of maize with appropriate management packages suitable for different agro-climatic regions.
- To spread the improved varieties with suitable management package among the maize growers either through 'Front Line demonstration' or government agencies or NGOs of the State.

Key production constraints in state for maize

- More than 90% of maize growing area of the state is rainfed and prone to vagaries of weather (drought, excess water situation), therefore it is considered as high risk prone crop by the farmers.
- Late arrival and early cessation of monsoon coupled with erratic and uneven distribution of rainfall adversely affects crop growth.
- Poor availability of quality seeds in the market.
- Typically short duration varieties and local land races used by the farmers have lower genetic yield potential.
- High yielding varieties/ hybrids which would require more inputs including high input management may not be affordable for poor/ marginal farmers of Jharkhand.
- Poor crop management with no plant protection measures for the control of insects and diseases.
- Lack of adequate irrigation facilities restricts even supplemental irrigation at critical crop stage.

- Soil erosion is a major problem due to lack of appropriate soil conservation measures under undulating topo-sequence.
- Soil acidity is a major constraint to obtain high crop productivity.
- Lack of rain water management in situ and its harvesting restricts efficient utilization of water.
- Shallow soil depth, poor water retention capacity, cation exchange capacity and soil fertility adversely affects crop production.
- Washing flow of various minerals like copper, manganese and iron ores from mining industries to the cultivated lands affects the soil fertility and also creates micro nutrient toxicities particularly in sub-zone VI.
- Weak linkages and interaction between the farmers, scientist and extension people from the government and the SAU specially in naxal affected areas.

AICRP technologies

Popular land races with their peculiar characteristics for their popularities

All together 59 land races have been collected by NBPGR- Center, Ranchi during last 15 years (Anonymous, 2002). The districts covered during exploration are Ranchi, Hazaribagh, Chatra, Koderma, Simdega, Gumla, Lohardaga, Giridih, Bokaro, East Singhbhum, W. Singhbhum and Saraikela. The characteristics of few lines are given below:

Collector No.	100 grains weight (g)	Important characters
VKG 22/17	16.78	60-70 days maturity
VKG 22/23	16.00	June/July sown, white irregular kernels
VKG 22/25	15.79	Medium size, white kernel
VKG 22/31	16.94	White irregular kernels
VKG 22/35	9.64	Small grain, white, irregular
VKG 22/58	14.03	Bold grain, deep orange, irregular shape
VKG 22/63	16.56	Orange, irregular shape
VKG 22/72	13.0	Dull orange, irregular shape
VKG 22/75	14.72	White orange, irregular shape
VKG 22/83	17.13	White, irregular shape

Popular hybrids under cultivation, HQPM-1, BIO 9637, 900M Gold, Isole, Allrounder, (Monsanto); 4794 (Proagro), NMH-909 (Nuzeeevidu).

Coverage of cultivars: No statistical data is available for coverage of area for the state under different categories of cultivars. Irrespective of any choice farmers generally procure the seeds from local markets where mostly single seed hybrid seeds (~70%) produced by private companies are available. Rarely (~10%) DCH seeds are marketed. In very remote areas people generally used their own seeds of OPVs or local landraces which may accounts maximum upto 20% of the total requirements of the state.

Cultivars released / developed for state by AICRP

Type	Name of cultivar	Year	Released by	Characteristics
Yellow Orange Maize	Suwan	1985	SVRC	Before inception of Jharkhand state BAU as voluntary centre of maize research worked in collaboration with RAU, Dholi for the development of Suwan. It is a medium maturity (in 95-100 days) composite with yield potential of 50-55 q/ha. Suitable for both *Kharif* and *Rabi* season.
Yellow	BM-1	1996	SVRC	It is an early maturity (maturing in 85-90 days) normal maize composite with yield potential of 40-45q/ha registered in PPVandFR Act in 2009.
Yellow maize	BVM-2	2005	SVRC	An extra early maturity.

Pipeline Hybrid/variety developed for state by AICRP

Type	Name	Yield potential (kg/ha)	Other Characteristics
SCH on yellow	BAUMH 2011-07	5625 (*Kharif* 2012)	DT- 51.8 days, DS- 54.6 days, DM-89.1 days, PH- 167 cm, EH- 79 cm
	BAUMH 2011-05	4760 (*Kharif* 2012)	DT- 48.3 days, DS- 50.9 days, DM- 86.8 days, PH-156.6 cm, EH-72 cm
	BAUMH 2011-04	5069 (*Kharif* 2012)	DT- 49.9 days, DS- 52.3 days, DM- 88.9 days, PH- 162.8 cm, EH- 71.5 cm
	BAUMH 2011-13	5602 (*Kharif* 2012)	DT- 53.8 days, DS- 56.3 days, DM- 93.8 days, PH- 184.0 cm EH- 84.4 cm
SCH on QPM	BAUQMH-17	7123 (*Rabi* 2012-13)	DT- 92.2 days, DS- 94.9 days, DM- 129.8 days, PH- 168.7 cm, EH- 82.8 cm
		4692 (*Kharif*-2014)	PH- 172.6cm, DT- 54.4 days,DS- 56.7 days, EH- 78 cm,DM- 92.8 days MR to TLB & MLB.
	BAUQMH-18	5812 (*Kharif*-2015)	PH- 179.2cm, EH- 75.3 cm, DT- 53 days, DS- 55.2 days, DM- 92.1, days, MR to TLB
Baby corn	BVM-2	2005 (cob without husk) 5893 with husk	PH- 171.8 cm, EH- 70.4 cm, CL- 9.1 cm, Cob Girth- 2.4 cm

DT- Days to tasseling, DS- Days to maturity, PH- Plant height, ER- Ear placement height, CL- Cob length, MR- Moderately resistant.

Characteristics of inbreds developed at Ranchi centre.

Trait	Inbreds							Remarks	Reference
	BAUIM-1	BAUIM-2	BAUIM-3	BAUIM-4	BAUIM-5	BQPM-2	BQPM-4		
Plant height (cm)	128.2-141.7	127.8-146.1	117.0-141.7	124.2-136.5	136.2-179.2	139.8-149.0	100.0-146.1	BAUIM-4suitable for both baby corn, green ears and grain yield. BQPM-2 for green ears and grain yield. BQPM-4 for baby corn yield.	Izhar and Chakraborty, 2013, 2014.
Ear placement height (cm)	55.5-60.7	58.0-60.7	57.8-65.5	50.0-60.7	89.7-96.8	55.4-65.8	54.3-62.8		
Weight of ears (with or without husk)/plant (g)	84.48/12.89	83.40/12.86	77.94/11.92	91.15/13.98		86.51/13.34	94.47/14.39	BQPM-4 found as was good combiner for grain yield and hybrid BQPM-4 x VQL-1 and BQPM-2 x VQL-1 has high tryptophan and starch content, respectively.	Rani *et al*., 2015
Length of baby corn with Husk (cm)	12.33	12.78	11.45	12.43		11.28	11.78	BQPM-4 was found to be most diverse with similarity coefficient of 0.49. Inbreds BAUIM-2 and BAUIM-5 were good for GCA effects.	Pandit *et al*., 2016
Length of baby corn without Husk (cm)	5.95	6.2	6	6.45		5.9	6.08		
Width of baby corn without Husk (cm)	1.32	1.22	1.12	1.17		1.32	1.28		
1000 grains weight (g)	143.6-185.3	171.4-201.8	168.6-170.0	151.5-211.0	195.2-227.9	204.3-248.7	160.7-241.4	Hybrids BAUIM-2 x HKI1532, BAUIM-2 x HKI335, BAUIM-5 x HKI1532 and BAUIM-5 x HKI335 had significant sca effects.	Sah *et al*., 2014
Grain yield/plant (g)	56.5-63.9	45.1-61.6	40.8-81.2	47.2-67.7	43.6-63.2	57.0-66.6	52.8-62.4	Hybrid BQPM-2 x BAUIM-2 had positive sca effects for quality traits and baby corn yield with highest estimates of standard heterosis.	
Iron (mg $100g^{-1}$)	2.31	2.23	2.2	2.44		2.52	2.39	BQPM-2 x BAUIM-2 good in protein, lysine and starch while BQPM-4 x K1105 good in tryptophan content.	Izhar and Chakraborty , 2013
Phosphorus (mg $100g^{-1}$)	172.7	171	167.7	176.3		170.7	173.7	BAUIM-2 recorded good yield with less disease infestation. Both gca and sca effects were significant for Maydisleaf blight disease.	Izhar and Chakraborty, 2013, 2013a;
Protein (%)	9.31	9.25	9.24	9.34		9.57	9.46		
Tryptophan (%)	0.4	0.41	0.38	0.48		0.63	0.66		
Lysine (%)	1.61	1.65	1.57	1.89		2.62	2.5		
Starch (%)	65.53	65.63	64.58	65.53		65.37	65.87		
BLSB	MR	R	R	MR		R	MR		
Grain yield(q/ha)	36.44	39.61	34.26	37.54		36.99	30.73		

Leaf Firing

BAUIM-I | BAUIM-5 | CM-111

HKI-1532 | BAUIM-4 X HKI-335 | BAUIM-1 X HKI-335

Stay Green

BAUIM-2 x HKI-1532 | HKI-1532 | BUIM-5 x HKI-1532

Tassel blast

BAUIM-2x HKI-335 | BQPM – 4 x HKI-335 | BAUIM-5 x HKI-335 | CM-111

Maize seed production in the state.

Year	Name of Cultivar/ Inbred	Quantity produced (q)				
			Breeder Seed	Foundation Seed	Certified Seed	Truthful Seed
2016	Suwan-1	157 cobs				0.83
	BVM-2	140 cobs				0.30
	BAUCM-3					0.50
2015	Suwan-1	350 cobs	2.40	-	18.41	4.34
	BVM-2	470 cobs	0.10	-	-	3.60
	BM-1	210 cobs				0.50
2014	Suwan-1	300 cobs		36		0.27
	BVM-2	350 cobs		3		0.26
	BM-1	12 cobs				
2013	Suwan-1	250 cobs	0.20	2.12		
	BVM-2	220 cobs	0.20	14.05		
	BM-1	51 cobs				
2012	Suwan-1			24.52		
	BVM-2			131.0		
2011	Suwan-1		0.60	16.54		
	BVM-2		1.02	14.0		0.30
	Priya					0.15
	HQPM-1				5.0	
	Male of HQPM-1		0.04			
	Female of HQPM-1		0.10			
2010	Suwan-1		1.68	46.8		1.57
	BVM-2		2.02			0.30
	BM-1	123 cobs				
	Priya	149 cobs				
	HQPM-1				3.73	
	Male of HQPM-1		0.25			
	Female of HQPM-1		0.38			
2009	Suwan-1		0.25	25.03	33.0	
	BVM-2		0.20			
	BM-1	130 cobs				
	Priya	165 cobs				
	HQPM-1				4.53	
	Male of HQPM-1				0.20	
	Female of HQPM-1				0.40	
2008	Suwan-1	198 cobs	0.25	21.90		16.0
	BVM-2	166 cobs	0.20			
	HQPM-1				2.0	
2007	Suwan-1	225 cobs	0.75	12.66	25.0	0.25
	BVM-2	153 cobs	0.92			0.49
	BM-1	355 cobs	0.80			0.61
	African Tall		2.00			

Changes in cropping system over the years

Despite good amount of rainfall received in Jharkhand, most of the fields are mono-cropped usually with upland rice are poor in productivity (<+1 ha). As per the analysis carried out by the Council for Social Development Delhi, Jharkhand agriculture sector is witnessing a slow shift from agriculture to horticulture species. The area under cereal crop has been dropping over time and so is the share of cereal crops in the total output of the sector (in value terms). On the other hand the rural focus on fruits and vegetables is growing rapidly. So, not only the state has become self-sufficient in vegetable cultivation, the productivity of the for most of the vegetables is better in comparison to all India average.

Recommended production technologies

Soil

Upland soils of Jharkhand are most suitable for cultivation of maize. Maize can be grown successfully on soils ranging from well drained sandy loam to silty loam with pH of 5.5 to 7.5 supports good crop growth. Jharkhand soils are acidic in nature, at pH lower than 6.0 application of 3 quintal of lime/ha in furrows should be done at the time of sowing. Soil have adequate water holding capacity and proper drainage to minimize the damage due to water logging specially at seedling stages.

Land Preparation

Field should be well levelled, 2-3 ploughings followed by planking. Fields should be free from weeds, clods and stubbles with drainage channel should be made all around the field.

Selection of cultivars

Full season Maturity (100-110 days):NMH-920, NK 6240 (NECH-131), SMH-3904, Pinnacle (MON 25), HQPM-1 (QPM Hybrid), Suwan (Composite) suited for timely sowing in adequate rainfall areas.

Medium Maturity (85-95 days): KMH-218 Plus, KMH-3426, NMH-803, HM-12, KMH 3712, P3441 (X8B691), HM-9, Malviya hybrid makka 2 (Hybrids) suited for rainfed areas and under delayed sowing.

Early Maturity (80-85 days): Parkash, X 3342 (Hybrid), Dewaki, BM-1 (composite) suited for intercropping in pigeonpea and sugarcane.

Extra Early Maturity (75-80 days): Vivek Maize Hybrid 43, Vivek Maize Hybrid-27 (Hybrid), D 994 (composite), BVM-2 (composite) suitable for growing as short duration catch crop or an intercrop.

Babycorn: BVM-2 and HM4

Sweetcorn: Madhuri and Priya

Popcorn: VL Amber Popcorn

Time of Sowing

Season	Optimum time of Sowing
Kharif	15 June to 15 July (sowing time coincides with the monsoon)
Rabi	15 October to 15 November (for inter-cropping sowing should be completed upto 30th October)
Summer	20 Jan to 15 Feb (Late sowing resulted in pollen desiccation and poor seed setting)

Seed rate, plant geometry and plant population

Purpose	Seed rate (kg/ha)	Plant geometry (cm)	No. of Plant/ha
Grain and green cob (*Kharif*)	20	70 x 20	71428
Grain and green cob (*Rabi*)	25	65 x 20	83333
Sweet corn	8	75 x 20	66666
Baby corn	25	60 x 15	111111
Pop corn	12	60 x 20	83333
Fodder	50	30 x 10	333333

Seed treatment

To protect seed from soil borne diseases and some insect-pests

Fungicide/ Pesticide	Rate (g/kg seed)	Disease/ Insect-pest
Bavistin and Captan	2.0	TLB, BLSB, MLB, stalk rot

Sowing

Furrows are opened with spade at required distance and seeds are planted in furrows as per recommended distance. Planting seeds in ridge and furrow method is most suitable for *Kharif* sowing.

Manures and fertilizer

Apply 5-10 t/ha of farmyard manure at the time of final land preparation.

Cultivar	Nutrient applied (kg/ha)		
	N	P_2O_5	K_2O
Hybrids	150	60	40
Composite	120	60	40

Full dose of phosphorus and potassium along with 1/3rd of nitrogen is applied as basal dose. The remaining nitrogen is applied in two equal splits at knee high and tasseling stage. However, the splitting of nitrogen at 5 stages: at basal (10%), 4 leaf stage (20%), 8 leaf stage (30%), flowering (30%) and grain filling stage (10%) is found more beneficial.

Water management

The annual rainfall in Jharkhand is 1400 mm, of which 82.1% is received during the periods of June to September and the rest 17.9% in remaining months. Maize is Mostly grown in *Kharif* as rainfed crop but irrigation should be given if moisture stress occurs at critical stages *i.e.* young seedling, knee high, tasseling, silking and grain filling stage.

Weed management

Maize crop is infested with grassy weeds (*Eleusine* sp., *Eragrostis* sp.), broad leaf (*Trianthema.*sp., *Digitaria* sp.) and perennials (*Cyperus* sp., *Cynodon* sp., *Sorghum helepense*). The critical period of crop-weed competition is 40-45 days after sowing. Application of atrazine @ 2.0 kg/ha as pre-emergence is effective for the control of weeds. Two hoeing at 20 and 40 days after sowing may take care of weeds.

Harvesting

Harvesting of cobs is done when husk around the cob turns brown. The moisture content in grain at harvesting can be up to 30%. Harvesting of baby corn is done at 1-2 days after silk emergence in morning or evening. Harvesting of sweet corn is done at 21-22 days after silk emergence.

Threshing

The harvested cobs should be spread evenly instead of making heap to avoid aflotoxin and should be kept on Tarpaulin/cemented floor to avoid infestation from pest and diseases. After sun drying, when moisture content in grain reached 14% threshing should be done by manual maize sheller or through maize sheller machine.The important insect pests that affect maize crop in Jharkhand are.

Insect-pest	Scientific name	Symptoms	Control
Stem borer	*Chilo partellus*	The damaging stage of the pest is larvae. The eggs hatch in about two to five days. The freshly hatched caterpillars migrate towards central shoot where they feed on the tender leaves and top internode. In younger plants, the growing point and base of central whorl gets badly damaged and drying up of the central shoot commonly known as 'dead heart'.	• Application of 6-8 Carbofuran 3G granules inside the maize whorls at 4 leaf stage. • Application of Imidachloprid 17.8 SL @ 2 ml/ liter of water • Use of 8 trichocards/ha at 10 days after sowing and again at 22 days after sowing • Intercropping of maize with cowpea

Pink borer	*Sesamia inferens*	The moth of the *Sesamia*sp.is nocturnal and lays eggs on lower leaf sheath. The larvae of the *Sesamia* sp.enter the plant near the base and cause damage to stem	
Shoot Fly	*Atherigona* sp.	It is a very serious pest of maize in spring and summer season. The tiny maggots creep down under the leaf sheaths till they reach the base of the seedlings. After this they cut the growing point or central shoot which results in the formation of characteristic dead hearts.	• Seed treatment with Imidacloprid 17.8 SL @ 6ml/kg of seed. • Furrow application of Carbofuran 3G @ 33 kg/ha before sowing int shoot fly prone areas.
Termite	*Odontotermes obesus*	Termite can cause substantial damage to the maize crop. They establish colonies much deep into the soil, it is very difficult to g et rid of the problem completely.	• Frequent irrigation before land preparation and during the crop growth reduces its infestation. • Application of Fepronil granules @ 20 kg/ha followed by light irrigation controls termites to a reasonable extent.

Major diseases of maize

Maize suffers from several diseases caused by fungi, bacteria and viruses. Disease attack in *Rabi* maize is comparatively lesser than the *Kharif* maize. The major diseases prevalent in Jharkhand are as under:

Diseases	Causal Organism	Management
Maydis leaf blight	*Helminthosporium maydis*	• Field Sanitation: Ploughing down of crop debris may reduce early infection. • Growing of resistant variety. • Spray of Diethane M – 45, Captan, Zineb @ 2 gm/lit 2 to 4 application at 7- 10 days interval. • Crop rotation for 2 – 3 years. • Use of healthy and certified seed.
Banded leaf and sheath blight	*Rhizoctonia solani f. sp. sasakii*	• Deep summer ploughing • Field sanitation • Seed treatment with Pseudomonas fluroscense @ 16gm/kg of seed • Rhizolex 50WP @1-2 g/litre when the crop is 30-40 days old
Downy mildew	The most common and dangerous pathogen of this disease are: *Perenosclerospora*	• Eradication of collateral host and disease crop debris. • Long crop rotation should be followed. • Deep ploughing of fields • Seed treatment with Metalaxyl @ 4gm/kg of seed is very effective.

	sorghi, Sclerophthora rayssiae, P. phllipinensis, P. sacchari	• Single spray of Ridomil MZ@2 g/litreof water is very useful. • Diathane M45 @0.3% after 10 days of sowing at weekly intrval, 4-6 spray give good control of disease
Bacterial Stalk Rot	*Erwinia carotovora, Erwinia chrysanthemi*	**Cultural practices** • Avoidance of water logging • Field should have a proper drainage • Planting of crop on ridges **Chemical control** • Bleaching powder containing 33% cl @ 10 kg/ ha as soil drenching at pre flowering stage • Application of calcium hypochlorite (klorocin) containing 22% cl @ 25kg/ha first before flowering second 10 days after 1st application
Common Rust	*Puccinia sorghi*	The only method of control is to grow only r esistant variety
Turcicum Leaf Blight	*Helminthsporium turcicum*	• Field sanitation • Growing of resistant variety • Spraying with Zineb/Maneb @ 2 gm/l at 8-10 days interval, 2-4 times depending upon severity

Changing diseases scenario and pest and disease management

In Jharkhand, *Chilo* is the major pest in *Kharif* damaging the crop severely. Among diseases, Maydis leaf blight and banded sheath and leaf blight are the major diseases affecting the maize crop. A survey conducted in *Kharif* 2005 and 2006, revealed that banded leaf and sheath blight is wide spread with disease severity ranging from 30.30 to 80.46% in and around Ranchi district of Jharkhand (Akhtar *et al.,* 2009). Common rust in maize caused by *Pucinia sorghi* was also reported during the first week of September, 2001 in *Kharif* while occurrence of common rust has been reported in *Rabi* from other parts of India (Lal *et al.*, 2011).

Table 1. Occurrence of Banded leaf and sheath blight disease in Ranchi, Jharkhand.

Location/Block	**Disease severity (%)**	
	2004	**2005**
Itki	35.10	45.50
Bero	52.80	57.20
HisriChauli	79.00	81.92
Jirabar	48.00	52.60
RAC farm	28.40	32.20

arts of India (Lal *et al.,* 2011).

Rainfall pattern in Jharkhand

The climate of the Jharkhand is from dry semi-humid to humid semi-arid type located on an elevation of 300 to 610 meter above sea level. The state receives rainfall from both the streams of monsoon *i.e.* Southwest monsoon and Northeast monsoon. The mean annual rainfall in the plateau and sub-plateau region is 1400 mm, on an average of which 82.1% is received during the period June to September and the rest 17.9 % in remaining months. Monsoon usually onset during the second to third week of June and continues up to second week of October.

Distribution of rainfall in different subzones of Jharkhand

Sub-zones	Annual rainfall (mm)	*Kharif* (June to October)
Sub zone (iv)	1320	(80.82)*
Sub zone (v)	1246	(70.00)
Sub zone (vi)	1400	(80.85)

There are on an average 130 rainy days in a year and in 75 days, rainfall is below 2.5 mm. Late arrival and early cessation of rainfall are not uncommon features. In normal years, pre-monsoon rains are received in the month of May, which helps for summer ploughing.

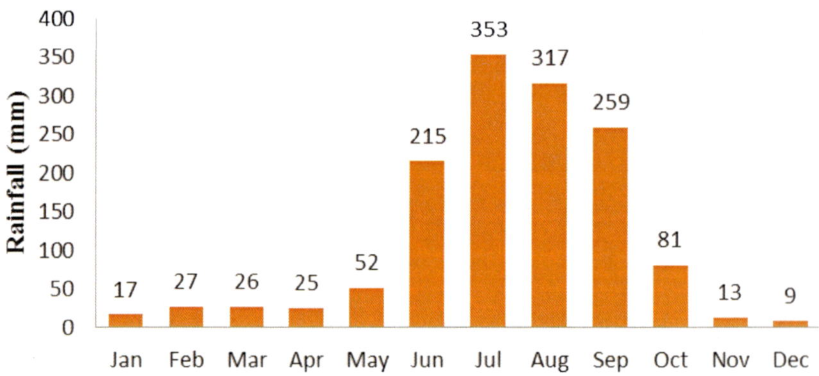

Fig. 3. Pattern of rainfall over the years (mean of 48 years).

Distribution of rainfall is also uneven and erratic. Dry spells (10-15 days) usually occur in July-August. Failure of *Hathia* rain (late September to early October) is observed once in four years, which not only adversely influence the grain development of standing crop but also affects establishment of second crop in winter season. Occasional winter rains during December/January is very helpful to winter crops.The monthly rainfall pattern at Kanke, Ranchi is given in Fig. 3.

Key R&D challenges and their probable solutions in maize for the state

1. Adoption of single cross hybrid technology - bringing more area under SCH
2. Development of biotic and abiotic stress tolerant genotypes
3. Application of modern technologies *viz.,* MAS, biotechnology to develop herbicide and insect tolerant maize
4. Intensification and commercialization of current maize production systems
5. Activating extension services to promote viable and acceptable technology
6. Favorable government policy
7. Integration of irrigation, nutrient and post harvest management practices to sustain the productivity
8. Identification of sources of tolerance to cold and heat stress
9. Development of efficient genotypes which can withstand low input stress
10. Intensification and commercialization of current maize production systems
    a. New hybrids with hybrid specific scientific crop management
    b. Targeting potential non-traditional areas
    c. Adoption of improved soil and resource conservation practices
11. New crop establishment methods like zero till maize after rice and maize on on permanent beds after wheat

Important factors responsible for increase/decrease of maize in the state

a. Mostly rain fed cultivation.
b. Local varieties and uncertified seed.
c. Imbalanced and inadequate use of fertilizers.
d. Inadequate plant population.
e. Broadcast sowing.
f. Delayed sowing.
g. Improper attention towards weed management and plant protection measures.

Future maize research and development strategies in the state

- Breeding for varieties with better adaptability for abiotic stresses (water-logging drought).
- Identification of major QTLs linked to different abiotic stresses and their pyramiding in different lines.
- Molecular genetic dissection of the genomic regions to understand the physiological mechanism.

- Marker assisted selection and transgenic approaches.
- Development of precision input management technologies in maize.
- Standardization of crop establishment and plant protection measures.
- Suitable post emergence weeds control measures.
- Development of moisture conservation practices for stabilizing yield and sustainable maize productivity.

Summary

Jharkhand state forms a part of Agro - climate zone - VII of the country known as Eastern Plateau and Hills Region with average annual rainfall of 1400 mm, of which 82.1% is received during the period June to September and the rest 17.9% in remaining months. Maize is the second most important crop after rice in *Kharif* and is grown in all the 24 district of Jharkhand specially in the northern belt of Jharkhand (Garhwa, Palamu, Latehar, Chatra, Hazaribag, Giridih, Dumka and Godda). The cropis mainly cultivated in *Kharif* as a rainfed crop in upland soil as only 15-18% of the total cropped area is irrigated in Jharkhand. Upland soils are usually red and acidic (pH 5.5-5.9), lighter in texture, poor in water holding capacity, low in organic matter, nitrogen, available phosphorus, sulphur, calcium and magnesium so, application of lime, organic manure and better nutrient management is the key to enhance the productivity the region. The composite varieties Suwan (Yellow Orange Maize), BM-1 (Yellow), BVM-2 (Yellow maize) released for maize cultivation in the state by Ranchi centre. After the initiation of All India Co-ordinated Maize Improvement Project at Birsa Agricultural University, Ranchi from September, 2005 work on single cross hybrid has been started and today, several hybrids/varieties are in pipeline The major factors responsible for low productivity of maize in the state are use of local varieties and uncertified seed, rainfed cultivation, imbalanced and inadequate use of fertilizers, inadequate plant population, improper attention towards weed management and plant protection measures. The frontline demonstration showed that there is a huge yield gap in the state and by bridging the constraints related to maize production the productivity can easily be enhanced by more than 100%. The specialty corn *i.e.* baby corn, sweet corn and pop corn cultivation has not become much popular in the state so, their popularization and bringing more and more area under single cross hybrid can improve the socio-economic status and bring prosperity to the farmers of Jharkhand.

References

Akhtar, J., Kumar, J., Kumar, V.A. and Lal, H.C. (2009). Occurrence of banded leaf and sheath blight of maize in Jharkhand with reference to diversity in *Rhizoctonia solani. Asian Agri. Sci.* 1(2): 32-35.

Anonymous. (2002). Annual Report, National Bureau of Plant Genetic Resources: 104-106.

Izhar, T. and Chakraborty, M. (2013). Evaluation of maize genotypes for chemical composition at silky and hard stage. *Green Farming*. 4(1): 11-14.

Izhar, T. and Chakraborty, M. (2013a). Genetic analysis of yield and quality parameters in baby corn (Zea mays L.) *Current Advances Agri Sci*. 5(1): 27-31.

Izhar, T. and Chakraborty, M. (2014). Genetic Analysis of Maize (*Zea mays* L.) Genotypes for Baby Corn, Green Ear and Grain Yield. *Maize Genomics Gen*. 5(1): 1-6

Lal, H.C., Kumar, Binay, Singh, C.S. and Chakraborthy, M. (2011). First report of common rust of maize in *Kharif* from Jharkhand. *Mycology Plant Pathology*. 41(4): 648-649.

Pandit, M., Chakraborty, M., Haider, Z.A., Pande, A., Sah, R.P. and Kumar, S. (2016). Genetic diversity assay of maize (*Zea mays* L.) inbreds based on morphometric traits and SSR markers. *African Agri Res*. 11(24): 2118-2128.

Rani, P., Chakraborty, M. and Sah, R.P. (2015). Identification and genetic estimation of nutritional parameters of QPM hybrids suitable for animal feed purpose. *Range Manage Agroforestry*. 36(2): 175-182.

Sah, R.P., Chakraborty, M., Prasad, K. and Pandit, M. (2014). Combining ability and genetic estimates of maize hybrids (*Zea mays* L.) developed using drought tolerant testers. *Maize J*. 3(1 and 2): 9-17.

10

Maize Research and Development in West Bengal

Srabani Debnath, Sonali Biswas and Anirban Maji

AICRP on Maize Bidhan Chandra Krishi Viswa Vidyalaya, West Bengal

**Corresponding Author's Email: srabanidebnath72@gmail.com*

West Bengal is the fifth largest state in terms of (GSDP) Gross State Domestic Product after Maharastra, UP, TN and AP and is the third largest contributor in agriculture sector of real GDP. West Bengal has 3% of India land mass and 8% population with population density of 1029 per km^2.

Agriculture accounts for about 17% (2013-14) GSDP and provides employment to over 55% of workers. About 64% of the geographical area of West Bengal is under cultivation. There is little scope for further increase the current cultivable area. Because it is difficult to maintain the growth rate in the absence of adequate infrastructure support, crop diversification and market access.

Food production remained stagnant as the green revolution by-passed the state. However, there has been a significant increase in food production since 1980's and the state is now one of the few Indian states with a surplus in food production, producing nearly 20% of the rice and 33% of potato. Agriculture accounts the largest share of the labour force. Rice, potato, jute, sugarcane and wheat are the top five crops of the state, while other major food crops include maize, pulses, oilseeds, wheat, barley and vegetables. In recent years, significant changes have occurred in maize production and utilization due to increasing commercial orientation of maize crop and rising demand for diversified end users, specially for feed and industrial uses. A sizable number of districts in West Bengal have potential for growing winter maize. Though maize favourably responds to better crop management both in *Kharif* and *Rabi* season, the erratic rainfall pattern of the south-west monsoon comes in the way of timely field operations of *Kharif* season. In absence of any major environmental impediments in *Rabi,* the desired field operations can be planned and executed at the most desired time. Moreover, the various environmental factors including absence of any major disease and insect pests in this season, helps in realizing the better profits from every additional unit of monetary inputs. Hence, future of *Rabi* maize in West Bengal is bright.

With the changing climate scenario, maize being a photo-insensitive crop has better options for adaptation and mitigation. The limitations of rising temperature during grain filling of wheat particularly in eastern India and declining yield of boro rice in West Bengal has shown a path for maize as a better option. Therefore, it is emerging as a potential driving force for diversification of rice-rice with rice-maize and other maize based high value cropping systems in water scarcity/ lowering of water table areas.

The upland rice in this state has very low productivity, therefore maize is the only suitable alternative crop and more area is likely to shift towards maize cultivation in near future in these non-traditional areas of West Bengal. Wheat crop is adversely affected with terminal heat stress due to sudden rise in temperature during crop growth and maturity which favours maize crop positively for suitable candid crop for diversification. Rice- maize system is practiced in West Bengal with acerage of more than 0.5 milion hectare.

Because of so many uses of maize crop it is getting importance in West Bengal with time. Animal feed is the largest end use segment for maize in West Bengal with more than 70% of total volumes used by feed industry. It is estimated that, the demand for maize will be fueled by population growth and increasing inclination towards higher protein consumption in the form of meat and eggs.

Maize as poultry feed is more acceptable than rice and wheat both in terms of price and nutrition. The nutritional value of maize is higher (3,365 kcal/kg) compared to rice (3,320 kcal/kg). Apart from feed and industrial applications, food-processing industry is a crucial end use segment for maize as it is being used for making food additives and sweeteners. Pop corn is very popular in common people. With the changing life-style corn flakes and baby corn is also getting popularity in West Bengal.

Poultry industry is heavily dependent on maize as it forms 50-60% of the input required for broiler feed and 25-35% of the input required for layer feed. Jowar is the closest substitute but its availability is constrained while wheat contains high non-starch polysaccharides which are indigestible by broilers and have to be depolymerised with an enzyme for release of energy. Hence, demand for maize in this segment will only grow in tandem with industry growth which is being driven by various factors like higher demand for meat and poultry with increasing income and strong growth in food services.

Besides, maize starch, an excellent source of carbohydrate, is a highly versatile industrial raw material and finds extensive applications in the textile, food, pharmaceutical and food segments that is expected to grow 12-15% (vol) in the next five years. Use of maize for biodegradable plastics, car-parts, edible oil and sweeteners is already picking up.

There are so many industries producing maize based products in West Bengal, such as, well dried yellow maize for poultry feed, produced by Sri Shyam Traders in Kolkata. Maize grits used on making extruded snacks like cheese balls, corn rings and others produced by Vitarich Agro Food Ltd. of Kolkata. Export quality superb maize and local quality maize (raw) produced by S.G. Traders, Pallishree Ltd., Sen Agrotech international of Kolkata. Animal feed, raw material for the production of paper, packaging materials, textiles, chemicals produced by Pallishree Ltd. of Kolkata.

Table 1. Trends in district-wise production of maize in West Bengal.

District	Production ('000 tonnes)		
	1990 - 91	**2000 - 01**	**2009 – 10**
Darjeeling	41.3	39.9	39.6
Malda	6.6	10.4	19.9
Jalpaiguri	5.9	5.1	23.7
Coochbehar	Trace	Trace	62.5
Dinajpur	Trace	0.1	21.37
Nadia	2.9	0.7	8.5
Purulia	14.2	26.3	9.8
Bankura	9.0	0.7	0.4
Birbhum	0.1	1.0	0.8
Midnapur (west)	1.5	-	2.9
Murshidabad	0.1	1.0	18.1

Table 2. Seasonal trends in yield (district-wise) of maize in West Bengal.

District	Yield (kg/ha)		
	1990 - 91	**2000 - 01**	**2009 – 10**
Darjeeling	1240	3096	2500
Malda	1240	3096	2314
Jalpaiguri	1463	3096	2008
Coochbehar	1240	3096	5799
Dinajpur	1240	3096	5321
Nadia	1139	2405	2362
Purulia	991	1874	1323
Bankura	2857	1835	2256
Birbhum	251	1835	1361
Midnapur (west)	1083	1835	2491
Murshidabad	1139	2405	4782

Major cropping system and production ecologies of maize in West Bengal

Most of the Lower Gangetic Plain is located in the state of West Bengal. Depending upon climate and soil, planning commission, Government of India redefined West Bengal into six agro-climatic sub- regions:

i) Northern hilly zone, ii) Tarai-Teesta flood plain, iii) Vindhya old flood plain, iv) Gangetic flood plain v) Undulating lateritic sub-region of the Eastern Plateau Region and vi) Coastal flood plain (Sen, Gupta, 2001). Depending upon the land situations and rainfall, a number of cropping systems followed in sub-region of West Bengal. Cropping systems followed in different agro-ecological regions of West Bengal are as below:

Land elevation	Ecology	Pre-*kharif*	*Kharif*	*Rabi*
Zone-I: Hill Zone				
High	Rainfed	Potato/ginger	Cabbage	Fallow
	Irrigated	Potato/orange	Cabbage	Vegetables
Low	Rainfed	Maize, soybean, ragi, finger millet, ginger	Rice, vegetables	Mustard, vegetables
	Irrigated	Maize, vegetables	Rice	Potato, vegetables, spices, wheat
Zone-II: Terai zone				
Upland	Rainfed	Jute	Rice	Vegetables, niger, toria, vegetables, pulse
	Irrigated	Jute, rice, tea, pine Apple, vegetables	Rice, vegetables, Tobacco	Vegetables, potato
Medium	Rainfed	Jute	Rice	Wheat, vegetables
Land	Irrigated	Jute, vegetables, Rice	Rice, Tobacco	Potato, vegetables, wheat
Lowland	Rainfed	Jute/rice	Rice	Vegetables, lentil
	Irrigated	Jute/rice/green manuring crop	Rice	Vegetables/lentil/rice/ linseed
Zone-III: Old Alluvial (Vindhya Alluvial zone)				
Upland	Rainfed Irrigated	Jute, sesame	Rice, blackgram	Toria, mustard, lentil
Medium	Rainfed	Jute	Rice	Pulse
Land	Irrigated	Jute, rice, sesame, vegetables, sugarcane	Rice	Pulse, wheat, vegetables, mustard
Lowland	Rainfed	Jute, blackgram, greengram, Vegetables	Rice	Linseed, khesari
	Irrigated	Green manure, sesame, vegetables	Rice	Vegetables, rice
Zone IV: New Alluvial Zone				
Upland	Rainfed	Jute, sesame, pulse	Blackgram, greengram, vegetables, rice	Vegetables, rapseed, lentil, mustard

	Irrigated	Jute, rice, vegetables, sesame	Rice	Wheat, vegetables, potato, mustard
Medium Land	Rainfed	Jute, greengram	Rice	Rai, pulse
	Irrigated	Jute, sesame, rice, vegetables, Sugarcane, maize	Rice	Potato, wheat, maize, vegetable, mustard
Low land	Rainfed	jute, green Manure, vegetables	Rice	Khesari, lentil, linseed, pulses
	Irrigated	greengram, sesame, vegetables	Rice	Rice
Zone-V: Red and Laterite Zone				
Upland	Rainfed	Amaranthus	Groundnut, Soybean, Rice, Maize, Arhar, Millet	Pulse, safflower, Oilseed
	Irrigated	Vegetables, Maize	Rice, Vegetables	Vegetables, potato, Wheat, toria, pulse
Medium Land	Rainfed	-----	Rice	Pulse, oilseed
	Irrigated	Rice, vegetables, Sesame, Maize	Rice	Potato, wheat, oilseed
Low Land	Rainfed	Green manure	Rice	Pulse, linseed
	Irrigated	Pulse, sesame, vegetables	Rice	Rice, mustard
Zone-V: Coastal Saline Zone				
Upland	Rainfed	-------	Vegetables, Rice	Chilli
	Irrigated	Vegetables, betelvine	Rice	Vegetables, water melon, mustard, barley
Medium Land	Rainfed	Vegetables	Rice	Khesari, linseed, chilli, sugarbeet
	Irrigated	Rice	Rice	Potato, mustard
Low Land	Rainfed	Green manure	Rice	Khesari, cotton
	Irrigated	Green manure	Rice	Rice

All India Coordinated Research Project for Maize (AICRP on Maize) which was established in 1957 under the name All India Coordinated Crop Improvement Project (AICCIP) on Maize organizes interdisciplinary and inter-institutional co-operative research programmes.

In 1957, research work was started under West Bengal government at Kalimpong farm under AICCIP on Maize. In 1973, the controlling power of this project was taken over by the University of Kalyani and in 1974 by Bidhan Chandra Krishi Viswavidyalaya. From 1974 both *Kharif* and *Rabi* season trials on maize crop were conducted at BCKV, Kalyani up to 1990. Very recently AICRP on Maize has restarted at BCKV, Kalyani centre during 12th plan in 2015-16 and the experiments are being conducted at district seed farm (AB Block) at Kalyani. There is only one regular center of AICRP on maize in this state at Kalyani (District – Nadia) under Bidhan Chandra Krishi Viswavidyalya. A voluntary center of AICRP on Maize is also working at Midnapur District under the supervision of State Government, department of Agriculture.

The mandate of Kalyani center are

- To identify and release zone specific or region specific high yield potential varieties which have superior grain yield, quality, resistance to diseases and pests with adaptability over wide range of cultural and regional environments.
- To develop improved agronomical practices and to find out low cost maize production technologies.
- To study the pattern of disease and pest occurrence and to devise ecologically sound and economically viable pest and disease management strategies in maize crop.
- To transfer and popularize the proven technologies to the farmers like replacement of old varieties with newly released high yielding ones, judicious application of inputs, etc. through front line demonstrations and other Extension Education Programmes.

Key maize production constraints in West Bengal are:

- It has six agro-climatic zone out of which Gangetic alluvium zone is largest (73.06% of total cultivable land) and most fertile but frequently faces natural vagaries like hailstorm, flood, drought, moisture stress, etc.
- Average labour cost has increased by 20-25% in the past 4-5 years. Moreover, the average age of the current farmer is 40+ years and the next generation is gravitating towards other profession. Going ahead, the labour availability is expected to reduce, resulting in increasing production costs across crop.
- Although maize requires less water as compared to paddy and sugarcane, but still the requisite irrigation infrastructure is grossly suboptimal. With drought cycle becoming shorter and increasing rainfall variation, development of irrigation infrastructure is an imperative to assure consistent improvement in productivity.
- There is lack of proper knowledge regarding the relevant chemical uses, integrated pest management, soil maintenance and other production technologies. Besides there is lack of knowledge transfer about information system (such as commodity exchanges, future contracts etc.). So, there is lack of better inputs availability and uses and delivering better farm level infrastructure.
- There is a lack of proper storage facilities due to which around 6% of the produce is lost. Absence of quality storage infrastructure for rodents and fungal attack proof resulting in minor losses.
- Due to improper drying of corn grain the grain quality is reduced.

- Lack of awareness among producer on new seed technologies and benefits.
- Lack of promotion of hybrid adoption and development of better hybrid seeds.
- Improper development in the field of mechanization which can solve labour problem in maize cultivation.

AICRP technologies

Popular land races with their peculiar characteristics (Sikkim – adjacent to West Bengal):

Name	Peculiar characters
Murli makai (Sikkim Primitive)	NA
Kaali makai	dark purplish black kernel type
Rathi makai	dark red kernels
Paheli makai	yellow/orange flint kernel type
Seti makai	white kernel type
Putali makai	transposon-induced pericarp variegation
Chaptey makai	white, dent type kernels
Gadbade makai	a mix of white and purple flint kernels
Bancharey makai	a high altitude maize with yellow, flint kernel type
Kukharey makai	Short statured plants
Kuchungdari	orange colored popcorn type kernels
Kuchungtakmar	a mix of yellow, white, purple and red kernels.

These landraces were collected from Sikkim by the author under the ICAR National Fellow Project in 2005, and characterized at both phenotypic and molecular levels (Prasanna 2010; Singode and Prasanna 2010; Sharma *et al.*, 2010).

Popular hybrids

Table 3. The popular maize hybrids suitable for West Bengal.

Extra early maturity	Early maturity	Medium maturity	Late maturity
Vivek 21 and 25, PEEH 5	Parkash , JKMH 1701, X 3342	Pratap Makka 4	Pro 311, Bio 9681, Seed Tech 2324

Table 4. Hybrid and Composites suitable for West Bengal.

Hybrids	Composites
DHM 103, Rajendra Makka 22, DHM 15	Megha, Pragati

Table 5. Some of the hybrid and composites which are released at national level that are suitable for West Bengal.

Hybrids	Composites
Ganga safed 2, Ganga 5, DHM-103, DHM-1, Ganga-11, DHM 105, Trishulata, Prakash, Deccan 107	Vijay, Ageti 76, Navjot, D 765, Diara 3, MCU 508, Kiran, Surya, Shakti 1, Madhuri (sweet corn)

Table 6. Some of the hybrids which are released by private seed industries for West Bengal.

Hybrid	Company	Maturity	Yield (q/ha)
P3522 (X35A019)	Pioneer OverseasCorporation, Karnataka	Late	91.0
P1864 (X35A019)	-do-	Late	91.0

Table 7. According to the duration some suitable cultivars of maize for West Bengal.

Duration	Composite	Hybrid
Early (78-85 days)	Diara-3, Pusa Composite-1	Deccan 109, Prakash
Medium (85-95 days)	Pusa Composite-2, Navjyot Kinap, Azad Uttam	Deccan-107, Ranjit, Ganga safed,
Late (95-115 days)	Suan	D KC 9081, 900 M Gold, PAC 740

Changes in cropping system in West Bengal over the years

The cropping pattern usually changes over time with the development of agriculture in West Bengal. A shift from traditional varieties to news HYVs and relatively more remunerative crops has added a new dimension to the West Bengal agriculture. This change is associated with the redistribution of land resources to different crops, which ultimately has a bearing on accelerating growth and efficiency of agriculture in West Bengal. In general, the farmers are guided by the relative profitability of crops in the system of commercialized agriculture. Thus, area under crops changes depending upon the past prices and changes in technology, expansion of irrigation and possibility of productivity change in the respective area and lastly the change of preference in the consumption basket of food habits of the West Bengal people (De, 2002; 2003).

Main cropping system of the state is rice-rice-mustard, jute/rice-mustard/pulses, jute-wheat, rice-potato, jute-rice-potato, maize-winter vegetables and greengram-mustard. According to the cropping pattern in most of the districts of West Bengal has noticeably changed in favour of high-value non-food grain crops such as potato, oilseeds and other non-food grain *Rabi* crops. More specifically, a shift in the crop diversification towards the high value crops can become a handy solution of the slowing down the pace of agricultural growth in the state. During the post-Green Revolution period (particularly after 1970) cropping pattern in West Bengal has changed in favour of high remunerative crops at the cost of the lower value crops. The analysis reveals that during the period 1970-71 to 1994-95, the area and production of *boro* rice, potato and mustard have increased rapidly and the development of irrigation and technology in other fields are the main factors behind the relatively rapid expansion of cultivation of the above mentioned crops. The farmers prefer that combination of crops from which they can derive maximum possible net revenue at least possible risk, if there is no dearth of essential factors of cultivation of those crops.

Mruthyunjaya and Kumar (1989) have found that during the period 1972 to 1983, in West Bengal, paddy and total oilseeds are the two crops which gained area allocation under gross cropped area (GCA) and there have been declined in area allocation under total pulses both in absolute and relative terms. The diversification of crop in terms of variation in acreage allocation, among the crops has taken place remarkably, in West Bengal during the post-green revolution period. The changes have taken place largely in favour of *boro* rice, potato, oilseed as a whole (especially mustard). Though in the early years of post-green revolution period, area allocation had moved in favour of wheat which later turned gradually from wheat to other *Rabi*-crops like potato and mustard. Crop intensification and/or diversification has now further increased with inclusion of short duration rapeseed and potato in between wet season rice or jute and dry season rice, resulting in higher production per unit area per unit time, higher nutrient removal, and varying changes in soil fertility as compared with rice–rice and rice–wheat systems, the two predominant cropping systems of the IGP. The limitation of rising temperature during grain filling of wheat particularly in West Bengal affecting yield of *Rabi* rice has shown a path to maize as better crop. Changing of rice-rice with rice-maize and other maize based high value cropping systems in winter on lowering water table is a major concern in West Bengal. The medium and uplands where subsistence yield of wheat, *Rabi* rice and other winter is obtained, could be substituted by winter maize in West Bengal. Winter maize has the clear cut comparative advantages of low incidence of diseases and insect pest, is not affected by temperature rise during winter (as the wheat is) and do not suffer on account of heavy rainfall.

Maize production technologies in West Bengal

Soil

Sandy Loam/clay loam

Cultivars

Hybrids-HQPM-1, DHM-103, DHM-105, Vivek 27, Prakash, X3342, Malviya Hybrid Makka2, Pro 311, Bio 9681, Seed Tech2324.

Composites-Navjot, Vijay composite

Sowing Time and Method

Kharif: Last week of June to first fortnight July.
Rabi: Last week of October to 15th of November.
Spring: First week of February.

Mostly seed to be sown in flat land sometimes on ridges in *Kharif* season.

Seed Rate (kg/ha)

Grain-20 kg/ha, Sweet corn-8 kg/ha, Baby corn-25 kg/ha, Popcorn-12 kg/ha, Green cob-20 kg/ha, Fodder-50 kg/ha.

Seed Treatment

Bavistin + Captan in 1:1 ratio @ 2 g/kg of seed

Imidachloprid @ 4 g/kg of seed.

Water Management

Rainfed in *Kharif* throughout the state, rainfed in *pre-Kharif* in Terai regions. Irrigated in *Rabi* season throughout the state and number of irrigation 5-7 depending upon soil type should be given at critical crop growth stages as and when needed.

Weed Management

Pre-emergence application of Atrazine @ of 1.0-1.5 kg a.i./ha or Alachlor (Lasso) @ 2-2.5 kg a.i/ ha or, Metolachlor (Dual) @ 1.5-2.0 kg a.i./ha, Pendimethalin (Stomp) @ 1-1.5 kg a.i./ha are effective way for control of many annual and broad leaved weeds.

Crop Protection

Stem borer- control by application of Fepronil @ 2 ml/l of water. Leaf Blight- control by Mancozeb @ 2.5 g/ l.

Yield

Kharif- 3.5 - 4 ton/ha

Rabi- 6-7 ton/ha

Important Diseases

1. Northern leaf blight caused by the fungus *Exserohilum turcicum*. Long elliptical, grayish green or tan lesions (2.5-15 cm) appear on lower leaves progressing upward in the affected maize plants.
2. Southern leaf blight: the causal agent is *Helminthosporium maydis*. Numerous small round to oval spots are found on the whole leaf surface.
3. Banded leaf and sheath blight: The fungal pathogen *Rhizoctonia solani* is the causal agent of this disease. The disease appears on plants at pre-flowering stage (40-50 days old plant) and within a period of 15-20 days spreads under favourable conditions from the lowermost sheath to the ear shoot. The magnitude of grain loss may reach as high as 100%.

4. Common rust (*Puccinia sorghi).*
5. Brown stripe downy mildew: This disease has been most severe in hilly parts of West Bengal special in areas that receive 100-200 cm of rains.

Presently, the disease is considerd as a major disease in West Bengal specially in warm and humid condition.

In near future some virus diseases may be occurred in severe form such as Maize mosaic and Maize Mosaic Dwarf diseases which are not getting importance presently.

Major insect pests

1. Pink Borer (*Sesamia inference*): It occurs during winter season.
2. Stem borer (*Chilo partellus*): It is a major pest of maize in this state during monsoon season.
3. Termites (*Odontotermes obesus*): It is also occurring in West Bengal in maize crop.

Management of pests and diseases

- Growing of PEMH-5, Vivek 21, Vivek 25, Pratap and Kanchan varities in recommended areas and need based spray of Mancozeb @ 2.5 g/l at 8-10 days interval can manage leaf blight diseases in West Bengal.
- Adoption of promising hybrids like Sheetal, and HQPM-1 helps in minimizing the common rust of maize.
- Pink borer and stem borer of maize can be managed by releasing 8 Trichocards (*Trichogramma chilonis*)/ha at 10 days after germination.
- Intercropping of maize with suitable varieties of cowpea is an effective eco-friendly option for reducing the incidence of *Chilo* in maize.
- For better management of termites in West Bengal Fepronil granules are applied @ 20 kg/ha followed by light irrigation. In case of termite incidence in patches application of Fepronil @ 2-3 granules/plant is done. Besides clean cultivation delays termite attack.

Post harvest management

- After the harvesting, maize requires drying for 10-15 days to reduce the moisture level upto 13-14%. The higher the moisture level, lower will be the price of produce. In WB, currently a large part of the maize production is sun dried, where the drying is not uniform. Also, farmers in certain areas of West Bengal sale spring maize (March-June) with high moisture just after harvesting, so as to use the field for the *Kharif* crop. Proper drying of grains by using driers improve the quality of produce. Moisture content is maintained to avoid cracking of kernels to maintain grain quality.

- Proper fumigation of the produce during storage should be done.
- Storage of the grain should be largely done in the large silos/bins which are monitored and treated regularly to prevent losses from pest attacks.

To minimize post harvest losses, the following measures should be followed

- Maize cultivated for kernels (grains) should be harvested when the kernels are matured with 25 to 30% moisture.
- Use proper method of harvesting.
- Dry the cobs immediately before threshing and kernels should be dried sufficiently before storage.
- Losses in threshing and winnowing should be avoided by using proper machineries.
- Follow sanitation during drying, packing and handling to avoid contamination of kernels and to protect from insects, rodents and birds, etc.
- Use proper techniques for cleaning and further processing.
- Adopt grading practices for proper evaluation and obtaining better price.
- Use strong, and free from infestation packaging material for storage and transport.
- Use proper scientific technique in storage for maintaining optimum moisture content.
- Use pest control measures (fumigation) before storage.
- Provide aeration to stored the grain and stir grain bulk occasionally.

Maize extension programmes in West Bengal

In West Bengal, the poultry industry alone requires about 2100 tonnes maize/day and for satisfying this demand, more than 2.5 lakh hectare is to be brought out under quality protein maize considering average productivity of 30q/ha. Considering the importance of QPM for poultry sector in West Bengal CIMMYT-India took initiative during 2004-05 to introduce QPM (HQPM-1) in West Bengal in collaboration with State Department of Agriculture, and QPM hybrid (HQPM-1) was first evaluated in Adaptive trial in all the agro-climatic zones of the state except the Coastal-Saline areas and performance of the hybrid was found to be excellent in every locations and average yield of 70q/ha was recorded. During 2007-08, total area under maize was about 1.0 lakh ha having production of about 30 lakh tonnes.

Considering the satisfactory performance and wide adaptability, decision was taken to popularize the hybrid across the state through demonstrations. Programme of conducting front line demonstration (FLD) in different seasons was started since *Kharif*, 2005 and continued upto *Kharif*, 2007.

Average yield of QPM (HQPM-1) under front line demonstration in West Bengal

Season	FLDs (ha)	Yield (q/ha)
Kharif, 2005	32	46.0
Rabi, 2005-06	368	52.0
Kharif, 2006	678	42.0
Rabi, 2006-07	532	53.5
Kharif, 2007 (in special problem areas)	920	37.0

Rice is the main cereal crop in West Bengal but with changing scenario maize crop is getting importance in recent years. So, there should be much more effort to popularize this crop through vigrous extension programmes.

Rainfall Pattern in West Bengal.

Month	Rainfall (mm)
June	502.38
July	688.56
August	554.22
September	511.65
October	265.74
November	33.50
December	9.65
January	17.14
February	34.03
March	51.94

Key R & D challenges for maize in West Bengal

- Changing climatic conditions resulting in drought/excess water associated with increased pressure of diseases/pests.
- Rainfed cultivation in *Kharif* on marginal lands with inadequacy of irrigation facilities.
- Lesser cultivation of single cross hybrid technology, which is a key to higher productivity gains like USA, China and other countries.
- Limited adoption of improved production-protection technology.
- Production and distribution system of quality seed at affordable prices.
- During *Kharif* season land preparation and timely sowing becomes so much difficult due to continuous rainfall.
- In *Kharif* season yield of maize is drastically reduced due to waterlogging particularly at initial stages, waterlogging for more than 24 hours can kill the

crop and later in the crop period, waterlogging can be tolerated for periods of up to one week, but yield will be reduced drastically.

- Weeds infestation is the serious problem in maize, particularly during *Kharif/* monsoon season.
- Fertilizer use efficiency is less due to leaching, runoff, excessive weeds and waterlogging in *Kharif*, is most common problem in different parts of the state.
- Lodging results in significant yield losses in *Kharif* season due to heavy rains coupled with high wind and in *Rabi* crop is also lodged due to high wind speed in February-March month.

Probable solutions for maize R & D

- Strategic maize research
- Development of climate resilient cultivars
- Conservation agriculture technologies in *Rabi* maize
- Policy intervention for bridging yield gaps
- Retaining local youth in agriculture
- Bridging yield gaps: Involving local youth

For land preparation and sowing

Perform tillage operations when the soil is in the optimum moisture range and land preparation and sowing method is done 10 to 15 days before onset of monsoon with one pre-sowing irrigation.

For waterlogging

1. Level the field, or avoid planting maize in the low lying areas.
2. Break up the hardpan by subsoiling
3. Install drainage channels.

For weed management

1. Adoption of recommended improved weed management practices
2. Increase population density of crop and/or N application rate to get more shading of weeds.
3. Move to another site or rotate with another crop that will allow better control of problematic weeds. Plant the crop on wide raised beds, or on ridges.

For proper nutrition

1. Balanced fertilizer application rate
2. Change fertilizer application method or timing
3. Improve drainage
4. Reduce weed competition
5. Apply micronutrient
6. Green manuring and use of vermicompost.

For lodging

1. Development of a shorter variety with lower ear placement
2. Alter density or spatial arrangement and reduce the density of the intercrop
3. Increase soil depth by subsoiling. The effective soil depth can also be increased by ridging or hilling up around the plants
4. Control insect pests which lead to lodging
5. Reduce unscrupulous N application to a non-responsive variety
6. Correction of K deficiency by fertilizer application

Factors important for increase/decrease of maize area/production/ productivity in the state

In West Bengal, the declining trend of area under maize is reversing (> 41.8%) and increasing the productivity (> 7.3%) as on 2013-14. Area, yield & production under Maize have increased enormously in last few years by adequate irrigation.

Year	Area (lakh ha)	Production (lakh ton)	Yield (kg/ ha)
2014-15	1.52	6.63	4351
Increase over 2010-11 (%)	71.93	88.21	9.49

Rice-rice and rice-wheat has been the traditional crop for a large part of West Bengal. However, extensive uses of water for paddy has lead to an increasing water demand resulting in severe water crisis. As a result area of maize is increasing.

In West Bengal the declining yield of boro rice, *Rabi* rice has shown a path to maize as a better option.

With the changing climate scenario maize being thermo and a photo insensitive is the only suitable alternative crop and more area is likely to shift towards maize cultivation in near future in non-traditional areas.

- Public-Private partnership could potentially be a route which could assist in improving various segment of the value chain – improved seeds, technical knowledge, storage infrastructure

- Currently large part of the value chain of maize production to delivery rank lower in quality compare to large global producers. About 35-37% of the maize acerage using low productivity OPV seeds primarily supplied by the government. There remains a gap in high quality knowledge transfer to farmers to enhance the quality of produce. There is a shortage of farm level infrastructure and quality storage facilities, resulting in quality degradation.
- For promoting maize government can introduce initiatives such as increasing MSP, seed and pesticide subsidy, installation of large scale driers at different locations, farmer's awareness programme/ fairs, extension programmes through KVK and universities and creation of efficient marketing infrastructures.
- Capacity building of farmers providing technology including mechanization and training to the farmers.

So, the areas of key considerations are

- The scalability of value chain initiatives
- Value impact both at end-user level and farm level
- Continued participation of private players
- Shift from top down to bottom up approach to meet new demands for innovations, products, information and extension services
- Maximize complementarity of public sector policies and dynamism of private sector to play major role for production of single cross hybrids (25-30% to 80%)
- Seed production of public bred hybrids and marketing by private sector by paying royalty
- Premium pricing for value added products
- Progressive policies for value chain/processing industries
- Assist and educate farmers in managing climate change/ market, etc.
- Provide farmers with affordable hybrids and other techniques
- Simultaneous scaling up of outputs through KVKs and the National Mission on Sustainable Agriculture for wider adoption by the farmers

Support required for promoting maize cultivation

- Access and support for inputs (seeds, fertilizer, etc.).
- Access to credit and crop insurance.
- Assistance in obtaining new seed varieties with higher yield/biotic and abiotic stress resistance, etc.
- Technology enabled knowhow support.

- Unified market platform offering better market access to corn producers.
- In order to support the increasing growth of value added corn uses / initiatives will have to be taken to encourage growing of better quality corn as well as specialty corn. Direct farmer linkage with corporate can help in building better procurement pathway for end-users company.
- Increase in thrust on promotion and adaption of mechanical harvesting could potentially assist in improving productivity and control farm losses.
- Harvesting process should be completely mechanized, which reduces losses due to post harvest farm activities (threshing, separating, etc.).

Summary

In West Bengal maize crop is also getting importance in recent years. The limitation of rising temperature during grain filling of wheat particularly in eastern India and the declining yield of *boro* rice in West Bengal and water scarce areas in western part of West Bengal affecting yield of *Rabi* rice has shown a path to maize as a better option. The medium and uplands where subsistence yield of wheat, *Rabi* rice and other winter crops is obtained, could be substituted by winter maize in West Bengal.

References

De, U.K. (2002). Economics of Crop Diversification. Akansha Publishing House, New Delhi.

De, U.K. (2003). Changing cropping system in theory and practice: An economic insight into the agrarian West Bengal. *Indian J Agric Econ*. 58: 64-83.

Mruthyunjaya and Kumar, P. (1989). Crop economics and cropping pattern changes. *Econ Political Weekly*. 24: A159-A161.

Prasanna, B.M. (2010). Phenotypic and molecular diversity of maize landraces: characterization and utilization. *Indian J Genet.* 70: 315–327.

SenGupta, P.K. (2001). Monograph on West Bengal. Directorate of Agriculture, Government of West Bengal, Kolkata, India.

Sharma, L., Prasanna, B.M. and Ramesh, B. (2010). Phenotypic and microsatellite-based diversity and population genetic structure of maize landraces in India, especially from the North East Himalayan region. *Genetica.* 138: 619–631.

Singode, A. and Prasanna, B.M. (2010). Analysis of genetic diversity in the North Eastern Himalayan (NEH) maize landraces of India using microsatellite markers. *J Plant Biochem Biotech.* 19: 33–41.

11

Maize Research and Development in Odisha

Digbijaya Swain, Pramila Naik, Devraj Lenka, B. Baisakh and Sanjay Pani [1]

Orissa University of Agriculture & Technology, Bhubaneswar, Odisha

[1]Directorate of Agriculture & Food Production, Odisha

Corresponding Author's Email: digbijaya72@gmail.com

Maize is the only cereal crop which is being used as human food, animal feed and as a source of large number of industrial products. In Odisha traditionally maize is a *Kharif* season crop and second most important crop next to paddy during *Kharif* season in terms of both area and production. The *Rabi* maize is cultivated under irrigated condition, and accounts only 7% of total area and production of the State. While only 24% of the *Kharif* maize area have irrigation facilities. Irrespective of the season, it is that is the fourth most important food grain crop with respect to area in the state which is next to rice, mungbean and urdbean. Even though maize is grown as a sole crop in the state but, some farmers are also adopting maize based inter-cropping and the major recommended/adopted inter-crops are arhar, cowpea, yam, runner bean, greengram and blackgram. The comparative analysis of compound annual growth rate (CAGR) of area, production and yield of the major crops is the state was done for the years 2004-05 to 2013-14. The CAGR study of major crops revealed that no other major crop including rice has the impressive growth rate a like maize in the state (Fig. 1).

Trend in area production and productivity of maize during 1950-2015

The maize area of the state has increased from 0.23 lakh ha in 1950-51 to 2.80 lakh ha in 2013-14, while the production has gone up from 0.09 lakh tons to 7.80 lakh tons. The yield has also been observed an unprecedented increasing trend of more than seven times higher during the period (Table 1). Thus maize is emerging as one of the major crops in Odisha over years. During 2014-15, maize was cultivated in an area of about 2.69 lakh hectares with an average yield of 2.79 t/ha and produced 7.51 lakh tons.

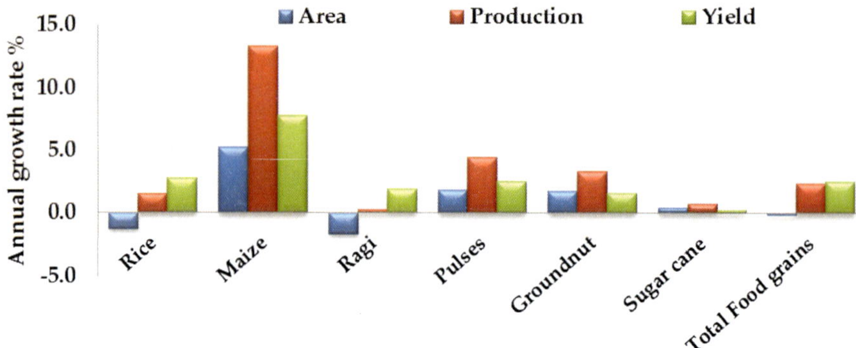

Fig. 1. Compound annual growth rates in area, production and yield of major crops during (2004-05 to 2013-14) in Odisha.

Source: Odisha Agriculture Statistics; Official website of Agriculture & Farmers' Empowerment, Govt. of Odisha.

Table 1. The trend in maize area (lakh ha), production (lakh tonnes) and yieid (kg/ha) during 1950-2014.

Year	*Kharif*			*Rabi*			Total		
	Area	Production	Yield	Area	Production	Yield	Area	Production	Yield
1950-51	-	-	-	-	-	-	0.23	0.09	390
1951-52	-	-	-	-	-	-	0.25	0.10	393
1952-53	-	-	-	-	-	-	0.25	0.09	373
1953-54	-	-	-	-	-	-	0.25	0.10	383
1954-55	-	-	-	-	-	-	0.25	0.10	383
1955-56	-	-	-	-	-	-	0.25	0.10	383
1956-57	-	-	-	-	-	-	0.24	0.09	395
1957-58	-	-	-	-	-	-	0.23	0.09	395
1958-59	-	-	-	-	-	-	0.22	0.09	422
1959-60	-	-	-	-	-	-	0.22	0.09	403
1960-61	-	-	-	-	-	-	0.22	0.09	417
1961-62	-	-	-	-	-	-	0.34	0.13	392
1962-63	-	-	-	-	-	-	0.31	0.20	648
1963-64	-	-	-	-	-	-	0.14	0.07	525
1964-65	-	-	-	-	-	-	0.22	0.18	818
1965-66	-	-	-	-	-	-	0.48	0.33	690
1966-67	-	-	-	-	-	-	0.53	0.49	933
1967-68	-	-	-	-	-	-	0.66	0.55	832
1968-69	-	-	-	-	-	-	0.68	0.56	820
1969-70	0.68	0.57	841	0.02	0.03	1183	0.70	0.60	851
1970-71	0.69	0.56	814	0.03	0.03	986	0.72	0.59	821
1971-72	0.72	0.60	830	0.03	0.03	1199	0.75	0.63	843
1972-73	0.75	0.57	769	0.03	0.04	1139	0.78	0.61	783
1973-74	0.82	0.65	795	0.03	0.05	1419	0.85	0.70	820
1974-75	0.93	0.70	752	0.04	0.04	1178	0.97	0.74	767

1975-76	1.15	1.12	976	0.04	0.05	1250	1.19	1.17	985
1976-77	1.21	0.98	815	0.03	0.05	1401	1.24	1.03	830
1977-78	1.26	1.06	842	0.04	0.05	1358	1.30	1.11	857
1978-79	1.26	1.15	910	0.05	0.06	1320	1.31	1.21	925
1979-80	1.24	0.72	580	0.05	0.07	1279	1.29	0.79	609
1980-81	1.74	1.67	956	0.07	0.08	1177	1.81	1.74	964
1981-82	1.47	1.60	1090	0.07	0.11	1551	1.54	1.71	1112
1982-83	1.53	1.81	1179	0.06	0.11	1807	1.60	1.92	1203
1983-84	1.85	1.84	992	0.07	0.12	1654	1.92	1.95	1017
1984-85	1.59	1.64	1034	0.06	0.09	1405	1.65	1.73	1048
1985-86	1.59	1.66	1042	0.06	0.09	1397	1.65	1.74	1055
1986-87	1.57	1.20	769	0.09	0.11	1216	1.65	1.31	792
1987-88	1.57	1.62	1035	0.06	0.07	1167	1.63	1.69	1040
1988-89	1.59	1.84	1157	0.08	0.11	1315	1.67	1.95	1165
1989-90	1.61	1.86	1156	0.08	0.11	1410	1.69	1.98	1168
1990-91	1.60	1.96	1224	0.07	0.11	1559	1.67	2.07	1238
1991-92	1.69	1.54	914	0.09	0.11	1199	1.78	1.65	928
1992-93	1.62	1.67	1033	0.08	0.11	1331	1.70	1.78	1047
1993-94	1.68	1.83	1092	0.08	0.11	1262	1.76	1.94	1100
1994-95	1.64	1.98	1209	0.11	0.13	1220	1.74	2.11	1209
1995-96	1.44	1.92	1337	0.10	0.14	1388	1.53	2.06	1340
1996-97	1.57	1.79	1141	0.08	0.10	1167	1.66	1.89	1142
1997-98	1.60	2.06	1281	0.09	0.10	1160	1.69	2.16	1275
1998-99	1.56	1.74	1116	0.08	0.09	1122	1.64	1.83	1117
1999-2000	1.65	2.06	1248	0.08	0.10	1243	1.74	2.17	1248
2000-01	1.70	2.10	1234	0.06	0.08	1273	1.76	2.17	1235
2001-02	1.56	1.75	1119	0.08	0.11	1287	1.64	1.85	1125
2002-03	1.52	1.70	1118	0.06	0.07	1244	1.58	1.77	1123
2003-04	1.66	1.83	1101	0.09	0.12	1389	1.75	1.96	1116
2004-05	1.76	2.30	1305	0.09	0.15	1649	1.85	2.44	1322
2005-06	1.79	2.63	1468	0.08	0.17	2092	1.87	2.80	1496
2006-07	1.90	2.99	1575	0.09	0.20	2147	1.99	3.19	1602
2007-08	2.04	4.53	2216	0.10	0.28	2824	2.14	4.82	2245
2008-09	2.11	4.76	2252	0.13	0.38	2935	2.24	5.14	2291
2009-10	2.17	4.75	2192	0.11	0.25	2169	2.28	4.99	2191
2010-11	2.36	6.03	2552	0.16	0.46	2820	2.53	6.49	2570
2011-12	2.45	5.59	2280	0.17	0.49	2912	2.62	6.08	2321
2012-13	2.62	6.2	2368	0.19	0.56	2948	2.81	6.76	2407
2013-14	2.60	7.21	2772	0.20	0.59	2954	2.80	7.79	2785
2014-15	2.51	6.96	2774	0.19	0.56	2981	2.69	7.51	2788

Source: Official website of Agriculture & Farmers' Empowerment, Govt. of Odisha

The decadal study on increase in maize area, production and productivity of the state since 1950-51 indicated that there is a steady increase till 2009-10. In the present decade (2010-11 onwards), there was sudden increase in area, production and productivity in the state (Fig. 2). However, the rate of increase i.e. growth rate in maize area, production and productivity in different decades during 1950-2014 was not consistent (Table 2).

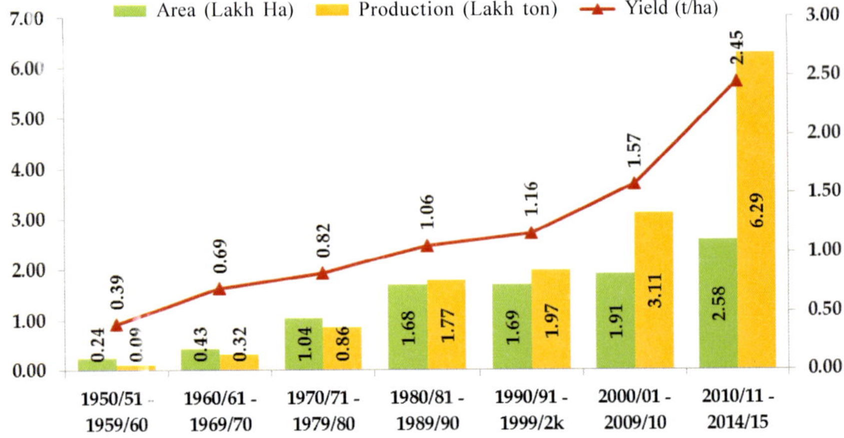

Fig. 2. Maize area, production and yield in Odisha over years (1950/51 – 2014/15).
Source: Odisha Agriculture Statistics; Official website of Agriculture & Farmers' Empowerment, Govt. of Odisha

Table 2. Compound annual growth rates (%) of maize in Odisha over decades (1950-51 to 2014-15)

Period	Compound Annual Growth Rate (%)		
	Area	**Production**	**Yield**
1950-51 to 1959-60	-1.11	-0.64	0.71
1960-61 to 1969-70	15.53	26.51	9.11
1970-71 to 1979-80	8.23	7.42	-0.79
1980-81 to 1989-90	-0.25	0.16	0.35
1990-91 to 1999-2k	-0.33	1.23	1.57
2000-01 to 2009-10	3.97	13.75	9.39
2010-11 to 2014-15	1.91	5.55	3.51

However, during the last ten years (2004-05 to 2014-15), there was significant increase in the maize area (1.5 times), production (2.7 time) and productivity (1.9 times) in the state (Fig. 3). The average maize yield has increased from 1.50 tons/ha in 2005-06 to 2.79 tons/ ha in 2014-15.

Major growing districts of Odisha

Maize is being cultivated more or less in all the districts of Odisha. However, out of thirty districts it is cultivated in more than five thousand hectares each in sixteen districts (Fig. 4). Almost all maize area (94%) and production (96%) comes from 16 districts (Fig. 5), among these, Nabarangapur district alone constitutes more than 26% of maize area in the state with a contribution of more

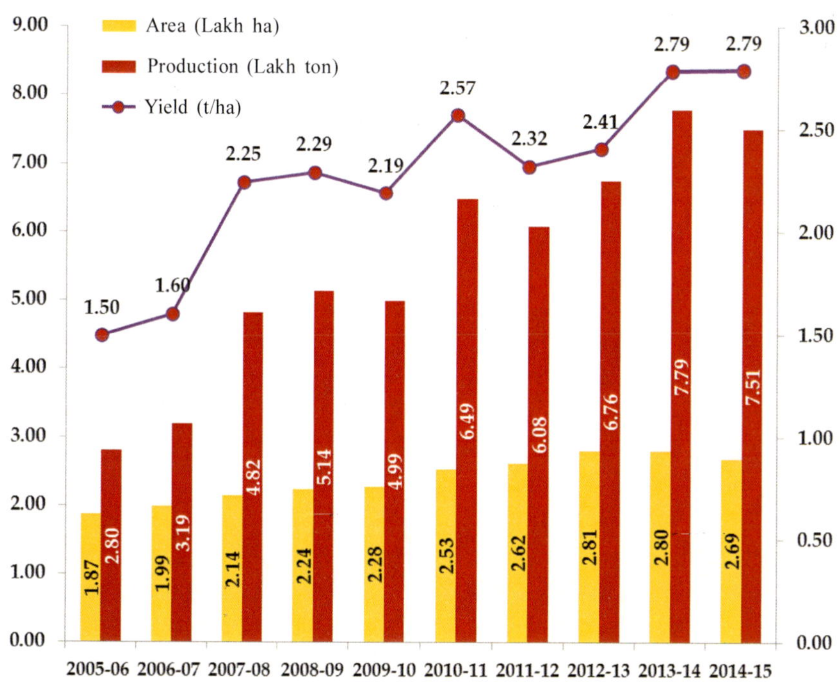

Fig. 3. Maize area, production and yield in Odisha during last ten years (2005-06 to 2014-15). *Source:* Odisha Agriculture Statistics; Official website of Agriculture & Farmers' Empowerment, Govt. of Odisha

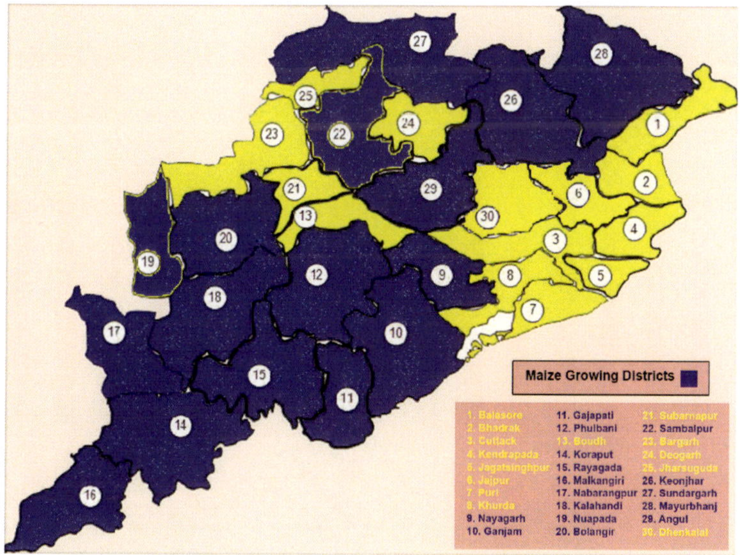

Fig. 4. Map of maize growing districts of Odisha.

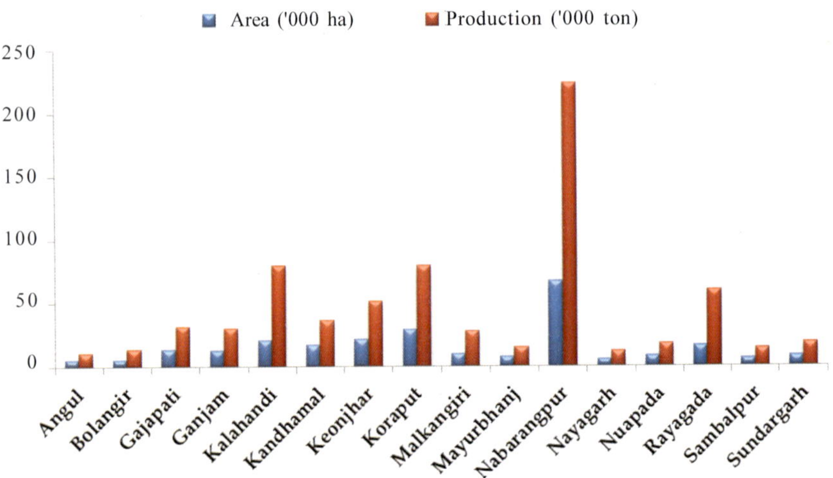

Fig. 5. Area & production in major maize growing districts of Odisha (2014-15).

than 30% of total production with a productivity of 3.3 tons/ha. Among other districts, Kalahandi has the highest productivity followed by Rayagada and Nabarangapur (Fig. 6). The steady growth in maize production and productivity over years is due to adoption of high yielding cultivars (composites/ hybrids) and scientific crop management practices. The majority of the maize produced

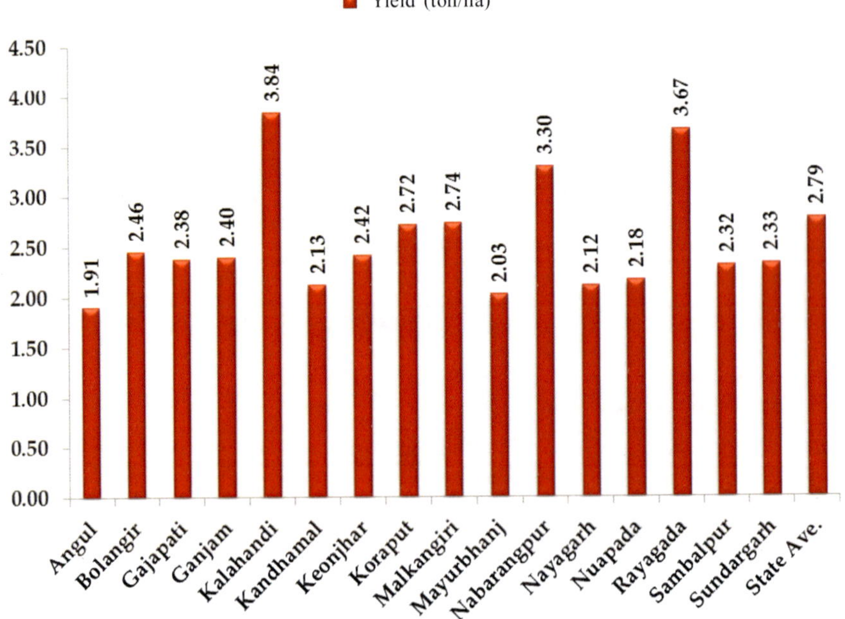

Fig. 6. Yield in Major Maize Growing Districts of Odisha (2014-15).

is of high yielding variety, which covers around 90% of the total area under maize in Odisha and contributes about 93% of the total maize production. Besides, the public sector, private sector companies like Pioneer India Seeds, Monsanto, Syngenta Seeds, Niziveedu Seeds, Advanta, Bio-seeds, Mahyco seeds, Rasi Seed etc. have played prominent role in supply of hybrid seeds to the farmers.

Current maize utilization pattern in Odisha

Despite of the steady growth in maize cultivation in Odisha over years, the maize industry is yet to witness its potential in terms of value addition. At present about 75–80% of maize production is exported to the neighbouring states such as West Bengal, Chhattisgarh, Andhra Pradesh and Karnatak for value addition and processing and only 20-25% is utilized within the state. Chhattisgarh and Andhra Pradesh have major share in import from Odisha. Out of 20-25% utilized within the state around 5% is processed within the state, while around 15-20% is retained by the farmers for their own consumption and for seed purpose. The consumers of Odisha get processed and value added products from outside the state. One of the main reasons for direct export of maize from Odisha to other states is the lack of primary processing/ storage facility at village level resulting in loss of quality of maize after harvesting due to high moisture content and hence results in immediate sale of maize from farmer's field even at a lower price. Presently there is a requirement of around 2.98 Lakh tons of maize grain per annum in Odisha for poultry feed and this demand is increasing year by year. The major feed producing industries in the state are annually utilizing around 50,000 MT tons of maize in processing to produce animal feed. There are only 4 units involved in corn flake manufacturing and 5-6 units in corn oil production in the state. There are no industrial units producing products like alcohol, starches, sweeteners, ethanol from maize within the state. Therefore there is high potential for establishing processing and value addition units for production of industrial/ consumer products from maize.

Based on the maize productivity data for the year 2014-15, the major maize-producing districts of Odisha are categorized and presented in Table 4. So far as the operational holdings of maize farmers of Odisha are concerned more than 86 % are small and marginal farmers (Table 5).

Genesis of AICRP centre

The All India Coordinated Maize Improvement Project (AICMIP) was initiated by ICAR in 1957 and AICMIP centre was established at OUAT during 1987-88. The mandate of AICMIP centre is to conduct multi-disciplinary and multi-location coordinated research trials, to identify appropriate technologies, to generate materials/ technologies for enhancement of productivity and production of maize and to disseminate the improved technologies through demonstrations. Maize hybrids like Deccan 103, Ganga II and composites like Navjot, Pratap, Megha were recommended for cultivation and are popular in the state. A number of high yielding varieties (composites/ hybrids) of different maturity group developed both by private and public sectors and crop management technology have been recommended for the state from time to time and adoption of these technologies by the farmers is the instrumentals in enhanced production and productivity of the state. At present some of the hybrids recommended in Odisha are HM 9, Vivek maize hybrid-27, HM 10, Vivek maize hybrid-17, P 3441, P 3501, Bio 9681, DMH 115, Bio 9544, Bio 9637, LG 32.81 (Yuvaraj Gold), Vivek maize hybrid-43, HM 12, Seed Tech 2324, NK 6240, Hishell, DKC 9081, NMH 713, etc.

Major maize based cropping systems in the state

The brief details of major maize growing districts with agro-climatic zones of Odisha along with maize based cropping system and intercropping system existed or being practiced in the state are given in Table 3.

Table 3. The maize based cropping and intercropping systems recommended across different districts of Odisha.

Agro-climatic zone	District	Annual rainfall (mm)	Maize based cropping sequence	Maize based intercropping
Eastern Ghat High Land	Major parts of Koraput, Nawarangpur	1522	Maize- horsegram/ blackgram/ toria/ potato/ Rapeseed mustard	Maize + Arhar/ blackgram/ greengram cowpea
South Eastern Ghat	Malkangiri & part of Koraput	1710		
Western Undulating Zone	Kalahandi & Nuapada	1352		
North Eastern Ghat	Kandhamal, Rayagada, Gajapati, part of Ganjam & small patches of Koraput	1597		
North Central Plateau	Mayurbhanj, major parts of Keonjhar	1534	Maize-horsegram/ greengram/ castor/ ground nut/ toria/	Maize + blackgram/ greengram
North Western Plateau	Sundargarh, parts of Sambalpur	1600		
East and South Eastern Coastal Plain	Nayagarh & part of Ganjam	1577	Maize-blackgram/ greengram/ horsegram	Maize + blackgram/ greengram/ arhar/ ragi/ mesta
Western Central Table Land	Bolangir, parts of Sambalpur	1614	Maize-ground nut/ potato/ Rape seed mustard-sesame/ vegetable	Maize + blackgram/ greengram/ cowpea
Mid Central Table Land	Angul	1421		

Major production constraints

The normal rainfall of the state of Odisha is 1451.2 mm and about more than 80% of total rainfall is received during June to September, monsoon period (Table 4).

Table 4. The rainfall pattern (mm) in Odisha over different months of the year.

District	Jan	Feb	Mar	Apr	May	Jun	Jul	Aug	Sep	Oct	Nov	Dec	Total
Angul	12.6	27.1	24.3	27.2	52.9	225.1	347.7	357.5	217.5	86.3	20.4	3.3	1401.9
Balasore	14.7	31.8	34.4	62.2	108.5	221.5	308.6	332.1	267.6	170.5	34.6	5.5	1592.0
Bargarh	12.5	19.1	22.0	20.0	25.6	205.6	397.2	374.4	222.6	52.8	10.4	5.1	1367.3
Bhadrak	11.8	29.6	36.1	51.2	91.4	198.2	293.6	311.7	216.8	145.3	37.3	4.9	1427.9
Bolangir	8.6	13.2	14.6	13.4	27.8	202.8	360.5	333.6	237.4	68.4	7.2	2.3	1289.8
Boudh	17.0	26.8	22.4	16.5	38.4	233.8	418.5	488.8	244.8	90.5	21.1	4.4	1623.1
Cuttack	9.9	28.6	24.7	28.3	71.5	210.0	308.3	339.1	229.2	125.6	45.0	4.1	1424.3
Deogarh	14.1	27.0	20.6	21.3	41.9	242.4	447.7	443.4	228.2	84.4	8.0	3.5	1582.5
Dhenkanal	10.5	21.9	33.7	41.3	69.6	225.7	317.9	344.8	220.6	104.4	36.2	2.2	1428.8
Gajapati	8.1	23.0	41.3	65.2	107.8	199.7	230.6	253.6	237.9	168.6	61.8	5.7	1403.3
Ganjam	9.4	24.1	32.6	36.6	65.4	168.3	220.8	246.8	216.3	177.7	71.1	7.1	1276.2
Jagatsinghpur	14.5	21.7	35.5	25.4	78.0	202.3	277.3	379.1	241.4	151.1	80.4	7.9	1514.6
Jajpur	12.9	25.1	28.4	46.1	93.7	238.5	350.9	341.2	238.0	140.2	41.0	3.9	1559,9
Jharsuguda	14.1	22.9	17.6	15.1	27.9	218.8	385.8	382.9	210.7	54.9	7.7	4.4	1362.8
Kalahandi	10.3	14.4	23.7	25.7	41.8	240.4	327.7	355.4	204.6	74.0	10.9	1.6	1330.5
Kandhamal	10.6	29.7	28.0	35.7	67.8	207.9	325.1	330.8	239.1	117.7	31.0	4.5	1427.9
Kendrapara	10.6	30.2	35.0	33.9	94.2	208.3	317.1	333.3	237.3	183.7	67.2	5.2	1556.0
Keonjhar	14.6	33.8	33.1	42.1	94.8	241.4	318.0	343.6	241.1	101.3	20.5	3.4	1487.7
Khordha	12.4	24.3	22.1	28.0	60.7	196.2	304.9	320.6	234.5	149.3	50.4	5.0	1408.4
Koraput	5.7	8.6	18.3	55.2	81.9	206.8	375.6	393.6	256.3	126.1	32.6	6.5	1567.2
Malkangiri	2.7	4.1	8.9	34.8	49.1	212.2	465.7	472.8	281.2.	109.5	23.6	3.0	1667.6
Mayurbhanj	10.3	28.0	40.2	52.5	101.2	265.8	337.3	359.9	262.0	114.1	21.6	7.3	1600.6
Nabarangpur	6.7	14.1	15.1	34.1	66.1	251.8	356.6	407.5	225.6	168.6	18.7	4.6	1569.5
Nayagarh	11.7	28.6	30.0	39.9	58.8	203.5	288.5	288.2	226.8	134.5	39.0	4.8	1354.3
Nuapada	12.0	14.6	19.7	21.8	31.2	210.3	347.1	327.8	214.5	68.9	15.3	3.2	1286.4
Puri	10.9	25.4	15.5	18.5	62.1	188.0	292.0	297.9	243.2	181.6	67.3	6.4	1408.8
Rayagada	9.6	22.9	38.0	52.7	87.8	195.3	259.5	273.7	199.1	109.7	32.9	4.7	1285.9
Sambalpur	14.8	24.5	18.2	16.5	32.3	221.0	429.5	442.4	224.7	54.7	12.7	4.4	1495.7
Subarnapur	12.5	17.2	15.7	15.4	29.3	217.4	399.6	408.9	228.0	59.8	11.4	3.3	1418.5
Sundargarh	15.1	24.9	16.0	16.0	40.6	237.4	386.4	393.9	211.5	67.7	8.7	4.2	1422.4
Odisha	**11.4**	**22.9**	**25.5**	**33.1**	**63.3**	**216.5**	**339.9**	**356.0**	**232.0**	**114.7**	**31.5**	**4.5**	**1451.2**

Source: Odisha Agriculture Statistics. 2013-14. Directorate of Agriculture & Food Production.

Though the total amount of rainfall is more or less constant over years, there is large spatial and temporal variation resulting prolonged dry (drought) or wet condition (water logging). About two-third of the total cultivated area being rain-dependent, drought poses a serious threat at regular intervals due to failure of monsoon. During last 20 years (1996-2015), drought has affected Odisha ten times (Table 5).

Table 5. Details of number of districts affected due to drought in Odisha during last twenty years.

Year	Out of 30 districts
1996	28
1997	16
1998	26
2000	29
2002	29
2009	18
2010	17
2011	21
2012	4
2015	25

During 2015 there was a deficit of 16.1% in the cumulative average rainfall of the state from June to October (Table 6).

Table 6. Rainfall deficit during June to October, 2015 in Odisha.

Month		Rainfall	
	Normal (mm)	Actual (mm)	Deficiency (%)
June	216.5	234.6	8.4
July	339.9	308.4	- 9.3
August	356.0	266.5	- 25.1
September	231.9	221.5	- 4.5
October	114.7	25.3	- 77.9

Predominantly the farmers of Odisha traditionally cultivate paddy in up- and mid-land during *Kharif* season under rainfed condition. In fact, 8.06 lakh ha up-land out of total 29.14 lakh ha and 15.80 lakh ha mid-land out of total 17.55 lakh ha is under paddy cultivation during *Kharif* season. But due to erratic pattern of monsoon rain, prolonged dry spell, absence of sufficient irrigation facility and recent climatic changes, is affected the most, especially in up- and mid-land conditions. Thus, there is a need for substitution of rice in up- and mid-land by non-paddy crops. In this context, maize is considered a promising option for diversifying agriculture in up- and mid-land areas of Odisha during *Kharif* rainfed condition as maize is less water intensive than rice and considered as a climate-change-ready crop. Thus, the farmers of Odisha are opting for maize cultivation in up- or mid-land. Thus, the maize is becoming an important crop over years in Odisha and will play a prominent role in state agricultural scenario in the coming years. Though 90% of total maize area of the state is under high yielding varieties with only 10% area under local cultivars, the productivity of maize in Odisha is far behind other states like Tamil Nadu, Andhra Pradesh, West Bengal, Punjab, Haryana and Karnatak (Fig. 7).

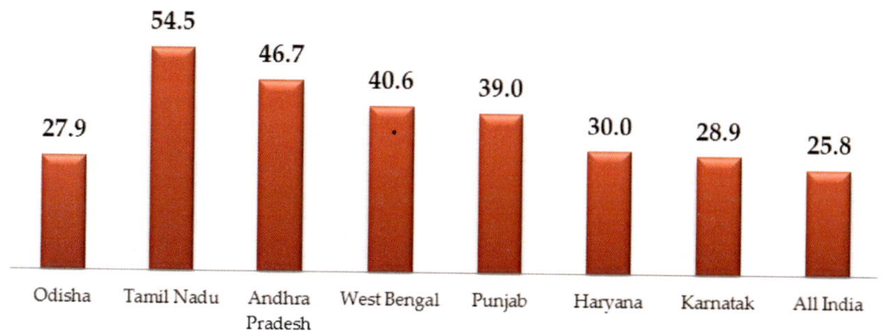

Fig. 7. Productivity (q/ha) of Odisha in comparison to other major producing states. *Source:* Odisha Agriculture statistics, 2014-15.

Based on the maize productivity data for the year 2014-15, the major maize-producing districts of Odisha are categorized and presented in Table 7.

Table 7. Categorization of maize producing districts

Yield	District	Yield (t/ha)
High (:> 4.0 t/ha)	Nil	
Medium (3.0 – 4.0 t /ha)	Kalahandi	3.84
	Rayagada	3.67
	Nabarangapur	3.30
Low (2.5 – 3.0 t/ ha)	Koraput	2.72
	Malkanagiri	2.74
Very low (<2.5 t/ ha)	Bolangir	2.46
	Keonjhar	2.42
	Gajapati	2.38
	Sundargarh	2.33
	Sambalpur	2.32
	Ganjam	2.40
	Nuapada	2.18
	Kandhamal	2.13
	Nayagarh	2.12
	Mayurbhanja	2.03
	Angul	1.91

The main reasons for low production/ productivity of the state are due to use of traditional local cultivars with low yield potential & use of F_2 seeds. In addition the operational holdings of maize farmers of Odisha are concerned more than 86% are small and marginal farmers who cannot practice for high-input agriculture (Table 8).

Table 8. The operational holdings of maize farmers in Odisha.

Particulars	Marginal (< 1.0 ha)	Small (1.0 – 2.0 ha)	Semi-medium (2.0 – 4.0 ha)	Medium (4.0 – 10. h)	Large (> 10.0 ha)
Holdings (%)	57.9	28.4	11.9	2.6	0.2
Total area (%)	38.6	32.6	20.4	7.1	1.3

Source: Odisha Agriculture Statistics; Official website of Agriculture & Farmers' Empowerment, Govt. of Odisha; Agricultural Census, 2010-11.

Further, hybrids do not express their potential due to non-suitability to agro-climatic condition under cultivation or non-suitability of available HYVs to biotic stresses of rainfed condition such as drought due to prolonged dry spell or water logging due to excessive rainfall. In addition, lack of technological know-how of the farmers about other improved scientific technologies such as herbicides/ in time weed management, nutrient management, inter cropping system, pest/ disease management and post harvest management are other reasons for low productivity. The government of Odisha has launched special projects from time to time for increasing maize production and productivity by considering its importance. To promote maize hybrids in the state a project in public-private partnership was implemented during 2010 (known as Project Golden Days) under the "Bringing Green Revolution in Eastern India" programme in 20 districts by involving seven leading seed companies of local, national and multinational level. In 2012, after its success in improving the crop yield, the state government re-launched the project under RKVY. Subsequently after 2015 drought, the government of Odisha has signed a MoU with CIMMYT and has sanctioned a RKVY funded project entitled "Stress-resilient Maize for Odisha" in which OUAT (AICMIP Centre) is a major partner and implementing organization. The main objective of the project is to identify high yielding stress-resilient maize cultivars and crop management practices, to strengthen maize seed system for timely availability of quality seed and to build the capacity of the maize growing farmers for a profitable and sustainable maize production system in Odisha. Besides, the Government of Odisha is promoting private sector investments on priority basis in the areas of maize value chain (animal feed processing unit, value added product units like corn flakes, popcorn, corn sweeteners etc., corn based starch producing units, corn oil units and rural warehousing units) across through different types of incentives, capital subsidy and tax exemptions in its Industrial Policy Resolution.

The gap between productivity at farmers' field and yield levels in FLDs and gap between average yield of Odisha and neighbouring states (Fig. 7) clearly indicates that there is wide scope for yield improvement of maize in the state. The yield gap among the productivity of different districts even though they belong to same agro-climatic zone (Fig. 6) may be narrowed down. Thus, there is lot of opportunities for increasing the maize production in Odisha through yield improvement, area expansion and other institutional measures. The strategies like promotion of hybrids in place of local traditional cultivars/ F_2 seeds, arrangement of timely availability of the hybrid seeds at lower price, identification of suitable stress resilient hybrids for different agro-climatic zones, district-wise/ agro-climatic zone-wise planning through a cluster approach, enhancement of productivity of medium productivity districts (yield: 3 - 4 t/ha) to high productivity (> 4.0 t/ha), special attention to low productivity districts for enhancement of productivity to medium level (3.0 – 4.0 t /ha) and capacity building of the farmers through technology transfer for adoption of scientific crop and post harvest management practices will go a long way to enhance the productivity. In order to expand maize area the following strategies like substitution of up- and mid-land paddy with maize during *Kharif* and increase in *Rabi* area under maize as part of diversification of rice-rice with rice-maize cropping system would help, just like in Andhra Pradesh, Tamil Nadu, Bihar and West Bengal due to favourable growing condition (more sun shine hours, mild temperature, less weeds), better water management, higher plant stand and low pest/ disease incidence. In Odisha the *Rabi* maize area is around 0.2 lakh ha and still 3.0 lakh ha is under rice cultivation. So there is lot of opportunity for diversification of rice-rice with rice-maize cropping system.

The profitability through maize cultivation may be enhanced through various institutional measures like promotion of farm mechanization in maize cultivation, inter-cropping in maize system, assured market linkage to ensure better price to the farmers, promotion of primary processing and high quality protein maize production, sweet corn and baby corn cultivation in peri-urban areas, development of storage and value addition infrastructure through private sector involvement and increasing irrigation facilities particularly through promotion of bore wells, dug wells, farm ponds, PLIP etc., providing credit and crop insurance facilities to ensure livelihood security of the maize farmers.

Future outlook

The conventional breeding methods should be supplemented with modern biotechnological tools such as molecular marker technology for heterotic grouping and assisting in phenotypic selection to accelerate the breeding process should be given importance. The doubled haploid technology for inbred development will help in development of hybrids suitable for agro-ecological conditions of the state with resistant to biotic and abiotic stresses such as drought, water logging and heat. The standardization of maize based intercropping for different agro-ecological conditions of the state along with development of crop management practices suitable for sole crop and inter crop to exploit maximum potential of the maize hybrids is also required.

References

Agricultural Census, (2010-11).

Annual Report on Natural Calamities. (2012-13). Revenue & Disaster Management Department Govt. of Odisha.

Investing in maize processing – Opportunities Abound. Developed under Odisha Inclusive Growth Partnership (OIGP).

Odisha Agriculture Statistics Directorate of Agriculture & Food Production, Govt. of Odisha.

Official website of Department of Agriculture & Farmers' Empowerment, Govt. of Odisha.

12

Maize Research and Development in Assam

Nabajyoti Bhuyan and Binod Kalita*
AICRP Maize, RARS, AAU, Gossaigaon
**Corresponding Author's Email: bnabajyoti@rediffmail.com*

Maize is presently the crop with growing importance in the state. Earlier it was primarily grown for human consumption only. In view of rapidly increasing demand of poultry feeds and fodder in Assam presently it has more use in some small industries. It is the raw material for most of the feed industries. Presently the crop has state acreage of 27953 hectare and production of 93179 tonnes with a productivity of 3333 kg/ha (2014-15).

While most other states in India are gradually moving away from their traditional agriculture based economy toward industry or service oriented economy, Assam is still predominantely dependent on the agricultural sector. While the sociopolitical problems afflicting the state since the last few decades are partly to blame for a lack of conducive environment for economic development of the state, particularly in industry or service oriented areas, there are various economic reasons (e.g. fragmented land) responsible for the lagging agricultural sector in the state. Economy of Assam is fundamentally based on agriculture. Over 70% of the state's population relies on agriculture as farmers, agricultural laborers, or both for their livelihood.

Assam produces both food and cash crops. The principal food crops produced in the state are rice (paddy), maize (corn), pulses, potato, wheat, etc., while the principal cash crops are tea, jute, oilseeds, sugarcane, cotton, and tobacco. Although rice is the most important and staple crop of Assam, its productivity over the years has not increased while other crops have seen a slight rise in both productivity and acreage.

Area and production scenario of maize

The crop acreage in the state is growing by year in the present scenario. The area increased by 52%, production by a massive 600% and the yield by 360%

from 2007-08 to 2014-15 in the state. It is mainly because of shifting of maize cultivation from OPVs and composites to Hybrids. Table 1 shows the change in area, production and yield of the crop year wise in the major districts of the state.

Table 1. District wise Area, Production and yield of Maize over years.

District		2007-08	2008-09	2009-10	2010-11	2011-12	2012-13	2013-14	2014-15
Dhubri	A	95	131	146	423	1148	1140	1176	1195
	P	56	77	88	254	752	841	5098	3209
	Y	600	600	600	600	655	737	4334	2685
Kokrajhar	A	624	668	1055	768	925	935	930	935
	P	324	346	549	399	512	613	2753	2230
	Y	520	520	520	520	553	656	2960	2385
Bongaigaon	A	157	175	172	177	188	206	421	894
	P	95	105	105	107	132	160	1839	2598
	Y	609	608	609	608	697	782	4366	2906
Chirang	A	329	418	406	478	356	516	397	297
	P	200	254	247	291	250	410	770	864
	Y	609	609	609	609	703	794	1939	2908
Goalpara	A	125	130	145	150	114	59	195	245
	P	73	75	84	87	75	39	410	582
	Y	580	580	580	580	658	653	2100	2375
Kamrup (r)	A	292	198	208	217	241	246	245	246
	P	181	122	129	134	172	197	889	713
	Y	620	620	620	620	711	801	3628	2898
Kamrup (m)	A	34	48	45	58	60	60	64	66
	P	21	30	28	36	42	42	228	167
	Y	620	620	620	620	700	697	3562	2535
Nalbari	A	85	30	35	35	46	110	115	70
	P	64	22	26	26	39	93	496	215
	Y	750	750	750	750	848	844	4316	3070
Barpeta	A	46	64	331	76	58	63	67	70
	P	29	41	216	49	44	55	123	221
	Y	650	650	650	650	775	867	1829	3156
Baksa	A	205	197	193	225	555	285	510	340
	P	154	148	145	169	472	272	2351	1183
	Y	750	750	750	750	850	956	4610	3478
Sonitpur	A	901	642	444	504	521	1151	866	1420
	P	524	373	253	287	338	828	4560	7323
	Y	582	582	569	571	648	720	5266	5157
Darrang	A	271	277	1250	1745	2237	2500	1647	5085
	P	163	167	754	1052	1528	1920	9882	14207
	Y	603	603	603	603	683	768	6000	2794
Udalguri	A	333	447	559	527	530	507	605	595
	P	208	279	349	329	375	404	1885	1723
	Y	625	625	625	625	708	796	3115	2896
Nagaon	A	398	88	415	435	428	453	529	593
	P	269	50	278	294	331	392	2074	1826
	Y	676	584	671	675	770	864	3920	3079

*A: Area ('000 ha); P: Production ('000 t) and Y: Yield: kg/ha.

In addition to some of the major problems such as land fragmentation, lack of modern technologies, or continued reliance on rain for irrigation, there are several other problems that hinder the development of agricultural sector in Assam are as follows:

1. *Natural calamities:* Floods and dry spells are the principal natural disasters faced by the farmers in Assam every year. The principal source of floods is the Brahmaputra river and its tributaries. Although it has been decades since the proposal to dredge the Brahmaputra came out, its progress and impacts are unknown. The loss of crop, livestock, house, cultivable land, and human lives are common during the yearly flood, which also takes a toll on human spirit.
2. *Capital deficiency:* Commercial capital, *i.e.* loans from banks or other credit agencies, generally is not accessible to farmers in Assam. As a result, borrowing from unscrupulous lenders, who are not regulated by the state, at an extremely high interest rate is common in the state. Tragically, in many instances some borrowers lose their livelihood, i.e. their cultivable land, to these devious lenders. However, the need for agricultural loan remains largely unfulfilled in Assam.
3. *Marketing problems:* Agricultural markets in Assam are under developed. Farmers sell to the nearest dealer/buyer immediately after harvesting when the price is at the lowest, instead of trying to find the best market for their products. Geographical isolation, weak transportation and lack of communication, poor marketing facilities, poor or nonexistent market intelligence (e.g. information on price and place to sell) are some of the principal marketing related problems. In terms of the role of government, earlier studies have shown that government efforts are more focused on collecting revenue (in the form of tax in the daily or weekly markets or through check gates) than facilitating the marketing of agricultural products in Assam (Bhuyan, 1990; Bhuyan *et al.,* 1990; Bhuyan *et al.*, 1988). A reorientation of the government's focus from revenue collection to marketing facilitation will be necessary if farmers are to be benefitted.
4. *Noneconomic factors:* Lack of education, ignorance about the changing economic conditions, outdated thinking, prejudiced cultural values, disturbed law and order situation, and lack of rigorous legislative and administrative machinery are some of the principal noneconomic factors that hinder agricultural development in Assam. While the first four factors are mostly socio-cultural, which may take time to change for the better, the later two factors are political. All these factors adversely affect agricultural as well as economic development of the state.

Key production problems of maize in the state: As mentioned above, the natural calamities and marketing problem are two main key constrains for maize production in Assam. The crop is mainly grown in the *Kharif* season, and sown during June-August. During this time the state receives most of the rainfalls and the areas of low to medium lands get inundated. Hence, sowing time gets delayed. Even if the sowing is done, the continuous rainfall makes the crop vulnerable for early damage during germination to seedling stage due to water logging. Again in some parts though *Rabi* maize is practiced, germination got affected due to low temperature and drought.

As there is no regulated market for maize, the growers have to sell their produce in local market or to some middle man. Thus the farmers do not receive the appropriate price for the crop. Now a days some small industries are coming up (like feed industries), which are making the marketing a bit easy for the growers. Still, the cases of middle agents are there where the farmers have to sell their products at lower price in the process.

AICRP centre and mandate

AICRP on Maize centre in Assam was stared on 1st November, 1979 at Assam Agricultural University Campus, Jorhat. Later on it shifted to Gossaingaon in the year 2010.

The mandates of the centre are as follows:

i. To undertake research for genetic enhancement in maize and its important attributes

ii. To develop efficient package of practices for increasing maize production in the state

iii. To popularize the crop in the region through frontline demonstration of high yielding varieties

Popular hybrids: The popular varieties presently cultivated in the region are from private sectors only. The Monsanto varieties under DKC series are performing and are the front-runners in the popularity chart. Maize hybrids from Adavanta seed company Ltd. are also doing good in the region. Some of the popular cultivars are: DKC 9081, DKC 9035, DKC 9120, Jwala, PAC 751, PAC 712, PAC 740, Kavari, Hishel, Bioseed 9662, CP 808 etc.

Changes in cropping system over the years: The cropping system in a region has changed gradually with the change of environmental factors, peoples' choice of food materials, availability of resources and ease of cultivation process etc. The cropping system of the state has been changed with time in a gradual manner over the years. The growing population and limited resources compelled people to grow more than one crop in a single plot of land. As per the crop maize, it is now being grown in place of summer rice in many areas of the state. The *Kharif* maize is also getting popularity as it needs less water than some other major crops. In places people are practicing *Kharif* maize followed by rapeseed and mustard and then *Rabi* maize. As per land situation, maize crop is being cultivated in upland to medium land situation rather than the typical rice growing low land areas. Thus in the changing scenario maize is replacing the rice-rice cropping system in recent days.

Table 2. Production technology and package of practices.

Production practice/technology	**Recommendation**					
Selection of Site and Land Preparation	Well drained sandy and sandy loam soils .					
Seed Rate	18-22.5 kg /ha					
Seed Treatment	Seeds should be dressed with carbendazim @ 2g/kg of seed or captan @ 2.5-3 g/kg of seed					
Sowing Time (*kharif*)	**LBVZ**	**UBVZ**	**CBVZ**	**NBPZ**	**BVZ**	**HZ**
	Feb-Apr	Mar-Apr	Mar-May	Feb-Apr	Mar-May	Apr-May
Sowing Time (*rabi*)	September-October and may extend to middle of November					
Method of Sowing	Seeds should be dibbled at a depth of 3-4 cm in rows at 65-75 cm apart and at a distance of 20-25 cm from seed to seed					
Manures and Fertilizers	Compost or FYM @ 5t/ha should be applied					
	Nutrient	**Requirement (kg/ha)**	**From**	**Fertilizer requirement (kg/ha)**		
	N	60	Urea	134		
	P_2O_5	40	SSP	250		
	K_2O	40	MOP	67		
	For Hills Zone, 90:40:40 kg / NPK per hectare.					
Method of Fertilizer Application	FYM or compost should be applied during land preparation. The entire quantity of SSP and MOP and half of the total urea is to be applied in furrows (8-10 cm deep) and covered with 4-5cm of soil. Sowing of seed should be done at least 2days after fertilizer application.The remaining half of urea should be top dressed in two equal doses followed by earthing up in each case. The first ¼ at 35 days after germination or when the plants are at knee high stage and the second ¼ at the tassel initiation stage of the crop i.e. 45-60 days after germination or at the time of elongation of the flag leaf whichever is earlier.					
Weeding and Interculture	Light hoeing and earthing up should be done. For further and adequate control of weed, atrazine should be applied @ 0.5–1.0 kg a.i/ha in 900 litres of water as pre-emergence spray.					
Irrigation and Drainage	Application of irrigation at the time of soil moisture deficit ensures optimum yield. Silking stages and grain filling period are the two critical stage of the					

	maize crop. In *Rabi* irrigation is needed at grand growth, tasselling and grain setting stage. Excess water can be drained out through surface drainage of 25 cm with, 15 cm deep spaced at 6 m distance.
Harvesting	Harvesting should be done as soon as the husk covers turn brown and the silks are completely dry. Ears (cobs) should be thoroughly dried in the sun before shelling.

*LBVZ: Lower Brahmaputra Valley Zone; UBVZ: Upper Brahmaputra Valley Zone; CVZ: Central Valley Zone; NBPZ: North Bank Plain Zone; BVZ: Barak Valley Zone; HZ: Hill Zone

Pest and disease management

Stem borer

When stem borer infestation is noticed any one of the following insecticides can be sprayed.

Insecticide	Quantity	Volume of water (l /h)	
		Power sprayer	Hand sprayer
Phosphamidon 100 EC	500 ml	200-250	400-500
Fenitrothion 50 EC	1.0 l	200-500	400-500

In addition to these all the dead hearts should be pulled out and destroyed to kill the lingering stage, of the pest in the stubbles.

Termites

Termites take a heavy toll of the crop especially in the *Rabi* season.

Aphids, cob borers, jassids and mites

Methyldematon 25 EC, monocrotophos 36 WSC, phosphamidon 100 EC should be sprayed @ 1ml/l against these pests.

Maydis leaf blight

Use of protective fungicides of thiocarbamate group (Mancozeb, zineb) @ 2.0-3.0 kg a.i./ha in 500-600 litres of water at the time of appearance of the disease symptoms or at knee high stages.

Sclerotial disease

In areas where banded sclerotial disease (*Rhizoctonia solani*) is noticed, the plants should be sprayed with carbendazim 0.05% (0.5 g/lit water) @ 700 l/ha of spray solution at an interval of 12-15 days. Matured cobs can be protected from bird damage by tying cobs with leaves of the same plant.

Turcicum leaf blight (southern corn blight)

The disease can be controlled by 2-3 foliar sprays with 0.25% to 0.3% mancozeb or zineb at 10-15 days interval. Application of urea followed by a light irrigation also helps to minimize the disease as it sometimes becomes serious due to poor management.

Seed rot and seedling blight

The best recourse to these problems is to sow certified seeds from a reliable source, which is pretreated with desired fungicides. If one wants to use his own seed, seed treatment is a must.

Pythium stalk rot

Apply Captan or Thiram @ 2.5 kg/100 litres of water/ha at the lower internodes of plants at 30-35 days after planting. Drain out excess rain water from the field.

Banded leaf and sheath blight

In areas where this disease is noticed the plant should be sprayed with carbendazim @ 0.05% @ 700 l/ha at an interval of 12-15 days.

Bird damage

The mature cobs can be protected from bird damage by tying the cobs with leaves of the same plant. Shinning, coloured strips may also be used to scare away the birds.

Post harvest management

Harvesting immediately after a shower should be avoided. Ears (cobs) should be thoroughly dried in the sun before shelling or storing unshelled. Keep the grain as clean as possible. Dry maize on cement floor or use tarpaulin to reduce chance of contamination.

At home, do not first heap the cobs in any room, kitchen or in the yard because this will expose them to all the dangers that cause post harvest losses. Transfer them to the drying place (like the crib) immediately. Dry on concrete or canvas not directly on bare soil.

During storage, the grain must remain dry and clean. Grain storage can be extended for up to 2 years without any significant reduction in quantity and quality. However, the majority of farmers sells off their maize grains cheaply soon after harvesting due to anticipated losses in storage and later buy food at exorbitant prices. There are improved storage structures that can prolong the storage duration until market prices for grains are favorable.

Frontline Demonstration in Maize

On an average more yield was observed in *Kharif* compared to *Rabi* season.

Table 3. Information of FLDs in maize conducted during *Kharif* seasons.

Year	FLD (ha)	Variety	Yield (q/ha)
1999	19	Vijay	24.4
2000	22	Vijay	20.89
2001	41	Ganga-11, Deccan-105	25.28
2002	19	Navjot	25.22
2003	47	Mahi Kanchan	33.36
2004	33	Vijay, Navjot	15.93
2007	15	Vijay	37.00
2008	10	Vijay	35.00
2009	60	PEHM 2 and Shaktiman 4	68.00
2010	15	Navjot	36.98

Table 4. Information of FLDs in maize conducted during *Rabi* seasons.

Year	FLD (ha)	Variety	Yield (q/ha)
1998-99	16	Vijay	11.94
1999-00	17	G-11, D-105	26.71
2000-01	40	Vijay	28.38
2001-02	1	Vijay	29.15
2013	04	Bio seed Hybrid	30.24
2014	80	HQPM-5 and PAC-740	35.7

Key R&D challenges and probable solutions: The crop is an emerging crop in the state and constant increase in acreage is being observed in recent years. Main factors determining its growth as a major crop can be outlined as followed.

Lack of suitable hybrids/composites: People are growing different varieties in the region. But a proper and suitable variety (hybrids/composites) is not established yet. The farmers are sowing whatever is available in the market.

Presently some initiatives are being taken by the Agricultural University to find out suitable hybrids for the region. Some private sector companies are also coming up with some good hybrids.

Improper crop management: The farmers are not following a definite package of practices. The fertilizer dose, protection measures etc. are not adhered to and thus the yield potential is not exploited to the fullest extent.

Awareness programmes/trainings have now been organized by the University Research Stations, KVKs and also by state Agricultural department to educate the people in maize growing.

Seasonal suitability: As the major season is *Kharif*, choice of varieties/ technologies suitable for the season is limited. The crop is affected with continuous rainfall during sowing time and with drought during cob formation stage.

Calibration regarding sowing time and other agronomic practices is necessary to solve the problem. Concurrently, *Rabi* maize also can be popularized in irrigated areas.

Researchable issues in maize

Maize being cultivated mainly in the Lower Brahmaputra Valley Zone, continuous feedback from farmers is received time to time. The researchable issues thus formulated are as followed:

i. Suitable hybrid for the area
ii. An effective fertilizer dose for the hybrids.
iii. Hybrids resistant or tolerant to pest and diseases.
iv. Technologies to protect the harvested products from stored grain pests.

Summary

Maize being a crop of high potential can give a boost to the economy of the state. The main limiting factors like suitable hybrids, seasonal heavy rainfall and proper agronomic practices could be dealt by a systematic study. The industrial use of the produce is helping the farmers to boost their cultivation.

References

Bhuyan, S.A. (1990). State intervention in agricultural marketing: Is it necessary? *Agric Marketing*. 33(1): 213.

Bhuyan, S., Demaine, H. and Weber, K.E. (1990). "Market regulation or regulated market? The case of Assam, India," HSD Monograph no. 19, Asian Institute of Technology, Bangkok, Thailand.

Bhuyan, S., Urs, S.D. and Weber, K.E. (1988). "Marketing farm produce: An efficiency test of traditional and regulated markets based on evidence from Assam, India."*Eco Bull Asia Pacific*. 39(2): 4655.

13

Maize Research and Development in Maharashtra

S.R. Kulkarni, S.S. Mahadik, V.M. Londhe, P.K. Pawar S.A. Patil and M.S. Pilane*

Mahatma Phule Krishi Vidyapeeth, Rahuri-413 722, Ahmednagar, Maharashtra
Corresponding Authors' Email: mipkop@yahoo.com

Maharashtra is the third largest state in terms of area in the country located in the north-centre of peninsular India. It links the southern part of the northern plains to the southern peninsula. Maharashtra occupies the western and central part of the country with total 3.08 lakh sq. km geographical area and has a long coastline stretching nearly 720 kilometers along the A*Rabi*an Sea. The prominent physical trait of the state is its Plateau character, which has up to 1350 meters high altitude. Physiographically this state may be divided into three natural divisions-the coastal strip (the Konkan), the Sahyadri or the Western Ghats and the Plateau. The Sahyadri mountain ranges provide a physical backbone to the state on the west, while the Satpuda hills along the north and Bhamragad-Chiroli-Gaikhuri ranges on the east serve as its natural borders. It has a geographical area of 307713 sq. km and is bounded by north latitude 15°40' and 22°00' and east longitudes 72°30' and 80°30'.

NBSS & LUP has divided the state into 356 soil mapping units and published a map of the soils of Maharashtra. It has broadly categorized as soils of Konkan coast, soils of Western Ghats, soils of upper Maharashtra, soils of lower Maharashtra. The soil status of Maharashtra is residual, derived from the underlying basalts. In the semi-dry Plateau, the regur (black-cotton soil) is clayey, rich in iron and moisture-retentive, though poor in nitrogen and organic matter. The rainy Konkan and the Sahyadri Range, the same basalts give rise to the brick-red laterites, which are productive under a forest cover, but readily stripped into a sterile varkas when devoid of vegetative cover. The soil in the Deccan Plateau is made up of black basalt soil. This type of soil is rich in humus. The soil is commonly known as the black cotton soil because it is best suited for the cultivation of cotton. The volcanic action which had taken place in the Deccan region has given rise to the soil texture and composition. These igneous rocks break down into the black soil which is very fertile. The Wardha-Waliganga river valley has old crystalline rocks and saline soils which make the soil infertile. This type of soil has a natural resistance to wind and water erosion because it is rich in iron and granular in structure. A very important advantage of this type of soil is that it can retain moisture. This makes the soil very reactive to irrigation.

Southwest monsoon is the main source of rainfall for the state. However, rainfall in state varies considerably; heavy rainfall in the coastal region to scanty rains in rain shadow areas in the central part to moderate rains in eastern parts of the state. The Rivers in Maharashtra are eastward flowing towards Bay of Bengal except Tapi which is flowing towards A*Rabi*an Sea at west.

For administrative purpose, Maharashtra with thirty-six districts is divided into six divisions (Konkan, Pune, Nashik, Aurangabad, Amravati and Nagpur) with each division consisting of minimum five and maximum eight districts.

Although Maharashtra is a highly industrialized state of India, agriculture continues to be the main occupation of the people. About 61% of the people directly or indirectly depend on agriculture and allied activities for their livelihood. Principal food crops of the state are rice, jowar, bajra, wheat, maize, tur, mung, urad, gram and other pulses. Maharashtra is a major producer of oilseeds like groundnut, sunflower, soybean etc. Major cash crops of the state are cotton, sugarcane, turmeric and vegetables. The growth rate of different crops in the state with respect to area, production and productivity varied depending on availability of resources and adoption of improved technologies. Horticulture has a very important place in this state, varieties of fruits like mango, orange, banana, grape, cashew nut etc. are produced in this state. The Nagpur oranges and Alphonso mangoes are very famous. There were about 10.91 lakh hectares of land under horticulture. Fishing is an important activity of this state.

In India, maize is cultivated throughout the year in one or other parts of the country. Earlier in Maharashtra, maize has been considered as poor man's crop, but now Maharashtra is one of the emerging predominant maize growing states in India. Presently, it accounts about 9 per cent of the total maize area with almost equal contribution to the total maize production in the country. Due to its multifarious uses as industrial food and feed crop maize is reached in every home of the society. The specialty corn like baby corn, sweet corn and popcorn area getting popular and the maize production scenario is changing from a neglected crop to a commercial crop. Thus, in Maharashtra maize is now occupying the place in the rich communities. As a result, the area under maize crop is increasing in Maharashtra.

Trends in APY of maize in the Maharashtra state from 2000 to 2015

Maize is mainly grown in *Kharif* season (8.06 lakh ha), *Rabi* season (2.40 lakh ha) and *Summer* season (0.36 lakh ha) in Maharashtra (2014-15). In the last fifteen years, the area of maize in Maharashtra has increased from 3.29 lakh ha to 10.59 lakh ha and the production increased to the extent of 22.03 lakh tones compared to 2.99 lakh tones before fifteen years (Table 1). Similarly, the productivity is also reached to 2080 kg/ha from 955 kg/ha.

Table 1. The trend in maize area, production and productivity in Maharashtra (2001-2015).

Year	Area (lakh ha)	Production (lakh ton)	Yield (kg/ha)
2000-01	3.29	2.99	955
2001-02	3.26	5.87	1800
2002-03	3.71	7.43	2003
2003-04	3.62	7.40	2045
2004-05	4.18	7.37	1863
2005-06	4.73	9.96	2106
2006-07	5.81	11.27	1941
2007-08	6.72	17.89	2665
2008-09	6.35	14.5	2258
2009-10	7.94	18.28	2302
2010-11	8.62	25.71	2961
2011-12	8.81	24.33	2762
2012-13	8.22	18.24	2219
2013-14	10.01	27.29	2727
2014-15	10.59	22.03	2080

District-wise decadal trends in area, production and productivity of maize (2000-2015)

Maize is grown in all the districts of the state in varying degrees. During 2001-02 to 2014-15, the area under maize crop has increased almost 2.4 times in the state (Table 2). The major maize growing districts are Ahmednagar, Aurangabad, Buldhana, Dhule, Jalna, Jalgaon, Nandurbar and Nasik. The maize yield in these districts has increased consistently during the past 10 years due to increasing adoption of hybrids. The highest increase in maize area was observed in Aurangabad, Buldhana, Dhule, Jalgaon, and Nasik districts.

Table 2. The district-wise trend in area, production and productivity of maize in Maharashtra.

District	2001-2010			2010-15		
	Area (00' ha)	Production (00' ton)	Yield (kg/ha)	Area (00' ha)	Production (00' ton)	Yield (kg/ha)
Nasik	659	1524	2177	1561	4980	2999
Dhule	186	377	1933	711	1919	2541
Nandurbar	224	397	1649	322	741	2411
Jalgaon	378	973	2439	983	3061	2962
Ahmednagar	291	579	1785	714	2052	2495
Pune	168	330	2069	295	907	2833
Solapur	283	455	1581	453	985	1972
Satara	143	263	1820	237	488	1960

Sangli	237	461	1933	486	1264	2460
Kolhapur	90	196	2114	90	232	2449
Aurangabad	974	2311	2304	1641	4538	2566
Jalna	584	1219	2025	661	1405	1877
Beed	74	89	1206	120	124	1035
Latur	53	56	1057	93	152	1467
Osmanabad	133	114	825	218	192	979
Nanded	12	16	1301	19	22	1270
Parbhani	23	27	1148	29	33	1141
Hingoli	9	11	1192	18	19	1210
Buldana	414	915	2048	490	1098	2094
Akola	15	32	1955	7	18	1934
Washim	1	3	2070	5	8	1504
Amaravati	14	29	2009	19	34	1876
Yavatmal	3	7	2041	9	9	1178
Wardha	1	2	992	3	3	273
Nagpur	4	8	1929	4	6	1481
Bhandara	1	1	1201	0	0	0
Gondia	1	3	1979	3	3	1150
Chandrapur	0	1	271	0	0	267
Gadchiroli	25	47	1886	29	55	1361
State Total	5000	10446	1688	9184	24353	1715

About 65 per cent of the total crop area contributing 70 per cent to the total maize production in the state showed high growth in yield during the past 10 years. During the period 2000-01 to 2009-10, the maize area in Aurangabad, Buldhana, Jalgaon and Nasik districts increased exponentially from 75 to 126 thousand ha, 21 to 85 thousand ha, 10 to 86 thousand ha and 23 to 144 thousand ha, respectively. The major maize growing districts in Maharashtra from last five years are Aurangabad (1.64 lakh ha), Nasik (1.56 lakh ha), Jalgaon (0.98 lakh ha), Ahmednagar (0.72 lakh ha), Dhule (0.71 lakh ha), Jalna (0.66 lakh ha), Buldhana (0.49 lakh ha), Sangli (0.49 lakh ha), Solapur (0.45 lakh ha) and Nandurbar (0.32 lakh ha). The brief details of production and productivity of these districts are given in Table 3.

Table 3. The major maize growing districts during last five years

District	Area (lakh ha)	Production (lakh tones)	Yield (kg/ha)
Aurangabad	1.64	4.54	2566
Nasik	1.56	4.98	2999
Jalgaon	0.98	3.06	2962
Ahmednagar	0.74	2.05	2495
Dhule	0.71	1.92	2541
Jalna	0.66	1.41	1877
Sangli	0.49	1.26	2460
Buldhana	0.49	1.10	2094
Solapur	0.45	0.99	1972
Nandurbar	0.32	0.74	2411

Considering the rising interest of farmers in maize cultivation, the State Agriculture Department has identified some private companies in 2013 to start a value chain development programme in 13 districts (Aurangabad, Jalgaon, Pune, Ahmednagar, Nashik, Buldana, Sangli, Jalna, Dhule, Nandurbar, Satara, Solapur and Osmanabad) for training of farmers on adoption of agricultural practices and for providing market services.

Utilization of maize in the state

In Maharashtra about 55 per cent of the grain produce concurrently is used for food purposes, about 14 per cent for livestock, 18 per cent for poultry feed, 12 per cent for starch and 1 per cent for seed. The starch is the main product of a maize processing unit, which is consumed in various other industries like food, pharmaceuticals, textiles, paper, hotels and restaurants, etc. The other products of starch industry include Gluten, Germ, Fibre (husk) and Corn Steep Liquor. Gluten has great demand in animal feed industry because of its high protein content (70%). Germ is used to extract edible germ oil which contains low cholesterol. The animal feed manufacturers uses fiber, mainly the husk; it has demand as animal feed in wet form itself. Corn Steep Liquor is one of the substitutes for culture media for manufacturing of antibiotics and other microbial production systems. However, there are very few private wet milling industries are currently running in the state with 150-200 tonnes per day capacity. Maize processors directly market their products to the consumers like pharmaceutical industries, hotels, textiles, paper industries, etc. and through traders as well and Mumbai is the major market for processed maize products. Most of the starch manufacturers of Gujarat, Maharashtra, Punjab, etc., have their marketing offices in Mumbai. Hence, Mumbai is the major trading centre for corn starch in Maharashtra.

Major cropping systems and production ecologies

Maharashtra state has nine agro-climatic zones *viz.*, South Konkan Coastal Zone, North Konkan Coastal Zone, Western Ghat Zone, Transition Zone-1, Transition Zone-2, Scarcity Zone, Assured Rainfall Zone, Moderate Rainfall Zone, and Eastern Vidarbha Zone. Several cropping systems are being practiced by the farmers of Maharashtra depending on agro-climatic zones. The major cropping systems are rice-wheat, rice-rice, rice-groundnut, rice-pulses, pearl millet-wheat/Pearl millet mustard, maize-wheat, sorghum-wheat, sugarcane-wheat/sugarcane-maize, cotton-wheat, soybean-wheat etc.

Genesis of AICRP centre/s

The AICRP centre at Kolhapur was established with the following mandate namely development of high yielding new maize hybrids and composites with desirable agronomic traits, resistance to biotic and abiotic stresses, evaluate the newly developed maize hybrids and composites from IIMR and find its suitability for the state, development of suitable agro-techniques for cultivation of newly developed maize hybrids and composites, study the resistance of newly developed maize genotypes to the stem borer, *Chilo partellus* (Swinhoe) under artificial condition, find out suitable control measures for insects and pests prevailing in the state and development of suitable maize based cropping systems.

Key production problems in state for maize

The increase in maize productivity depends on several factors, which are more than simply increasing the adoption of hybrids. The increase in profitability, though slow, maize cultivation has helped thousands of small and marginal farmers in ensuring their livelihood security. It is one of the reasons that the marketed surplus of maize crop has increased in the states, and slowly this crop is moving towards becoming a cash crop. However, maize crop faces several challenges namely dependency on rainfed conditions, insufficient irrigation facilities, non-availability of quality seeds on time, non-availability of stress tolerant varieties/ hybrids. The dependency on rain along with aberrant weather conditions reduces the maize yields in different regions. In addition, improper soil selection (mostly grown on marginal and sub-marginal soils), poor weed management, lack of farm mechanization, improper insect pests and disease management, indiscriminate use of chemical fertilizers and low use of FYM also add to increased cost of cultivation and reduced profitability due to lower yields. The sudden emergence of biotic and abiotic stresses in the state under changing climate scenario and insufficient transfer of technology for maize farmer is other problems which hamper enhancement of production. However, there is scope to increase processing to guard against market price fluctuation through networking with industries.

AICRP technologies

The following cultivars are released and notified for commercial cultivation. The brief details of the cultivars released by AICRP centre located in the state are given below (Table 4).

Table 4. The cultivars released by AICRP on maize centre, MPKV, Kolhapur.

Type	Cultivar	Year	Released by	Other Characteristics
Yellow	Hunis	1977	SVRC	Selection from 30 parents from Hungeri (USA) and India. Yield: 45-50 q/ha.
	Manjri	1980	SVRC	Reddish yellow grain. Yield: 40-50 q/ha
	Karveer	2005	SVRC	Bold orange flint grain. Yield: 52-55 *(Kharif)*; 65-68 *(Rabi)*
	Rajarshi	2009	SVRC	High yielding, orange yellow, semi-flint, mid-late maturity, suitable for *Kharif* and *Rabi*, Yield: 70-75 *(Kharif)*; 95-100 *(Rabi)*

	Phule Maharshi	2016	SVRC	High yielding orange yellow, semi-flint hybrid. Yield: 76.05 *(Kharif)*, 87.61*(Rabi)*
White	African tall	1982	SVRC	Green fodder: 60-70 MT,Grain: 40-50 q/ha
	Panchganga	1986	SVRC	White grain, short duration, suitable for intercrop
Sweet corn	Phule Madhu	2016	SVRC	Brix: 14.89

Seed production

AICRP on maize, Kolhapur, Maharashtra, produces breeder seed of the cultivars released by the centre for augmenting the certified seed production and distribution to farmers. The major agencies involved in augmenting the seed production are Maharashtra State Seed Corporation, National Seed Corporation, private agencies (through MoUs). The brief details of the seed produces and distributed during the period are given below (Table 5).

Table 5. The seed production report during 2011 to 2016.

Season	Seed produced by MPKV, Kolhapur (q)		Seed distribution in the state			
			MSSC	NSC	Private	Total
Kharif 2011			5196	100	92341	97637
Kharif 2012			2447	121	94625	97193
Rabi 2012			6693	0	11240	17933
Kharif 2013	0.45	GPM-456 (Female)	2396	0	140693	143089
Rabi 2013-14	0.3	GPM-342 (Male)	13233	0	20050	33283
	125.19	KMH 22168 (Rajarshi)				
	180.2	African tall				
Kharif 2014	3.95	GPM-456 (Female)	3099	824	107197	111120
Rabi-2014	2.65	GPM-342 (Male)	38	3420	30833	34291
	96.82	African tall				
Kharif 2015	3	GPM-456 (Female)	66	3614	111295	114975
Rabi 2015-16	3.5	GPM-342 (Male)	3787	0	28058	31845
	2.4	QMI 1403 (Female)				
	4.9	QMI 1401 (Male)				
	104.1	QMH 1025				
	60.35	African tall				
	3.1	QMHSC 1182				
Kharif 2016 (Tentative)	4.9	QMI 1401 (Male)	414	3708	132291	136413
	1.66	QMI 1403 (Female)				
	1.3	QMH 1025				

Changes in cropping system

There was declining trend of area under food crops (Table 6). However, rice, bajra, tur showed higher growth in area under cultivation, but growth in area of rice was higher during 1971-80; the same condition was of tur, pulses and total food grains (Table 7). It may because of the green revolution in Maharashtra.

Table 6. The decadal average area ('00 ha), production ('00 MT) and yield (kg/ha) of main crops of Maharashtra

Crop	1961-1970			1971-1980			1981-1990			1991-2000			2001-2010			1961-2010		
	Area	Prod.	Yield	Area	Prod.	Yield	Area	Prod.	Yield	Area	Prod.	Yield	Area	Prod.	Yield	Area	Prod.	Yield
Rice	13367	13078	978	14002	17591	1249	15084	22488	1494	15318	24374	1593	15328	23730	1555	14620	20252	1374
Wheat	8662	3909	226	10228	7503	707	9522	8483	889	8072	10020	1230	9002	12976	1357	9097	8578	882
Kharif Jowar	25550	15621	613	27634	21387	772	28988	27861	969	23145	24752	1025	15891	16947	1000	24242	21313	876
Rabi Jowar	31588	16733	478	33594	14238	424	35971	16259	452	32692	18654	569	31314	16960	540	33032	16568	493
Bajara	18360	5232	281	17050	5736	329	18178	7659	422	18114	12647	696	15430	9867	650	17426	8228	476
Other cereals	134	4935	1010	2475	4666	936	4922	4357	486	4723	5803	1221	6870	8304	1218	3814	5613	1054
Total cereals	105493	60484	1795	107079	68467	1734	116065	89396	1325	102065	104443	995	92593	92819	1018	104659	83122	1381
Gram	3563	1086	521	4044	1304	498	5811	2174	714	7269	4124	950	9926	5452	1143	6123	2828	765
Tur	5841	3386	583	6141	3213	521	7780	4833	616	10216	6017	588	10790	7229	699	8153	4936	602
Other Pulses	16814	5826	516	16978	5368	476	16843	5230	461	16433	7341	654	16006	2618	246	16615	5277	471
Total Pulses	24350	8255	518	26611	7204	409	30460	10408	490	34313	17274	684	34661	18999	723	30079	12428	565
Total Food grains	130033	65559	2016	132626	77684	1962	139499	101222	1409	135744	121173	1145	127377	112290	1164	133056	95586	1539
Sugarcane	1629	10893	142	2527	18240	123	3369	27683	118	4930	324362	118	5688	468988	132	3628	NA	NA
Cotton	27019	12486	109	27269	11304	93	27109	14585	128	26991	23355	128	26536	31640	191	26208	18674	129
Groundnut	10437	6960	269	8263	5442	662	6978	4945	120	6243	4987	814	5433	4734	924	7471	5414	757
Safflower	5896	3004	NA	4633	2296	NA	2972	1523	713	NA	NA	497	NA	NA	514	4500	2274	504
Soybean	NA	0	NA	NA	0	NA	3307	3910	1071	6416	8528	1085	16961	18543	1138	10291	11931	1106
Sunflower	NA	NA	NA	NA	NA	NA	4061	1960	469	4670	2134	447	3359	1825	673	4022	1976	545
Total Oilseed	18274	17449	958	16184	16710	1123	22878	16591	731	26275	19625	742	29381	28058	941	22598	19686	899

Sources: Crop Report of Maharashtra, Agricultural Statistics Information, Epitome-II, Govt. of Maharashtra

Table 7. The decadal cumulative growth rate (%) of area production and yield of main crops of Maharashtra

Crop	1961-1970			1971-1980			1981-1990			1991-2000			2001-2010			1961-2010		
	Area	Prod.	Yield	Area	Prod.	Yield	Area	Prod.	Yield	Area	Prod.	Yield	Area	Prod.	Yield	Area	Prod.	Yield
Rice	0.2	-0.6	-0.8	1.3	6.5	5.2	0.2	0.4	0.2	-0.7	1.2	2	-0.5	0.3	0.7	0.3	1.6	1.2
Wheat	-0.6	-1.4	2.8	4.7	12.5	7.9	-4.1	-2.2	2	3.3	5.1	1.9	3.1	4.2	1.1	-0.2	2.7	3
Kharif Jowar	-0.2	-3.2	-3.3	3.1	17.4	14.2	-0.7	-2.4	-0.8	-4.8	-4.1	0.8	-2.7	-1.3	-0.5	-1.2	0.4	1.3
Rabi Jowar	-1.8	-1.9	-0.2	0.7	8.7	8.1	0.1	2.3	2.2	0.2	0.6	0.5	-0.4	2.8	3.1	-0.2	0.4	0.6
Bajara	2.7	3.1	0.6	-1.1	4	5.1	2	5.1	3.1	-1.2	1.5	2.6	-0.7	-1.4	-0.7	0.2	2.1	2.4
Other cereals	0.3	1	0.7	-0.9	-5	-4.1	-0.3	0.1	0.3	2.9	5.1	2.1	2.2	3.5	1.3	0.6	1.2	0.6
Total cereals	0.1	-1.6	-1.6	1.5	11.2	9.6	-0.8	0.8	1.7	-1.1	0.7	1.7	-0.2	1.3	1.4	-0.3	1.4	1.7
Gram	-2.1	-5.7	-1.8	2.3	7.5	5.2	2.1	8.2	6.1	1.9	7	5.1	-5.2	1.4	6.6	2	4	2
Tur	1.8	-2.4	-4.2	2.5	6.3	3.8	2.8	5.4	2.6	0.1	6.1	6.1	-2	-2.1	-0.1	1.6	2.1	0.5
Other Pulses	-0.1	-0.4	-0.2	-0.2	-0.6	-0.3	2.7	7.5	4.8	-1	2.1	3.1	-0.6	-2.2	-1.6	-0.1	-1.8	-1.6
Total Pulses	1.3	-1.8	-3.1	1.4	2.2	0.8	2.4	11	8.7	0.9	4.4	3.5	-1.6	-1.6	-0.1	1.3	2.5	1.2
Total Food grains	0.2	-1.2	-1.3	1.5	10.9	9.4	0.3	1.7	1.4	-0.7	1	1.7	-1.8	0.6	2.4	-0.1	1.6	1.7
Sugarcane	3.4	2.4	0.9	5.9	6.6	0.4	-0.9	-3	-1.9	2.9	3.6	0.8	1.4	3	1.7	3.2	NA	NA
Cotton	0.7	-0.7	-1.4	-1.3	7.1	8.4	-0.5	2.8	3.3	1.7	6.4	4.7	-0.2	2.5	2.7	0.1	2.5	2.4
Groundnut	-2.8	-3.2	-0.4	-0.09	4.2	4.9	3	3.1	3.1	-4.8	-0.3	4.5	2.4	1.3	-1.1	-1.6	-0.7	0.9
Safflower	NA	NA	NA	NA	NA	NA	2.1	1.9	-0.2	-4.6	-5.4	-0.8	3.6	7.7	4.1	-0.2	-3	0.1
Soybean	NA	NA	NA	NA	NA	NA	11.5	41	29.8	18.3	20.4	2.1	-4.9	0.6	5.5	9.4	10.1	0.7
Sunflower	NA	NA	NA	NA	NA	NA	10.4	40.1	29.6	-1	-6.6	-5.7	-2	5	7.1	-2	-0.2	1.8
Total Oilseed	-1.5	0.6	2.1	-0.1	-3.5	-3.4	3.6	-1.1	-4.7	0.5	4.2	3.7	1.1	2.5	1.4	1.4	1.2	-0.4

Sources: Crop Report of Maharashtra, Agricultural Statistics Information, Epitome-II, Govt. of Maharashtra

After this decade, the low growth was recorded because the place of food grain was taken by the non-food crops. Among non-food crops, sugarcane and soybean recorded higher growth rate in area under cultivation. Except 1981-90, the sugarcane recorded higher growth rate among remaining decades. The growth of area of groundnut, safflower, sunflower and cotton observed mixed growth trend in decades. The agricultural economy of Maharashtra mainly depends upon the sugarcane and cotton. Sugarcane recorded higher growth but cotton crop growth was mixed. During 1961-70 and 1971-80, the growth in area of cotton was positive, but during the remaining three decades, it was negative growth. But in the overall period (1961-2010), it could record growth just only 0.1% per annum which is much lower than area under sugarcane which was 3.2% per annum. There was shift in cropping pattern of Maharashtra from food crops to non-food crops.

During last 50 years (1961-2010), few crops namely sugarcane, cotton, rice, soybean, groundnut, bajara and maize (other cereals), recorded higher growth in production in Maharashtra. During last ten years (2004-05 to 2014-15), the area under maize was increased from 4.18 to 10.59 lakh ha, which is >2.5 times. The maize crop has played a significant role in changing the cropping pattern of Maharashtra, particularly in the region of North-Maharashtra, Western-Maharashtra and Marathwada.

Production Technologies recommended and package of practices

1. For late and medium maturity varieties the plant population should be 60000 per hectare and for early types it should be 90000 per hectare.
2. The treatment of *Azotobactor* to maize seed @ 15 g/kg at the time of sowing is recommended.
3. For effective control of weed in maize, pre-emergence spraying of Atrataf on soil @ 2 to 2.5 kg/ha is recommended.
4. For obtaining higher green fodder yield of maize, cu. African Tall, 75 kg/ha and application of 210 kg N/ha are recommended for fodder maize yield in "D" and "E" types soils of sub-montane zone of Maharashtra under rainfed condition.
5. Sowing of maize from 22nd meteorological week (i.e. dry seeding from 28th May to 3rd June) is recommended for "D" and "E" types of soils in sub-montane Zone under rainfed condition.
6. Application of 120:60:40 NPK kg/ha are recommended for medium and late maturing varieties of maize and 'N' should be applied in three equal doses *i.e.* 30, 40 and 50 days after sowing in maize.
7. Cultivation of safflower (for vegetable) as intercrop in maize with application of 120 kg N/ha is recommended for higher monetary returns.

8. The cropping system of growing maize + Groundnut followed by Wheat with 100% recommended fertilizers are found beneficial.
9. New intercrops identified along with their cultivars for *Kharif* & *Rabi* season

S. No.	Particular	*Kharif*	*Rabi*
1	Name of intercrop	Groundnut	Safflower
2	Package of practices	Maize + Groundnut in paired row planting system.	Safflower in between rows of maize for vegetable purpose.
3	Additional benefit expected	Rs. 3 to 4 thousand/ha.	Rs. 2 to 3 thousand/ha.

10. For obtaining higher yield and monitory returns from the sequential cropping of Maize and Gram and to maintain fertility of soil. It is recommended that, the application of 75% of Recommended Dose of Fertilizer (RDF) (NPK kg ha^{-1}) through chemical fertilizer and remaining 25% N through green foliage of subhabul (46.15 q ha^{-1}) along with seed treatment of azotobacter to maize and application of 50% RDF to succeeding gram.
11. Soil test based targeted yield approach for balance fertilization of maize in inceptisol. Fertilizer equation without FYM (t ha^{-1}). Fertilizer equation with FYM (t/ha^{-1}) are as follows: FN = 4.51 × T – 0.65 × SN FN = 3.88 × T – 0.56-3.19 × FYM FP_2O_5 = 1.93 × T – 1.05 × SP FP_2O_5 = 1.91 × T – 0.99 × SP – 1.46 × FYM FK_2O = 2.57 × T – 0.16 × SK FK_2O = 2.09 × T – 0.13 × SK – 1.08 × FYM Where, FN, FP_2O_5 and FK_2O are fertilizers N, P_2O_5 and K_2O in kg ha^{-1}, T is yield target in q ha^{-1}, SN, SP and SK are soil available N, P and K in kg ha^{-1} and FYM is Farm yard manure in t ha^{-1}.
12. For effective control of armyworm in maize spraying of cypermethrin 0.01%, Fenvelrate 0.01%, Chlorpyriphos 0.05%, methyl parathion 0.05% or dusting with methyl parathion 2D @20 kg/ha immediately after incidence is recommended.
13. For effective control of stem borer in maize spraying of Quinolphos 25 EC @ 1000 ml in 500 of water or Phorate 10G @ 10 kg granule application in leaf sheath or spraying of 5% Neem extract is recommended.

Changes in Disease scenario

In Maharashtra, except *Turcicum* and *Maydis* Leaf Blight, no any serious disease on maize has been recorded. However, due to continuous cropping of maize, diseases like, post flowering stalk rot and banded leaf and sheath blight are recently noticed in the state.

Changes in Pest scenario over the years

Earlier, stem borer, *Chilo partellus* was restricted only to the assured rainfall zone. Though, it is a serious pest of maize, not any evidence of major loss in yield has been recorded. However, in maize growing area of Maharashtra *viz.*, Aurangabad, Jalgaon, Nasik, Dhule and Ahmednagar, farmers are facing problem of pink borer, *Sesamia inferens* on late *Kharif* as well as *Rabi* maize.

Pest and disease management

Endosulfan 35 EC has been banned for its use in agricultural crops, alternative for it, Dimethoate 30 EC, which has a label claim, is recommended for effective control of stem borer, *Chilo partellus*.

Post Harvest management

The harvested cobs should be sun dried to avoid storage pests. It is advisable, if possible keep ear intact with husk. It may prevent infestation of stored grain pests.

Extension and outreach

In order to transfer technology to farmers front line demonstrations have been conducted on 1460 and 934 ha land during *Kharif* and *Rabi* seasons, respectively during 2006-2016.

Key R&D challenges and their probable solutions in maize for the state

There is considerable scope to increase coverage of maize area with hybrids and high yielding varieties. The industrial utility of maize in food processing, feed and industrial production, like biofuels fast expanding. The high potentiality to export large quantities of hybrid maize seed due to low cost of seed production may be strengthened further. The rich genetic diversity of maize like prolific type landraces need to be utilized in the breeding. Considering the high yield potentiality of maize, there is wide scope for replacement of other crops due to its wider adaptability of maize in varied climatic change. Further, development of quality maize seed for higher production and research on post harvest technology and food processing and maize grain storage needs to be strengthened.

14

Maize Research and Development in Karnataka

Mruthunjaya C. Wali[1*], *Hulihalli U.K.*[1], *Harlapur S.I.*[1], *Kachapur R.M.*[1] *Talekar S.C.*[1], *Puttaramanaik*[2], *Shobha D.*[2] *and Mallikarjun N.*[2]

[1]*All India Coordinated Maize Improvement Project, University of Agricultural Sciences Dharwad-580 005,* [2]*AICRP, University of Agricultural Sciences, Bengaluru-560 065 Corresponding Author's Email: ars_arabhavi@rediffmail.com*

Karnataka is India's eighth largest state in geographical area covering 1.92 lakh sq km and accounting for 6.3% of the geographical area of the country. As per the population Census 2011, agriculture supports 13.74 million workers, of which 23.61 per cent are cultivators and 25.67% agricultural workers. A total of 123,100 km^2 of land is cultivated in Karnataka constituting 64.6% of the total geographical area of the state (Table 1). Of the total cultivated area in the state, about 33.0% is covered by cereals, 42% by pulses and 19% by oilseeds and 6.0% by commercial crops (Savitha and Kunnal, 2015). Agriculture employs more than 60% of Karnataka's workforce. The large portion of agricultural land in the state is exposed to the vagaries of monsoon with severe agro-climatic and resource constraints.

Table 1. Land use pattern in Karnataka.

Land Use Pattern	Area (Lakh ha)				
	1967-68	1977-78	1987-88	1997-98	2010-11
Non-Agri. purpose	9	1036	12	13	14
Permanent pasture	1676	14	11	10	9
Current fallows	11	13	11	14	13
Net Sown area	101	9940	106	104	104
Gross Irrigated area	13	17	24	30	41

Maize (*Zea mays* L.) is the third most important cereal in India after wheat and rice. Currently in Karnataka State it is cultivated over 11.87 lakh ha with 32.23 lakh tonnes production having average productivity of 2830 kg ha^{-1}. Maize is grown on very wide range of soils and climatic conditions in Karnataka (Table 2).

Table 2. Soil types and pattern in different districts of Karnataka.

Soil Type	Soil Characteristics	Districts
Black Soil Properties	In texture, soil varies from loam to clays. Generally they are neutral to alkaline in reaction, calcareous and well supplied with bases such as Ca, Mg, K. Black soils are known to get self ploughed due to their swelling and shrinking properties with changes in moisture content.	Districts are Belgaum, Bijapur, Gulbarga and Bidar; also parts of Raichur, Chitradurga and Bellary
Laterite soil	Laterite soils result from advanced stages of weathering; highly leached, they are poor in bases and very acidic in reaction. The moisture retentivity of the soil is very poor; soil contains adequate quantities of organic matter.	Malnad and coastal areas of UK, DK and parts of Dharwad, Chikmagalur, Hassan
Red and red loamy soil	They are light textured, from sandy to gravelled or loamy, with poor aggregating ability. They are poor in bases and acidic to neutral in reaction.	Shimoga, Chikmagalur, Hassan, Mysore and Kodagu
Coastal alluvial	The surface soil is generally grey, yellow or light brown; the intensity of the colour increases with depth. The soils are acidic in nature, low in cation exchange capacity and bases.	Dakshin Kannada, and UttarKannada
Dark brown clayey soil	They are clayey, low in bases, rich in organic matter as the surface soil receives the decomposition product of the virgin forest	Dakshin Kannada, Uttar Kannada, Kodagu & Mysore
Mixed red and black soil	Black soil seen in the low lands and valleys has properties resembling those of medium black soil. Soils are productive under good management practices	Belgaum, Bijapur, Dharwad, Raichur, Bellary

The trend in area, production, productivity, of maize during 1960-2016

Karnataka stands first in terms of area and 3rd in terms of production in India (Yadav *et al.*, 2016). Since 1950, the area and production of maize in the state is increasing continuously (Table 3 and Fig. 1).

Table 3. Decadal seasonal trends in APY of maize in the state from 1960 to 2016.

Year	Area(lakh ha)	Production(lakh t)	Yield (kg/ha)
1960-61	0.11	0.12	1090
1961-62	0.13	0.13	1000
1962-63	0.14	0.10	710
1963-64	0.14	0.11	790
1964-65	0.17	0.13	760
1965-66	0.18	0.10	560
1966-67	0.24	0.16	670

1967-68	0.37	0.47	1270
1968-69	0.51	1.04	2040
1969-70	0.57	1.69	2960
1970-71	0.63	2.17	3440
1971-72	0.74	2.43	3280
1972-73	0.75	2.20	2930
1973-74	0.87	3.24	3720
1974-75	0.94	2.51	2670
1975-76	1.27	3.78	2980
1976-77	1.26	3.36	2670
1977-78	1.39	4.13	2970
1978-79	1.51	4.29	2840
1979-80	1.15	3.70	3217
1980-81	1.57	3.81	2430
1981-82	1.58	4.19	2650
1982-83	1.56	3.57	2290
1983-84	1.66	4.66	2810
1984-85	1.88	4.77	2540
1985-86	1.67	3.98	2380
1986-87	2.27	5.76	2540
1987-88	2.05	5.10	2490
1988-89	2.55	6.77	2660
1989-90	2.53	7.09	2800
1990-91	2.46	6.17	2510
1991-92	2.80	8.46	3020
1992-93	3.10	10.21	3290
1993-94	3.10	9.50	3070
1994-95	3.50	10.40	2970
1995-96	3.31	12.01	3630
1996-97	4.46	13.85	3110
1997-98	5.61	15.11	2690
1998-99	5.12	16.71	3260
1999-00	6.09	16.88	2770
2000-01	6.69	21.12	3160
2001-02	5.80	26.09	4500
2002-03	6.50	21.29	3280
2003-04	6.20	12.71	2060
2004-05	8.50	25.12	3110
2005-06	9.36	58.07	3160
2006-07	9.61	27.19	2980
2007-08	11.73	35.79	3050
2008-09	10.69	30.29	2833
2009-10	12.00	31.76	2647
2010-11	12.87	44.44	3633
2011-12	13.49	40.85	3028
2012-13	13.22	34.75	2629
2013-14	13.30	34.80	3016
2014-15	13.70	43.18	3318
2015-16	11.87	32.23	2830

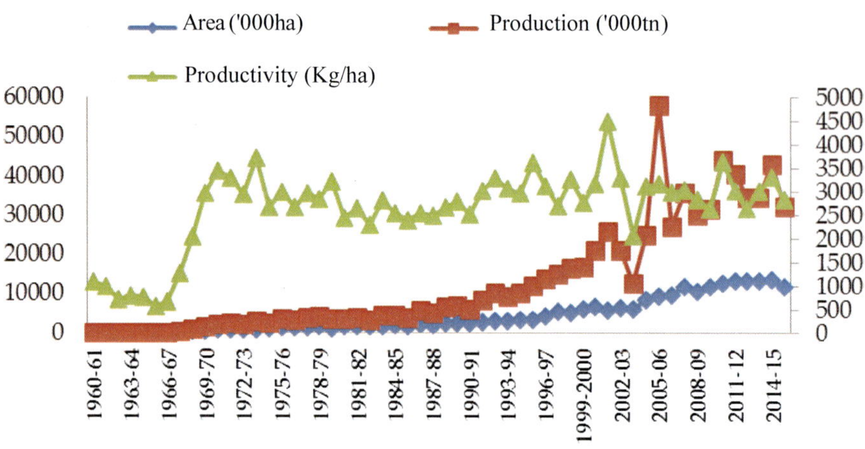

Fig. 1. The trend in area, production and productivity of the state.

The contribution of the state to the national maize production was around 10 per cent during 1995 and it rose to 15.3 per cent in 1999 and further to 19.0 per cent during 2011-12 from 15.0% area. During the triennum (TE) 2011-12 Karnataka had the highest share (14%) of the area and 18.32% of the production. The share of Karnataka has increased by five fold for areas and three fold in production. Three independent variables namely high yielding varieties (HYV), average annual rainfall and type of soil were considered to analyze for 12 years. Based on the analysis it was found that the impact of HYV seed to maize yield was significant. With a change in 1.0 unit in HYV area of maize in the state, there was a change of 4.7 per cent in maize yield. The factors credited for the development of maize production in Karnataka is attributed to high demand from the feed industry, introduction of high yield hybrid seeds, assured market price.

Maize cultivation has spread to irrigated areas of non-traditional belts of the state which is grown successfully in uplands of high rainfall areas in *malnad* where water is insufficient for paddy cultivation. The major maize growing districts of the state are Dharwad, Haveri, Davanagere, Chitradurga, Belgaum, Bellary, Bagalkot, Shimoga and Bijapur which grow more than 75,000 hectares of maize every year (Table 4 & Fig. 2).

Table 4. The district-wise trends in APY of maize during *Kharif*, *Rabi* and Summer season from 2015-16.

District	Area (hectare)			Production (tonnes)			Yield (kg/ha)		
	Kharif	*Rabi*	Summer	*Kharif*	*Rabi*	Summer	*Kharif*	*Rabi*	Summer
Bagalkote	27789	13864	2647	102932	45242	9261	3899	3435	3683
Bangalore-Urban	917	106	106	2503	268	298	2873	2661	2955
Bangalore-Rural	14695	82	89	41769	207	250	2992	2661	2955
Belgaum	96482	23620	3061	202197	52844	8715	2206	2355	2997
Bellary	82764	3829	2146	217872	10633	5578	2771	2923	2736
Bidar	1235	148	0	2330	374	0	1986	2661	0
Bijapur	27783	9884	2540	46189	17183	5084	1750	1830	2107
Chamarajanagar	36049	1968	552	149109	6267	2759	4354	3352	5261
Chickballapur	55407	456	514	185913	1432	1961	3532	3306	4015
Chikmagalur	25831	0	0	61668	0	0	2513	0	0
Chitradurga	89794	1524	875	269220	4034	2152	3156	2786	2589
Dakshina Kannada	0	0	0	0	0	0	0	0	0
Davanagere	163015	2352	1033	492004	6958	4317	3177	3114	4399
Dharwad	26421	3345	1134	49999	5984	2744	1992	1883	2547
Gadag	31263	2864	36	48233	4723	101	1624	1736	2955
Gulbarga	1803	538	295	2110	505	276	1232	988	985
Hassan	76828	2156	74	243191	6718	208	3332	3280	2955
Haveri	158681	8014	4001	361642	23655	10871	2399	3107	2860
Kodagu	3423	10	15	15401	25	42	4736	2661	2955
Kolar	550	38	52	1501	96	146	2873	2661	2955
Koppal	38602	5212	1396	43933	8823	2606	1198	1782	1965
Mandya	4376	351	166	15826	887	466	3807	2661	2955
Mysore	29061	7668	474	86358	25299	1331	3128	3473	2955
Raichur	135	811	170	171	2050	477	1337	2661	2955
Ramanagaram	1453	175	359	4525	399	1368	3278	2403	4011
Shimoga	48558	245	2495	194946	1049	7004	4226	4508	2955
Tumkur	24031	27	149	62073	68	418	2719	2661	2955
Udupi	0	7	1	0	18	3	0	2661	2955
Uttara Kannada	4756	127	343	22302	321	963	4936	2661	2955
Yadgir	885	45	31	1861	114	87	2214	2661	2955
Karnataka State	**1072587**	**226176**	**24754**	**2927778**	**2661**	**69486**	**2873**	**2661**	**2955**

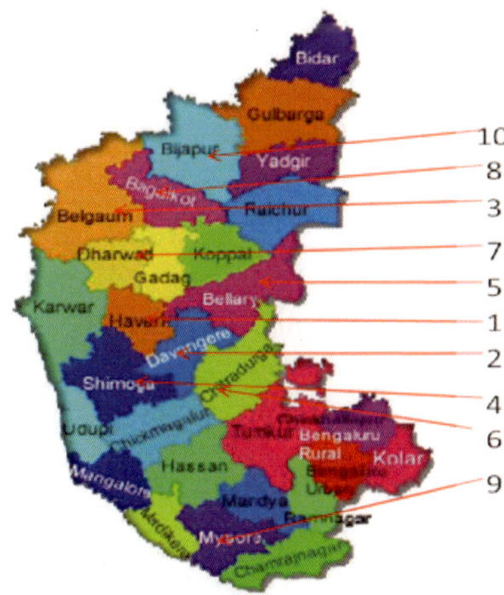

Fig. 2. Area wise major maize growing districts (in descending order) in Karnataka.

Growth rate of maize area, production and productivity

The growth rate of maize was the highest compared to other traditional crops (rice, tur) in the State. For instance, the growth rate (CAGR) of maize was 8.5 per cent and 7.2% for production and area expansion respectively in the last 30 years. It is much higher than the growth of rice, which is registered at 2.2% and 1.1% of production and area expansion respectively during the same period. The same holds true for the tur crop as well, the growth rate of area and production of tur was much lower than that of maize during the same period. As the growth rate of maize production is higher than the growth rate of its area expansion during the last three decades, it infers that the yield level of the crop was good during the period (Table 5).

Table 5. The growth rate of area (lakh ha), production (lakh MT) of major traditional crops of the state.

Crop		1980-81	1990-91	2000-01	2010-11	CAGR
Rice	Area	11.1	11.7	14.8	15.4	1.1
	Production	22.6	24.3	38.5	43.0	2.2
Maize	**Area**	**1.6**	**2.5**	**6.7**	**12.9**	**7.2**
	Production	**3.8**	**6.3**	**21.4**	**44.4**	**8.5**
Tur	Area	3.4	4.6	5.8	8.9	3.3
	Production	.3	1.8	2.6	5.3	4.9

During the period 1998-99 to 2009-10, except three districts *viz.,* Bangalore (Urban), Kolar and Udupi, all the districts where maize is being cultivated, have shown positive and significant growth rate that contributed to the state's growth rate. The highest growth rate in terms of area has been observed in Mandya followed by Uttara Kannada, Hassan and Chikkamanaguluru. However, their share to the states total production is very low (Singha and Chakravarthy, 2013). Based on the log linear regression, the growth trend among different districts it was found that Uttara Kannada and Chikkamanguluru witnessed significantly higher growth rate in terms of area and production followed by Bijapur, Hassan, Raichur and Shimoga.

In the past 15 years (1998-99 to 2012-13), substantial growth in maize area has been witnessed in the state especially in Davangere (3.89%), Haveri (6.49%), Belagavi (3.37%), Bellary (8.3%) and Chitradurg (8.14%) districts. During this period the overall growth of the state in terms of area was around 7.6%. Among the districts, Davanagere has highest area (1.88 lakh ha) with production 5.03 lakh tonnes and productivity of 3563 kg/ha. During this period the overall growth of the state in term of area was around 7.6%. The rapid growth was due to high yielding varieties (HYVs) adaptable to wide range of environments (Table 6). The area expansion in Karnataka was around 7.98% which was marginally lower from Maharashtra (9.19%).

Table 6. The area under high yielding variety of maize in Karnataka (000 ha).

2002-03	203-04	2004-05	2005-06	2006-07	2007-08	2008-09	2009-10	2010-11
624	591	843	919	958	1108	1059	1186	1230

The maize yields have also increased at the rate of 0.23 per cent per annum. The studies showed that with a change of one unit in area under HYVs of maize in the state, a change of 4.7 per cent in yield of maize was observed. Thus the impact of HYVs seed to enhance the maize yield of state is significant and clear. However, the yield improvement was more visible in the districts of Chitradurg, Hassan and Shiomga compared to the other. Similarly, the impact of average annual rainfall on the growth of yield is 2.4%. However, the impact of soil on the growth of yield is not significant. It was found that between 2001 and 2003 there was a significant growth in yield of maize compared to 1998. The production and yield level of the state have increased at the rate of 7.86 and 0.23% per annum. However, the highest growth in production was observed in Chitradurg (8.72%) district.

Major cropping systems and production ecologies

Maize-maize (double cropping systems practiced in command areas), Soybean-maize, maize-wheat, cotton-maize are major cropping systems practiced in the state. It is being cultivated in irrigated, rainfed, rainfed with protective irrigation ecologies. However, the rainfed area is nearly 55-60 per cent and this area is likely to increase in future. The area under irrigated ecosystem is around 40 to 45 per cent.

Genesis of AICRP centres

The agricultural research station, Arabhavi located in the district of Belgaum has been established during 1909 as a result of construction of weir across Ghataprabha river, Dupdhal village during 1889. It was the only station for scientific work of improvement of cereals, including maize in erstwhile state of Bombay. This station is located 10 km away from Gokak on Gokak-Sankeshwar road. Nearest Railway station is Ghataprabha (Miraj-Bangalore section). This research station is situated at an altitude of 640 m, latitude of 16°12' N and longitude of 74° 54' E. The soil is of medium black type. Source of irrigation is Ghataprabha canal water which flows from July to February making it possible to raise two crops in a year. This station receives an average rainfall of 566 mm. During early 1900s, "Arabhavi Local" variety of maize was being grown by the farmers in Belgaum district for longer time. The area was around 18000 acres; of this area, more or less half of the area (which includes both *Kharif* and *Rabi* seasons) was under irrigation provided by the Gokak canals. The canal provided irrigation for about eight months only, where the maize crop, by virtue of its short duration found a place in the system of mixed and double cropping within the available period of water and this was the major food crop in the canal area.

The research work on maize was initiated much before the inception of the All India Co ordinated Maize Improvement Project (AICMIP). In 1948, a section for improvement of maize was started to cater the needs of the cultivators of canal region in the then Bombay state, in particular, and other irrigated tracts in general, with the objective to evolve high yielding and early maturing superior strains of better quality. The work on hybrid maize production was started in 1947-48 at the agricultural school, Arabhavi, to cater the needs of the corn growing areas in the region. The selection for desirable plants from local and exotic materials, and continuance of their progenies by controlled pollinations resulted in six early, five mid-late, and five late, pure breeding and superior inbred lines in local materials, and four in exotic materials *viz.*, Sahara, Peruvian, Swali White and pride of Salin. With this as foundation stock, the programme of hybridization was initiated. All the indigenous inbreds were crossed inter se and with exotic inbreds as well. The F_1 crosses were tested for hybrid vigour and yield. The crosses with exotic lines *viz.*, S-23, originating from Sahara, an African variety showed promising. Thus, the crosses that showed promise were repeated and trials conducted. In 1955, due to constant research work and efforts of the Breeders, a three way hybrid S-23 (I-5 × L-5) was developed and released by this centre which recorded 30 per cent higher yield over "Arabhavi local" with two per cent higher protein and it was popular up to 1961. The production and distribution of hybrid maize seeds was in operation from 1955 to 1961 subsidized by ICAR under comprehensive scheme for scientific improvement of cereals and pulses. However, the research made in India revealed that the indigenous varieties have a very narrow genetic base and the hybrids developed with such material cannot make much headway since they did not registered significantly superior grain yields as compared to local varieties.

In 1957 the Indian Council of Agricultural Research in collaboration with Rockefeller foundation formulated a co-ordinated maize breeding scheme on all India basis. The scheme was operated at five main centres and 11 sub-centres representing principal maize growing zones of the country under the overall supervision of a Project Co ordinator and Associate Project Co ordinator. Agricultural Research Station at Arabhavi also happened to be one among the 11 sub-centres. With the inception of the AICMIP, Arabhavi centre was also recognized as one of the testing centres. In 1981, the resistance breeding work on sorghum downy mildew and Turcicum leaf blight disease was started functioning at Zonal Agricultural Research Station, VC farm, Mandya under the auspices of AICRP on Maize. In the beginning, the problem of identifying Turcicum leaf blight from downy mildew infected plants was felt difficult in view of the systemic nature of Sorghum downy mildew on maize resulting in complete killing of plants as a result, no plant foliage was available for the establishment ofTurcicum leaf blight. Even on surviving downy mildew affected plants there was overlapping of necrotic streaks developed on the foliage and Turcicum leaf blight spots. This mislead the observations of the two diseases. Repeated efforts were made to establish Turcicum leaf blight incidence in downy mildew disease free spots at Mandya with artificial inoculation which was also did not give fruitful results.

Thus a proposal was made for consideration during seventh five-year plan to open a sub-station with minimum staff strength at agricultural research station, Naganahalli, Mysore. Full pledged research work on Turcicum leaf blight disease resistance was started functioning from May, 1997. The centre was considered as 'hot spot' for Turcicum leaf blight incidence and the research works on TLB and polysora rust were carried out till *Kharif* 2008 at agriculture research station, Naganahalli. In view of the conversion of agricultural research station, Naganahalli into completely organic farming research station, the AICRP (maize) project was shifted from Naganahalli to zonal agricultural research station, VC farm, Mandya from December, 2008.

The sorghum downy mildew sick plot was created and maintained from 1981 and the germplasm is being continuously screened against this disease. Mandya center, being the only center in India identified as 'hot spot' location for sorghum downy mildew, the hot-spot location for TLB and polysora rust were also established at ZARS, Mandya from *Rabi* 2008 and work is continuing.

AICRP technologies

The AICRP centres located in the state has released several cultivars, from time to time to continuously enhance the yield levels of the states. The brief details of the cultivars released are given in Table 7.

Table 7. Cultivars released for state by AICRP.

Type	Cultivar	Year	Released by	Other Characteristics
Composite	Renuka (G-25)	1984	SVRC	Early maturing (85 to 90 days) suitable for growing with cotton as mixed cropping system
Composite	Prabha (G-57)	1989	SVRC	Full season, tolerant to stem borer, TLB, downy mildew, erwinia stalk rot, late wilt
Composite	NAC-6004	1998	SVRC	Late maturity: composite Grain color: Orange yellow Grain type: Semi-dent

				Recommended for both rainfed/irrigated ecosystem in southern Karnataka, Tolerant to Sorghum downy mildew, Turcicum leaf blight and Polysora rust diseases
Hybrid	DMH-1	1998	SVRC	Full season hybrid tolerant to TLB,SDM
Composite	NAC-6002	1999	SVRC	Early maturity composite Grain color: Orange yellow; Grain type : Semi-dent Recommended for inter cropping with pulses and oil seeds. Tolerant to Sorghum downy mildew, Turcicum leaf blight and Polysora rust diseases
Hybrid	DMH-2	2002	SVRC	Single cross hybrid tolerant to TLB, SDM
Hybrid	NAH-2049 (Nithyashree)	2006	SVRC	Single cross hybrid and late maturity. It is suitable for both Irrigated and Rainfed situation in southern Karnataka. Grain colour: Orange yellow; Grain type: Semi-dent Resistant to Sorghum downy mildew, Turcicum leaf blight and Polysora rust diseases.
Hybrid	EH-434042 (Arjun)	2009	SVRC	Three way cross, tolerant to TLB, SDM, PFSR, Rust
Hybrid	NAH-1137	2010	SVRC	Single cross hybrid. Stay green type; Late maturity Grain colour: Orange yellow; Grain type: Semident Suitable for both *Kharif* and *Rabi* Season.Tolerant to Sorghum downy mildew, Turcicum leaf blight and Polysora rust diseases
Hybrid	GH-0727 (Shrushti)	2014	SVRC	Single cross hybrid tolerant to TLB, PFSR and moderately tolerant to SDM

1. DMH-1

This is a double top cross hybrid released during 1997-98. This is a long duration (115-120 DAS) hybrid. The hybrid is a stay green type which remains green at the time of harvest. This is tolerant to turcicum leaf blight (TLB).

DMH-1

2. DMH-2

This is a first ever single cross hybrid developed in South India and released during the year 2002. This is a long duration (115-120 DAS) hybrid. The hybrid is tolerant to turcicum leaf blight (TLB) and moderately tolerant to post flowering stalk rot (PFSR). This hybrid can be cultivated both in *Kharif* and *Rabi* season.

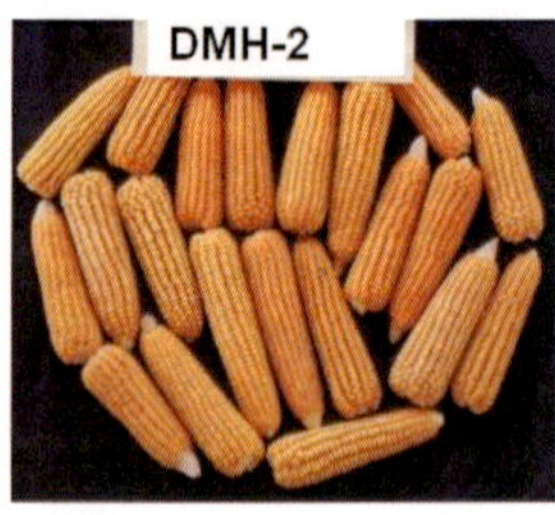

3. EH-434042 (Arjun)

This is a three way cross hybrid which takes 114-118 DAS for maturity. This hybrid is having multiple disease tolerance namely turcicum leaf blight (TLB), rust, maydis leaf blight (MLB), post flowering stalk rot and sorghum downy mildew (SDM). The hybrid grows tall and fodder yield realized is very high compare to any of the hybrids. This hybrid was released during 2008.

4. GH-0727 (Shrusti)

GH-0727 is a single cross long duration hybrid (115-120 DAS). It can be grown under irrigated ecosystem during *Kharif* season in Zone 3 and 8 of Karnataka. It is tolerant to Turcicum leaf blight (TLB) and rust. The hybrid is having attractive ear features with orange yellow colour grains. The hybrid is having 84.88 shelling percentage, with 100 grain weight and yield are 36.82 gm and 72.0 q/ha respectively.

Patents/Registration

Hybrid	Year of registration	Registration No.
DMH-2	2010	E59 ZM83 07 416No. 17/2010
DMH-1	2012	E60 ZM84 07 417No. 4/2012

Arabhavi local was one of the popular landrace which was grown during early 1900s to 1960s. It was white type, suitable for preparing roti, rawa, porridge and matures in 90-95 days. However, of late most of the maize area (95%) in the state is under hybrids. As per the informal estimates, the most popular hybrids among farming community in the state are CP-818, CP-999 developed by Charoen and Pokhpond Seeds Limited. The area under CP-818 is the highest in Karnataka state. It is followed by the hybrid NK-6240 from M/S Syngenta India Limited. The other popular hybrids are 900 M Gold, DKC 8101 etc. from Monsanto India Limited. Further, the hybrids developed by AICRP centres are also being cultivated but in limited area. The major hybrids developed by AICRP centres of the state are Arjun (EH-434042), GH-0727 (Shrushti), NAH-2049 (Nithyashree) and Hema. In Karnataka around 20 per cent area is under single cross hybrids, whereas the rest of the area (60-75%) is under either three-way or double cross hybrids. Only 5 to 10% area is under OPV.

Production technologies recommended

1. Early planting of maize gives higher yields. Maintenance of plant population of 55,000-66,000 per ha. is a must for achieving higher yields.
2. Planting of maize in flat bed followed by earthing up in case of medium to shallow red soils in moderate rainfall area and planting maize on ridges in heavy rainfall areas and irrigated heavy black soil is recommended for optimum yields.
3. Alternative furrow irrigation, alternatively in heavy black soil is found ideal.
4. Excess of moisture during early stages is quite harmful and optimum moisture after flowering is a must.
5. Weedicides like atrazine or simazine can be safely used in maize crop as pre-emergence spray at the rate of 2 kg/ha.
6. Detopping (removal of portion above the ear) 30 days after silk drying can be safely practiced without affecting grain yield which gives lot of green fodder at the time of fodder scarcity.
7. Intercropping of maize in cotton or freshly planted sugarcane with early maturing maize genotype is a handy cropping system for the small and marginal farmers.
8. Intercropping of pulses (Cowpea, Soybean, Redgram, Greengram and blackgram during *Kharif*) and oilseeds (Sunflower and Groundnut in *Kharif;* Safflower in *Rabi*) was found to be highly remunerative compared to sole crop of maize.
9. Combined application of Zn:Fe:Mn (20:20:20 kg/ha) with recommended dose NPK enhances the grain yield of maize.
10. Need based plant protection sprays with endosulphon for insect pests and Dithen M-45 for downy mildew and leaf blight gives optimum yields.
11. In-situ green manuring of two rows of fodder cowpea or sunhemp in between 90 cm rows of maize and incorporating at 45 days after sowing is found to enhance grain yield of maize and improves soil fertility. In-situ green manuring is found to have residual effect by increasing the grain yield of succeeding wheat crop.
12. October first fortnight to November first fortnight is found to be optimum sowing time for *Rabi* maize.
13. Intercropping of Renuka composite with Cotton (cv. ACP-71) is found to be profitable which is beneficial over sole crop.
14. Intercropping of two rows of soybean (JS 335) in between 90 cm rows of maize (DMH-1 and DMH-2) is found to be profitable and farmer could get additional profit of Rs. 3000 over sole maize.

15. In-situ green manuring of fodder Cowpea or Sunhemp with maize during the *Kharif* season followed by wheat in *Rabi* is found to be profitable cropping system in GLBC area.
16. The DMH-1 (Dharwad Maize Hybrid-1), the double top-cross hybrid developed from this centre was released during 1996 and it is extensively cultivated by the farmers of Karnataka State.
17. The single cross hybrid DMH-2 (Dharwad Maize Hybrid-2), developed at this centre is recommended for release in the state during 1999. This hybrid is 20 and 8 percent superior to Deccan-103 and DMH-1 respectively. The hybrid is also tolerant to turcicum leaf blight, fairly resistant to charcoal stalk rot.
18. DMH-1 a double top cross long duration hybrid is identified as moderately resistant to TLB, SDM and PFSR.
19. A multiple disease resistant single cross hybrid namely, DMH-2 is noted resistant to TLB, SDM and PFSR under artificial inoculations.
20. For management of TLB, it is recommended that spraying of Mancozeb at the rate of 2.5 grams/L of water after initiation of the disease should be done in susceptible maize cultivars at knee-high stage. Second and third spray to be done at intervals of 10 days, if necessary.
21. Spraying of propiconazole 25% EC (Tilt) @ 1 ml/L of water at 30 and 45 days after sowing most effective in controlling rust.
22. Early sowing during *Kharif*, first fortnight of June resulted in minimum foliar disease pressure and enhanced maize yields.

Value addition in maize

Crop	Technology Developed	Year
Maize	Nippattu, Laddu, Vermicelli, noodles, crispies and papad are added to Package of practice **(POP)**	2010
Maize	Nutrimix, vadamix and idlimix are added to Package of practice **(POP)**	2011
Popcorn	Popcorn burfi, bar and other popcorn value added products are added to Transfer of practice **(TOT)**	2013

Package of practices for maize in Karnataka

The crop can be sown throughout the year in all the seasons. However, May to June, September to October and January to February are the best time of sowing for *Kharif, Rabi* and summer season, respectively. The crop should not be sown in the month of September, in the Downey mildew disease infected regions in the previous year. The crop is not suitable for the places which cross 40.5°C temperature. Under such places, grain yield will be reduced due to insufficient grain filling in the ear and improper pollination. The seed rate of 22.5 to 27.5 kg/

ha with 10 tonnes of FYM or 1 tonnes of poultry manure. The manures should be incorporated 2-3 weeks before sowing. The chemical fertilizer dose should be used based on soil test result; under deficient soils, zinc sulphate should be mixed with soil before sowing. In iron and zinc deficient black clay soils, zinc sulphate and vermi compost mixture should be incorporated to soil at the time of sowing. However, the recommended fertilizer dose is 150:75:75:: N:P:K, along with 25 kg/ha each of zinc and ferrous sulphate.

The major maize based cropping systems are maize-wheat, maize-chickpea, maize + soybean intercropping system (1:2). Weeds can be controlled by using Simazine or Atrazine 50 WP herbicide at 2.5 kg/ha in sandy soils and 3.0 kg/ha in black soils by dissolving in 500 liters of water on the day of sowing or next day after sowing. At the time of herbicide application, there should be sufficient moisture available in the soil and the soil should be in good tilth condition, all the seed born weeds can be controlled by this herbicide.

Disease scenario

Maize is one of the most important cereal crops in Karnataka and is widely grown in various climatic conditions across the state. The occurrence and severity of diseases and resultant losses are dependent on the prevailing agro-climatic conditions and hybrids grown in an area. A total of twelve fungal diseases *viz.*, turcicum leaf blight, maydis leaf blight, curvularia leaf spot, brown spot, common rust, poysora rust, charcoal stalk rot and fusarium stalk rot, seedling blight, diplodia ear rot *(Syn.Stenocarpella maydis)*, sorghum downy mildew and banded leaf and sheath blight were recorded during the surveys (Fig. 3). Some of the diseases *viz.*, turcicum leaf blight (*Exserohilum taxicum*), maydis leaf blight (*Dreschlera maydis*), polysora rust (*Puccinia polysora*), common rust (*Puccinia sorgi*), and charcoal stalk rot (*Macrophomina phasealina*) were prevalent in all the districts of northern Karnataka with moderate to heavy incidence on almost all the hybrids grown. Whereas diseases like banded leaf and sheath blight (*Rhizoctonia solani* f. sp. *sasakii*), seedling blights (*Exserohilum turcium* and *Fusarium* sp.) curvularia leaf spot (*Curvularia lunata*) and ear rots (*Diplodia maydis (Syn. Stenocarpella maydis)*, *Fusarium* sp.) were predominantly prevalent in heavy rainfall areas of hilly zone and transitional zone. The banded leaf and sheath blight (BSLB) was very severe in Kalaghatagi taluks of Dharwad district and Mundagod taluk of Uttar Kannada district during *Kharif* season. The low to moderate incidence of seedling blight was observed in Dharwad, Haveri and Belgaum districts during *Kharif* season, whereas high severity of curvularia leaf spot was noticed in Kalaghatagi and Dharwad taluks of Dharwad district and also in Davanagere and Haveri districts. Ear rots noticed in heavy rainfall areas *viz.*, Hanagal, Hirekerur, Haliyal taluks in low to moderate incidence.

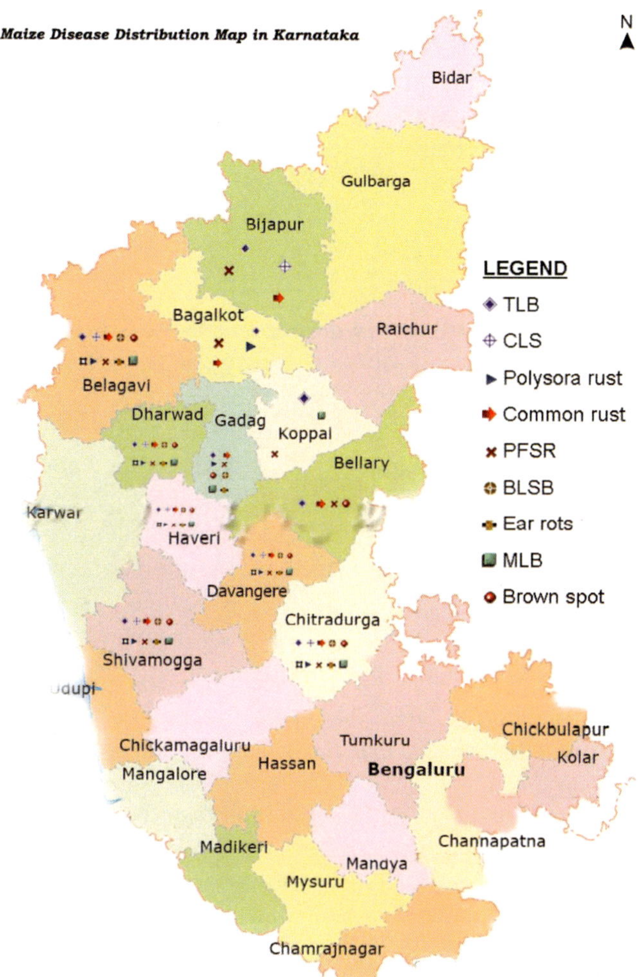

Fig. 3. The disease distribution map of Karnataka.

The major diseases like TLB, PFSR, common rust and polysora rust will continue to represent problems in certain conditions, whereas other diseases like, BLSB, CLS, MLB, diplodia ear rot and brown spot which were not previously known as limiting factors, but now identified as new challenges in maize production in Karnataka. The severity of TLB, MLB and polysora rust in Haveri, Dharwad and Belgaum districts may be attributed to extensive and continuous cropping of maize and cultivation of susceptible hybrids. The increasing trend in disease severity of maize can be attributed to late sowing (late July to August) due to irregular supply of water from command area canals and unpredictable start of rainy season and unawareness in application of suitable chemicals and also cultivation of highly susceptible hybrids. Some areas *viz.*, Haveri, Dharwad, Kalaghatagi, Bailhongal and Davanagere recorded high severity of foliar diseases and are considered as endemic for foliar diseases, whereas *viz.*, Ramdurga, Gokak, Mudhol, Bellary, Koppal, Harapanahalli, Jamakhandi, Navalgund and Nargund recorded more than 30% incidence of PFSR. Hence, these locations were identified as hot spots for PFSR disease(s). Moderate to severe incidence of diplodia ear rot and BLSB were noticed in Kalaghatagi taluk (Dharwad district), Haveri and Davanagere districts. However, the disease severity varied from locality to locality and was more in monocropping system.

Turcicum leaf blight (TLB) and maydis leaf blight (MLB) diseases were noticed during the second fortnight of July and severity progressed later on and reached maximum during September in Dharwad, Haveri and Gadag Districts. Maydis leaf blight was very severe in Kalaghatagi and Haveri taluks. Polysora rust appeared during second fortnight of August and recorded maximum severity during first fortnight of October in Dharwad, Bailhongal and Haveri taluks. Moderate to severe incidence of banded leaf and sheath blight (BLSB) was noticed during second fortnight of September and attained severe at grain filling stage in Kalaghatagi and Mundagod taluks. Moderate to severe incidence of *Curvularia* leaf spot was notice in Dharwad and Haveri Districts during September first fortnight.

In the scenario of climate change, the survey and surveillance of maize diseases during past five years revealed that, the furgal foliar diseases *viz.*, turcicum leaf blight, common rust, maydis leaf blight, curvularia leaf spot, diplodia ear rot, brown spot and banded leaf and sheath blight were found widespread during rainy season. The post flowering stalk rot diseases *viz.*, charcoal stalkrot and Fusarium stalk rot were noticed in *Rabi*/summer seasons. Turicum leaf blight (*Exserohium turcicum*) was prevalent at all the growth stages of the crop. Its severity was 5-15 per cent during vegetative stage, while it was 15-40 and 45-80 per cent during tasselling and post flowering stages respectively. The disease was noticed both under rainfed and irrigated situations, Severity of turcicum leaf blight was observed in severe form in high rainfall areas and also fields which had low nitrogen levels coupled with low moisture stress during early stages of crop growth. Maydis leaf blight (*Dreschlera maydis*) was recorded up to 16 per cent at tasselling stage and 50 per cent during post flowering stage. Common rust (*Puccinia sorghi*) was widespread in irrigated command areas and assured rainfall areas of transitional zone. The late sown crop was most affected by rust at flowering to grainfilling stage. The severity of common rust was 35-60 per cent. Curvularia leaf spot (*Curvulaira lunata*) severity was noticed during tasselling to post flowering stage of the crop up to 50 per cent. Hot and humid climate favored the development of Curvularia leaf spot. Banded leaf and sheath blight (*Rhizoctionia sloani* f sp. *saskii*) was recorded in the range of 5-25 per cent in heavy rainfall areas during flowering to grain filling stage. BLSB was predominant in paddy-maize and maize-maize cropping sequence. Brown spot (*Physoderma maydis*) intensity was recorded up to 25% during vegetative to post flowering stage. Brown spot severity was high in areas where mono cropping was practiced. In some locations, stalk breaking and stalk rot induced by *Physoderma maydis* was noticed besides foliar infection. Sporadic incidence of diploida ear rot (*Stenocorpella maydis*) was also noticed at grain filling to maturity stage in heavy rainfall areas. The bird's damage and earworm infestation aggravated the ear rot intensity.

A survey conducted recently, revealed that, the severity of curvularia leaf spot was prevalent in moderate to severe intensities on most of the hybrids grown in Karnataka during rainy season and cause extensive damage to the crop thus lowering yields. The disease severity ranged from 22.50% to 68.4%. High disease severity was recorded in Dharward, Haveri and Belagum districts during tasselling to post flowering stage of the crop. Hot and humid climate favoured the development of disease during flowering to grain filling stage. The diseased samples were analyzed. Microscopic examination of the cultures of different locations revealed association of two species of curvularia namely, *Curvularia lunata* and *Curvularia eragrostidis*. The most prominent was *Curvularia lunata* followed *by Curvularia eragrostidis*. The loss estimation studies showed that, in susceptible hybrids the avoidable yield loss ranged form 23.8 to 35.0%. In addition the yield losses due to TLB was estimated to the tune of 56% in susceptible cultivar; CM-202. In case of moderately resistant genotypes namely; DMH-1, Prabha and Deccan-103 losses were 13.6, 16.5 and 20% respectively.

The continuous survey and surveillance of maize diseases along with identification of germplasm tolerant to post flowering stalk rots is required to tackle the challenges of diseases in future. In addition there is need for proper diagnosis of viral diseases which are transmitting through seeds since major hybrids of multinational companies are ruling.

Extension and outreach

In order to demonstrate the worth of improved technology, during *Kharif* 2014 season, there was a target to cover 1,55,256 ha of maize cultivated area in the Davanagere district with improved management interventions. However beyond target, 110% (1,72,144 ha) of the maize cultivated area was covered under improved management. The results of CCEs (crop cutting experiments) showed huge benefits through adoption of improved management in enhancing maize grain and fodder production across all taluks in the district (Table 8). The results were compared with the farmers' management, it was observed that the grain productivity increased by 18% to 29% and fodder productivity by 24 to 42%.

On an average, there was 22% increase in grain yield and 31% increase in fodder yield in the district which translates to additional 1,210 kg/ha grain production and 1,077 kg/ha fodder production.

Table 8. Yield improvement through improved practices v/s farmer's practices - a case study of Davangere district.

Study area	**Farmers' management**			**Improved management**			**% increase over FP**		
Davanagere **District**	**Grain**	**Fodder**	**TDM**	**Grain**	**Fodder**	**TDM**	**Grain**	**Fodder**	**TDM**
Channagiri	5950	3430	11400	7050	4290	13800	18	25	21
Davanagere	5540	2750	9950	7130	3900	13300	29	42	34
Harihar	5740	4220	11840	6900	5240	14250	20	24	20
Honnali	4310	3370	8840	5300	4650	11240	23	38	27
Mean	5385	3443	10508	6595	4520	13148	22	31	25

Source: Bhoochetana Report 2014-15.

Similarly, the technologies were also demonstrated through front line demonstrations and tribal sub-plan across different districts of the state. The glimpses of yield gap between global maize productivity vis-β-vis other are given in Fig. 4.

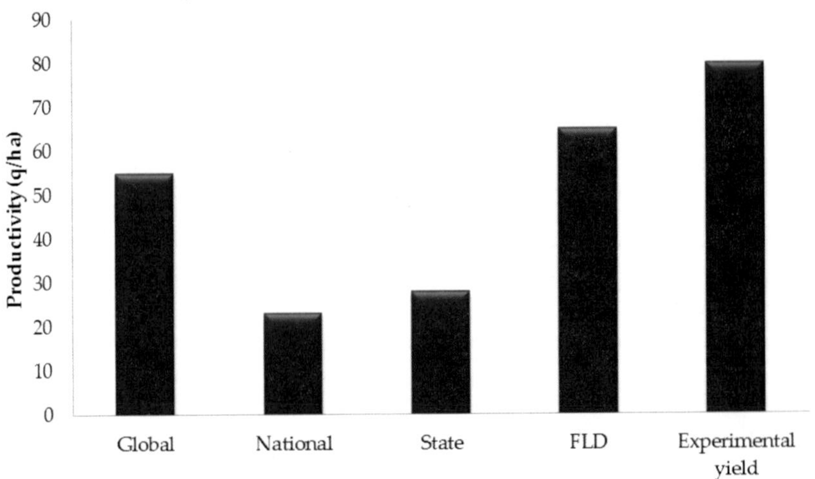

Fig. 4. The comparison of yield level of maize

Key production problems in the state for maize

The shift in traditional crops and cropping systems with maize based systems are gaining importance in view of changing resource base under the current farming scenario (Table 9). The major production constraints are due to rainfed cropping system. The incidence of terminal stress especially the moisture stress reduces the yield due to low rains as monsoon recedes (Table 10, Fig. 5 and 6).

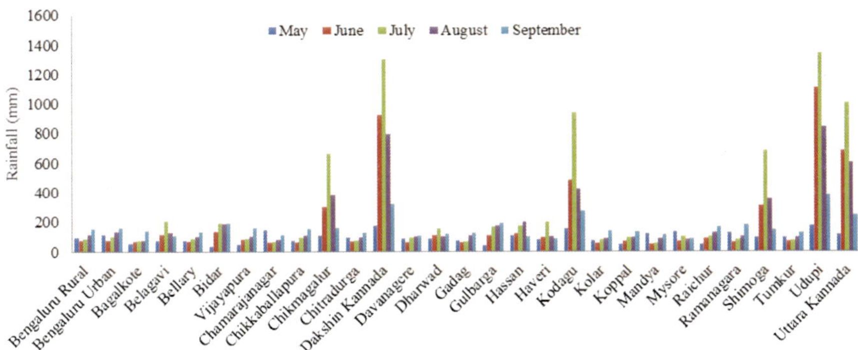

Fig. 5. Monthly rainfall pattern during *Kharif* season in different districts of Karnataka (1951-2008).

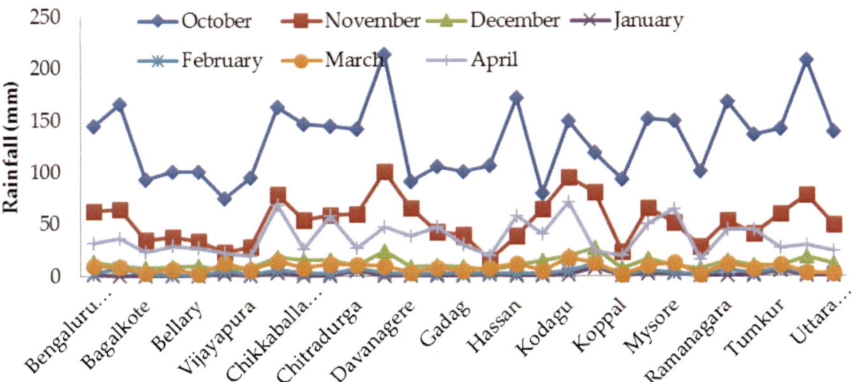

Fig. 6. Monthly rainfall pattern during *Rabi* season in different districts of Karnataka (1951-2008).

Thus development of early maturing terminal drought stress resilience will substantially help in sustaining the yield levels. Whereas in irrigated ecosystems (Table 11), hybrids with tolerance to high-density planting will help in further enhancement. The efficient nutrient management techniques along with water use efficient hybrids will bring down the cost of production thus enhances the profit margin of maze farmers. Finally, the incidence of diseases like TLB, PFSR and rust also act as major constraints, which can be reduced by developing disease resistant hybrids. In addition some abiotic stresses like deficiency of zinc, iron and phosphorus were also noticed in Dharwad, Haveri and Belgaum districts.

Table 9. Changes in cropping pattern in Karnataka.

Crop	Share in GCA (%)					
	1960-63	**1970-73**	**1980-83**	**1990-93**	**2000-03**	**2007-10**
Rice	9.9	10.7	10.3	10.3	11.4	11.6
Sorghum	28.0	21.8	19.2	18.0	15.1	10.9
Pearl millet	4.8	4.6	5.4	3.3	2.7	2.6
Maize	**0.1**	**0.7**	**1.4**	**2.3**	**5.3**	**9.0**
Finger millet	9.6	9.8	9.8	8.8	7.7	6.4
Wheat	2.9	2.9	3.0	1.7	2.2	2.2
Small millets	4.2	4.1	3.2	1.1	0.6	0.3
Total cereals	59.7	55.4	52.4	45.5	45.0	43.1
Pigeon pea	2.7	2.5	3.3	3.9	4.4	5.0
Chick pea	2.5	1.4	1.3	1.7	3.7	6.1
Total pulses	11.9	11.0	13.2	13.8	16.9	18.3
Food grains	71.9	68.3	66.6	59.4	61.9	61.4
Groundnut	8.4	9.2	7.6	10.5	7.8	6.8

Rainfall pattern

Drought is recognized as the most important constraint across the rainfed environments (Fig. 7). The changing climate not only affects the frequency and intensity of abiotic stresses like moisture stress (Table 12), but also affects intensity of biotic stresses on crop plants. Maize in Karnataka is affected by an array of diseases and insect-pests; these include the post-flowering stalk rots (PFSR), turcicum leaf blight (TLB), downy mildews, ear rots, stem borers and weevils, besides high incidence of mycotoxins in maize grain harvested in some parts of the country. Which causes post-harvest losses due to poor drying facilities is yet another challenge. Maize is particularly vulnerable to the reproductive stage heat stress. Climate projections also suggest that elevated temperatures. Poor soil fertility (including micronutrient deficiencies) in degraded lands and low nutrient use efficiency are also among the most important factors limiting maize productivity and yield stability. Low adoption of improved production technology continues to be a major challenge for improving maize productivity in India. Poor agronomic management of crop causes sub-optimal realization of yield potential of cultivars. Weed infestation in the *Kharif* season is the single biggest challenge in maize-based cropping systems. Lack of systematic regulated marketing, negligible processing and price glut during harvest are some of the other challenges the state farmers are facing.

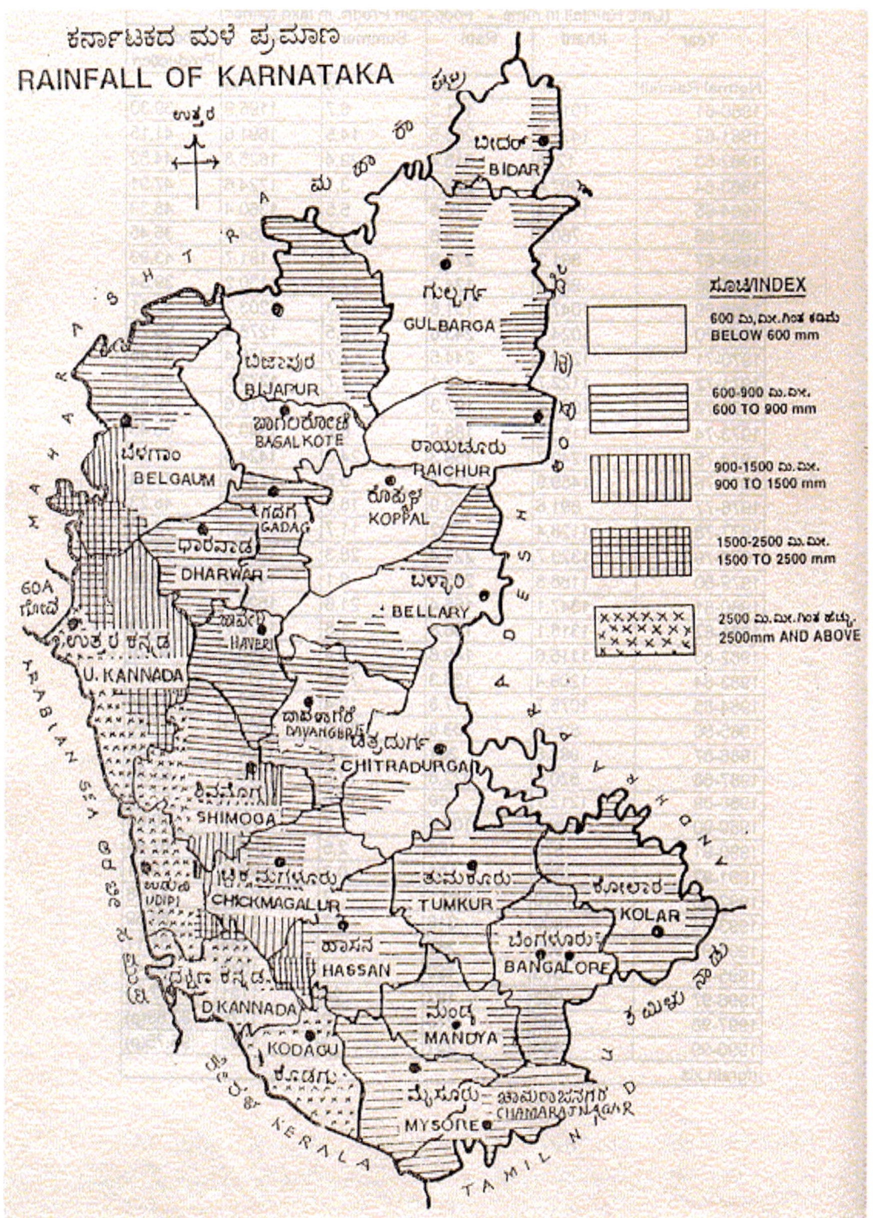

Fig. 7. Rainfall map of Karanataka.

Table 10. Mean monthly rainfall and number of rainy days during *Kharif* and *Rabi* seasons for Karnataka state from 1951 to 2008.

District	Rainfall	May	Jun	Jul	Aug	Sep	*Kharif*	Oct	Nov	Dec	Jan	Feb	Mar	Apr	*Rabi /Summer*
Bengaluru	A	94	71	87	115	154	521	145	63	14	2	3	10	32	269
Rural	B	6.5	4.9	7	8.1	8.3	34.8	8.1	4.3	1	0.5	0.5	0.6	2.9	17.9
Bengaluru	A	113	73	103	132	158	579	166	65	10	1	8	9	37	296
Urban	B	6.5	4.9	7.1	8	8.1	34.6	8	4	1.1	0.6	0.6	0.5	2.8	17.6
Bagalkote	A	57	68	75	74	143	417	93	35	8	1	2	4	24	167
	B	3.1	5	5.7	5.1	8	26.9	5	2	0.5	0.2	0.3	0.5	1.8	10.3
Belagavi	A	76	114	208	128	110	636	101	38	9	1	2	7	29	187
	B	3.8	7.4	11.8	9	7.5	39.5	6.4	2.4	0.6	0.2	0.1	0.6	2.4	12.7
Bellary	A	71	69	88	99	132	459	101	34	10	1	1	3	27	177
	B	3.6	4.3	6.2	6.3	7.3	27.7	5.4	2.4	0.5	0.2	0.3	0.3	1.8	10.9
Bidar	A	31	134	192	184	195	736	75	24	6	3	5	13	23	149
	B	2.4	7.4	11.6	10.1	10.7	42.2	3.6	1.6	0.4	0.5	0.8	1.1	2.1	10.1
Vijayapura	A	44	78	89	103	157	471	95	28	8	1	2	6	20	160
	B	3.1	5	5.7	5.1	8	26.9	5	2	0.5	0.2	0.3	0.5	1.8	10.3
Chamaraja	A	146	58	64	82	112	462	163	79	19	4	6	15	68	354
nagar	B	6.5	4.9	7	8.1	8.3	34.8	8.1	4.3	1	0.5	0.5	0.6	2.9	17.9
Chikkaballa	A	74	59	92	107	154	486	147	54	16	1	4	8	26	256
Pur	B	6.5	4.9	7	8.1	8.3	34.8	8.1	4.3	1	0.5	0.5	0.6	2.9	17.9
Chikmagal	A	105	301	662	382	160	1610	145	59	16	1	3	12	58	294
Ur	B	6.1	15	21.4	18.5	11.6	72.6	9.1	4.1	1	0.3	0.3	0.8	3.9	19.5
Chitradurga	A	94	69	74	92	128	457	142	60	10	5	7	11	27	262
	B	4.3	4.2	6.9	6.1	6.2	27.7	6	2.9	0.6	0.3	0.3	0.3	1.8	12.2
Dakshin	A	171	921	1301	796	320	3509	214	101	25	1	4	10	47	402
Kannada	B	6.4	24.8	28.6	26.3	17.2	103.3	11	4.8	1.1	0.3	0.1	0.4	2.2	19.9
Davangere	A	84	63	92	98	108	445	91	66	9	1	2	4	39	212
	B	4.3	4.2	6.9	6.1	6.2	27.7	6	2.9	0.6	0.3	0.3	0.3	1.8	12.2
Dharwad	A	88	105	152	100	123	568	106	43	10	1	3	8	47	218
	B	4.4	7.5	12	9.2	7.3	40.4	6.6	2.5	0.6	0.2	0.2	0.5	2.6	13.2
Gadag	A	76	62	67	108	128	441	101	40	9	1	3	5	30	189
	B	4.4	7.5	12	9.2	7.3	40.4	6.6	2.5	0.6	0.2	0.2	0.5	2.6	13.2

Gulbarga	A	38	108	163	175	194	678	107	16	5	2	3	8	20	161
	B	2.2	7.3	10.5	8.7	9.7	38.4	3.8	1.4	0.2	0.2	0.5	0.7	1.7	8.5
Hassan	A	107	118	172	196	98	691	172	39	11	1	3	12	58	296
	B	7.1	8.9	13.8	10.6	7.7	48.1	8.9	4.6	1.1	0.4	0.4	0.7	3.6	19.7
Haveri	A	82	96	201	103	85	567	80	65	15	1	3	5	41	210
	B	4.4	7.5	12	9.2	7.3	40.4	6.6	2.5	0.6	0.2	0.2	0.5	2.6	13.2
Kodagu	A	151	486	938	426	270	2271	150	95	20	2	5	18	71	361
	B	8.9	19.6	25.6	21.5	14.4	90	12.4	6	1.4	0.5	0.4	1.4	5.9	28
Kolar	A	75	55	79	87	142	438	119	81	27	8	12	13	25	285
	B	4.9	4.3	6.4	7.2	7.6	30.4	7.2	4.7	1.3	0.7	0.5	0.6	3.1	18.1
Koppal	A	50	65	91	96	133	435	93	24	7	1	1	2	20	148
	B	2.9	5.9	7.4	7.2	8.1	31.5	4.9	1.9	0.3	0.2	0.3	0.5	1.5	9.6
Mandya	A	123	48	51	85	112	419	152	66	17	3	5	10	50	303
	B	7.2	3.6	3.9	5.1	6.2	26	8.5	4.2	1	0.3	0.4	0.6	3.6	18.6
Mysore	A	135	65	100	81	89	470	150	51	10	3	4	13	65	296
	B	7.8	5.2	7.3	6.5	6.2	33	8.7	4.3	1	0.4	0.4	0.9	4.4	20.1
Raichur	A	45	85	101	126	168	525	101	28	7	1	1	2	16	156
	B	2.9	5.9	7.4	7.2	8.1	31.5	4.9	1.9	0.3	0.2	0.3	0.5	1.5	9.6
Ramanagar	A	127	63	82	102	178	552	168	53	15	1	6	12	45	300
	B	6.5	4.9	7	8.1	8.3	34.8	8.1	4.3	1	0.5	0.5	0.6	2.9	17.9
Shimoga	A	92	307	681	352	144	1576	136	41	10	1	3	7	45	243
	B	4.5	13.2	20.5	16.9	9.8	64.9	7.6	3	0.7	0.2	0.2	0.6	2.7	15
Tumkur	A	94	69	74	92	128	457	142	60	10	5	7	11	27	262
	B	5.5	4.3	6	6.6	7.2	29.6	7.3	3.8	0.8	0.3	0.4	0.5	2.3	15.4
Udupi	A	170	1105	1341	840	383	3839	208	78	19	1	3	4	30	343
	B	6.4	24.8	28.6	26.3	17.2	103.3	11	4.8	1.1	0.3	0.1	0.4	2.2	19.9
Uttara Kannada	A	116	681	1006	602	248	2653	139	49	12	1	3	3	25	232

Table 11. Source of irrigation in Karnataka (triennium averages) area in lakh ha

Triennium	Canal	Tanks	Tube wells	Wells	Other sources	Net Irrigated Area	Gross Irrigated Area
1960-63	2.56	3.58	0.0	1.46	1.46	9.06	9.96
1970-73	4.38	3.67	0.04	3.11	1.00	12.20	15.02
1980-83	6.11	3.17	0.07	4.02	1.71	15.08	18.59
1990-93	8.94	2.65	2.11	5.32	3.04	22.05	27.45
2000-03	8.81	2.29	6.17	4.69	3.57	25.53	30.67
2004-07	10.11	1.86	9.36	3.93	3.86	29.12	35.21
2008-11	11.08	1.99	12.24	4.23	4.19	34.90	41.87

Table 12. District wise rainfall (mm) pattern in Karnataka state

District	Normal Rainfall (mm)	Days	1988-2000	2001-2003	2004-2006	2007-2009	Mean 1998-2010	2011	2012	2013	2014
Bagalkot	584	40	634	377	489	689	554	516	372	528	637
Bangalore (Rural)	740	41	893	571	766	792	756	721	487	744	830
Bangalore (Urban)	835	49	1001	604	895	921	855	954	533	821	821
Belgaum	842	53	899	630	1010	1023	901	1153	734	825	1124
Bellary	604	40	566	422	524	648	550	380	418	482	566
Bidar	886	49	893	730	799	749	812	719	727	970	606
Bijapur	632	40	583	405	511	668	547	356	440	564	884
Chamarajanagar	730	43	930	673	948	831	936	698	438	536	868
Chikamagalur	2073	87	2362	1777	2445	2690	2321	1772	1542	2125	1978
Chitradurga	495	31	586	422	602	722	607	363	528	548	769
Dakshina	3519	117	4161	3568	3978	4008	3969	4293	3392	4223	3724
Davanagere	623	43	638	471	657	832	678	535	542	701	901
Dharwad	787	58	679	474	702	793	676	744	526	696	980
Gadag	631	43	586	417	545	758	587	397	403	518	698
Gulbarga	839	45	765	548	649	723	690	593	594	776	702
Hassan	1148	58	1137	836	1348	1347	1186	1516	1141	1499	1500
Haveri	782	61	730	529	730	908	751	772	550	764	1095
Kodagu	2692	111	2806	2292	3251	2953	2814	3077	2001	3429	3015
Kolar	614	32	675	546	699	819	696	643	760	579	493
Koppal	587	36	602	417	470	736	571	417	392	560	722
Mandya	648	37	816	572	801	696	732	664	446	569	782
Mysore	730	47	894	631	804	721	769	782	491	681	833
Raichur	654	37	664	473	540	663	590	358	353	639	549
Shimoga	2421	89	2452	1764	2407	2582	2308	2869	2359	3059	2991
Tumkur	585	32	716	484	630	707	650	549	473	584	691
Udupi	4252	121	4783	4002	4250	4667	4471	5113	4265	4815	4010

Summary

The maize improvement need to be strengthened through development high yielding hybrids combined with drought tolerance along with adoption of farm practices namely disease control, weed control and irrigation management techniques to enhance yield levels. The development of GM maize may act as effective and sustainable solutions to control insect pests, manage abiotic stresses like drought and also prevent early stage losses due to weeds. The technology may have positive impact on social life of farmers and environment. The increase in thrust on promotion and adoption of mechanical harvesting could potentially assist in improving productivity and control farm losses. However, access to financial support, information and technological knowhow are areas where cultivators require support from governments and private organizations. Apart from the above a multipronged improvement is required in different aspects of maize production, processing, marketing and consumption for enhancing maize production in Karnataka. The public bred hybrids have on par yield levels with private bred hybrids. Besides, the cost of cultivation will be relatively less compared to private bred hybrids. But as the extension functionaries and popularization programmes with regard to public bred hybrids is week, the present programme envisaged as an important tool in bringing limelight into public bred hybrids. It is well known fact that the corporate sector always concentrate favorable environment for their commercial market, but it is the public sector, which needs to tackle all faculties of crop improvement programmes. The public-private partnership (PPP), needs to be strengthened to harness one another strengths.

References

Savitha, M.G. and Kunnal, L.B. (2015). Growth performance of cereals in Karnataka : A district wise analysis. *Agri Update*. 10(4): 288-293.

Singha, K. and Chakravorty, A. (2013). Crop diversification and growth of maize in karnataka: An assessment. Working paper series 299.

Yadav, O.P., Prasanna, B.M., Yadava, P., Jat, S.L., Kumar, D., Dhillon, B.S., Solanki, S. and Sandhu, J.S. (2016). Doubling maize (*Zea mays*) production of India by 2025-Challenges and opportunities. *Indian J Agric Sci.* 86(4): 427–34.

15

Maize Research and Development in Andhra Pradesh and Telangana

Lavakumar Reddy M., Narsimha Reddy V., Sreelatha D. Swarnalatha V. and Ameer Basha S.

Maize Research Centre, ARI, Rajendranagar, Hyderabad-500030
Corresponding Author's Email: mrcari@rediffmail.com

Maize production in India has been dominated by erstwhile Andhra Pradesh and Karnataka which together contribute around 38 per cent of the total maize production of the country. Erstwhile Andhra Pradesh has achieved the highest mean yield of 4441 kg/ha in the country during the last decade (2001-02 to 2010-11) by surpassing Karnataka which had the highest productivity in the country, earlier. The main reason for achieving the highest productivity in the country was due to adoption of improved production technology along with 100% area grown under hybrids. In Telangana, maize is the third most important crop after rice and cotton. It is cultivated in 5.74 lakh hectares, predominantly in *Kharif* season (4.54 lakh ha), with a productivity of 3.69 t/ha (2015-16).

Trends in maize area, production and productivity in the state (1950 to 2016)

The erstwhile Andhra Pradesh (including Telangana) is the non-traditional maize growing state. But, the climate of the state is very favorable for the crop and hence maize can be grown in any season in the state. However, it is mainly grown as a rainfed crop during monsoon season, concentrated in Telangana. The trends in area, production and productivity of maize in erstwhile Andhra Pradesh have shown a remarkable increase during the past six decades. During 1955-56, the maize acreage in the state was very less (1.89 lakh ha), which has been up by 5.3 times to the present (2014-15) level of 9.95 lakh hectares. The production has also increased dramatically from merely 1.14 lakh tonnes during 1955-56 to 42.36 lakh tones during 2014-15, which is nearly 37 times higher. Similarly, the productivity of maize in the state was merely 601 kg/ha during 1955-56, which has been increased to 4257 kg/ha during 2014-15 (Fig. 1)

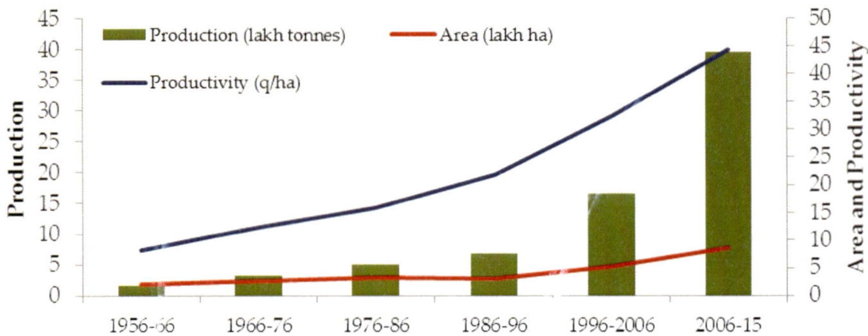

Fig. 1. Trends in area, production and productivity in erstwhile Andhra Pradesh.

Trends show that till 1990s the growth in area, production and productivity of maize was very slow but thereafter it got the momentum. The maize area, production and productivity have increased by 426%, 3616% and 608%, respectively, over six decades (Table 1).

Table 1. The area, production and productivity of maize in erstwhile Andhra Pradesh during 1955-56 to 2015-16.

Year	Area (lakh ha)	Production (lakh tonnes)	Yield (kg/ha)
1955-56	1.89	1.14	601
Mean	**1.89**	**1.14**	**601**
1956-57	1.88	1.05	557
1957-58	2.06	1.17	583
1958-59	1.91	1.61	842
1959-60	1.85	1.00	540
1960-61	1.82	1.55	852
1961-62	2.05	2.15	1050
1962-63	2.06	1.63	790
1963-64	2.02	1.63	806
1964-65	2.07	2.57	1244
1965-66	2.16	1.83	846
Mean	**1.99**	**1.62**	**811**
1966-67	2.27	1.73	761
1967-68	2.27	2.57	1135
1968-69	2.37	2.47	1045
1969-70	2.47	2.97	1205
1970-71	2.56	3.44	1344
1971-72	2.71	2.94	1084
1972-73	2.87	2.92	1018
1973-74	2.81	4.20	1493
1974-75	2.97	4.84	1632
1975-76	3.04	4.98	1636
Mean	**2.63**	**3.31**	**1235**
1976-77	2.95	3.01	1024
1977-78	3.05	4.74	1553
1978-79	3.16	3.60	1141

1979-80	2.90	3.94	1359
1980-81	3.21	7.25	2262
1981-82	3.31	6.31	1906
1982-83	3.33	7.44	2234
1983-84	3.41	5.22	1531
1984-85	3.13	4.33	1383
1985-86	2.89	4.14	1433
Mean	**3.13**	**5.00**	**1583**
1986-87	3.07	4.59	1494
1987-88	3.03	5.33	1756
1988-89	2.97	4.91	1653
1989-90	2.98	6.70	2244
1990-91	3.09	6.46	2086
1991-92	3.17	6.35	2001
1992-93	3.22	8.56	2660
1993-94	3.04	7.76	2554
1994-95	3.21	8.58	2670
1995-96	3.33	8.76	2630
Mean	**3.11**	**6.80**	**2175**
1996-97	3.61	7.90	3299
1997-98	3.96	10.83	2737
1998-99	3.99	13.83	3470
1999-00	4.39	13.97	3182
2000-01	5.23	14.27	2727
2001-02	4.28	14.59	3401
2002-03	5.26	14.86	2825
2003-04	7.21	24.77	3437
2004-05	6.57	20.64	3142
2005-06	7.58	30.87	4073
Mean	**5.21**	**16.65**	**3229**
2006-07	7.25	24.62	3396
2007-08	7.86	36.21	4607
2008-09	8.52	41.52	4873
2009-10	7.83	27.62	3527
2010-11	7.44	39.56	5317
2011-12	8.64	36.58	4234
2012-13	9.72	48.55	4995
2013-14	10.63	59.68	4673
2014-15	9.95	42.36	4257
Mean	**8.65**	**39.63**	**4431**

**Source*: Directorate of Economics and statistics.

However, the major increase was achieved during the past few years mainly due to adoption of single cross hybrids and shift in acreage under non-traditional areas. The quantum jump in maize area, production and productivity of the state itself indicates a revolution, which has been possible in India due to the combined efforts of public and private maize researchers, seed companies, farmers and the policies of central and state governments.

District-wise decadal and seasonal maize in erstwhile Andhra Pradesh

The area under maize has increased in all the districts of erstwhile Andhra Pradesh i.e., Mahaboobnagar, Karimnagar, Medak, Warangal, Nizamabad, Guntur, West Godavari, Khammam, Ranga Reddy, Kurnool, Adilabad, Anantapur, Vizianagaram, Prakasham, East Godavari, Vishakapatnam and Srikakulam (Fig. 2). However, during 1955-56 to 1994-95, the maize area came more in Karimnagar, Medak, Warangal, Nizamabad and Adilabad districts of Telangana. After 1995-96, the maize area has increased dramatically in Rayalaseema region of present Andhra Pradesh. Similarly, the productivity of maize also increased over six decades (1955-56 to 20014-15) in all the districts of erstwhile Andhra Pradesh. However, the highest productivity was achieved in Guntur (8890 kg/ ha), East Godavari (7247 kg/ha), Krishna (7106 kg/ha), Nellore (6928 kg/ha) and West Godavari (6729 kg/ha) (Table 2). This was mainly due to cultivation of maize in irrigated areas and use of recommended package of practices.

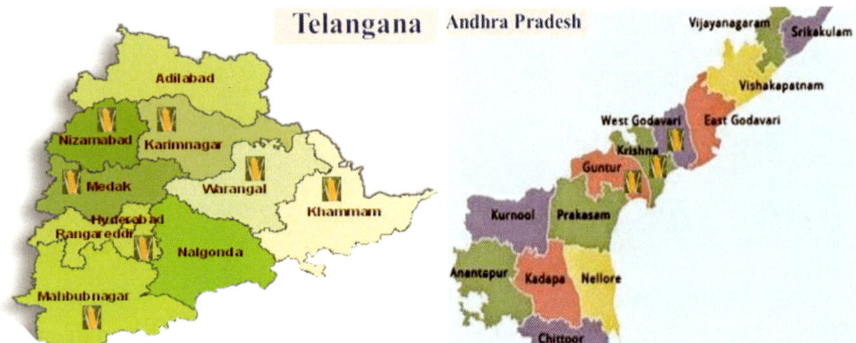

Fig. 2. State maps showing major maize growing districts.

The potentiality of maize hybrids can be fully exploited under assured irrigated conditions. Though the area under maize was highest in Telangana districts, its cultivation is mainly under rainfed conditions. As a result, the productivity was comparatively low in Khammam (5324 kg/ha), Nizamabad (4623 kg/ha), Karimnagar (4400 kg/ha), Warangal (4237 kg/ha), Adilabad (3432 kg/ha), Ranga Reddy (2979 kg/ha), Medak (2858 kg/ha) and Mahaboobnagar (2595 kg/ha) districts (Table 2).

AICRP Centers' genesis and their mandate

Maize Research Centre, ARI, Rajendranagar, Hyderabad, Telangana

The All India Coordinated Maize Improvement Project (ACMIP) established Maize Research Station at Amberpet, Hyderabad in 1957 with the following mandate:

Table 2. District-wise decadal seasonal trends in area, production and productivity of maize in erstwhile Andhra Pradesh.

District	1st decade 1955-56 to 1964-65			2nd decade 1965-66 to 1974-75			3rd decade 1975-76 to 1984-85			4th decade 1985-86 to 1994-95			5th decade 1995-96 to 2004-05			6th decade 2005-06 to 2014-15		
	A	P	Y	A	P	Y	A	P	Y	A	P	Y	A	P	Y	A	P	Y
Srikakulam	1	1	846	0	0	1107	0	0	2167	0	1	1812	1	3	3303	6	35	5562
Vizianagaram	0	0	0	0	0	0	0	1	2636	1	1	2009	4	18	4059	16	69	4016
Vishakapatnam	2	2	862	2	2	1115	6	12	3065	7	13	1609	8	16	1951	7	22	2906
East Godavari	1	1	778	2	2	1174	2	4	2109	3	7	2331	6	28	4535	9	65	7247
West Godavari	0	0	859	1	1	1217	2	3	2197	3	8	2650	15	67	5036	43	294	6729
Krishna	1	1	880	2	2	1160	2	4	2113	2	4	2056	9	33	3654	22	154	7106
Guntur	2	2	885	2	2	1135	2	3	2116	4	8	2037	13	70	4608	67	598	8890
Prakasam	0	0	0	0	0	1441	0	0	1954	7	18	2311	11	39	3403	11	72	6572
Nellore	0	0	855	0	0	1144	0	0	2145	0	1	1903	0	1	3517	1	8	6928
Kurnool	1	1	860	0	0	1214	0	1	2093	1	1	1905	3	7	2819	25	133	5429
Anantapur	0	0	871	1	1	1228	1	2	2193	1	2	1928	4	14	3024	18	84	4859
Kadapa	0	0	849	0	0	1166	0	0	2145	0	0	1697	0	0	2843	3	11	6418
Chittoor	0	0	843	0	0	1193	0	1	2099	1	1	1940	1	2	3343	2	9	5668
Rangareddy	0	0	0	0	0	0	6	7	1089	5	7	1572	13	36	2589	31	104	2979
Hyderabad	6	4	779	7	6	874	7	7	1491	0	0	2053	0	0	0	0	0	0
Nizamabad	22	23	997	41	62	1474	60	113	1868	61	142	2141	52	184	3504	74	345	4623
Medak	41	27	659	48	39	819	64	67	1035	68	98	1454	86	194	2257	115	336	2858
Mahabubnagar	0	0	754	0	0	1195	0	1	1670	2	4	2208	45	86	2146	133	344	2595
Nalgonda	2	1	752	2	2	1124	2	2	1460	1	2	2032	3	8	2862	3	8	2351
Warangal	33	22	701	41	51	1175	45	72	1562	30	54	1836	54	164	2954	81	352	4237
Khammam	6	6	884	7	7	980	6	7	1244	9	22	2152	21	63	3013	33	175	5324
Karimnagar	59	50	847	79	101	1529	92	181	1959	82	216	2651	107	382	3542	123	567	4400
Adilabad	17	13	777	20	19	940	23	27	1145	24	31	1184	29	86	2983	24	81	3432

Note: A: Area in '000 ha; P: Production in '000 tonnes and Y: Yield in kg/ha **Source*: Directorate of Economics and statistics

- Development of high yielding – early, medium and long duration hybrids tolerant to pests and diseases
- Development of specialty corn hybrids in sweet corn, QPM, baby corn and popcorn
- Development of maize hybrids tolerant to abiotic stresses
- Maintenance and production of nucleus and breeder seed of parental lines of released hybrids and composites
- Development of suitable seed production technologies for maize parental lines and hybrids released from the centre
- Standardization of population densities and fertilizer schedules for pre-release maize hybrids
- Studies on weed management with post emergence herbicides and cropping systems in maize
- Management of pests and diseases in maize

The centre was the head quarters for peninsular India. In 2006, by considering the demand of winter nursery and maize research of the Acharya N. G. Ranga Agricultural University (now Professor Jayashankar Telangana State Agriculture University), the center was permanently shifted to ARI, Rajendranagar, Hyderabad with allocation of 11.04 ha land. Recently in 2015, an additional 3.40 ha land has been allotted to Maize Research Centre to cater the increasing needs of maize research and seed production.

Agricultural Research Station, Karimnagar, Telangana

The Agricultural Research Station, Karimnagar was established in the year 1955 to conduct research on all aspects of agricultural crops like maize, pulses and oilseeds with a special emphasis on maize under Acharya N. G. Ranga Agricultural University (now Professor Jayashankar Telangana State Agriculture University), Rajendranagar, Hyderabad. However, the All India Coordinated Maize Improvement Project was started in the year 1993 with the following mandate.

- Development of medium and short duration high yielding hybrids for rainfed and irrigated conditions
- Development of drought tolerant high yielding hybrids of maize for different situations
- Development of high yielding hybrids with resistance to important diseases

- Improvement of special types of maize, *viz.*, popcorn, sweet corn and fodder maize
- Re-stabilize the package of practices of maize to increase productivity
- Maintain inbred lines and germplasm of maize for future use in seed production and breeding programmes.

AICRP technologies

Landraces and their utilization in breeding

During the pre-independence era, farmers used to grow landraces in maize called as Warangal local (Chinna makka) in Warangal, Karimnagar and Nizamabad area, Khammam local (Pedda makka/Nakka jonna) in Khammam tribal area and Vishakapatnam local (Potta jonna) in agency area of Vishakapatnam of Andhra Pradesh. These were poor yielders and mainly grown for family consumption. These landraces were used in Indian maize breeding programmes which was started in year 1957 as All India Coordinated Maize Breeding Schemes having sub-centres in different zones of India. The main objective of AICMIP was to improve maize yields by replacing the poor yielding landraces and local varieties in order to improve the maize productivity in the farmers' fields. Maize breeders derived improved inbred lines like CM 100, CM 101, CM 102, CM 103, CM 104, CM 105, CM 1106, CM 200, CM 201 and CM 202 from indigenous germplasm/landraces.

Cultivars developed by MRC

The first double cross maize hybrid, "Deccan" was developed by the breeders by using inbred lines - CM 104, CM 105, CM 201 and CM 202 (Vital Rao, 1983). The centre has released six single cross hybrids, *viz.*, Amber Shakti (2004, 1st QPM), DHM 115 (2004), DHM 111 (2009), DHM 113 (2009), DHM 117 (2009), DHM 119 (2010) and DHM 121 (2013) (Table 3); five double cross hybrids, *viz.*, DHM 101 (1975), DHM 103 (1982), DHM 1 (1988) and DHM 105 (1990); three three-way cross hybrids which were 1st in the country *viz.*, Trishulatha (1991, long duration), DHM 107 (1992, medium duration) and DHM 109 (1993, short duration); three composites, *viz.*, Ashwini (1987), Harsha (1988) and Varun (1990, 1st medium duration & drought tolerant). In addition, it has developed popcorn composite variety, Amber popcorn in 1975 and popcorn hybrid, BPCH 6 (1st in the country from public & private sector) in 2012, which were released for cultivation across India. Further, sweet corn composite variety, Madhuri (1990, 1st in the country) and Priya (2002), developed by the centre, were also released for cultivation all over India. The details of the hybrids developed by the centre during 1957 to 2016 are given in Table 4.

Table 3. Single cross hybrids released by AICRP Centre.

Hybrid	Year of release	Specific characters	Photograph
Released from Maize Research Centre, Hyderabad			
DHM 115	2005	• Short duration • Tolerant to TLB, rust & PFSR • Tolerant to *Chilo partellus*. • Yield – 6.5-7.0 t/ha	
DHM 111	2009	• Medium duration • Non-lodging • Tolerant to TLB, MLB & BSDM • Yield – 7.0-7.5 t/ha	
DHM 113	2009	• Long duration • Non- lodging • Stay green at physiological maturity • Tolerant to TLB, MLB and BSDM. • Yield – 7.5-8.0 t/ha	
DHM 117	2009	• Medium duration • Non-lodging with sturdy stem • Stay green at maturity • Tolerant to TLB, MLB & PFSR • Yield – 8.0-8.5 t/ha	
DHM 119	2010	• Medium duration • *Kharif* season • Moderately tolerant to stalk borer • Yield – 7.5-8.0 t/ha	
BPCH 6	2012	• Popcorn hybrid • Medium duration • Yield 4.0-4.5 t/ha	
DHM 121	2014	• Medium duration • Tolerant to TLB, MLB & PFSR • Yield 7.0-7.5 t/ha	

Released from ARS, Karimnagar, Telangana			
Karimnagar makka	2016	• Short duration • Yield 6.0-7.0 t/ha	
Karimnagar makka -1	2016	• Medium duration • Tolerant to wilt • Yield 7.0-8.0 t/ha	

Registered lines and hybrids

Seventeen normal inbred lines developed from MRC, Hyderabad, *viz.*, BML 2, BML 3, BML 4, BML 5, BML 6, BML 7, BML 8, BML 10, BML 11, BML 12, BML 13, BML 14, BML 15, BML 16, BML 20, BML 21 and BML 22; seven inbreds tolerant to post flowering stalk rot *viz.*, BPPTI 28, BPPTI 32, BPPTI 34, BPPTI 35, BPPTI 37, BPPTI 38 and BPPTI 44 and three inbred lines resistant to stem borer *Sesamia inferens*, *viz.*, DMR E62, E63 and DMR-ENT-2-3 were registered with NBPGR. In addition, Priya sweet corn composite and four single cross hybrids, *viz.*, DHM 117, DHM 111, DHM 113 and DHM 119 were registered with PPV& FRA. Further, DHM 121 & BPCH 6 are under the process of registration with PPV& FRA.

Maize seed production and MoUs

Andhra Pradesh and Telangana states are the seed bowl of India. About 40,000 ha area is under hybrid maize seed production in *Rabi* i.e. more than 3/4th of India's hybrid maize seed produced. About 400 seed companies are doing their business in the states, as the climate of the states is congenial for seed production and storage. The state has experienced proliferation of seed growers, irrigation facilities, drying & processing plants and cold storage units. Since 2013-14, about 65 tons of DHM 117 and 5 tons of DHM 121 certified seed were produced and distributed to the farmers for cultivation. The state agricultural universities are involved in seed production of nucleus and breeder seeds of released hybrids. The university supplies breeder seed to various public seed corporations and private seed companies to make use of them in producing hybrid seed for commercial cultivation. During the period 2010 to 2016, thirteen MoUs (Table 5) were made with 12 private seed producing companies to produce and market DHM 117 (10 MoUs) and DHM 121 (3 MoUs).

Since 2010, about 30 q of BML 6, 15 q of BML 7 and 5 q of BML 45, the parental lines seed of DHM 117 and DHM 121 hybrids were produced and supplied to NSC, APSSDC, TSSDC, Rajasthan Seed Corporation, Maharashtra Seed Corporation and Private seed companies.

Table 4. Cultivars released for state by AICRP centres.

S. No.	Hybrid	Year	Released by	Type	Pedigree	Other Characteristics
Released by MRC, Hyderabad (ANGRAU/PJTSAU)						
Normal Hybrids: Yellow/orange						
1	DHM	1961	CVRC	Double cross	(CM 104 X CM 105) X (CM 202 X CM 201)	Long duration, yellow, semi-flint
2	DHM 101	1975	CVRC	Double cross	(CM 115 X CM 114) X (CM 202 X CM 206)	Long duration, yellow, semi-flint
3	DHM 103	1982	CVRC	Double cross	(CM 120 X CM 118) X (CM 119 X CM 208)	Long duration, tolerant to leaf blight and stalk rot
4	DHM 1	1988	CVRC	Double cross	(CM 209 X CM 120) X (CM 211 X CM 130)	Short duration, resistant to leaf blight
5	DHM 105	1990	CVRC	Double cross	(CM 119 X CM 120) X (CM 211 X CM 131)	Short duration, tolerant to leaf blight and stalk rot
6	Trishulatha	1991	CVRC	Three-way cross	(CM 131 X CM 211) X CM 120	Long duration, yellow-orange, semi-flint, tolerant to TLB, MLB and stalk rot
7	DHM 107	1992	CVRC	Three-way cross	(CM 119 X CM 211) X CM 132	Medium duration, yellow, semi-flint, tolerant to leaf blights, SDM and *Erwinia* stalk rot
8	DHM 109	1993	CVRC	Three-way cross	(CM 119 X CM 211) X CM 133	Short duration, yellow-orange, semi-dent, tolerant to leaf blights, SDM and *Erwinia* stalk rot
9	DHM 115	2004	CVRC	Single cross	BML 13 (CM 148) X BML 10 (CM 149)	Short duration, orange, flint, tolerant to leaf blight, rust and stalk rot
10	DHM 111	2009	SVRC	Single cross	BML 6 X BML 15	Medium duration, yellow, semi -dent, nutrient responsive, tolerant to lodging and stay green
11	DHM 113	2009	SVRC	Single cross	BML 2 X BML 7	Long duration, orange, semi-flint, nutrient responsive and tolerant to lodging
12	DHM 117	2009	SVRC	Single cross	BML 6 X BML 7	Medium duration, orange-yellow, flint, nutrient responsive, tolerant to stalk rots and lodging and stay green type
13	DHM 119	2010	CVRC	Single cross	BML 2 X BML 15	Medium duration, yellow, flint, nutrient responsive, resistant to lodging

14	DHM 121	2013	CVRC	Single cross	BML 45 X BML 6	Medium duration, non-lodging, tolerant to stem borer, leaf blights and post flowering stalk rots
Varieties: Yellow/orange						
1	Ashwini	1987	CVRC	Composite	-	Composite, tolerant to drought, medium duration
2	Harsha	1988	CVRC	Composite	-	Composite, yellow orange, semi-flint, tolerant to drought and TLB
3	Varun	1990	CVRC	Composite	-	Composite, tolerant to drought, medium duration
Specialty corns: Yellow/orange						
1	Amber	1975	CVRC	Composite	-	Popcorn composite, medium duration
2	Madhuri sweet corn	1990	CVRC	Composite	-	Yellow, dent, short duration suitable for table purpose, sugar 24%, fodder yield 20-25 t/ha
3	Priya sweet corn	2002	CVRC	Composite	-	Yellow, dent, medium durationtolerant to TLB, stalk rots, sugar 20-22%, fodder yield 20-25 t/ha
4	Amber Shakti	2004	CVRC	Single cross	BML 23 X BML 15	QPM hybrid, medium duration
5	BPCH 6	2012	CVRC	Single cross	BPCL-3 X BPCL-6	Popcorn, orange, medium duration, suitable for *Kharif* and *Rabbi*
Fodder maize:						
1	APFM 8	1997	CVRC	-	Varun x Palampur local	Leafy, non-lodging
Released by ARS, Karimnagar (PJTSAU) :						
Yellow/orange						
1	Karimnagar Makka 1	2016	SVRC	Single cross	PFSR 3 X BML 7	Short duration, wilt tolerant
2	Karimnagar Makka	2016	CVRC	Single cross	KML 225 X BML 7	Medium duration, suitable for *Kharif*

Table 5. Particulars of MoUs entered by private seed companies with the University for production and marketing rights.

Name of the Company	Date of Commencement	Date of Termination	Hybrid	Fee Paid (Rs. in lakhs)
M/s Vikky Agri-sciences Pvt. Ltd, Hyderabad	2010	Three years	DHM 117	5% royalty on company sales
M/s Sampoorna Seeds Yemiganur, Kurnool	2010		DHM 117	
M/s ABC Seeds India Pvt. Ltd, Hyderabad	2011		DHM 117	
M/s Shakthi Seeds Pvt. Ltd., Hyderabad, T.S.	24-02-2014	23-02-2018	DHM 117	3.00
M/s Sonam Seed Technologies Pvt. Ltd., Karimnagar, T.S.	24-02-2014	23-02-2018	DHM 117	3.00
M/s Genesis Agro Seeds Pvt. Ltd., Reg. Office, H.No. EWS 86/1, KPHP Colony, Kukatpally, Hyderabad	01-09-2014	31-08-2018	DHM 117	3.00
M/s Naveen Seeds, Adoni, Kurnool, A.P.	16-09-2015	15-09-2019	DHM 117	3.00
M/s Naveen Seeds, Adoni, Kurnool, A.P.	23-11-2015	22-11-2019	DHM 121	3.00
M/s Sri Laxmi Venkateswara Seeds, Peddapadu, Kurnool, A.P.	19-05-2016	18-05-2020	DHM 117	3.00
M/s Muralidhar Seeds Corporation, Kurnool, A.P.	19-05-2016	18-05-2020	DHM 121	3.00
M/s Sayaji Seeds, Kathwada, Ahmedabad, Gujarat	29-06-2016	28-06-2020	DHM 117	3.00
M/s Chakra Seeds, Kurnool, A.P.	06-12-2016	05-12-2020	DHM 117	3.00
M/s Sri Sai Laxmi Seeds, Kurnool, A.P.	06-12-2016	05-12-2020	DHM 121	3.00

Technology demonstrations and outreach

The organization of field day and training programmes are adopted to popularize the technology among the farmers. Maize hybrids/varieties and technologies were popularized in maize farming community by conducting various minikits, front line demonstrations, kisan melas, and exhibitions through university extension wing. During 2008-16, 1611 FLDs were conducted in farmers' fields in *Kharif* (895), under rainfed; and *Rabi* (716), under irrigated conditions. The mean yield in FLDs was 23.0 to 60.0 per cent during *Kharif,* and 4.3 to 41.6 per cent during *Rabi* season was higher over the state average (Table 6). The combined efforts of FLDs, extension work and availability of high yielding hybrids like DHM 117 and DHM 121 have led to increase in maize area under single cross hybrids in the state. As a consequence, the maize productivity of the state has increased to 5317 kg/ha during 2010-11 from merely 275 kg/ha during 1950-51. In addition, the zero-tillage technology was also popularized through frontline demonstrations in the state. Further, in order to reach remote tribal farmers, ICAR has implemented tribal sub-plan programme (TSP) to popularize maize hybrids, production technologies for normal maize, sweet corn, popcorn and baby corn. During 2011-16, 310 demonstrations were conducted under TSP programme in *Kharif* and *Rabi* seasons. The mean increase in yield over control plot ranged from 12.22 to 28.7 per cent (Table 7).

The farmers were also demonstrated with the production technologies like timely management of weeds by using Atrazine as pre-emergence, and timely control of stem borer by using Moncrotophos at 12-15 days after germination followed by whorl application of Carbofuran 3 G granules. The earlier practices of farmers like application of only nitrogen in the form of urea/DAP was replaced with balanced use of fertilizers, emphasizing on potash application. Majority of the farmers have adopted improved technology for achieving higher yields in maize.

Table 6. Demonstrations conducted under FLD Programme.

Year	FLDs in *Kharif* (No.)	Mean yield (q/acre) in FLD plots	State Yield (q/acre)	Increase over state yield (%)	FLDs in *Rabi* (No.)	Mean yield (q/acre) in FLD plots	State Yield (q/ acre)	Increase over state yield (%)
2008-09	-	--	-	-	270	62.8	48.7	28.0
2009-10	90	61.2	49.6	23.0	180	70.2	49.6	41.6
2010-11	50	59.6	37.1	60.0	65	73.0	76.0	4.0
2011-12	150	39.8	30.2	24.1	35	75.3	66.3	11.9
2012-13	30	50.0	28.1	43.8	66	67.8	62.8	4.3
2013-14	500	62.5	34.4	55.0	100	65.45	60.76	7.8
2014-15	25	62.5	49.8	25.5	--	--	--	-
2015-16	25	61.5	40.1	53.3	--	--	--	--
2016-17	25	49.5	39.5	25.3	--	--	--	--

Table 7. Demonstrations conducted in our state under Tribal Sub-Plan Programme.

Year	FLDs	Yield (q/acre) in FLD plots	Yield (q/acre) in control plots	Increase overcontrol plots (%)
2011-12	10	29.8	25.8	15.5
2012-13	50	32.6	28.1	13.7
2013-14	50	30.1	26.5	12.2
2014-15	40	26.2	21.5	18.2
2015-16	100	21.9	17.5	20.2
2016-17	60	13.7	10.8	28.7

Impact of AICRP technologies

During 1950-51, the maize area in the state was predominantly under low yielding local cultivars/landraces. As a result, the productivity of maize in the state was merely 275 kg/ha. However, due to continued efforts of our AICRP centre, new and high yielding hybrids were developed and released for the state for commercial cultivation. Similarly, private sector also concentrated more on commercializing maize hybrids. The farmers realized the impact of new cultivars with respect to higher productivity over local varieties (poor yielders) and large numbers of farmers have started adopting improved hybrids. As a result, the local varieties and landraces were completely replaced in the state and 100 per cent maize area in the state got covered under hybrids. Another reason for complete replacement of landraces was due to shift in acreage under non-traditional areas. Presently, large number of double, triple and single cross hybrids released by public and private sector are available to the farmers. The popular hybrids which are currently cultivated in the state are namely, DHM 117, DHM 121, P 30V92, P 3396, NK 6240, 900 M Gold, Pinnacle, Kaveri 50, NK 30, HT 5205, CP 808, CP818, DOW 4683, 25K60 and 25K45. During the last three years (2013-2015), the centre has produced and sold about 65 tons of DHM 117 and 5 tons of DHM 121 seed to farmers for cultivation. The state has witnessed a remarkable increase in production, productivity as well as area due to adoption of single cross hybrids. Thus the productivity of maize was increased to 5317 kg/ha, which was the highest average state's productivity in the country during 2010-11.

Cropping systems and production ecologies

During 1950s, tribal farmers used to grow maize in shifting cultivation for their family consumption. In plain regions of Telangana and Andhra Pradesh, farmers grow maize as sole crop (land kept fallow till next *Kharif*) under rainfed situation during *Kharif* season. Maize possesses wide adaptability under diverse agro-climatic conditions and this flexibility makes it an attractive crop for cultivation in sequence or with different crops under various seasons and agro-ecologies. Under irrigated conditions, farmers grow rice-maize, maize-maize, rice-maize-pulses, groundnut-maize, maize-groundnut, maize-chickpea, maize-coriander, maize-sesame, sugarcane-maize and turmeric-maize. In *peri*-urban areas, maize-vegetables (tomato, brinjal, bhendi, potato, field bean etc.) and maize-flower crops (like marigold) cropping system are followed in irrigated conditions. Under rainfed, farmers grow redgram and cowpea as intercrops. Some farmers earn good profits by growing beans as intercrop with maize. Some of the farmers grow maize as mixed crop in turmeric for their family consumption. Maize is also grown as guard crop in summer vegetables in fixed rows to protect the tomato and leafy vegetables from heat stresses. Maize is considered to be very nutritious fodder for livestock and can be easily fitted into forage cropping system. Hence, farmers grow maize-berseem, maize-lucern and sorghum-maize. It is also grown as intercrop in forage crops particularly in pulses (Kumar *et al.,* 2013).

Production technologies recommended and package of practices

Particular	Package of practices
Normal maize	
Variety/Hybrids	DHM 113 (long), DHM 117, DHM 121 & DHM 119 (medium duration) and DHM 115 (short duration) and private hybrids released for state.
Suitable soils	Red sandy loams to medium black soils with good drainage facilities are preferable. Light red sandy soils are not suitable for cultivation under rainfed conditions.
Seed treatment	Seed treatment with Thiram/ Captan/ Mancozeb @ 3 g/kg seed. Metalaxyl 4 g/kg seed for downy mildew.
Season and sowing time	*Kharif*: June to July 15 (in case of delayed monsoon, sowings may be extended up to July end using short duration hybrids). *Rabi*: October to November.
Seed rate/sowing method	8-10 kg per acre, 60x20 cm^2 spacing with 33,333 plants per acre. Under rainfed situations, sowing can be taken up after overnight seed soaking after rainfall of 50 mm. **Method of sowing**: Sowing one side of ridge at the distance of 1/3rd from top facilitates irrigation as well as drainage. Excess seedlings should be thinned 10 days after emergence to have one seedling/ hill.
Fertilizer doses & time of application	For *Kharif* crop, a dose of 70-80 kg N, 24 kg P_2O_5, 20 kg K_2O per acre is recommended. Nitrogen may be applied in three equal splits i.e., at sowing, knee high and flowering stages. For *Rabi* crop, 80-100 kg N, 32 kg P_2O_5, 32 kg K_2O per acre is recommended. Nitrogen has to be applied in four equal splits *viz.*, at sowing, knee high (30-35 DAS), flag leaf emergence (50-55 DAS) and tasseling-silking stage (60-65 DAS). Apply 20 kg of Zinc Sulphate per acre for every 2-3 seasons. If symptoms appear on standing crop, spray Zinc Sulphate @ 2 g/l of water.
Weed management	• Pre-emergence spraying with Atrazine 50 W.P. @ 800-1200 g/ ac depending on soil type effectively controls most of the broad leaved weeds. • In intercropping system involving pulse and oilseed crops, pre-emergence application of Pendimethalin 30% EC @ 1.0 l/ acre in 200 litre of water is recommended. • Post-emergence application of Atrazine (same dose) or 2,4-D Sodium salt 80 WP @ 400 g or Tembotrione @ 115 ml +

Atrazin 400 g or Sempra @ 36 g/ac at 20-25 days crop or 4 leaf stage of weeds.

- Intercultivation followed by earthing up is done at 30-35 days.

Water management

Adequate moisture should be present in the soil at the time of sowing for germination. Care should be taken to avoid water logging up to 30 days after sowing. Flowering, grain filling and dough stages are sensitive to moisture. For *Rabi* crop, heavy soils may require 5-6 irrigations, whereas light soils need 8 irrigations. Irrigation can be given once in 8 days in sandy loam soils and once in 15 days in heavy soils depending on the weather.

Pest and Disease management

a) Pests:

The stem borer, *Chilo partellus* infests the crop during *Kharif* and pink borer *Sesamia inferens* infests during *Rabi* season. The borers cause dead hearts in early stage of crop. The pest incidence is recognized by the presence of shot holes in the leaf blades as well as exit holes on the stem.

Management of Stem borers:

- Farm sanitation: Removal of infested plants, ploughing the field soon after harvest and collection and burning of stubbles to kill the hibernating larvae and pupae.
- Growing stem borer tolerant hybrids (DHM 117).
- Adjusting the sowing time to avoid peak activity of the borers to coincide with critical stage of the plant (10-12 days old).
- Maintaining optimum plant density in the field (33,000 plants/ac).
- Intercropping with legumes such as cowpea, soybean, redgram and green gram in 2:1 ratio encourages the buildup of natural enemy population.
- Trap crop: Sorghum is the preferred host for *C. partellus*. Sowing 2-3 rows of trap crop on all sides of maize and uprooting it after 45 days.
- Clipping of lower leaves of maize on which most of the eggs are laid.
- Crop rotation with non-host crops destroys the buildup of pest due to non-availability of host.
- Release of egg parasitoid, *Trichogramma chilonis* @ 8 cards per ha twice i.e., at 12 and 22 days after germination.
- *Cotesia flavipes* is the dominant and most widely distributed larval parasitoid.
- Use of recommended dose of fertilizers, avoiding excessive use of nitrogen that increases pest attack.
- Prophylactic spray of Monocrotophos 36 SC @ 1.6 ml/l or Chlorantriniliprole 20 SC @ 0.3 ml/l of water at 10-15 DAG followed by Carbofuran 3G @ 7.5 kg/ha in plant whorls in case of severity.
- **Aphids:** Spray Dimethoate 30 EC @ 2 ml/l or Monocrotophos 36 SC @ 1.6 ml/l of water.

b) Diseases:

The important diseases of maize are leaf blight (*Exserohilum turcicum*), late wilt (*Cephalosporium maydis*) and charcoal rot (*Macrophomina phaselina*). One to two sprayings of Mancozeb @ 2.5 g/l of water at 8-10 days interval starting from knee high stage of the crop controls the leaf blight. Banded leaf and sheath blight is observed in some of the districts and when the symptoms are noticed, stripping of the affected bottom 2-3 leaves along with their sheath and spraying of Propiconazole @1 ml/l is recommended. For late wilt & charcoal rot, crop rotation, removing plant debris, summer ploughing, avoiding moisture stress after flowering and growing tolerant hybrids should be followed.

Rust (*Puccinia sorghi*): Spraying of Mancozeb @ 2.5 g/l as soon as pustules appear and repeat at 10 days interval till flowering.

Downy mildew: Seed treatment with Metalaxyl @ 7 g/kg seed, plant debris and collateral host should be removed spray Mancozeb @ 2.5 g/l or Metalaxyl MZ @ 2 g/l of water.

Harvesting

Harvesting should be done when the crop attains physiological maturity i.e., 7-8 weeks after flowering or when the grain moisture is 25-30 %.

Physiological maturity can be known by yellowing of most of leaves which start drying, ear husks turn brown and papery, ears begin to droop on stalk, and black layer formation at the bottom of kernels. In the absence of irrigation facilities and failure of rains, maize can be harvested at any stage of its growth for different uses. At pre-flowering stage for fodder, flowering stage for baby corn and milky to early dough stage for green ears. Stalks/ Stover can also be used for silage making or to feed cattle.

Zero-till maize

Sowing time

Sowing in rice fallows during *Rabi* from October to November which may extend up to December or January in coastal districts.

Sowing method

Depending on the soil type, one last irrigation should be given to the standing paddy crop so that sufficient moisture would be available for the germination of subsequent maize crop. Spacing of 60 x 20 cm can be maintained using a rope and seeds should be dibbled with a gap of 20 cm at a depth of 5.0 cm. Zero till drill machine can also be utilized for dibbling the seed.

Weed control

Pre-emergence application of Paraquat 24% @ 1.0 litre / acre to arrest re-growth of paddy stubbles followed by Atrazine @ 1.0 kg/ acre in 200 litre water. Combination of Paraquat and Atrazine was also found to give effective weed control.

Pest and disease management

Ear borers: In the coastal areas of Andhra Pradesh, incidence of cob borers like *Sesamia inferens*, *Spodoptera spp.*, *Helicoverpa spp.* and hairy caterpillars were observed in zero-tillage maize.

Management: Setting up of Light traps and Pheromone traps in the field can attract the adults of *H. armigera* and *S. litura*/ Hand picking and destruction of larvae/ Dusting with Folidol 2 % on the silks/ Spraying of Carbaryl 50 % WP @ 3 g/l or Monocrotophos @ 1.6 ml/litre water can manage the pest.

	Pink borer *Sesamia inferens* infests during *Rabi* season. The control measures are similar to that of normal maize.
	Diseases: Common diseases are late wilt and sheath blight. Follow the recommended control measures as given for normal maize.
Water management	Irrigation should be given at 30, 45, 60, 80 and 90 days depending on the soil type.
Specialty corn	
Variety/Hybrid	**Sweet corn:** Madhuri, Priya, Win Orange, Sugar 75 (private hybrid).
	Popcorn: Amber popcorn, Pearl popcorn, VL popcorn and BPCH 6 (hybrid).
	Baby corn: VL 42, Him 123, Him 129, Madhuri, VL78, JH 3459 & VL Baby corn 1, DHM 115.
	Quality Protein Maize: HQPM-1, HQPM-4, HQPM-5, HQPM-7, Vivek QPM 9.
Seed treatment	Same as for normal maize.
Sowing time	Baby corn can be grown throughout the year under assured irrigation.
Seed rate/sowing method	4.0 kg per acre for sweet corn, 5 kg for Popcorn, 7-8 kg for Quality Protein maize and 10 kg for baby corn.
	Spacing of 60 x 20 cm for sweet corn, popcorn and QPM and 45 x 20 cm for baby corn.
Fertilizer doses & time of application	For sweet corn and baby corn- 60-72, 24, 20; for popcorn 32, 24, 20; for QPM 72-80, 24, 20 Kg N-P_2O_5 and K_20 /acre. For sweet corn, popcorn and QPM, three splits (basal, knee high and tasseling and silking); for baby corn in two splits (basal and knee high stage) are recommended.
Pest management	Similar to that of normal maize except that granules applications for the control of stem borers is not recommended for sweet corn and baby corn.
Harvesting	Sweet corn is ready for harvest 20-25 days after completion of pollination (soft white lustrous kernels turning yellow that produce milky liquid when punctured). For baby corn, the ears are harvested when the silks are 1-2 cm long i.e., within 1-2 days after silk emergence, otherwise ears quickly become too long and tough. For popcorn, harvesting should be done when seed moisture is 20-25% and ears dried up to 18% moisture in shade. The dried grain should contain 13-14% moisture for good popping.

Key production problems in state for maize

Maize is grown mainly under rainfed situation during *Kharif* season (76%) and irrigated conditions during *Rabi* season (24%). The mean annual rainfall of the state is 939 mm, of which, 742 mm (79%) from south-west monsoon and 197 mm (21%) from north-east monsoon. Generally SWM enters in the state during the month of June and recedes in the September; whereas, NEM enters the state in 2nd fortnight of October and continues up to December. During 2007-16, there was a deficit rainfall of 23 per cent over normal rainfall (742 mm). Rainfall distribution of SWM was also erratic in most of the years; as a result many drought periods were experienced during the seasons. Deficit rainfall was reported in the years 2006, 2007, 2009, 2011, 2012, 2013, 2014 & 2015. Therefore, area under maize was decreased in the years 2009, 2010, 2011, 2012 & 2015 during *Kharif* season. The mean productivity of maize during *Kharif* season is 3.22 t/ha, whereas in *Rabi* it is 5.07 t/ha. A yield gap of 1.85 t/ha is there between *Kharif* and *Rabi* seasons. This yield gap can be bridged by providing protective irrigations to maize in drought periods during *Kharif* season. Thus, drought is recognized as the most important constraint across the rainfed growing areas.

In addition to maize being grown predominantly under rainfed conditions by small holder, resource-poor farmers, they face other constraints like overdependence on rainfall, prolonged droughts during reproductive stage, poor agronomic management, lack of access to quality seed and yield losses due to pre- and post-harvest pathogens, insect pests and weeds. This situation is likely to be worsened in the coming decades due to climate change, *viz.*, temperature changes and inadequate and/or uneven pattern of rainfall distribution during the cropping season (Table 8 & 9). Climate projections also indicate elevated temperatures, especially in the drought prone areas. Maize is particularly vulnerable during reproductive stage to heat stress. However, alleviating the effects of drought alone could increase average maize yields by 35 per cent. These highlight the importance of developing improved maize varieties with tolerance to both drought and heat stress.

During the last 4-5 years, unexpected rains are coming at the time of harvest (cyclones, North-east monsoons) as a result, maize produce gets wet under rains. Due to lack of threshing floors at village level, farmers are not able to avoid wetting and proper drying of produce in time. Hence, post harvest losses are increasing during *Kharif* season due to fungi, insects, rodents and birds etc. These post-harvest losses are classified into three main categories; quantitative loss, qualitative loss and economic or commercial loss. Quantitative loss indicates the reduction in physical weight and can be readily quantified and valued, example a portion of grain damage by pests or losses during transportation. A qualitative loss is the contamination of grain by molds; such loss is often aggravated by inappropriate handling, poor storage facilities, insects and other pests. Mycotoxins are toxic secondary metabolites of fungi that frequently contaminate the maize in the field and/or during storage. It poses health risk to humans and domesticated animals if not properly managed because of their acute and chronic effects. The most important mycotoxins in maize are the aflatoxins, fumonisins, deoxynivalenol and ochratoxi. Common stored grain insects that cause damage to maize are maize weevil, lesser grain borer, red flour beetle, khapra beetle,

Table 8. Rainfall pattern during *Kharif* seasons (month wise).

Month/ Year	2006		2007		2008		2009		2010		2011		2012		2013		2014		2015		2016	
	Rf	RD	Rf	RD	Rf	RD	Rf	RD	Rf	RD	Rf	RD	Rf	RD	Rf	RD	Rf	RD	Rf	RD	Rf	RD
JUN	44.5	7	134.3	10	66.7	6	82.0	5	113.7	6	40.7	4	139.5	6	145.7	8.0	53.6	3	160.0	8	90.0	8
JUL	87.0	7	50.8	4	83.7	8	54.0	4	278.9	16	191.0	11	261.6	11	150.2	9.0	108.0	8	25.2	3	144.7	11
AUG	211.2	11	172.5	13	506.3	15	203.7	9	203.1	13	147.3	9	99.4	9	158.1	11.0	184.1	10	126.8	12	180.6	6
SEP	165.3	14	146.3	13	204.6	7	165.5	9	243.8	13	68.1	5	117.9	10	110.6	8.0	60.6	7	168.2	9	391.6	17
	508.0	39.0	503.9	40.0	861.3	36.0	505.2	27.0	839.5	48.0	447.1	29.0	618.4	36.0	564[illegible]	36.0	406.3	28.0	480.2	32.0	806.9	42.0

Source: Agro climate Research Centre, PJTSAU, Rajendranagar, Hyderabad; Rf: Rainfall in mm; RD: Rainy days.

Table 9. Rainfall pattern during *Rabi* seasons (month wise).

Month/ Year	2006-07		2007-08		2008-09		2009-10		2010-11		2011-12		2012-13		2013-14		2014-15		2015-16		2016-17	
	Rf	RD	Rf	RD	Rf	RD	Rf	RD	Rf	RD	Rf	RD	Rf	RD	Rf	RD	Rf	RD	Rf	RD	Rf	RD
OCT	17.5	2	14.8	3	53.6	2	96.0	6	108.0	7	47.9	4	58.9	5	253.2	9.0	69.2	3	36.4	2	32.2	4
NOV	27.3	2	15.8	1	12.6	2	30.2	2	26.8	4	27.5	1	47.0	2	31.0	2.0	10.6	1	17.3	1	0.0	0
DEC	0.0	0	0.0	0	0.0	0	5.0	1	18.2	2	0.0	0	0.0	0	0.0	0.0	0.0	0	1.4	0	2.0	0
JAN	0.0	0	0.0	0	0.0	0	9.6	1	0.0	0	0.0	0	5.0	1.0	0.0	0.0	0.0	0	0.0	0	--	
	44.8	4.0	30.6	4.0	66.2	4.0	140.8	10.0	153.0	13.0	75.4	5.0	110.9	8.0	284.2	11.0	79.8	4.0	55.1	3.0	34.2	4.0

Source: Agro climate Research Centre, PJTSAU, Rajendranagar, Hyderabad; Rf: Rainfall in mm; RD: Rainy days.

Angoumois grain moth, rice moth and saw toothed beetle. Damage during storage leads to grain discoloration, weight loss and reduction in quality due to development of mycotoxins. The losses are estimated to be 10 per cent (Reddy *et al.*, 2012).

Changes in diseases scenario over the years

In India under the aegis of All India Coordinated Maize Improvement projects, 16 out of the 61 disease that occur in India have been identified as the major ones. The economic yield losses due to diseases have been estimated to be 13.2 per cent. Various diseases attack maize crop in different stages of crop growth and reported to cause significant yield losses. These diseases not only affect quantity of the produce but also quality of the produce. Maize diseases can be broadly classified in to foliar diseases affecting the foliage, stalk rots affecting the stalks and ear rots affecting the ear. The major foliar diseases are leaf blights (TLB and MLB). However, in recent years banded leaf and sheath blight and downy mildew are reported with increased incidence. Among the post flowering stalk rots, charcoal rot causes maximum yield loss followed by late wilt in the state. Ear rot is observed particularly during rainy season caused by saprophytic fungi (Reddy *et al.*, 2012).

The changing climate not only affects the frequency of abiotic stresses but also biotic stresses on crop plants. Raising temperatures and variations in humidity could potentially affect the diversity and responsiveness of pathogens and insect pests. Maize is affected by an array of diseases and insect pests; these include downy mildews, post flowering stalk rots (PFSR), banded leaf and sheath blight (BLSB), *Turcicum* leaf blight (TLB), ear rots, stem borers, aphids and weevils.

Changes in pest scenario over the years

The major stalk borers are spotted or striped stem borer and pink stem borer, which cause economic yield losses due to dead hearts if not controlled. Among all the insect pests of maize, stalk borers, *Chilo partellus* (Swinhoe) during *Kharif* and *Sesamia inferens* (Walker) during *Rabi* are the most serious pests in our state and cause yield loss of 25-80% during severe conditions. Now and then, aphids are also reported to cause considerable damage due to infestation of tassel. Among sucking pests, incidence of aphids is observed in maize particularly in sweet corn (Reddy *et al.*, 2012). Various pests cause damage to leaves, stem, tassel and ear and result in significant yield losses. These pests not only affect quantify of the produce but also quality of the produce. Maize pests are broadly classified in to stalk borers, sucking pests, leaf feeders, root feeders and ear borers.

In recent years, often termites cause severe damage by lodging crop, particularly in light soils under rainfed situation. Sometimes, *Spodoptera* (*litura/exigua*) causes severe damage to maize seedlings during *Rabi* season. Cob borer damage is reported in sweet corn by *Helicoverpa armigera,* however its incidence on commercial crop is 1-2 per cent only. Recently, Silk cutter (*Euproctis*) incidence is increasing in *Kharif* maize. Often *Chiloloba* beetles are observed on tassel, feeding on anthers as a result grain formation gets affected.

Changes in cropping systems over the years

Maize is cultivated in 4.54 lakh ha (2015-16) during *Kharif* season under rainfed conditions as a sole crop (fields are kept fallow till next season) in Telangana State. However, farmers adopt intercrop under rainfed situation with redgram. In *peri*-urban areas, farmers follow crop rotation in maize with vegetables. Whereas in turmeric growing areas, maize is grown along with turmeric as intercrop or mixed crop to meet their family requirements. Maize is grown as a guard crop in fixed rows in summer vegetables like tomato, French bean etc. During the last decade, area under maize is increasing in non-traditional areas due to suitability of maize in rice-maize cropping system under zero tillage, since the cultivation of *Rabi* rice has become difficult due to shortage of water and power. Maize (African tall) fits well to forage crops' cropping system due to its high nutritive value for milk production, because of growing dairy industry to cater the demand of milk production.

Changes in cropping season over the years

Maize has emerged as one of the potential alternative *Rabi* crop after rice in erstwhile Andhra Pradesh and a quantum jump (7.4 t/ha) on productivity front was also witnessed during 2008-09 (Fig. 3). The phenomenal increase in area (34%), production (49%) and productivity (31%) in *Rabi* during last decade could be attributed to the cultivation of single cross maize hybrids in non-traditional area and also increasing area under seed production (Table 10) (Sreelatha *et al.,* 2016).

Table 10. Trends in area, production and productivity in erstwhile Andhra Pradesh during *Rabi* season.

Year	Area (lakh ha)	Production (lakh tons)	Yield (q/ha)
2005-06	1.65	9.89	59.98
2006-07	1.90	11.77	61.89
2007-08	2.48	17.58	65.90
2008-09	3.3	26.52	74.09
2009-10	2.81	17.65	62.81
2010-11	3.04	23.12	72.00
2011-12	3.32	21.36	66.32
2012-13	3.98	27.29	68.57
2013-14	3.43	18.62	54.28
2014-15	3.26	16.96	52.02
Mean	**2.91**	**19.08**	**63.79**

Recent trends in seasonal acreage of maize in Telangana state showed a dramatic change during the past one decade (Table 11). During 2006-07, the maize acreage during monsoon season was 5.5 lakh ha that has decreased to 4.54 lakh ha during 2015-16; whereas, the acreage during *Rabi* season has grown up from 1.07 lakh ha during 2006-07 to 1.12 lakh ha during 2015-16.

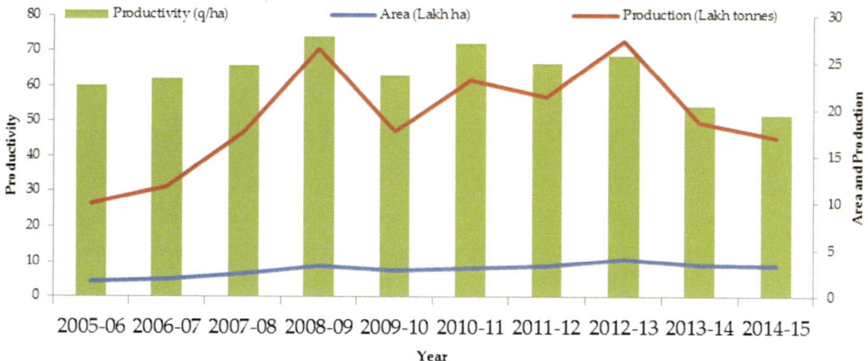

Fig. 3. Trends in area, production and productivity in erstwhile Andhra Pradesh during *Rabi* season.

Table 11. Area, production and productivity of Telangana.

Year	**Area (Lakh ha)**			**Production (Lakh tonnes)**			**Productivity (Tonnes/ha)**		
	Kharif	*Rabi*	**Total**	*Kharif*	*Rabi*	**Total**	*Kharif*	*Rabi*	**Mean**
2006-07	5.50	1.07	6.57	2.29	5.21	7.50	0.42	4.87	1.14
2007-08	5.22	1.49	6.71	2.39	7.99	10.38	0.46	5.36	1.55
2008-09	4.93	1.94	6.87	1.51	8.27	9.78	0.31	4.26	1.42
2009-10	4.57	1.12	5.69	8.31	5.34	13.65	1.82	4.77	2.40
2010-11	3.91	1.20	5.11	14.36	6.32	20.68	3.67	5.27	4.05
2011-12	4.49	1.42	5.91	12.09	6.83	18.92	2.69	4.81	3.20
2012-13	4.74	1.89	6.63	19.78	9.66	29.44	4.17	5.11	4.44
2013-14	5.53	1.99	7.52	24.38	10.87	35.25	4.41	5.46	4.69
2014-15	5.22	1.70	6.92	14.09	9.03	23.12	2.70	5.31	3.34
2015-16	4.54	1.12	5.66	11.96	4.25	16.21	2.63	3.79	2.86
Average	4.87	1.49	6.36	11.12	7.38	18.49	2.33	4.90	2.91

Source: Directorate of Economics and statistics

The mean productivity of maize was 2.33 t/ha in *Kharif* and 4.90 t/ha in *Rabi* during the decade (Fig. 4). This shows that maize is occupying more acreage under *Rabi* season as well as non-traditional areas that indicates that maize is emerging as one of the potential driver for crop diversification in the state. Moreover, the *Rabi* season maize is more assured crop with higher productivity potential compared to monsoon season. Therefore, in areas where *Rabi* season rice crop suffers due to water scarcity, the maize has emerged as potential alternative crop.

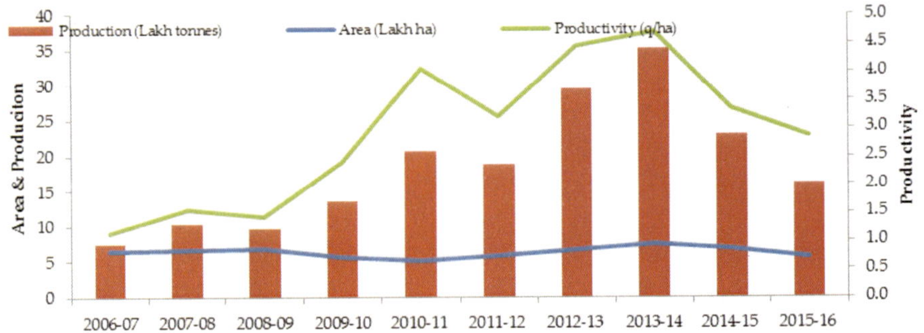

Fig. 4. Trends Area, production and productivity of Telangana State.

Future research and development strategy

Maize is predominantly grown under rainfed conditions by small holder, resource-poor farmers, despite constraints including changes in climate, overdependence on rainfall, prolonged droughts during reproductive stage and yield losses due to pests, diseases and weeds. In order to overcome the challenges there is need to develop high yielding short, medium and long duration hybrids of normal maize, QPM, popcorn and sweet corn with tolerant to biotic (pests and diseases) and aboitic stresses along with lodging resistance (dwarf plant type). In addition, water management studies for irrigated and rainfed maize in view of climate change, development of cropping systems suitable to rainfed maize, nutrient management with reference to micro-nutrient studies in maize are also important to sustain the yield levels in long-run. Further, application of marker assisted selection for traits like drought, PFSR, BLSB and QPM will accelerate the breeding process.

Summary

Maize, being a versatile crop, can be cultivated in all agro-ecologies. There is a clear cut increase in area, production and productivity of maize in erstwhile Andhra Pradesh. This indicates that there is a maize revolution over six decades. The efforts of IIMR and SAUs under umbrella of All India Coordinated Research Project on Maize were realized with the advent of single cross hybrids for commercial maize production. Maize can be fitted in multiple cropping systems and helps in crop intensification, thereby farmers prefer to grow maize in all seasons. At the same time, Govt. of India, ICAR and SAUs have brought changes in policies to encourage the private partnership in maize research and seed production. IIMR has oriented research towards the development of high yielding hybrids and improved technologies over six decades to achieve the goal. Maize has been exploited industrially in the last two decades because of the development of poultry, starch and other by-products.

Acknowledgement

Special thanks to Dr. G. Manjulatha, Sr. Scientist, ARS, Karimnagar, Dr. T. Pradeep, Director (Seeds), SRTC, Hyd, Dr. J.C. Sekhar, PS, Dr. S. Sunil, Sr. Scientist, Dr. P. Lakshmi Soujanya, Scientist, WNC, IIMR, Hyderabad for help rendered by them to bring out this chapter.

References

Agricultural statistics at a glance (Ed) by Directorate of Economics and Statistics, Govt. of Andhra Pradesh.

Kumar, A., Jat, S.L., Parihar, C.M., Singh, A.K. and Kumar, V. (2013). Crop diversification through maize based cropping system. In *Maize production systems for improving resource-use efficiency and livelihood security*. (Ed) by Kumar *et al.* Directorate of Maize Research, New Delhi. pp. 4-8.

Reddy, N.V., Pradeep, P., Ranga Reddy, R., Sudarshan, M.R., Sreelatha D., MuraliKrishna, K., SivaLakshmi, Y. and Sekar, J.C. (2012) “Maize-The Miracle Crop”, (Ed) by Anuradha M, University Press, Acharya N.G. Ranga Agricultural University, Rajendranagar, Hyderabad. pp. 1-91.

Sreelatha, D., Lavakumar, R.M., Narsimha R.V., Swarna Latha, V. and Sekhar, J.C. (2016). “Production technology for zero-tillage maize”. Maize Research Centre, Professor Jayashankar Telangana State Agricultural University, Rajendranagar, Hyderabad. pp. 1-10.

Vital-Rao, S. (1983). Maize Research in India in Hybrid Maize. The Indian Council of Agricultural Research, New Delhi 110 001. pp. 21-31.

16

Maize Research and Development in Tamil Nadu

Ravikesavan R[1]., G. Nallathambi, A.P. Sivamurugan[1]
N. Kumari Vinothana[2], P. Renukadevi[1], S. Manimegalai[3]
P. Thukkaiyanan[2], K. Sethuraman[2] and R. Radhajeyalakshmi[2]
[1]Department of Millets, TNAU, Coimbatore
[2]Maize Research Station, Vagarai
[3]Department of Entomology, TNAU, Coimbatore
Corresponding Author's Email: chithuragul@gmail.com

Maize (*Zea mays* L.) is the third most important grain crop in India after rice and wheat with respect to area and production. It has wide ecological adaptability and is grown in almost all parts of the country extending from extreme semi-arid to sub-humid and humid regions. The crop is also popular in low and mid-hill areas of Western and North Eastern regions. It has the highest yield potential and used as human food, animal feed and as a source of large number of industrial by-products. Hence, it is called as "Queen of cereals". In India, about 59% of the total production is used as feed, while the remaining is used as industrial raw material (17%), food (10%), exports (10%), and other purposes (4%).

Tamil Nadu has a total geographical area of 13 million hectares with a coastal line of 922 km and is endowed with 3 per cent of water resources of the country. Out of 13 million hectares of geographical area, the cultivable area in Tamil Nadu is around 7 million hectares and 55 per cent of which is dryland. The State has gross irrigated area of 33.94 lakh ha out of which 79% is under food crops and 21% is under non-food crops. In Tamil Nadu, maize was grown in an area of 0.14 lakh ha during 1970-71 with the annual production of 0.16 lakh tonnes, mainly concentrated in Tanjore, Pudukkottai and Trichy districts. Due to rapid increase in the demand of maize for poultry and animal feed, industrial uses and introduction of high yielding varieties and hybrids, the area under maize has gone up to 3.8 lakh ha with the production of 22.45 lakh tonnes during 2013-14. The major maize growing districts are Perambalur, Tuticorin, Dindigul, Erode, Karur, Ariyalur, Salem and Southern districts of Tamil Nadu. The demand and production of maize is increasing more rapidly as compared to other major commodities. It is estimated that the demand for maize will continue to increase in coming days (Yadav *et al.*, 2016). Thus, in the next 10 years there is a necessity and opportunity for doubling India's maize production from the current level of approximately 26 million MT.

Decadal trends in APY of maize in Tamil Nadu from 1965 to 2014

Maize has emerged as an important crop in Tamil Nadu mainly for feed and industrial purposes. The analysis revealed that the area under maize cultivation increased from 12700 ha (1965-66 to 1974-75) to 92300 ha (1995-96 to 2004-05) which was 627.8% increase over 1965-66 to 1974-75 period. The highest area of 259700 ha was observed in 2005-06 to 2013-14 period, which was 20 times more than 1965-66 to 1974-75 period. The remarkable achievement in production (1124500 ton) was made during 2005-06 to 2013-14, which was 87 times more than the average production achieved during 1965-66 to 1974-75. The maize productivity has increased from 1029 kg/ha (1965-66 to 1974-75) to 4151 kg/ha (2005-06 to 2014-15), which was 303.4% increase over 1965-66 to 1974-75 period (Table 1).

Table 1. The decadal trend in maize area, production and productivity in Tamil Nadu

Block years	Area ('000 ha)	Production ('000 t)	Yield (kg/ha)	Shift in productivity
1965-66 to 1974-75	12.7	12.9	1029	-
1975-76 to 1984-85	20.0	26.9	1282	253 (24.6%)
1985-86 to 1994-95	31.9	50.1	1576	294 (22.9%)
1995-96 to 2004-05	92.3	147.2	1611	35 (2.2%)
2005-06 to 2013-14	259.7	1124.5	4151	2540 (157.7%)

Source: Season and Crop Report, 2013-14

Decadal seasonal trends in APY of maize in Tamil Nadu from 1995 to 2014

The area under maize during *Kharif* has increased from 45500 ha (1995-96 to 2004-05) to 158600 ha (2005-06 to 2013-14). The highest increase in maize area (248.6%) was achieved during 1995-96 to 2004-05 (Table 2). The area has also increased during *Rabi* season also. With respect to production, *Kharif* season contributed more compared to *Rabi* season. Nevertheless, the productivity enhancement during *Rabi* (189%) was higher than *Kharif* (139.1%).

Decadal trends in APY of maize district-wise from 1995 to 2014

Maize area in the state has increased almost three times in past ten years and maize yield had increased significantly in almost all maize growing districts in states. (Kumar *et al.,* 2013). The decadal period of 1995-96 to 2004-05 has shown that maize was grown by large number of farmers in Dindigul (26035 ha), Coimbatore, Perambalur, Salem and Viruthunagar districts. The highest production was achieved in Dindigul district (51627 MT) which was followed by Coimbatore district. Nevertheless, the highest productivity was attained in Villupuram district. The increase in area, production and productivity was remarkable during 2005-06 to 2013-14 decadal period over 1995-96 to 2004-05 period. Higher area under maize was observed in Perambalur (46006 ha), Dindigul, Salem and Thoothukudi districts. However, Dindigul (184710 MT), Perambalur, Salem and Erode districts contributed more to the total production as compared to other districts. The higher productivity was achieved in Pudukkottai (6195), Theni and Namakkal districts (Table 3).

Table 2. The decadal seasonal trend fo maize in Tamil Nadu

Period	*Kharif*				*Rabi*			
	Area ('000 ha)	Production ('000 t)	Yield (kg/ha)	Shift in yield	Area ('000 ha)	Production ('000 t)	Yield (kg/ha)	Shift in Yield
1995-96 to 2004-05	45.5	70.9	1584	-	49.6	80.8	1633	-
2005-06 to 2013-14	158.6	625.1	3788	2204 (139.1 %)	97.2	456.8	4719	3086 (189%)

Table 3. The district-wise decadal trends in APY of maize during 1995 to 2014

District	1995-96 to 2004-05			2005-06 to 2013-14		
	Area ('000 ha)	Production ('000 t)	Yield (kg/ha)	Area ('000 ha)	Production ('000 t)	Yield (kg/ha)
Kancheepuram	0.0049	0.0019	397	0.0119	0.03	2518
Thiruvallur	0.0006	0.0001	203	1.33	3.1388	2360
Cuddalore	0.7843	1.4768	1883	11.751	23.936	2037
Villupuram	1.2085	3.3077	2737	6.3474	29.008	4570
Vellore	0.6708	1.1135	1660	1.5046	6.9332	4608
Thiruvannamalai	0.2225	0.3769	1694	4.2464	17.364	4089
Salem	6.3031	11.724	1860	21.41	96.217	4494
Namakkal	0.6613	1.0144	1534	5.482	29.373	5358
Dharmapuri	0.5526	0.6095	1103	2.3946	10.974	4583
Krishnagiri	0.0215	0.0059	276	1.3002	5.5415	4262
Coimbatore	19.015	24.168	1271	12.929	67.476	5219
Thiruppur	0	0	0	13.13	50.014	3809
Erode	3.9945	8.5842	2149	15.318	81.233	5303
Tiruchirapalli	4.6598	5.9972	1287	7.5388	27.456	3642
Karur	0.0278	0.0507	1823	0.8243	3.528	4280
Perambalur	15.38	15.949	1037	46.006	133.83	2909
Ariyalur	0	0	0	8.2687	14.867	1798
Pudukkottai	0.2836	0.5848	2062	3.5238	21.83	6195

Thanjavur	0.343	0.9162	2671	1.2492	5.8375	4673
Thiruvarur	0.0203	0.0521	2568	0.0032	0.0082	2550
Nagapattinum	0.0269	0.0707	2629	0.017	0.0798	4696
Madurai	0.8584	1.722	2006	4.6262	14.263	3083
Theni	5.3954	9.4689	1755	10.351	59.444	5743
Dindigul	26.035	51.627	1983	37.919	184.71	4871
Ramanathapuram	0.0428	0.038	888	0.388	1.6409	4229
Viruthunagar	6.1235	12.425	2029	16.12	69.268	4297
Sivagangai	0.0014	0.0014	1016	0.0687	0.2735	3981
Thirunelveli	4.8221	3.877	804	9.0107	32.285	3583
Thoothukudi	3.722	2.6352	708	16.649	57.821	3473
Nilgiris	0	0	0	0	0	0
Kanniyakumari	0	0	0	0	0	0
State	**101.18**	**163.0**	**1611**	**259.72**	**1078.10**	**4151**

The average maize productivity in Tamil Nadu is estimated to have risen by 98 per cent between 1986 and 2014. The success in breeding good-quality, high-yielding and disease-tolerant hybrids with better adaptability are believed to have played a significant role in the steady expansion of corn farming. According to state government record, the major maize producing districts can be categorized into high-, medium-, and low-productivity zones (Table 4).

Table 4. The grouping of districts based on their productivity level.

High maize productivity districts (maize grain yield > 4t/ha)	Coimbatore, Dindigul, Erode (with highest yield of 7.2 MT/ha), Namakkal, Pudukottai, Salem, Theni, Tiruppur, Trichirapalli, Thirunelveli and Villupuram together constitute 61 per cent of total maize area in the state.
Medium maize productivity	Cuddalore, Madurai, Perambalur, Thoothukodi, Virudhunagar
districts (maize grain yield 2-4t/ha)	and Ariyalur together constitute about 37 percent of total maize area in the state
Low maize productivity districts (maize grain yield < 2t/ha)	Nil

Utilization of maize in Tamil Nadu

Poultry feed

India is the third largest egg-producer and the ninth largest producer of poultry meat in the world. The earlier estimates suggested that egg production in India has surged to about 95 billion numbers by 2015 at a CAGR of over 8 per cent, with erstwhile Andhra Pradesh alone contributing over 30 per cent followed by Tamil Nadu (20%). Namakkal in Tamil Nadu is India's egg export hub and accounts for over 90 per cent of total egg exports from the country. With the assembly lines for broilers reviving up ever more every year and turning out 42 million birds a week, the pressure from this industry, which contributes an estimated Rs. 450 billion to the national income, which is actually fuelling investment in maize cultivation in the state. Tamil Nadu has been a pioneer in implementing 'noon meal programme' (NMP) in the state. From June 1998 onwards, one boiled egg was supplied to children along with meal once in a fortnight. Currently, under the scheme, 2.5 million students of 2-5 years age are provided one egg each day from Monday to Friday along with the meal. If any other Indian state with a large number of children under malnutrition adopts it in the on-going mid-day meal scheme (MDMS), then the demand for egg may increase tremendously, ultimately resulting into higher maize demand. Such pressures will accelerate in the wake of campaigns by the national egg coordination committee (NECC) to raise egg consumption of Indians. Therefore, maize would continue to provide livelihood security to millions of small and marginal farmers by giving higher income from the improved productivity, and in the long-run, with increased per-capita income, it would help in improving the food security through higher consumption of livestock based products.

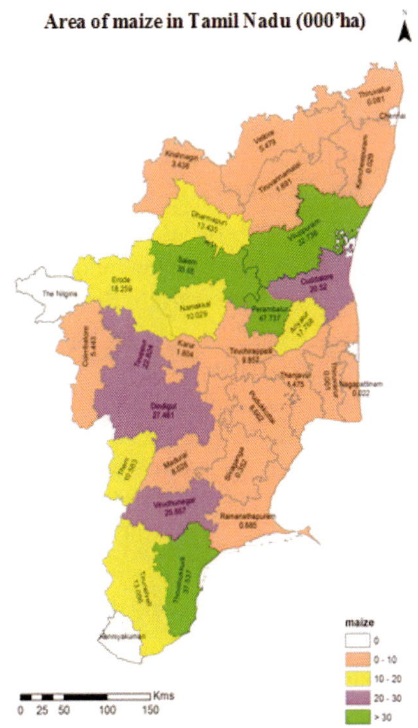

Fig. 1. State map showing major maize growing districts

Livestock feed

Dairy farming is a fast emerging potential business in Tamil Nadu. Dry stalks, ear sheath and shanks of maize can be used as fodder and feed. These can also be used in the preparation of silage for lean season. The quality of maize fodder is considered better than that of sorghum and pearl millet, as the latter crops possess anti quality components such as HCN and oxalate, respectively. The demand for value added dairy products in India is expected to grow by 20-30 per cent in the next five years which is attracting many foreign dairy players to enter into its supply chain. It would further boost the demand for maize based feed and fodder for dairying in future.

Corn starch

Starch is the major industrial product of maize manufactured by a process known as wet milling. The demand for starch is increasing across range of industries such as manufacture of paper, plastics, cellophane, films and food processing; it is likely that the requirement of maize for industrial usage will also increase.

Major industries utilizing maize in Tamil Nadu

Santhosh Limited is the one of the largest corn wet millers located in Athur, Salem district of Tamil Nadu having more than 60 years of experience in agro processing field. It has the crushing capacity of more than 225000 MT per annum. The success of the industry depends on the efficient recovery of the co-products *viz.*, corn starch, dextrin, sweeteners, speciality starches, corn germ, corn gluten meal, corn fibre, liquid glucose syrup, high maltose corn syrup etc. SPAC starch products (INDIA) Ltd in Erode, Tamil Nadu is one of the successfully long run industry for past 20 years which deals with manufacturing of tapioca and maize based co-products *viz.*, maize starch, dextrin, carboxyl methyl starch, corn germ, corn gluten meal, maize mixed fibre, thin boiling starch and dextrin. Suresh Enterprise in Salem, Tamil Nadu made a humble beginning in the year 1986 as a firm dealing only in Tapioca starch. Recently, the firm has started expanding its business supplying maize starch and modified starch. Suguna's Corn Procurement Division, which embarked upon an ambitious program of centralized procurement of maize has succeeded in creating an efficient, dependable and viable system in Karnataka, Tamil Nadu and Andhra Pradesh and is planning an expansion into Maharashtra. Suguna has setup an objective to create backward integration by directly procuring corn from the farmers at the farm gate. SKM Animal Feeds & Foods (India) Limited entered into feed manufacturing business in 1981 and built a feed manufacturing unit at Nanjaiuthukuli, a rural area in the district of Erode-State of Tamil Nadu. The factory began its commercial production in 1983 with the production capacity of 36000 MT of animal feed per year. The present production capacity is 90000 MT per month. Presently SKM occupies a pre-dominant position in the Cattle and Poultry Feed business in the states of Tamil Nadu, Kerala, Karnataka and Andhra Pradesh due to its continued improvement and innovation. However, there is ample opportunity for the development of maize industries, as the support of Government of India is growing for processing industries and improved market linkages.

Major cropping pattern and production ecologies

Tamil Nadu is geographically located between 8°5' and 13°35' North latitude and between 76°14' and 80°21' East longitude. As a result of this geographical position, Tamil Nadu enjoys semi arid climate, which favours higher crop productivity under irrigation. The cropping intensity is 125%. Tamil Nadu receives a total rainfall of 911 mm, of which the north-east monsoon (NEM, Oct-Dec) receives 47 per cent followed by the south-west monsoon (SWM, Jun-Sep) which contributes 35 per cent. There is very less rainfall of 4 per cent during winter (Jan-Feb) and 14 per cent during summer (Mar-May). The cropping systems are highly specific and vary from farm to farm. These are decided by and large by a number of soil and environmental factors, which determine over all agro-ecological setting (Table 5). But the potential productivity and monetary benefits acts as guiding principles while selecting a particular cropping system at farmer's level. The predominant cropping pattern adopted in different agro-climatic zones of Tamil Nadu is as follows.

Genesis of AICRP centre(s)

Coimbatore centre was the first centre which was established in Tamil Nadu under AICRP system. However, after the establishment of exclusive maize research station at Vagarai by TNAU to concentrate on rainfed ecosystem, the AICRP centre on maize was also sanctioned to this centre from the year 2011. The mandate of the AICRP centres is to conduct coordinated trials under different management technologies *viz.*, spacing, weed management and fertilizer dose requirement, to screen maize genotypes received through coordinated trials against different diseases and also to organize front line demonstrations and tribal sub-plan to transfer the technologies to the farming community.

Key production problems in Tamil Nadu

In Tamil Nadu, out of 7 million ha of cultivable land, the net irrigated area accounts for only 27.26 lakh ha, whereas the remaining area is under rainfed (60%). Thus the farming depends solely on rainfall. The rainfall distribution during north-east monsoon (October-December), south-west monsoon (June-September), summer (March-May) and winter (January-February) is 47%, 35%, 14% and 4% with 26, 12, 8, 2 rainy days, respectively. The average annual rainfall of the state is around 912 mm which is less than the national average of 1200 mm. The table below depicts the average rainfall over 50 year period in the state.

Month	Rainfall (mm)
February-May (Summer)	
February	13.4
March	18.4
April	42.6
May	67.5
June to September (*Kharif*)	
June	45.9
July	67.9
August	87.3
September	115.9
October-January (*Rabi*)	
October	179.9
November	169.1
December	87.1
January	17.4
Annual	**912.4**

Table 5. Agro-climatic zones with soil types, rainfall and cropping pattern of Tamil Nadu.

Agro-climatic zone	Districts	Soil Types	Normal rainfall (mm)		Cropping pattern	
			SWM	NEM	Irrigated	Rainfed
North Eastern Zone	Kancheepuram, Tiruvallur, Cuddalore, Vellore and Tiruvannamalai	Red sandy loam, clay loam, saline coastal-alluvium	443.9	319.3	1. Sugarcane-ratoon sugarcane 2. Rice-pulses 3. Rice-groundnut 4. Maize-vegetables-pulses	1. Pearl millet-groundnut 2. Groundnut-sesamum 3. Sesamum-groundnut
NorthWestern Zone	Dharmapuri, Salem and Namakkal	Non-Calcareous red or brown, calcareous black	298.7	557.6	1. Tapioca 2. Castor + groundnut 3. Sorghum-sugarcane 4. Rice-rice-pulses 5. Cotton-groundnut-pulses 6. Maize-groundnut/sesamum/turmeric	1. Tapioca 2. Castor + groundnut-horse gram 3. Minor millets-horse gram
Western Zone	Erode, Coimbatore, Tiruppur, Karur, Namakkal, Dindigul and Theni	Red loam, black	191.5	320.4	1. Maize-pulses 2. Rice-sesamum/groundnut/pulses 3. Sugarcane 4. Turmeric-cotton-pulses	1. Groundnut-maize/pulses
Cauvery Delta Zone	Trichy, Perambalur, Pudukkottai, Thanjavur, Nagapattinam, Tiruvarur and Part of Cuddalore	Red loam, alluvium	393.1	573.3	1. Rice-rice-pulses 2. Rice-rice-sesamum 3. Rice-pulses/sesamum/cotton/groundnut 4. Rice-pulses/sesamum/cotton 5. Sugarcane	-
Southern Zone	Madurai, Sivaganga, Ramanathapuram, Virudhunagar, Tirunelveli and Thoothukudi	Coastal alluvium, black, red sandy soil, deep red soil.	222.7	427.5	1. Maize-pulses/groundnut 2. Cotton-pulses 3. Rice-rice-pulses 4. Sugarcane 5. Groundnut-cotton	1. Semidry rice-minor millets/chillies
High Rainfall Zone	Kanyakumari	Saline coastal alluvium, deep red loam	477.4	496.4	1. Rice-pulses 2. Tapioca	1. Tapioca

The frequent occurrence of drought and complete crop failure may occur due to erratic distribution of rainfall with long dry spells during rainy season. The inadequate soil moisture storage due to improper soil and moisture conservation practices, optimum plant population is not maintained by farmers. Soils in the dry farming regions are eroded, degraded or marginal leading to poor soil fertility. Further, imbalanced application of manures and fertilizers results in widespread occurrence of nutrient deficiencies in soils. Thus, low water and nutrient use efficiency under irrigated condition resulting in increased cost of cultivation. The mechanical and chemical methods of weed control are practiced by less percentage of farmers. In fact indiscriminate use of pesticides and fungicides by the farmers results in resurgence of pests. The shortage of labour during peak season and insurgence of poor quality seeds at the time of heavy demand are also a matter of concern. Majority of the farmers in Tamil Nadu are small (15%) and marginal farmers (77%) with limited resources which permits only subsistence farming resulting in low and unstable crop yields. The low productivity is attributed to the cumulative effect of many constraints for crop production.

AICRP technologies

The maize cultivation in Tamil Nadu is dominated by hybrids. Different types of hybrids like single cross, three-way cross, double cross were predominant. However, after the potential of single cross maize hybrids were noticed, breeding for single cross maize hybrids took the driver seat. TNAU has also strengthened its efforts for developing single cross hybrids (Table 6). During 2012, a promising single cross TNAU maize hybrid, COH(M) 6 suited both for irrigated and rainfed condition was released by CVRC both for state and central. It is being popularized for commercial cultivation. Subsequently four hybrids namely COH(M) 7, COH(M) 8, COH(M) 9 and COH(M) 10 were released during 2013 at national level. In addition, private sector also plays a vital role in the maize seed industry in the state.

Table 6. The brief details of cultivars released by TNAU

Type	Cultivar	Year	Released by	Other Characteristics
Yellow	COH 1	1982	SVRC	Resistant to downy mildew disease
Yellow	CO 1	1985	SVRC	Resistant to downy mildew and tolerant to drought
Yellow	COH 2	1989	SVRC	High yield and resistant to downy mildew diseases
Yellow	COH 3	1996	SVRC	Highly resistant to downy mildew disease and stem borer. High starch and high protein content
Yellow	COH (M) 4	2002	SVRC	Dark yellow colour seeds, Resistant to powdery mildew and stem borer, high protein content
Yellow	COH(M) 5	2006	SVRC	Grains yellow in colour and flint-semi flint in texture Suitable for cultivation both under

				irrigated (throughout the year) and rainfed (Sept.-Oct. sowing) situations
Yellow	COH(M) 6	2012	SVRC/CVRC	Grains are bold, orange yellow in colour and semi dent type. High shelling (81%) with high test weight (400 g/1000 seeds).
Yellow	COH(M) 7	2013	CVRC	It is a long duration (110 d), single cross normal hybrid; grains are bold, orange yellow in color with dent type
Yellow	COH(M) 8	2013	CVRC	Multiple disease resistance *viz.* MLB, TLB, RDM, DM and moderately resistant to PFSR
Yellow	COH(M) 9	2013	CVRC	Highly stable, high yielding hybrid, high test weight (38 g/100 kernels), high shelling (82%) and wider adaptability across the Zones 3 and 5 in India.
Yellow	COH(M) 10	2013	CVRC	It is a modified single cross normal hybrid with medium duration (90-95 days), grains bold, orange yellow in colour and semi dent type
Babycorn	COBC 1	1998	SVRC	It has a duration of 50-65 days for baby corn production (6660 kg/ha), 65-70 days for green fodder (32.2 tonnes/ha) and 100-105 days for seed production (2000 kg/ha).

COH(M) 6 COH(M) 8

Changes in cropping system over the years

The state has faced sudden price rise in feed cost in the recent past. Therefore, the maize crop is being grown in different regions of the state depending on the quantity of rainfall received and its distribution, availability of irrigation water, climate of the locality and soil type. Presently, Tamil Nadu is one of the fastest emerging maize producing state in India because of growing demand from the large poultry sector in the state. Farmers have adopted high-yielding long-duration hybrids with improved production technologies like drip fertigation in large scale. The adoption of latest package of practices in maize by farmers has resulted in good harvest which is expected to continue in future also. The major maize based cropping systems being followed in different ecologies are given below.

Zone	Cropping system
North-eastern zone	Maize-vegetables-pulses (irrigated)
North-western zone	Maize-groundnut/sesamum/turmeric (irrigated)
Western zone	Maize-pulses (irrigated) Groundnut-maize/pulses (rainfed)
Southern zone	Maize-pulses/groundnut (irrigated)

In recent years, other factors *viz.*, demand and supply of the produce, availability and cost of inputs, labour wages, consumer preference, monetary benefits, crop production techniques also influenced the change of crops and season in a particular cropping pattern in a particular region. The district-wise decadal analysis revealed that in all the zones except high rainfall zone, the maize is accommodated as an important crop in the cropping pattern. The districts *viz.*, Cuddalore, Villupuram, Namakkal, Erode, Coimbatore, Theni, Dindigul, Trichy, Perambalur, Ariyalur, Pudukkottai, Thanjavur, Virudhunagar, and Thoothukudi have more area under maize compared to previous decade.

Rabi maize

In recent years Tamil Nadu has become one of the major states where *Rabi* maize has emerged as an important crop in the non-traditional season and non-traditional areas. The other states where predominant *Rabi* maize is being grown are Andhra Pradesh, Bihar, Karnataka, Maharashtra and West Bengal. In Tamil Nadu, there is potential to increase the production of maize by increasing the production of *Rabi* maize in the coming years as *Rabi* maize has a higher yield as compared to *Kharif* maize. Though maize favourably responds to better crop management both in *Kharif* and *Rabi* season, the erratic rainfall pattern of the south-west monsoon comes in the way of timely field operations of *Kharif* season. In the absence of any major environmental impediments in *Rabi*, the desired field operations can be planned and executed at the most desired time. The area under *Rabi* maize is on the increase due to enhanced area coverage in Southern part of Tamil Nadu. The role played by the private sector is enormous in the maize seed industry of Tamil Nadu. Many popular hybrids from private sector includes, Pro 311, KH 510, Deccan 103, M 900 gold, Pioneer 30 V 92, NK-6240, S-6668, NK-6850, Hi-shell, Bio 9681 and Seed tech 2324.

Production technologies recommended and package of practices

Season	1. Adipattam (July-August) 2. Purattasipattam (September-October) 3. Thaipattam (January-February)
Varieties/Hybrids	CO H (M) 6 and CO H (M) 8
Soils	Maize can be grown in variety of soils ranging from loamy sand to clay loam. However, soils with good organic matter content having high water holding capacity with neutral pH and well drained are considered good for higher productivity.

Preparatory cultivation	Plough 2-3 times to obtain a good tilth. Form ridges and furrows of 6 m long and 60 cm apart or form beds of size 10 m^2 or 20 m^2. Spread 12.5 tonnes of FYM/ha before ploughing and incorporate well.
Seed rate	20 kg/ha
Spacing	60 x 25 cm
Fertilizers	Blanket recommendation
	TNAU micronutrient mixture: 30 kg/ha
	Varieties
	135:62.5:50 kg NPK/ha
	Basal: 33.8 kg N, 62.5 kg P and 50 kg K
	I Top dressing: 67.5 kg N at 25 DAS
	II Top dressing: 33.8 kg N at 45 DAS
	Hybrids
	250:75:75 kg NPK/ha
	Basal: 62.5 kg N, 75 kg P and 75 kg K
	I Top dressing: 125 kg N at 25 DAS
	II Top dressing: 62.5 kg N at 45 DAS
Irrigation	Irrigate immediately after sowing and give life irrigation on the third day and thereafter once in 7-10 days depending upon soil condition.
After cultivation	Apply pre-emergence herbicide, Atrazine 50 WP @500 g/ha on 3 DAS as spray on the soil surface followed by one hand weeding on 40-45 DAS.
Plant protection	**Shoot fly:** Methyl dematon 25 EC @500 ml/ha or Dimethoate 30 EC @500 ml/ha or Neem seed kernel extract @ 5%
	Stem borer: Apply in the leaf whorls on 20 DAS
	Quinalphos 5 G @ 15 kg/ha or Carbaryl 4 G @ 20 kg/ha.
	Downy mildew or Crazy top: Metalaxyl 72 WP @ 1000 g/ha or Mancozeb @
	1000 g/ha on 20 DAS
	Leaf spot: Mancozeb or Captan 1000 g/ha
Harvest	Harvest the ears when the sheath covering the ear turn yellow and dry and the seeds become fairly hard and dry.
Yield	Irrigated: 7.4 tonnes/ha
	Rainfed: 5 tonnes/ha

Changing scenario of pest status in Tamil Nadu

Maize is an important crop grown for food, livestock and poultry feed, beverages and starch. More than 95 per cent of the maize area in the state is dominated by single cross hybrids. It was only after the introduction of hybrids; maize growers became more conscious of the fact that maize also needs better plant protection measures because of the escalated pest problems. The prevalence of insect pests is one of the important factors that adversely affect maize production both in pre-harvest stage and during storage. Insect outbreak cause serious problems and could reduce the yield up to 75% and even total crop failure in case of severe infestation. In maize, 139 insect-pests causes varying degrees of damage to maize crop and among them a dozen of insect-pests *viz.,* maize stem borer, *Chilo partellus* (Swinhoe), pink stem borers, *Sesamia calamistis* (Hampson) and *Sesamia inferens* (Walker), cutworm, *Agrotis ipsilon* (Hufnagel), shoofly, *Atherigona orientalis* (Schin), aphid, *Rhapalosiphum maidis* (Fitch), sugarcane leafhopper, *Pyrilla perpusilla* (Walker), armyworm, *Mythimna separata* (Walker), cob borer, *Helicoverpa armigera* (Hubner) and shoot bug, *Peregrinus maidis* (Ashm), have been found responsible for huge economic loss to maize cultivation. The yield losses due to pests vary greatly depending upon the region, season, maize variety and fertilization (De Groote *et al*., 2003).

The survey conducted by Thangachamy (2013) on maize in six major districts of the state *viz.,* Bhavanisagar, Dindigul, Madurai, Tiruchi and Theni, revealed that the stalk borer, *C. partellus*, cob borer, *H. armigera*, aeroplane bug, *P. perpusill*, whorl aphid. *R. maidis*, ash weevil, *Myllocerus maculosus* and grasshoppers, *Oxya velox*, *Acanthoplus* sp. and *Oedaleus senegalensis* were the major pests. Whereas the rice weevil, *Sitophilus oryzae* (L.) is the most serious pest among the insect pests infesting stored maize. Infestation during storage results in deterioration in the nutritional status of maize grains (Sridhar, 2008). *C. partellus*, known as spotted stem borer is the most notorious pest found throughout India. The population dynamics and seasonal abundance studies at Coimbatore have revealed that adult activity is higher in January than in other months, with highest damage on the crop sown during March and lowest on the crop sown in June or October (Mahadevan and Chelliah, 1986). The dead heart incidence was found to be at its peak resulting in 21.30 per cent damage in crop sown during last week of February and first week of March (Elangovan, 1987). Pink borer, *S. inferens* affects maize crop during the *Rabi* season and mainly restricted to Peninsular India. Shoot fly, *A. orientalis,* a serious pest in Peninsular India causes damage to maize plants at the seedling stage and leads to drying of the seedlings or 'dead heart'. The early sowing during first fortnight of February avoids build up of shoot fly population. Termite, *Odontotermes obesus (*Rambur) is a major problem in certain areas. If not controlled, it can cause substantial damage to the maize crop, since they establish colonies much deep into the soil.

In order to manage insect pests in maize, broad-spectrum insecticides like organochlorines, organophosporous compounds and carbamates dominated integrated pest management in maize before 1990's. Besides granular application, organophosporous insecticide sprays are also recommended to avoid pest insect outbreak. In the late 2000's, neonicotinoid pesticides replaced many organophosporous and carbamate pesticides in maize crop protection system. Neonicotinoids like imidacloprid and thiamethoxam, because of their systemic activity find place for seed treatment and also for soil application. Thiamethoxam 70 WS @ 4.0 g/kg seeds was found better for seed treatment against aphids and shoot fly and did not produce any phytotoxicity or ill effects to predatory coccinellids (Dineshraja, 2013). Thiamethoxam 30 FS at 5.0-8.0 ml/kg was found to be effective in reducing the incidence of shoot fly, stem borer, aphids and termite in maize (Baskaran *et al.,* 2011). Other newer molecules like Fipronil, Chlorantranilipole and Spinosad are also used for insect pest management. Application of Fipronil granules @ 20 kg/ha followed by light irrigation controls termites to a reasonable extent. If the infestation occurs in patches, applying few granules of Fipronil on and around the patches control termite infestation. The recent trends of chemical less agriculture, that is reduced pesticide usage and ecological sustainability have led to increased interest in other methods of management such as cultural methods, use of botanicals, promoting natural control by conservation of predators and parasitoids and cultivation of *Bt* maize.

Intercropping of maize with lab lab and cowpea reduced the incidence of *C. partellus* and increased the yield of maize (Elangovan, 1987). Saranya (2015) reported that application of neem oil 5% was found to inhibit hatching of *C. partellus* eggs by 95 per cent followed by 0.2% chlorpyriphos (86%), neem oil 1% (79%), NSKE 5% (78%) and *Jatropha* leaf extract 5% (78%). Boomathi (2003) found that NSKE 0.18% significantly reduced the egg hatchability of pigeon pea pod borer, *H. armigera.* Increasing CO_2 concentration in storage atmosphere to 60% resulted in 100% mortality of larvae of *Sitophilus oryzae* at 4 days of exposure time on maize seeds. In the long term storage study, no adults emerged in treated seeds up to 120 days after treatment with 60 and 70 per cent CO_2 concentrations and seed germination remained unaffected. The mortality of larvae of *S. oryzae* due to seed treatment with botanicals and inert dusts was above 90% in maize seeds. Two hymenopteran parasitoids belonging to families Pteromalidae and Bethylidae, were recorded from insect cultures of *S. oryzae* (Sridhar, 2008). Thangachamy (2013) reported that under experimental conditions, transgenic corn recorded the lowest infestation by *Chilo sp.* (1.4%) while it was maximum of 12.82% in non transgenic crops. *H. armigera* damage was only to the tune of 3.85% in *Bt* corn when compared to 25.53 per cent in non transgenic crops. Temperature treatment of maize seeds was attempted by Nandhini (2015) against *S. oryzae.* If for seed purpose, the harvested maize can be exposed to 65°C for 30 min and treated with ball milled turmeric powder at the rate of 5 g/kg where in the germination was in optimum requirement besides reducing the infestation. In case of seeds to be used for grain purpose, the seeds can be heated to 70°C for 30 min and treated with either plain or ball milled turmeric consideration on the cost of ball milling.

Several natural enemies have been reported to attack *C. partelllus*. Among the parasitoids reported, egg parasitoid *Trichogramma* spp. and larval parasitoids,

Aphanteles favipes and *T. chilonis* are important. *T. chilonis* has been found parasitizing egg masses of *C. partelllus* up to 70%. *A. flavipes* is the most dominant and most widely distributed larval parasitoid of *C. partellus* in India with a parasitisation ranging from 5.3-42.8%. The larval parasitoids recovered included *Cotesia sesamiae*, *Cotesia flavipes* and *Goniozus indicus*, and the pupal parasitoid was *Pediobius furvus; the f*ungal pathogen *Beauveria bassiana* and *Metarhizium anisopliae* were also found to be effective in controlling *C. partellus*. The predatory spider species dominating in the maize ecosystem of Coimbatore are *Lycosa barnesi, Pardosa birmanica*, *Lycosa psedoannulata, Salticus* sp. and *Hippasa lycosina.* The dendrogram analysis of spiders revealed that the families, Clubionidae and Gnaphosidae were closely related when compared to other predatory spider families. (Saranya, 2014). Predators like Coccinellids, Chrysoperla, Argiope, Salticus were found in more number under natural conditions of maize cultivation. Even though several natural enemies have been reported on maize borers, but in general, the efficiency of natural enemies in particular farming environments is not known. The scope for successfully controlling maize stem borers with natural enemies is limited by the short cropping period and the lack of continuous habitat for natural enemies. They are rendered less effective due to increase in temperature and ultra violet light. In this context, in order to achieve higher crop productivity along with quality, it is highly essential to manage the pest by selecting and adopting an ideal pest management strategy taking into consideration the pest population dynamics, abundance and ecology.

Changes in disease scenario over the years in Tamil Nadu

In Tamil Nadu, the major diseases recorded for the past one decade are sorghum downy mildew (SDM) caused by *Pernosclerospora sorghi,* TLB (Turcicum Leaf Blight), MLB (Maydis Leaf Blight), Post Flowering Stalk Rot (PFSR/ Charcoal rot) caused by *Macrophomina phaseolina* and very recently Curvularia leaf spot caused by *Curvularia lunata* (Renukadevi, Unpublished, 2017).

Sorghum Downy mildew (SDM)

The important pathogen causing downy mildew in Tamil Nadu is *Peronosclerospora sorghi*. Payak (1975) reported that the sorghum downy mildew disease causes yield loss of 30-70%. The genetic variability among the isolates of *P. sorghi* causing downy mildew studied by Mathiyazhagan *et al.* (2008) and Sireesha and Velazhahan (2015) have confirmed the heterogeneity among the *P. sorghi* isolates of sorghum and corn. Ladhalakshmi *et al.,* 2009 developed SCAR marker for *Peronosclerospora sorghi* infecting maize. This markers help in detecting maize isolate of *P. sorghi* from sorghum isolate. The sorghum downy mildew incidence was found to be in higher side (>50%) in western districts of Tamil Nadu up to 2011. But for the past five years the incidence was found to be <10% due to seasonal vagaries. In southern districts, the SDM incidence was found to be <1%.

Leaf blight

In Tamil Nadu, the major leaf blights *viz.,* turcicum leaf blight (*Exserohilum turcicum),* maydis leaf blight (*Drechslera maydis*) and banded leaf sheath blight (BLSB) *(Rhizoctonia solani.)* are more prevalent in cooler condition with high humidity. The incidence was found to be 10-25% in the southern, western and central districts during *Kharif* season. Lesion sizes of blights may vary in inbreds and hybrids due to different genetic backgrounds.

Common rust (*Puccinia sorghi*)

The common rust was recorded in maize growing areas with subtropical temperate and high-land environment of Tamil Nadu. The circular to elongate, golden brown to cinnamon brown pustules are visible over both leaf surfaces changing to brownish black at plant maturity. It appears at the time of tasseling and the incidence was found to be only <5%.

Charcoal rot/ post flowering stalk rot

The pathogen *Macrophomina phaseolina*, affects the plant mostly after flowering and the disease is named as post flowering stalk rot. The stalk of the infected plants can be recognized by greyish streak. The pith becomes shredded and greyish black minute sclerotia found on the vascular bundles. Shredding of the interior of the stalk often causes stalks to break in the region of the crown. The crown region of the infected plant becomes dark in colour. Shredding of root bark and disintegration of root system are the common features. High temperature and low soil moisture (drought) favours the disease. This disease is now becoming epidemic in central districts of Tamil Nadu.

However, there is a shift in disease pattern during the past six years as the major disease SDM is gradually becoming disease of lesser economic importance. Presently, due to the seasonal vagaries such as change in minimum temperature and monsoon rainfall pattern, the diseases such as TLB, MLB, Charcoal rot and CLS may assume epidemic proportion in coming years.

Management of diseases

Various approaches were made to manage diseases, rogueing and destroying of infected plants as they appear in the field is the most common management measure of SDM. However, avoiding maize-sorghum crop rotation in field where

SDM has occurred previously as well as avoiding of sowing of maize adjacent to a field of maize or sorghum will further prevent spread of SDM through secondary infection. The seed treatment with Carbendazim @ 2 g/kg or Thiram @ 4 g/kg or Metalaxyl @ 3 g/kg of seed and need based foliar sprays of Metalaxyl + Mancozeb @ 1000 g/ha, Mancozeb 1000 g/ha at 20 days after sowing will effectively check the disease. The multiple resistant hybrid COH6 is recommended to overcome all the diseases. However, chemical control is practically feasible but non-judicious and untimely application has made this approach ecologically and economically unsuitable. Ecologically and socially acceptable methods of biological control of SDM were also reported by Kamalakannan and Shanmugam, (2009), who reported that the foliar spray of talc based formulation of *Pseudomonas fluorescens* at the rate of 2 g/l and plant extract of *Prosophis chilensis, Azadirachta indica* at 10% concentration were found to be significantly reducing downy mildew disease and Sireesha and Velazhahan, (2016), who reported that the seed treatment with talc based formulation of *Bacillus subtilis* G1 at the rate of 10 g/kg of seed reduced the downy mildew incidence up to 54% and increased germination percentage and seedling vigour of maize under green house condition which could be further exploited for commercial scale-up for eco-friendly management of downy mildew. The leaf blights are effectively controlled by foliar spray of Mancozeb or Zineb at 2-4 g/L at 10 days interval after first appearance of the disease. The PFSR is effectively managed by following the integrated disease management practices such as crop rotation with non-host crop. Avoiding water stress at flowering time reduces disease incidence. Further, application of potash at 80 kg/ha in endemic areas and soil application of *P. fluorescens* (or) *T. viride* @ 2.5 kg/ha + 50 kg of well decomposed FYM or sand at 30 days after sowing also reduces disease incidence.

Extension

Front line demonstrations are being organised every year to popularise the adaptable technologies to the farmers.

Year	FLDs (No.)	Area	Varieties /Hybrids	Districts covered	FLD yield (kg/ha)	Farmers practice/ state's yield (kg/ha)
2012-13	400	400	COH(M) 6 (CMH 08-282)	Permbalur, Tirupur, Tiruvannamalai, Villupuram, Dindigul, Prambalur, Tanjore, Karur, Cudalore, Tirunelveli, Salem and Erode	7945 (K) 7975 (R)	6252

2013-14	400	400	COH(M)6	Salem, Tuticorin, Coimbatore, Erode, Tiruvarur, Karur, Tiruvannamalai, Tiruppur and Ariyalur	7888	6042
2014-15	50	50	COH(M)6	Perambalur and Dindigul	6843 (K) 7840 (R)	6281 (K) 6945 (R)
2015-16	50	50	COH(M)6 and COH(M) 7	Coimbatore and Tiruppur	6318 (K) 7003 (R)	5900 (K) 6545 (R)
2016-17	100	100	COH(M)6, COH(M) 7 and COH(M) 8	Coimbatore, Perambalur and Theni	6127 (K)	5601 (R)

K-*Kharif* season R-*Rabi* season

The aforementioned data revealed that there is a large yield gap exists between the yield obtained in front line demonstration conducted by the scientist and the yield obtained in the farmer's field. This is attributed to improper adoption of recent technological interventions by the farming community. The existing yield gap may be abridged by innovative extension methodologies to reach-out to farmers to enable them to adopt improved production technologies thus enhancing the productivity.

Tribal sub-plan trainings

Technological demonstrations and popularization of the maize hybrids among the tribal community paved a way for their financial upliftment and knowledge of maize cultivation. Tribal sub-plan scheme is being operated in MRS, Vagarai from 2013 in the identified tribal villages of Tamil Nadu. The tribal farmers were given training on "Maize Production Technologies" which involved the conduct of field demonstration and training lectures on package of practices of maize cultivation, single cross hybrid technology, value addition and integrated pest and disease management in maize. The trainings were conducted in the tribal villages of Parthasarathipuram, Perumalpudur, Thamaraikulam, and Valavichettipatty (2013-2014), Pachalur, Porupar Settlement and Sempirankulam (2014-15), Valichettipatti, Ponnurukki hamlet, Kodanthur hamlet, Manupatti village, Kilakkuchettipatti, (2015-16), Andipatti, Manjalaru,d Kuthiraiyaru and Ponthupuli (2016-17). Field days have also been organized to know the impact of this programme and performance of maize hybrid in the tribal holdings. The contact tribal farmer was distributed with 8 kg of TNAU maize hybrid CO6 seeds and inputs for one acre of land area. Till now a total of 750 tribal farmers (50 farmers/village/training) got benefitted.

Key R&D challenges

The relatively more variation exists in annual rainfall from year to year which leads to frequent crop failures in regions with lesser rainfall. Improper distribution of rainfall results in poor growth and development of crops. Prolonged dry spells especially at critical stages leads to reduction in crop yield. High atmospheric temperature, low relative humidity, hot dry winds, high potential evapo-transpiration are the major climatic factors which hamper the productivity. In addition, edaphic factors namely inadequate soil moisture availability, poor soil fertility, soil deterioration due to wind and water erosion are added challenges for enhancing the productivity. Finally, the socio economic factors like non-availability inputs timely, and high-cost of critical inputs, the higher labour wages, non-availability of credit facilities, low monetary returns and high cost of cultivation, fragmentation of landholdings, farmers attitude, skill and knowledge also poses serious challenges which influence the adoption of recent technological interventions.

R&D challenges	Probable solutions
Drought tolerance & productivity enhancement	Development of hybrids/varieties suited for rainfed condition Development of hybrids/varieties with tolerance to biotic and abiotic stress
Productivity enhancement	Technologies for yield maximization, rainfed ecosystem, intercropping and physiological disorders.
Mechanization	Identifying suitable crop geometry for complete mechanization
Documentation of newly emerging diseases	Management of diseases through chemical and bio-control agents and identification of resistant sources

Future thrust

The development of high yielding varieties/hybrids in maize to suit the different agro-ecological regions of Tamil Nadu will continue to be top priority of the AICRP centres. However, breeding hybrids with resistance to major diseases and stem borer will be given priority along with developing agro-techniques for improving the productivity. Further, the efforts will be made to manage aberrant weather conditions for mitigating early- mid- and late-season drought. The attention is required to establish research-extension-industrial linkages to further strengthen the breeding programmes. Presently, TNAU has strong linkage with department of agriculture for promotion of university varieties and hybrids. In order to promote maize hybrids, such kind of linkages is made through adaptive research trials and on-farm trials. The present linkages can be further strengthened in case of maize through hybrid seed production in state seed farms. Besides the research-extension linkage, the maize industries like feed industries, starch industries and other industrial stake holders have to be involved in the linkage, so as to have healthy buy-back arrangements with farmers and enjoy mutual benefits. Further training farmers, extension officers and seed producers is required to improve their skills. In this regard, substantial investments and improvements are needed in training components involving farmers, extension officers and seed producers to capitalize the large unexploited maize production potential. The farmers and extension officers have to be trained in ever changing maize production technologies and marketing. Training the seed producers to impart efficient hybrid seed production methodologies would also reveal rich dividends.

Summary

There is a significant opportunity for maximising maize yields to meet the ever increasing demand in the state. The following strategies *viz.*, replacement of low yielding varieties with high yielding hybrids, creation of awareness among the farmers in respect of production and value addition in maize through trainings, organize field demonstrations to show the yield potential of the high yielding varieties/hybrids with scientific package of practices, inclusion of legumes in crop rotation to enrich soil fertility and for sustainable productivity, public-private partnership for production and delivery of high quality seed, promotion of processing industries and value addition, strengthening post harvest infrastructure, establishing a chain of community based driers at producer level for improving quality, commodity group formation for higher market price, incentive for maize based processing industries will further enhance the maize yield further.

References

Baskaran, R. K.M., Rajavel, D. S., Manisegaran, S., Bagyaraj, K., Vathsala, V. and Palanisamy, N. (2011). Efficacy of thiamethoxam 30% FS as seed treatment against early stage insect pests in maize. *Pestology.* 35(7):45-47.

Boomathi, N. (2003). Eco-friendly management of gram pod borer, *Helicoverpa armigera* (Hub.) (Lepidoptera:Noctuidae) on pigeonpea (*Cajanus cajan*) (L.) Millsp.). M. Sc. (Ag.) Thesis, Tamil Nadu Agric. Univ., Coimbatore-3.

De Groote, H.W., Overholt, J.O., Ouma and Mugo, S. (2003). Assessing the impact of *Bt*maize in Kenya using a GIS model. *Paper presented at the Intl. Agric. Econ. Conf.,* Durban, August, 2003.

Dineshraja, R. (2013). Bioefficacy of thiamethoxam 70 WS as seed treatment against insect pests of maize and sunflower. M. Sc. (Ag.) Thesis, Tamil Nadu Agric. Univ., Coimbatore-3.

Elangovan. (1987). Epidemology and ecology of maize downy mildew, *Peranosclerophora sorghi* (Weston and Uppal Shaw) and stem borer *Chilo partellus* (Swinhoe) and their management. M. Sc. (Ag.) Thesis, Tamil Nadu Agric. Univ., Coimbatore-3. 92 p.

Kamalakannan, A and Shunmugam, V. (2009). Management Approaches of Maize Downy Mildew Using Biocontrol Agents and Plant Extracts. *Acta Phytopathologicaet Entomologica Hungarica 44 (2), pp. 255–266.*

Ladhalakshmi, D., Vijayasamundeeswari, A., Paranidharan, V., Samiyappan, R. and R. Velazhahan. (2009). Molecular identification of isolates of *Peronosclerospora sorghi* from maize using PCR-based SCAR marker. *World J MicrobiolBiotechnol*. 25: 2129–2135.

Mahadevan, N.R. and Chelliah, S. (1986). Influence of intercropping legumes with sorghum on the infestation of stem borer, *Chilo partellus* (Swinhoe) in Tamil Nadu in India. *Trop Pest Manage.* 32: 162-163.

Mathiyazhagan, S.M., Karthikeyan, M., Sandosskumar, R. and Velazhahan, R. (2008). Analysis of variability among the isolates of Peronosclerosporasorghi from sorghum and corn based on restriction fragment length polymorphismof ITS region of ribosomal DNA. *Archives Phytopath Plant Protection.* 41: 31-37.

Nandhini, P. (2015). Non-chemical management strategies against rice weevil on maize. M. Sc. (Ag.) Thesis, Tamil Nadu Agric. Univ., Coimbatore-3. pp. 114.

Payak, M.M. (1975). Downy mildews of maize in India. *Trop Agric Res*. 8: 13-18.

Saranya, V.S.L. (2015). Eco-friendly options for the management of *Chilo partellus* (Swinhoe) in maize at egg stage. M. Sc. (Ag.) Thesis, Tamil Nadu Agric. Univ., Coimbatore-3. 101p.

Sireesha, Y. and Velazhahan, R. (2016). Biological control of downy mildew of maize caused by *Peronosclerosporasorghi*under environmentally controlled conditions. *J Applied Natural Sci.* 8(1): 279-283.

Sireesha, Y. and Velazhahan, R. (2015). Assessing genetic diversity in *Peronosclerospora sorghi* causing downy mildew on maize and sorghum. *Indian Phytopath.* 68(1): 73-77.

Sridhar. N. (2008). Eco friendly strategies for management of *Sitophilus oryzae* (Curculionidae: Coleoptera) in stored product maize. M. Sc. (Ag.) Thesis, Tamil Nadu Agric. Univ., Coimbatore-3. pp. 94.

Thangachamy, P. (2013). Impact of *Bt* Maize and assessment of arthropod diversity in Maize ecosystem. Ph. D(Agrl. Entomology) Thesis, Tamil Nadu Agric. Univ., Coimbatore-3.

Yadav. O.P., Prasanna. B.M., Yadava, P., Jat, S.L., Kumar, D., Dhillon. B.S., Solanki, I.S. and Sandhu, J.S. (2016). Doubling maize production of India by 2025-Challenges and opportunities. *Indian J Agric Sci.* 86(4): 427-34.

17

Maize Research and Development in Madhya Pradesh

V. K. Paradkar, R.K. Sharma, Mahender Singh# and Narendra Kumawat#*

**AICRP on Maize, JNKVV, ZARS, Chhindwara. (MP)*
#AICRP on Maize, RVSKVV, ZARS, Jhabua (MP)
Corresponding Author's Email: paradkarvkp@yahoo.co.in

The state of Madhya Pradesh is one of the high profile maize growing state of the country cultivating maize in about 11.32 lakh hectare area which is ~12% of the total maize area of the country having 90-95% of total maize area having rainfed. Madhya Pradesh is situated in heart of India between 17°-26° N latitudes and 74°-84° E Longitude. The rainfall in maize growing belt varies from 800 to 1600 mm. Due to erratic and uneven rainfall in the state, the production and productivity of maize fluctuates every year. Maize cultivation in the area is associated with resource poor tribal farmers. As the demand for maize is growing globally due to its multiple uses for food, feed and industrial uses, we need to produce more from same or even less resources, which can be achieved through targeting technologies in potential regions.

Besides, monsoon crop due to mild winter conditions the climate allows the cultivation of maize during winter successfully under assured irrigation conditions. At present about 50 thousand ha area is under *Rabi* cultivation with high productivity levels as compared to monsoon crop. However, the major area increase was achieved during past few years mainly due to adoption of hybrids and shift in acreage under non-traditional areas, replacing soybean crop on farmers field. Uncertain climatic conditions and more disease and pest problem lead farmers to adopt maize in place of soybean.

The trends in area production and productivity of maize in Madhya Pradesh has shown a remarkable increase during four decades. During 1966-67 the maize acreage in the state was very less (5.59 lakh ha) that has grown up by 11.32 lakh ha in 2014-15. Moreover, the productivity also increased dramatically from mere 811 kg/ha during 1966-67 to 1513kg/ha in 2014-15.

Maize in Madhya Pradesh at a glance

- Dry spell occurred in different phases starting from germination to maturity affects the grain yield of maize.
- Improved hybrids/varieties of maize grown only 65 to 70% area of the state.
- The lead maize production districts are:-Chhindwara, Jhabua, Dhar, Khandwa, Ujjain, Narsingpur, Neemuch, Mandsor, Betul and Rajgarh.

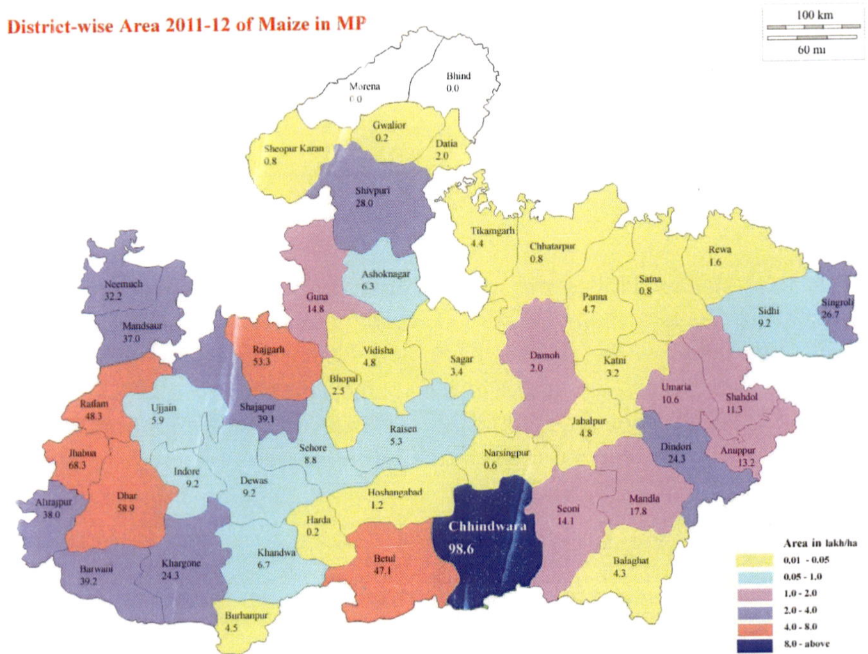

Soils: The Madhya Pradesh has diverse soil and climatic conditions and has been divided in to eleven agro- climatic zones. The soils in the state varied between and within agro- climatic zones with wider coverage under alluvial loamy and black soils. The maximum acreage under maize in having black soils where maize has emerged as a potential crop.

The main soil types found in Madhya Pradesh are alluvial deep black, medium black, shallow black, mixed red and black, mixed red and yellow and skeletal soil.

Rainfall: The annual rainfall received in the state varies from 1600 mm in the eastern districts to 800 mm in the northern and western regions. In some years rainfall goes much below the normal. Most of the rainfall is received in the monsoon period from June to September.

Spread and growth of area and productivity of maize in Madhya Pradesh

The district-wise mapping of maize area and productivity in the state showed that maximum area (2011-12) recorded in Chhindwara district. The highest productivity of maize has also been recorded in Chhindwara district (4248 kg/ha) followed by Ratlam district (2229 kg/ha)

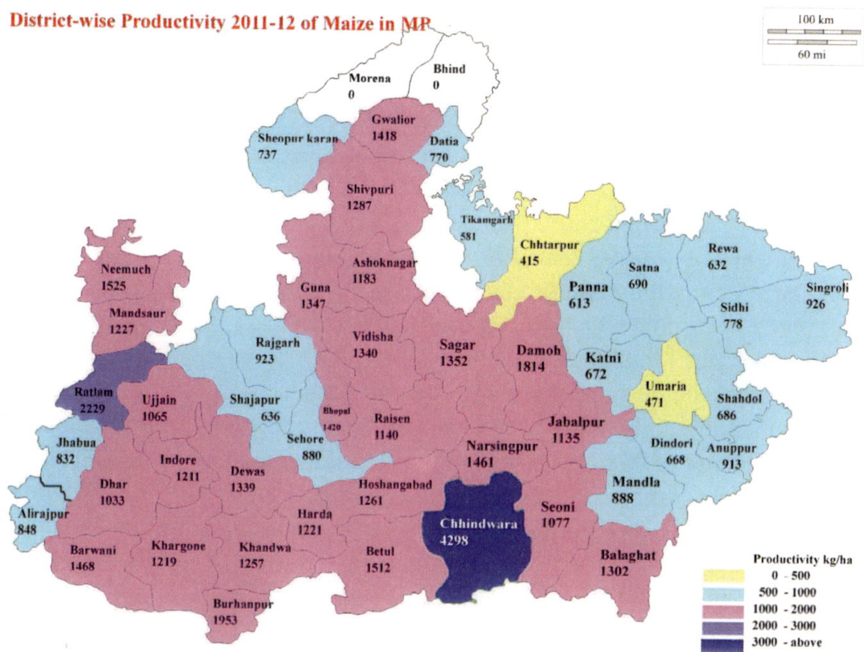

The farmers of MP are more advance in adoption of hybrid technology in maize. Most of the farmers are planting double cross hybrid seed of maize since past five years but now they are switching with the introduction of single cross hybrid technology and availability of hybrids that are more productive and having better adoptability. Resultantly, nearly 70% acreage of maize is covered under hybrid.

The map shows the spread of acreage under maize in different districts of Madhya Pradesh. The growth rate of maize acreage in state during 2014-15 varied from 0.2 to 98.6 ha. The Chhindwara, Seoni, Betul and Mandsour district showed higher growth rate increase but Chhindwara district showed remarkable higher growth rate of maize acreage. Therefore, the maize acreage is increasing in non-traditional maize growing areas of the state particularly is rainy season. Productivity growth rate of maize in the state has shown remarkable increase up to 2014-15 (Fig. 1).

Recent trends in seasonal acreage of maize in Madhya Pradesh showed dramatic change within 2-3 years. The acreage during *Kharif* season has grown upto 1132 thousand ha during 2014-15. However, the major increase was achieved during past few years mainly due to adoption of hybrids and shift in acreage under non-traditional areas.

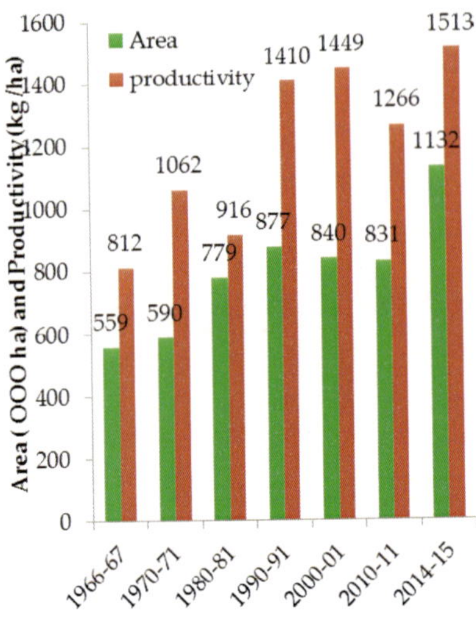

Fig. 1. Trends in area and productivity of maize in Madhya Pradesh.

There are three wet milling industries units in Madhya Pradesh with a crushing capacity of about 450 MT maize annually. In Madhya Pradesh, there is no plant with crushing capacity of less than 100 MT/ day. The starch in the main product of a maize processing unit, which is consumed in industries like food, pharmaceuticals, taxtiles, paper, hotel and restaurants. The other products include gluten, germ, fiber (husk) and corn steep liquor. Gluten has great demand in animal feed industry because of its high protein (70%). The data recorded from industry of Mandsaur district showed that maize grain purchase in the year 2013, 2014 and 2015 is 510764, 478993 and 487023 q per year, respectively while consumption was 509643, 495393 and 478128 q per year, respectively.

Indicated in Fig.2, the lowest purchase of maize grain by the industries was done in August month during 2013-16. However, the consumption was stable throughout the year but the lowest maize consumption recorded in June.

Agro-climatic regions identification of zones in Madhya Pradesh

Traditionally the state is divided into five major crop zones according to the predominance of different crops. These are:

1. **Rice zone:** consisting o the districts of Baster, Raipur, Durg, Rajnandgaon, Balaghat, Bilaspur, Mandla, Raigarh, Surguja, Sidhi and shahdol.
2. **Rice-wheat zone:** consisting of the districts of Rewa, Panna, Satna, Jabalpur and seoni.

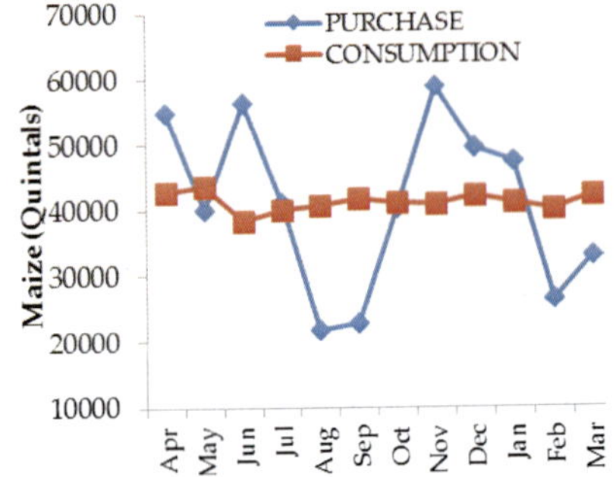

Fig.2. Month-wise mean maize purchase and consumption by Madhya Pradesh state based industries during 2013-16 (Source: data provided by Rajaram Maize Starch Industries, Mandsaur, Madhya Pradesh)

3. **Jowar-wheat zone:** consisting of the districts of Guna, Shivpuri, Gwalior, Bhind, Datia, Tikamgarh and Chhatarpur in the north and Betul and Chhindwara in the south.

4. **Wheat zone:** Wedged in between the two stretches of jowar-wheat zone are the Vindhya plateau and Narmada valley including the districts of Vidisha, Sagar, Damoh, Bhopal, Raisen, Sehore, Hoshangabad and Narsingpur which consist of a stretch of medium to heavy black soils, where the bulk of the wheat and gram crop grown with lower rainfall and medium black soil. Although, the division of the state into these five crop zones is fairly valid, it must be pointed out that these crops zones are not agro. Climatically homogenous except most parts of the Jhabua district adjoining Gujarat including the districts of Rajgarh, Shajapur, Dewas, Mandsaur, Ratlam, Ujjain, Indore, Dhar, Khandwa and khargone.

The different areas of the zone are agro-climatologically unrelated. It is therefore, necessary to recognize one or more agro-climatic regions in each of the crop zones.

Crops and cropping pattern

Out of the total cropped area of 208.26 lakh ha, in Madhya Pradesh, food crops (cereals and pulses) account for 172.76 lakh ha. The area under soybean is increasing fast and covered maximum area of 45.46 lakh ha in year 2007-08.

The main crops grown in *Kharif* season are soybean, paddy, maize, bajara and tur etc. and in *Rabi* season wheat, gram, mustered, cotton, jowar and vegetables. Sugarcane, custard, apple and banana are also grown in some districts. Among the individual crops soybean followed by paddy (16.78 lakh ha) and maize (8.69 lakh ha).

In Madhya Pradesh, in comparison to cereals and pulses, higher area of oilseed (5167.1 thousand ha) covered and second highest area is in cereals (3690.1 thousand ha) but area of pulses is decreasing day by day which is now 950.7 thousand ha.

AICRP centres in Madhya Pradesh

The first AICRP maize centre was established in at JNKVV, ZARS, Chindwara. Later on, in M.P. one more centre was established at RVSKVV, ZARS, Jhabua, in XI plan which was closed subsequently.

Constraints for maize production in Madhya Pradesh

- Maize grown in rainfed area under poor management and on marginal lands.
- Delayed planting of maize in heavy soil.

- The majority of the farmer not maintaining optimum plant population.
- Use of imbalance dose or no use of organic and inorganic fertilizer as per recommendations.
- Micronutrient deficient area especially for zinc.
- Water management practices not adopted properly.
- Improper management of insect pest is one of the causes of low production.
- There is a problem of weeds in early stage of crop growth and farmers are not properly managing this problem.
- The farmers does not uses certified seed in certain area.
- Low adaptation of improved technologies among the maize growers due to lack of education, poverty and other socio-economical constraints.
- The crop cannot tolerate the long dry spell during the flowering to grain filling stages, which resulted in heavy yield reduction.

Popular maize hybrid in Madhya Pradesh

Most of the farmers are planting double cross hybrid maize since past several years but recently switched to single cross hybrid (Table 1). Presently nearly 65.68% acreage of maize cover under hybrid.

Table 1. Area under improved maize varieties in M.P.(Source: Estimate obtained from expert elicitation (EC) workshop)

Top Maize Varieties indentified in EC	Area adopted (%)
JM-216	18.89
NK-30	5.88
P-3501	4.93
RASI-4212	4.51
JVM-421	4.51
Hishell	4.46
NK-6240	3.57
BIO-9637	3.45
HQPM-1	2.73
DKC-7074	2.62
DKC-8101	1.43
BIO-9544	1.43
Jawahar Maize-12	0.89
Bio-9720	0.82
Jawahar Maize-8	0.59
Ganga Safed-2	0.59
RASI-4794	0.59
BIO-4790	0.30
P-3501 and P-3502	0.30
Other hybrids >100-120	29.0
Others (Desi/Land races or F2-F3 generation)	8.41
Total	**99.9**

On the basis of seed of maize hybrid in Madhya Pradesh, prominent private sector hybrids are KMH-3712, NMH-803 and PRO-312. In public sector variety JM-216, HQPM-5 and HQPM-1 taken as 11500, 9000 and 8900 q, respectively (Table 2 & 6).

Table 2. Variety-wise production of certified seed (q) for *Kharif* 2016 (Estimated).

Variety	SSC	G Farm	BeejSangh	NSC	Private Agency	Total
Pusa Early hy-M-2	0	0	0	0	5250	5250
Pusa Early Hy M-1	0	0	0	0	5750	5750
BISCO-740	0	0	0	0	5500	5500
BISCO-855	0	0	0	0	6200	6200
NMH-803	0	0	0	8500	5600	14100
Bio-96-82	0	0	0	4000	0	4000
BIO-9637	0	0	0	0	0	9500
MM-7725(HISHELL)	0	0	0	0	100	1600
CANDY (KSCH-333)	0	0	0	140	19	159
PRO-311	0	0	0	0	9200	9200
KMH-3426	0	0	0	0	9500	9500
KMH-3712	0	0	0	1500	15200	16700
PRO-312	0	0	0	0	10550	10550
JKMH-175	0	0	0	0	5400	5400
DKC-7074	0	0	0	0	9500	9500
Bio-9544	0	0	0	3000	9800	12800
Total Maize Hybrid				**20440**	**107069**	**127509**
JM-8	0	0	0	0	0	0
JM-12	0	0	0	0	0	0
JM-216	61	9	305	0	11500	11875
Chandan-2	0	0	0	0	0	0
Chandan-3	0	0	0	0	0	0
NLD	0	0	0	0	0	0
HQPM-1	0	0	100	0	8900	9000
Jawahar Campo	0	0	0	0	0	0
Vijay Campo	0	0	0	0	0	0
Ajad Kamal	0	0	0	0	0	0
HQPM-5	0	0	0	0	9000	9000
Total Maize	**61**	**9**	**405**	**124169**	**136469**	**146084**

SSC- State Seed Corporation, Madhya Pradesh
G Farm- Government Farm, Madhya Pradesh
BeejSangh- BeejSangh, Bhopal, Madhya Pradesh
NSC - National Seed Corporation

Table 3. Composites released from the Chhindwara centre.

Cultivar	Maturity (days)	Year	Area of adoption	Remark, if any
Chandan Makka-1	100-105	1972	For whole M.P.	Yield 48-50 q/ha, Yellow seeded medium maturity.
Chandan Makka-2	80-85	1973	For whole M.P.	Yield 25-28 q/ha, extra early white seeded composite.
Chandan Makka-3	95-100	1976	For whole M.P.	Yield 45-48 q/ha, medium maturing yellow seeded
Jawahar Makka-8	82-83	1997	For whole M.P. rainfed	Extra early, white, yield 40-42 q/ha, for double cropping
Jawahar Makka-12	88-90	1999	For whole M.P. rainfed	Early, white, for double cropping, yield. 42-45 q/ha.
Jawahar Makka-216	96-98	2004	For whole M.P. rainfed	Early, yellow, yield50- 55 q/ha
Jawahar pop 11	100-105	2006	For whole M.P. rainfed	Medium and, yellow, yield 20- 25 q/ha

These varieties released by JNKVV, Jabalpur from Chhindwara centre by state variety release committee (SVRC) for the Madhya Pradesh (Table 3). Out of these, JM-8 is mostly adopted by the farmers of Jhabua district due to white grain colour but JM-216, preferred by farmers of the state due to high yield not only *Kharif* but in *Rabi* season also with the advantage of medium maturity drought resistance and full grain filling.

Table 4. Domains-wise area covered by different group of varieties in MP during 2014-15.

Particluars	Domain 1	Domain 2	Domain 3	MP State
Agro-climatic Zone	Khymore plateau and stapura hills and Satpura plateau	Malwa and Vindyan plateau	Jhabua Hills	
Districts	Rewa, Satna, Seoni, Jabalpur, Katni, Panna, Sidhi, Chhindwara, Betul	Ujjain, Dewas, Indore, Rajgarh, Shajapur, Ratlam, Mandsour, Neemach, Dhar	Jhabua, Alirajpur	
Local varieties (%)	4 %	3 %	12 %	6.4 %
Modern varieties				
Composites (%)	28 %	24 %	23 %	25.0 %
Hybrids (%)	68 %	73 %	65 %	68.6 %

The seed replacement rate (SRR) in 2013-14 was 73.91% which replaced upto 86.64% in 2016-17, it indicates farmers are aggressively adopting improved cultivars in maize (Table 5).

The farmers of Madhya Pradesh are more advance and adopting hybrid in maize cultivation as compared to composite and local varieties. The spread of different group of maize cultivars in different domains of Madhya Pradesh, shown that higher area covered by hybrid cultivars compared to other domain (Table 4).

Table 5. Achievements of seed replacement ratein M.P.

Year	Total area to be sown (000 ha)	Seed distributed (q)	SRR (%)
2013-14	1002	111089	73.91
2014-15	1132	117097	69.44
2015-16	1098	120789	73.34
2016-17	1211	120789 (Estimated)	86.64 (Estimated)

Table 6. Maize seed production in the state of M.P.

S. No.	Name of inbred/ composite	Quantity of seed produced in (kg)						
		2006	2007	2008	2009	2010	2015	2016
(a) Nucleus Seed								
1	Jawahar Makka-8	25	25	25	25	5	5	10
2	Jawahar Makka-12	25	25	25	25	5	5	10
3	Jawahar Makka-216	25	25	25	25	5	20	40
(b) Breeder Seed								
1	Jawahar Makka-8	900	1180	1040	-	-	-	-
2	Jawahar Makka-12	210	180	80	-	-	-	-
3	Jawahar Makka-216	3200	3800	3600	4200	2800	3200	5000
4	HKI-163	-	-	685	1416	2560		
5	HKI-193	-	-	1165	311	850		
6	HKI-161	-	-		50	2050		
(c) Foundation seed								
1	Jawahar Makka-216	-	-	-	-	-	28500	15000
(d) Hybrid Seed								
1	HQPM-1 (CWA)	-	-	1125	-	-		
2	HQPM-1 (JBP)	-	-	6900	-	-		
3	HQPM-1 (Farmer)	-	-	1000	275	-		
4	HQPM-5	-	-	100	275	350		
****(E) TL seed**								
1	Jawahar Makka-216	-	-	-	-	-	25000	25000

* Foundation seed produced by cooperative seed samiti.
** TL seed produced by farmers group for their own use.

Cropping system

The AICRP recommended various sequential (Table 7) and inter (Table 8) cropping systems for various agro-eco-regions of the state.

Table 7. Maize based sequential cropping systems in different ago-climatic zones of M.P.

Agroclimatic region	Cropping system	
	Irrigated	**Rainfed**
Chhattisgarh plain	Rice-maize-greengram	
Northen Hills	Maize-wheat, Maize-potato-wheat	Maize-legumes
Kaymo	Maize-mustard, Maize-wheat, Rice-maize	Maize-mustard, Maize-legumes
Central Narmada Valley	Maize-sugarcane, Maize-wheat,	Maize-mustard, Maize-legumes
Vindhya Malwa Nimar	Maize-potato-wheat, Maize-wheat-greengram	
Sathpura	Maize-wheat, Maize-potato-wheat, Maize-wheat-greengram	Maize-wheat, Maize-mustard, Maize-legumes
Gird	Maize-wheat, Maize-potato-wheat	Maize-mustard
Bundelkhand	Maize-wheat, Maize-potato-wheat Maize-wheat-greengram	Maize-wheat, Maize-mustard Maize-legumes
Jhabua	Maize-wheat, Maize-potato-wheat, Maize-wheat-greengram	Maize-mustard, Maize-legumes

Table 8. Maize based intercropping systems for M.P.

Intercropping systems	Suitable area/situation
Maize + Soybean	All maize growing areas
Maize + Pigeonpea	
Maize + Cowpea	
Maize + Mungbean	
Maize + Urdbean	
Maize + high value vegetables	*Peri*-urban interface
Maize + cut flowers plant	
Baby corn + vegetables	
Sweet corn + vegetables	

I. *Kharif* seasons AICRP agro-technologies

Date of sowing and crop management

1. Planting of maize at onset of monsoon gave higher yield and found optimum for sowing of maize crop. Dry sowing also suggested where palewa irrigation is available. Pre-monsoon sowing in first week of June gave an extra yield from 23 to 28% than the normal sowing.

2. Planting of maize in the third week of June gave higher yield. A reduction in yield by 25% to 60% were noticed if the sowing is delayed beyond second week of July to end of July.

Method of sowing

3. Sowing of maize in flat bed followed by earthing up after knee high to pre tesseling stage found superior over other prevailing methods *viz.* sowing on ridge and sowing in furrows by drilling.

Seed treatment

4. Seed soaked in tap water over night and then dried in shade for few hours followed by seed treatment with fungicide gave 10 % higher yield over dry seed fungicidal treatment.

Plant population

5. Maize plant population of 54,000 per ha (75*25 cm) for late maturing hybrid, 66,000 per ha (75*20 cm) for medium maturing hybrid and 83,000 per ha (60*20 cm) for composite found to be optimum.

Seeds per hill

6. Two seeds per hill followed by thinning at 15 days after sowing was found proper to maintain desired population.

Nutrient Management

Response of different doses of nutrients to maize varieties over the years shows increasing response of nitrogen doses up to 120 kg/ha.

Nitrogen (kg/ha)	2006	2007	2008	2009	2010	Mean
40	—	4363	—	—	—	4383
60	3079	4989	—	—	—	4516
100	—	5131	—	—	—	4905
120	4223	5587	—	—	—	5131
160	—	5505	—	—	—	5505
180	4619	5818				5818
$N:P_2O_5:K_2O$ (kg/ha)						
100:50:50	—	—	6896	4002	4735	5211
150:65:65	—	—	7856	5419	5478	6251
200:80:80	—	—	8011	6506	5777	6764

7. For Extra Early composite like JM-8, one can choose 70:30:20 N_2:P_2O_5 and K_2O kg/ha dose with a plant population not more than 77,000 to get maximum yield of with C:B ratio of 1:3.23.

8. While considering interaction effect of 90:40:30 N_2: P_2O_5 and K_2O kg/ha with 70,000 plants/ha, JM-12 (Early) has given maximum grain yield (40.5 q/ha).

9. Application of FYM in addition to NPK and biofertilizer significantly increase the yield of maize (i.e. Azotobacter + 75% N + full PK through fertilizer + FYM 10 t/ha + PSB +75% P full NK through fertilizer + 10 ton FYM /ha).

10. SSNM based on the nutrient expert gave 13.7% higher yield of maize over recommended fertilizer practices (RDF) at Chhindwara.

11. The bio-fertilizer coktail (Mix stain) techniques was superior over single stain and chemical fertilizer. Application of 200:75:75 kg/ha N: P_2O_5 : K_2O along with bio-fertilizer (Coktail) in maize was found economic in comparison to higher dose of fertility levels.

Intercropping

12. At Chhindwara, maize + soybean gave 30% while maize + black gram gave 50% extra remuneration over sole maize while maize + cowpea in 1:1 ratio in between maize rows gave 13% more remuneration then sole maize.
13. Intercropping of maize (H) for grain with forage maize (African tall) proved to be more profitable. Highest monetary can be obtained from intercropping as against sole forage maize (Rs. 6785/ha) when harvested at milk stage.
14. Inter-cropping with black gram in between the rows of early maize not only gave additional production (3.6 q/ha) of black gram but also increase the maize grain yield (45.2 q/ha) against sole maize (43.9 q/ha).
15. Maize + cowpea fodder can successfully be grown together in 1:1 ratio to harvest good yields of quality fodder rich in protein for meeting the nutritional requirement of cattle.
16. Suitable period of early harvest would be at physiological maturity (black point appears on the germ end of grain), which enables the farmers to grow the crop into early (rainfed) cropping systems during *Rabi*. It means after cessation of monsoon during September end, farmers can go for 10 days early planting of safflower, linseed and mustard under rainfed situation after taking produce from early maize composite.
17. Cowpea inter crop mulch in maize can conserve the moisture which gave 7% increase in maize grain yield over maize grown without mulch, in addition to 8 q/ha of green pods of cowpea.
18. It is quite feasible to grow vegetables as an intercrop with *Kharif* maize like Indian bean, sponge guard, okra and bottle guard in between two rows of medium duration composite maize (JM-216) for extra returns.
19. Crop grown for green manuring purpose as the intercrop with maize growing sesbania an intercrop of maize did not change value in the treatment of maize: sesbania (Harvest at 35 DAS) in 1:1 ratio (4130 kg/ha) and 1:2 ratio (4150 kg/ha.) in comparison to maize sole (4130 kg/ha).
20. Maize with sunhemp green manuring in 1:2 ratio was found good for minimum yield reduction of maize crop and maximum green foliage of green manuring sub crop followed by maize + sunhemp at 1:1 row ratio.

Weed management

21. Pre-emergence application of 1.0 kg/ha atrazine + followed by manual weeding at (30 DAS) was effective in controlling weeds in sole maize.
22. Study resulted that grassy weeds are highly undesirable in maize field and they causes much damage as compared to non-grassy weeds or sedges. Maximum yield loss in maize was noticed by grass weeds followed by broad leaved weeds and sedges.

Moisture conservation

23. Side ridge sowing of maize i.e. dibbling on side of the ridge was found superior over the other methods in initial water stress condition as well as excess water condition.
24. Maize sowing in dry condition before onset of mansoon with two seed per hill and row placement of FYM gave significantly higher i.e. 25% grain yield.

II. *Rabi* season AICRP agro-technologies

Planting time

1. The first fortnight of November is the optimum time for *Rabi* maize planting.

Plant population

2. 80000 plants / ha (60*20 cm) was found to be optimum to achieve maximum yield.

Response of NPK

3. 150 N: 72 P_2O_5:60 K_2O kg/ha was found economic in maize during winter.

Time of N application

4. Nitrogen is to be applied in three equal split dose in *Rabi* maize i.e. at sowing, knee-high stage and pre-tasseling stage for higher yield.

Intercrop during *Rabi*

5. Maize: fenugreek (1:1) was found superior in maize based inter cropping system along with fenugreek, coriander and radish for seed purpose as intercrop recommended.
6. Sweet corn: marigold (big) or gladiolus or marigold (small) or Chrysanthemum (1:1) intercropping system is also recommended for higher returns.
7. Maize: onion (1:1) was found superior in inter crop system.
8. In between two rows of *Rabi* maize (75 cm. apart) fenugreek, early pea (Arkel), coriander and spinach to be grown as green vegetable with higher monetary returns of 11-16%, over sole maize.

Herbicides

9. Higher grain yield of maize was recorded by hand weeding (manual) or Atrazine @1.0 kg/ha.

Recommend maize production technology

A. Kharif maize

Date of sowing : First week to second week of June.

Method of sowing : Sowing of maize in flat bed followed by earthing up after knee high to pre tesseling stage found superior over other prevailing methods *viz.* sowing on ridge and sowing in furrows by drilling.

Plant population : 54,000 per ha (75*25 cm) for late maturing hybrid, 66,000 per ha (75*20 cm) for medium maturing hybrid and 83,000 per ha (60*20 cm) for composite found to be optimum.

Seed Treatment : To protect the maize crop from seed and major soil borne diseases and insect-pests, seed treatment recommended as per the below given details:

Disease/insect-pest	Fungicide/ Pesticide	Use (g/kg seed)
TLB, BLSB, MLB	Carbandazim + Thiram in 1:2 ratio	3.0
BSMD	Apran 35 SD	4.0
Pythium Stalk Rot	Captan	2.5
Termite, shoot fly	Imidachlorpid	4.0

Nutrient management

Fertilizer : N : P_2O_5 : K_2O

Composite : 100 : 50 : 50

Hybrid : 150 : 60 : 60

Inter cropping : Maize intercropping with early soybean, blackgram, cowpea legume in 1:1 ratio. Intercrop like Indian bean, sponge guard, okra and bottle guard is also recommended between two rows of maize.

Weed management : Atrazine @1.0 kg/ha as pre-emergence application.

B. Rabi maize

Planting time : first fort night of November

Plant population : 80000 plant/ha (60 x 20 cm) nutrient management

N: P_2O_5: K_2O kg/ha : 150 : 70 : 60

P and K to be applied at the time of sowing and nitrogen to be applied in three equal split dose i.e. at sowing, knee high stage and pre-tasseling stage.

Intercropping

Rabi maize with vegetable for seed :

Maize : fenugreek

Maize : coriander

Maize : radish

Maize : onion (1:1)

Sweet corn maize with flower plant :

Sweet corn : marigold big (1:1)

Weed management : Atrazine @1 kg/ha or Pendimethaline @1.0 kg/ha as pre emergence application.

Seed treatment: A given in *Kharif* maize

Insect-pest management

Stem borer (*Chilo partellus*)

Major pest of maize is stem borer.

Pink borer (*Sesamia inference*)

It occurs during winter season particularly in peninsular India.

Borer control : For control of borer granular application of corbofuran 3G @ 7.5 kg/ha in plant whorls at 10-15 days after germination is very effective.

Termites *(Odontotermes obesus)*

Termite is also an important pest in many areas is also an important pest in many areas and for control of this clorpyriphos @ 2ml/l water should be applied.

Disease management

Several diseases occur during different seasons in various parts of the MP that leads to loss to yield if not managed properly in time.

Turcicum leaf blight (*Exserohilumturcicum*) (TLB)

The disease is prevalent in cooler conditions with high humidity in Madhya Pradesh. Long, elliptical, grayish green or tan lesions (2.5-15 cm) appear on lower leaves progressing upward. Sprays of mancozeb @ 2.5 g/lt (with adjuant @ 0.05%) at 8-10 days interval advised for its management.

Maydis leaf blight (*Drechsleramaydis*)

The symptoms of maydis leaf blight lesions on the leaves are elongated between the veins, tan with buff to brown or dark reddish brown borders. For control spray of Dithane 2-75 or Zineb @ 2.4– 4.0 g/liter of water (2-4 applications) at 8-10 days interval after first appearance of symptoms of disease.

Yield gaps in maize in Madhya Pradesh state

Contribution of production factors to maize yield in the state is as follows:

1. Variety /Hybrids 22.25%
2. Plant protection measures 13%
3. Agronomic management
   a. Plant population 8-10%
   b. Fertilizer 37%
   c. Weed management 16-18 %
   d. Irrigation 10-12%

Yield gap analysis

The average yield and of maize in last five year were lower in M.P. compared to the national average (Table 9). However, the state yields was lower than the farmers potential and potential research farm yield represented by demonstration yield (5205 kg/ha) and research station yield (6216 kg/ha). The total yield gap of 4703 kg/ha, which is the different between research station yield and the state average yield can be broken up in to two components *viz.* yield gap I or research gap (difference between research station and demonstration yield) of 1012 kg/ha and yield gap II or extension gap (difference between demonstration and state average yield) of 3691 kg/ha.

Table 9. Yield gap of maize in M.P. 2014-15.

Particulars	India	Madhya Pradesh
Area (Lakh/ha)	92.33	11.32
Productivity (kg/ha)	2564	1513
Mean yield of maize cultivars	5204	6216
Yield gap (kg/ha)		
a. Research gap (yield gap I)		1012 (16.28)
b. Extension gap (yield gap II)		36.91 (70.92)
c. Total yield gap (yield gap III)		4703 (67.82)

The extension gap can be minimized by motivating the farmers to adopt improved production practices.

- Develop variety suitable to the soil condition with short duration for minimum rainy day's as well as for rainfall areas.
- Adopt medium to long duration hybrid in high rainfall / irrigation facility available areas.
- Assessment of the variety- wise demands for seed for finally indents by state government.
- Motivate the farmers to produce foundation or certified seed as per guidance of scientist or seed certification officers.

Linkages between agricultural research and extension systems

The important features of these linkages are monthly workshop, seasonal workshop, of regional research planning committee, discipline level research planning committee, zonal research planning committee, and the central research planning committee. In these channels the production technology generated at regional level is discussed and finalized taking in to consideration the views and suggestion of the extension officers of the department of agriculture. In addition, research programme of the station scrutinized and approved after passing through above mentioned channels. The research and extension linkages are promoted through the training of extension staff by research scientists through fortnightly meeting, visit to farmers fields and the activities of the department of agriculture. The coordination between extension and research is ensured by:

a. Involving field officers in planning, execution and evaluation of technical programme.
b. The condition of adaptive research by extension staff under technical supervision of scientists.
c. By providing a system of feedback for field problems to research for finding solutions.
d. By evolving need based and problem oriented production technology so that farmers are tangibly benefitted and the farmers develop a sense of confidence in the extension workers.

Climatic pattern in Madhya Pradesh

The overall climate varies from semi-arid to sub-humid with hot summer (38 to 44 °C), cool and dry winter (7 to 13°C) with rainfall (during June to October) maximum 1200 to 1600 mm in Chhattisgarh plains and Northern hills region of Chhattisgarh plains, and minimum rain fall of 800 to 1500 mm in other agro-climatic zones. average crop growing period is 150 to 180 days.

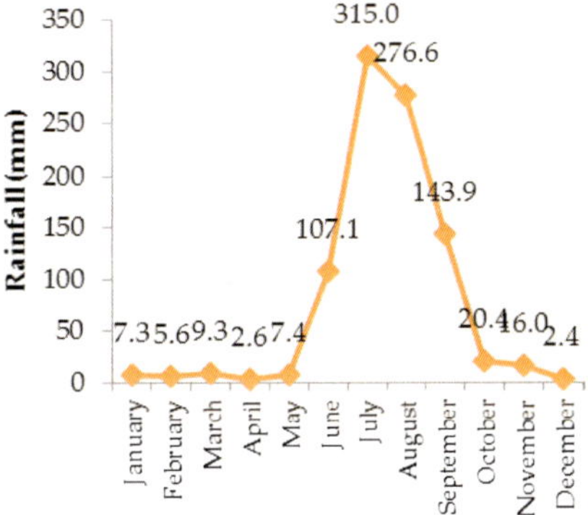

Fig. 3. Monthly rainfall pattern in M.P.

1. **Rainfall:** The average rainfall of this region is around 1194.0 mm out of which 40% rains receives during 28th July to 24th August, where July and September months jointly receives only 43% of rains (Fig. 3). Average number of rainy days varies from 62 in Chhindwara and 70 days in Betul districts with average being 66 days (Fig. 4).

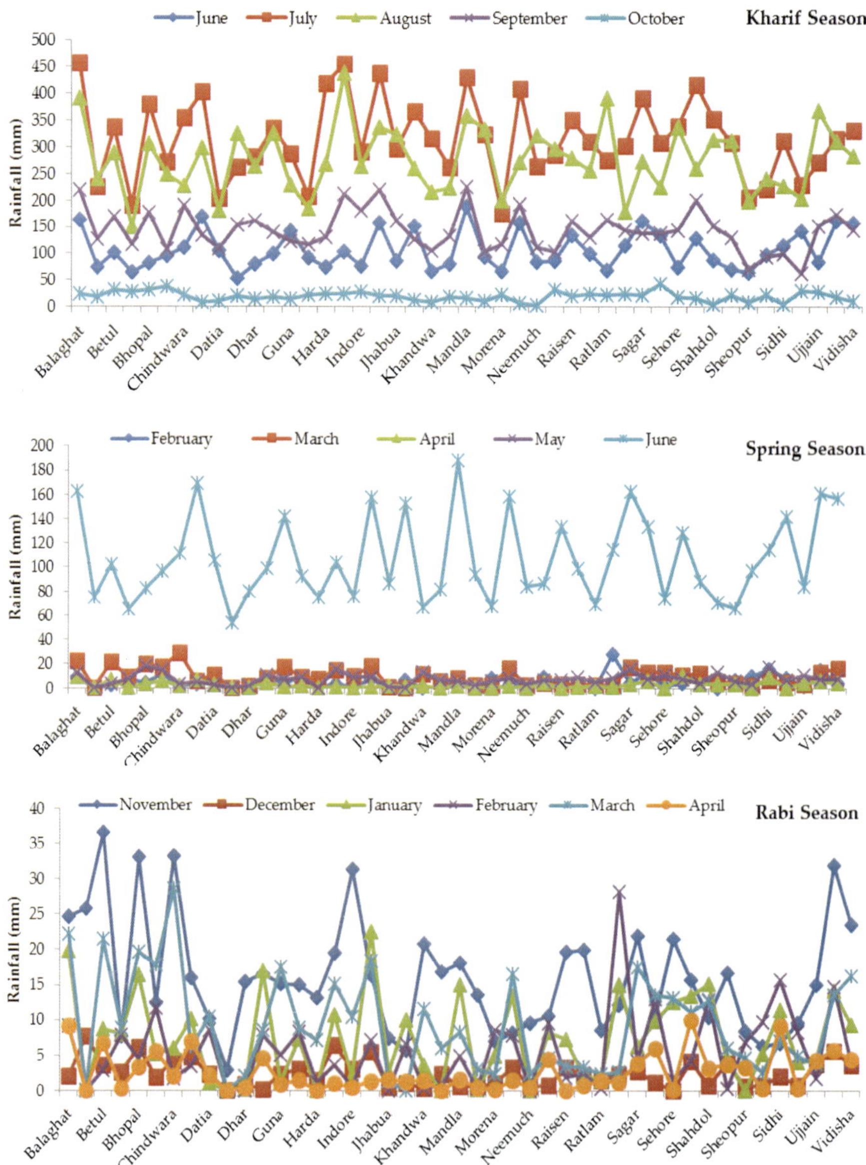

Fig. 4. Rainfall patterns in different districts of M.P. during all the three maize seasons.

2. **Temperature** : The maximum temperature which is the highest during the months April, May and June which varies from (38°-42°C). The temperature starts deceasing from Mid-June onwards or 25th week (38°C max.-25°C mini) and come down to 26°C max.-22°C mini. during middle of August (33rd week). The temperature remains constant till middle of December except the duration 15th September to 31st October, where it fluctuates from 28° to 30°C. The minimum temperature is the lowest of 5.8° and 4.6° C which can be observed during and of December to middle of January. During 1st April to 10th October, the minimum temperature is always greater than 20°C (Fig.5).

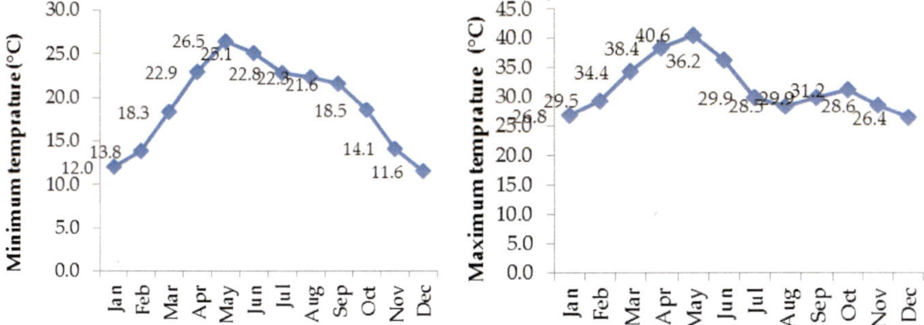

Fig. 5. Mean maximum and minimum temperature in last 100 years of different months in Chindwara.

Future research and development strategy for maize in Madhya Pradesh

The maize particularly during winter season becoming a potential alternative crop to winter wheat in MP. Two technology i.e. single cross hybrid and flatbed method are becoming very popular in maize system. Further strengthening of these technologies for farmer's profitability and long term suitability, the following actions/intervention are to be initiated :

- The seed replacement rate in maize in Madhya Pradesh is very high and almost all the farmers are taking hybrids of maize in the state. However, the seed is costly due to lack of seed production chain of the public bred single cross hybrids. Therefore, there is a need to strengthen the seed production by public private partnership so that the availability of SCH seed can be ensured in time at less cost.
- Most of the farmers of irrigated area are growing maize with high plant density/ha with inappropriate and imbalance nutrient application that needs to be optimized for profitability and long- term sustain ability.
- Massive efforts on training of the local artisans and farmers are essentially needed for realizing better productivity and farm profitability and the accelerate the adoption of the technology.

- Identification of the compatible genotypes that give higher yield under minimum or maximum rainy days conditions.
- The 'No- till' planting of maize-mustard or soybean- winter maize is some district of MP is accepted technology. However till now, manual dibbling is being practiced due to non-availability of planters in the region. Therefore, there is a need to strengthen the local manufacturing units for the no till planters that suitable for local requirements.
- Developing agro-technique i.e. nutrient, weed, and water and pest management, for emerging cropping system like soybean-maize or maize-mustard where in the existing management practices may not work due to varied edaphic requirement.
- More area (> 18000 ha) but less productivity (< 1000 kg/ha) districts like Jhabua, Dhar, Rajgarh, Alirajpur, Shajapur, Dindori, Mandla, Shahdol, Seoni, Sihor and Anuppur actually have tribal farmers and grow good crop under traditional practices. Need low inputs technology and organize field day and lecture and training on crop production practices.
- More rains and paddy area of Satna, Rewa, Panna, Murena, Umariya, Sidhi, Katni Jabalpur, Balaghat, Damoh, Hoshangabad and Narsingpur. In these paddy districts in future maize can be grown in *Rabi* and summer season successfully as popcorn, baby corn, sweet corn and maize grain. Some districts have low area of maize crop which may have possibilities to increase area with the help of maize fair, kissan gosthi, thesil wise.
- In some district having more soybean area but some past year decrease the production of soybean area due to disease and insect problem in this reason farmers needs to be educated to grow maize crop.
- Some district having minimum rainy days, uncertain and moisture ability problem. In these district, short duration varieties (60-92 DAS) needed for maize production. Maize crop as baby corn (60 day), sweet corn (75 day) and for green cob (80-85 day) can be grown successfully.

Future research strategy for maize in Madhya Pradesh

- Development of early and medium maturity single cross hybrids tolerant to biotic and abiotic stress for *Kharif* season.
- Development of full season maturity single cross hybrids for *Rabi* season.
- Development of cultivars for specific uses (popcorn, sweet corn, baby corn, starch, oil, high lysine etc.)
- Development of cultivar specific package of practices.

- Development of crop production technology for determination of minimum tillage, water management (low and excess) and low nitrogen input.
- Development of maize-based cropping system for maintenance of fertility / productivity of soils.
- Generation of information on agronomic aspects of maize crop used for specific purposes.
- Demonstration of improved technology in farmer's field through Minikit and front line demonstration (FLDs) in both the seasons.
- Popularization of non-traditional uses of maize (Pop corn, sweet corn and baby corn).
- Extension of maize cultivation at *Rabi* and *spring* season.
- Encouragement of maize cultivation by providing higher support price.
- Projection of maize as an alternative source of edible oil.
- Establishment of maize based industries in Madhya Pradesh.

18

Maize Research and Development in Gujarat

S.M. Khanorkar, P.K. Parmar, K.H. Patel and V.J. Patel*

Main Maize Research Station, Anand Agricultural University, Godhra-389 001, Gujarat
**Corresponding Author's Email: subhkhanorkar@yahoo.com*

Maize ranks fourth after rice, wheat and bajara with respect to area and production, and grown on 4.23 lakh ha average with 6.72 lakh tones production and yield 1589 kg/ha in the Gujarat (2014-15). The state has 38.0% lower than national yield. It mainly grown in Panchmahal, Dahod, Mahisagar, Vadodara, Chhota Udepur, Kheda, Sabarkantha, Banaskantha and Aravali districts as rainfed *Kharif* and irrigated winter/*Rabi* crop for food, feed and industrial raw materials.

It is staple food and feed of tribal and grown as sole and intercrops with pigeon pea, black gram and soybean in *Kharif* season. The white and yellow flints to semi-flint maize are preferred to prepare bread and other food preparations by

local inhabitants of the region. Maize is also one of the main constituent of good and palatable fodder in dairy industries. The specialty corns like, sweet corn and popcorn becoming popular to consume as snacks and use in vegetable preparations throughout the year. On account of non-availability of sufficient maize grain (~9 to 10 ton per day), the local starch industries are procuring it from Andhra Pradesh, Karnataka, Maharashtra and other states to make starch and its value added derivatives (sucrose, fructose, glucose, sweetener, sorbitol, dextrin etc.) along with other by-products like grits, poultry feed, corn oil etc., by dry and wet milling processing of maize. Being a good quality of maize starch than tapioca, the textile, pharmaceuticals and paper industries prefer to use corn starch and its value added derivatives for manufacturing quality cloth and paper. In addition, common households like to use starch for preparation of ketch-up, ice-cream, cake, biscuit, chocolate, water etc.

Trends in area, production and yield of maize (1950 to 2016)

In Gujarat, maize grown on 1.90 lakh ha area with production of 1.50 lakh tonnes and 754 kg/ha yield during 1951-60 (Fig. 1). Interventionof improved production technologies through crop improvement and production researches along with their extension education activities on farmers fields, the area, production and yield gradually improved and achieved 142, 387 and 111 per cent enhancement, respectively during period of 2011-15 over 1951-60. Among maize producing districts are Panchmahal, Dahod, Sabarkantha, Kheda, Banaskantha and Vadodara (Fig. 2).

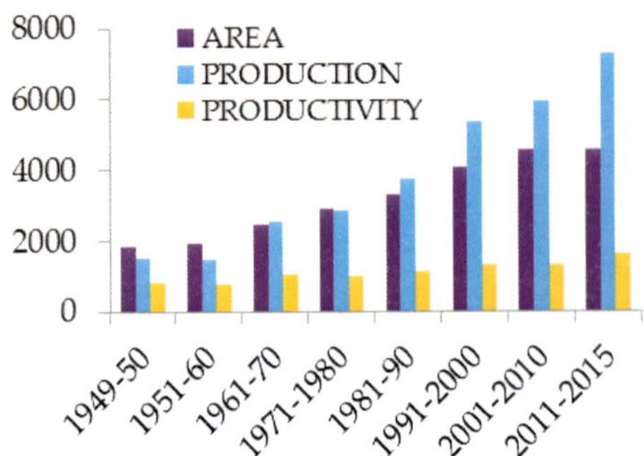

Fig. 1. Area (00 ha), production (00 t) and productivity (kg/ha) trends of maize in Gujarat (1950-15).

The state needs to cultivate more area under single cross hybrids over the existing around 20-25 per cent for further enhancement of grain productivity. Based on five and half decade rainfall pattern, the data revealed that the rainfall ranged from 795 mm (1965-70) to 1006 mm (2011-2016) with mean of 52 years as 892.2 mm. It indicates that rainfall is showing increasing trend in this agro-ecology situations (Fig. 3).

The grain yield of *Rabi* maize in Gujarat remarkably improved by 36.7 to 65.2 per cent over the *Kharif* maize productivity during 2009-15 (Table 1). It indicates better opportunity to enhance maize production and yield if area under winter/ *Rabi* maize will increase by sole crop and cropping system *viz.* paddy-maize and groundnut-maize in non-traditional area. This surplus maize production may meet demands of maize starch based industries of the state. There are also possibilities to introduce single cross hybrid and sweet corn for green cobs.

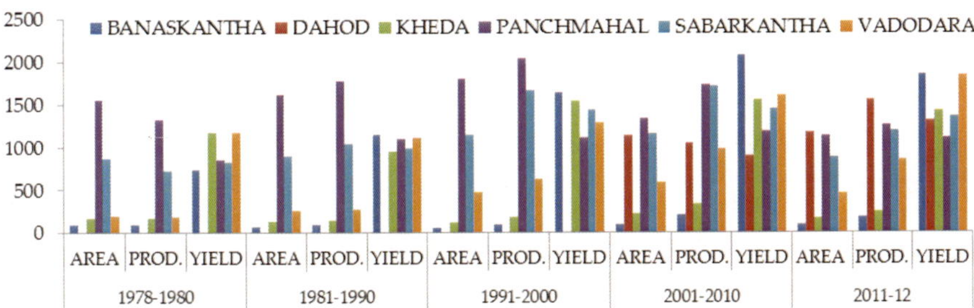

Fig. 2. Area (00 ha), production (00 t) and yield (kg/ha) in maize growing districts of Gujarat (*Source:* Department of Statistics, Govt. of Gujarat).

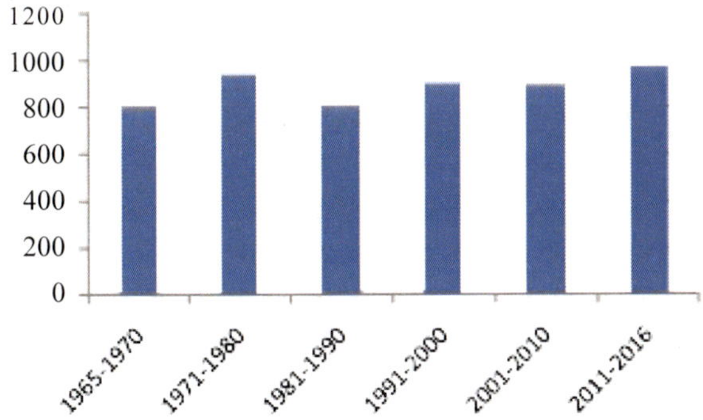

Fig. 3. Rainfall (mm) trends in last fifty years at the Godhra station (*Source:* MMRS, AAU, Godhra).

Table 1. Recent seasonal trends in area, production and yield of maize in Gujarat (Source: DACNET, GoI).

Year	Season	Area (000 ha)	Production(000 t)	Yield (kg/ha)
2009-10	*Kharif*	411	369	964
	Rabi	86	137	1593
2010-11	*Kharif*	423	692	1636
	Rabi	78	1281	1645
2011-12	*Kharif*	387	539	1393
	Rabi	129	247	1915
2012-13	*Kharif*	373	625	1676
	Rabi	85	166	1953
2013-14	*Kharif*	333	422	1267
	Rabi	128	259	2023
2014-15	*Kharif*	318	463	1456
	Rabi	105	209	1990

Major industries utilizing maize in the Gujarat

The state has well established 14 maize based starch industries listed below and needed 9-10 tonnes/day maize grain for production of starch and its value addition derivatives like, sucrose, fructose, glucose, sweetener, sorbitol, dextrin etc. for textile, paper, pharmaceutical and poultry industries. Corn oil is becoming popular and healthier source of vegetable oil now a days as cooking media is an additional byproduct which extracted from maize germs obtained during wet milling in starch industries. Most of these industries export starch and its by products to south Asian countries. In addition, maize food based entrepreneurs are also manufactures corn flakes, popcorn, nutritive biscuits and many house hold items for healthy human diet.

1. Everest Starch, Rajkot-Ahmadabad National Highway, **Brahmapuri**, Gujarat; **Phone:** 098985 24662
2. Anil Products Ltd., Anil Rd, Government Employees Colony, Bapunagar, **Ahmadabad,** Gujarat 380018 **Phone:** 07922202722
3. Mamta Starch Products, 57, Airabhai Market, Diwan Ballubhai Road, Kankaria, Kankaria, **Ahmadabad**, Gujarat 380008 **Phone:** 07922866691
4. Sayaji Industries Limited, Chinubhai Nagar, P O Kathahwada Maize Product, Odhav, **Ahmedabad**, Gujarat 382430 **Phone:** 079 2290 1581
5. Bharat Starch Industries, Patriot Complex, Race Course Circle, Opposite Amrakunj Society, Race Course Circle, **Vadodara,** Gujarat 390007 **Phone:** 0265 239 8095
6. Santosh Starch Products Limited, 55/78 NIDC Estate, Lambha Road, **Ahmedabad,** Gujarat 382405 **Phone:** 079 26651800
7. Mahalaxmi Maize Products, C-Block, 2nd Floor, B.G. Tower, Opposite Delhi Gate, Shahibaug Rd, Bhadreshwar Society, Kazipur Dariyapur, **Ahmedabad,** Gujarat 380004
8. Paramguru Agro Industries - Maize Cattle Feed, Hirpura, Post - Hirpura, Ta. Vijapur, Aaglod Road, **Hirpura,** Gujarat 382870 **Phone:** 09409199438
9. Gujarat Ambuja Exports Ltd, Ambuja Tower, Opposite Memnagar Fire Station ,Post Navjivan, Navarangpura Gam, Navrangpura, **Ahmedabad**, Gujarat 380014 **Phone:** 07926423316
10. Roquette Riddhi Siddhi Pvt. Ltd, 18th Floor, Shapath V, Opp. Karnavati Club, S.G. Highway, **Ahmedabad**, Gujarat 380015 **Phone:** 07940016000
11. Babji Food Industries, 23, Mega G.I.D.C, Kharedi, **Dahod**, Gujarat 389151 **Phone:** 02673 -293 456
12. Gigal Industries, Lakulesh Complex, 201, Near Char Rasta Deluxe Society, Nizampura, **Vadodara,** Gujarat 390002 **Phone:** 09426357833

13. Maize Products, Kathwada, **Ahmedabad,** Gujarat 382430 **Ph:** 07922901581
14. Rahi Agro Industries, 91/19-20, Phase: 1, G.I.D.C. Vatva, **Ahmedabad,** Gujarat 382445 **Phone:** 09426354047.

Major cropping systems and production ecologies

Because of climate change and residual rainfall change including insect pests' incidences and ecology pattern, the state witnessed changes and created alternative in cropping system suited to farmers in current situations. The contingency plan is also regularly prepared as and when needed, and executed by the Department of Agriculture, Government of Gujarat, Gandhinagar with the objectives to sustain and enhance maize production and yield under climate change in the state. Maize is usually cultivated more than 65-70 per cent as sole and intercrop under rainfed situations in the state. Looking to suitability, the maize growers adopt Maize-Chickpea, Maize-Wheat, Paddy- Maize in cropping system and Maize + Pigeon pea, Maize + Blackgram, Maize + Soybean, Maize + Castor and Maize + Cotton as inter-cropping systems under moisture conservation regimes. The tribal prefers to grow extra early and early maturity maize hybrids/varieties in undulated topography with low soil fertility and under residual soil moisture ecology in winter/*Rabi* season. They are still to stick to either 30 or 45 cm row to row distance with low utilization of chemical fertilizers and prefer high plant density. The farmers who have sufficient irrigation facility and good soil fertility conditions. Medium to late maturity hybrids with recommended crop production technology have estimated 25-30% of total maize area in the state. In addition, sweet corn farming for consumption of semi urban area as green cobs with extra green fodder for dairy industry is fetching more remuneration in short duration (75-80 days).

AICRP centre/s genesis and their mandate

In Gujarat, the AICRP on Maize was implemented in 1957 under administration of the State Department of Agriculture in first five year plan of the country at Dahod, a tribal dominated area adjoining to Madhya Pradesh and Rajasthan. Later on, the project shifted to Godhra which is 70 km away from Dahod in 1960 and being continued with the mandate to improve production and productivity of maize to meet demands of tribal as food and feed in the state. Later on, the project was executed and administered by the newly established Gujarat Agricultural University, Datiwada, Sardar Krishinagar, Palanpur with the revised mandate based on breeding and production technologies of maize, cropping systems for different agro-climatic zones in 1974. The project released and made availability of quality seed of Ganga Safed 2, Ganga-5, Ganga-11, Hi-starch, Trishulata, Decan 101, Decan-111, Madhuri (Sweetcorn), Amber popcorn,

African Tall (Forage) and production technologies developed to maize growers in the state. Ganga Safed-2, a double top cross white grain hybrid released in 1964, is still under cultivation in the region. African Tall, old fodder variety is also being cultivated to meet demands of dairy farmers in the state. In addition, Amber popcorn and other related specialty corns hybrids like, sweet corn (Madhuri, WOSC and SG-75) are also under cultivation in *peri*-urban area for consumption in the state. In 1983, the National Agriculture Research Project (NARP) implemented with objectives of multidisciplinary and local need based maize research to enhance further productivity in the state. In addition, the station has made collaborative research on heat and drought phenotyping under climate resilient in maize project with international institution CIMMYT, Mexico and CIMMYT (India). The station also has collaborations for seed village and transfer of technology in tribal belts of central India (Gujarat, Rajasthan and Madhya Pradesh) with 14 NGOs to enhance yield under rainfed cultivation with mandate of food and nutrition security since 1986. The early maturing, high yielding, biotic stress resistance, and drought tolerance/ avoidance variety/hybrids have been developed by the for *Kharif* and winter/*Rabi* seasons. The low cost production technologies of maize preferable to farmers developed and disseminated through extension education. The varieties/hybrids of specialty corns *viz.* sweet corn and popcorn and their production technologies have been also developed due to increased demand in the state. The awareness and dissemination of latest improved production technologies are being executed through effective extension education activities like, FLDs, TSP, NSP, Krishi Mahotsav, Field Day, literatures, Akashwani and Doordarshan programmes, trainings etc.

Key production problems for maize in Gujarat

- Non availability of sufficient quantity of certified seed of notified single cross hybrids at affordable rate during sowing time.
- Less availability of early maturity, high yielding and climate resilient single cross hybrids suitable to agro ecological situations.
- Erratic and improper distribution of rainfall in *Kharif* results in low productivity.
- Around 45-50 per cent maize area is undulated and hilly topography with shallow soil depth of low fertility having less soil moisture holding capacity in tribal belt adjoining to Madhya Pradesh and Rajasthan states, where maize production is more challenging.
- Low adoption of spacing, herbicide, integrated disease and insect pest management, integrated fertilizer and irrigation managements practices at critical stage of crop growth not adopted by tribal maize growers.

- On account of severe damage to maize fields by wild pigs and blue cows (roze), the maize cultivation is being discouraged and replacing by cash crops like castor, cotton, tobacco and eucalyptus plantation in plain.
- Less awareness of specialty corns production technology *viz.,* Quality Protein Maize, sweet corn, baby corn, popcorn and their farm mechanization for efficient cultivation and post harvest management to get optimum yield with quality.

AICRP generated maize technologies for Gujarat state

The Sameri, Dhudh Mogar and Kathudi are popular land races under cultivation in tribal belt. These land races are of white flint grain, with extra early to early maturity and low yielder but having good taste quality in bread making and hence preferred by consumers. Based on market survey, the private single cross hybrids, double cross hybrids, double top cross hybrids and three way cross hybrids tested in AICRP multi-location programme are under cultivation in the state *viz.* Rasi-4797, Advanta-781, Advanta-740, Advanta-751, Ganga Kaveri-777, Ganga Kaveri-3344, Ganga Kaveri Safed-2, Ganga Kaveri-3059, Dhanya-7314, Dhanya-1107, Dhanya-7705, Hitech-5801, Hitech-5402, Pioneer- 3502, Pioneer-3396, Prabal and All Rounder (Monsanto), Syngenta-6240, Swarna (Nuzivedu) and Kanak-51 (Seedtech).

Table 2. Hybrids recommended by AICRP on Maize for Gujarat in recent years.

Cultivar	Year
GAYMH-1	2013
KMH-3426, NMH-731, NMH-803, KMH-3712, P3441, P3502	2012
NK 6240 and GAWMH-2	
BIO-9682, HQPM-1, Co 6, SMH-3904, Vivek Maize Hybrid 43, P3501	2011
Bisco 111, PMH 5 and YM-9905 (PKVM-Shatak)	
DKC 7074 R	2009
HM-10, Chandramani and Pant Sankul Makka-3	2008
HQPM-5	2011
Pratap Makka-5	2006
Pratap Makka-3 and Azad Kamal (R 9803)	2005
Pratap Hybrid Maize-1	2004
Amar and Kohinoor	2001

In addition, many more hybrids recommended for commercial cultivation by (Table 2) AICRP on Maize in the Central Western Zone (Rajasthan, Gujarat, Madhya Pradesh and Chhattisgarh). Out of them, HQPM-1, GAYMH-1, GAWMH-2 and Co-6 single cross hybrids are being cultivated in the state. The seed availability observed to be major constraint for adoption of the rest hybrids (Table 2).

Table 3. Open pollinated varieties and single cross hybrids developed by Godhra centre.

Type	Cultivars	Year	Released by	Notification/ Registration
Yellow orange flint	Gujarat Makai-2	1995	SVRC	7-2/95-SD IVREG/2007/336
	GAYMH-1(SCH)	2013	SVRC	Under notification
White flint	Gujarat Makai-3	1999	SVRC	S.O.937(E)07-09-2002REG/ 2007/337
	Gujarat Makai-4	2000	SVRC	36-K,6,30-5-2001REG/2007/ 339
	Gujarat Makai-6	2002	SVRC	S.O.283 (E)12-3-2003REG/ 2007/338
	Narmada Moti	2002	SVRC	S.O.937(E04-09-2003)REG/ 2007/340
	GAWMH-2 (SCH)	2012	SVRC	Under notification
QPM (Yellow dent)	HQPM-1 (SCH)	2011	SVRC	Endorsed

Coverage of cultivars

The estimates of coverage of area under single cross hybrids, double cross hybrids and open pollinated varieties are not statistically available. However, the probable approximate area accounted to be 20.0 per cent in single cross hybrids, 30.0 per cent double cross including double top cross and three way cross hybrids and 50.0 per cent open pollinated varieties including land races under cultivation in the state. The station has accomplished to release open pollinated varieties and single cross hybrids for the state. The farmers have adopted all these cultivars in both *Kharif* and *Rabi* seasons (Table 3).

Based on breeder/certified seed indent and demand, the quality seed of respective varieties and single cross hybrids produced by Godhra centre was 1244.46 quintals which was supplied to seed producing agencies, NGOs and private companies in the state to reach to the farmers. The seed also sold to farmers and used in FLDs and adoptive trials to demonstrate on farmers fields during 2004-16. The seed production and its sale revealed that the farmers of central and North States of India adopted these for rainfed cultivation (Table 4). In addition, the state recent release single cross hybrids, GAYMH-1 and GAWMH-2 are also becoming popular in central Gujarat (Fig. 4).

Table 4. Quality seed production of released varieties and hybrids by Godhra centre.

Year	Variety / hybrids	Quantity (q)
2003-04	Gujarat Makai-2	0.72
	Gujarat Makai-3	5.60
	Gujarat Makai-4	1.20
2004-05	Gujarat Makai-2	0.56
	Gujarat Makai-3	7.00
	Gujarat Makai-4	19.40
	Narmada Moti	17.00
	WOSC	7.00
2005-06	Gujarat Makai-2	2.20
	Gujarat Makai-3	3.30
2006-07	Gujarat Makai-2	0.10
	Gujarat Makai-4	0.15
	Gujarat Makai-6	0.12
	WOSC	0.12
2007-08	Gujarat Makai-3	6.36
	Gujarat Makai-4	17.00
	Gujarat Makai-6	2.24
	Narmada Moti	21.34
	WOSC	0.48
2008-09	Gujarat Makai-2	5.70
	Gujarat Makai-3	21.00
	Gujarat Makai-6	58.79
	WOSC	0.30
2009-10	Gujarat Makai-4	14.40
	Gujarat Makai-6	62.56
	Narmada Moti	13.40
	WOSC	2.55
	HQPM-1 (SCH)	2.90
2010-11	Gujarat Makai-2	17.75
	Gujarat Makai-3	49.56
	Gujarat Makai-4	8.00
	Gujarat Makai-6	62.56
	Narmada Moti	23.28
	WOSC	0.80
	HQPM-1 (SCH)	4.34
2011-12	Gujarat Makai-3	30.20
	Gujarat Makai-6	131.97
	Narmada Moti	17.07
	Amber Popcorn	0.20
	WOSC	4.22
	HQPM-1(SCH)	2.80

2012-13	Gujarat Makai-2	10.64
	Gujarat Makai-3	1.00
	Gujarat Makai-6	78.56
	Narmada Moti	15.68
	Amber Popcorn	1.00
	WOSC	1.66
	HQPM-1(SCH)	23.00
2013-14	Gujarat Makai-2	3.45
	Gujarat Makai-3	31.00
	Gujarat Makai-4	33.40
	Gujarat Makai-6	57.00
	Narmada Moti	30.20
	Amber Popcorn	6.00
	WOSC	1.15
	HQPM-1(SCH)	10.96
2014-15	Gujarat Makai-2	9.84
	Gujarat Makai-3	30.48
	Gujarat Makai-4	4.82
	Gujarat Makai-6	38.64
	Narmada Moti	15.70
	Amber Popcorn	2.80
	WOSC	0.97
	HQPM-1 (SCH)	3.77
	GAYMH-1(SCH)	11.60
	GAWMH-2 (SCH)	3.28
2015-16	Gujarat Makai-2	10.14
	Gujarat Makai-3	33.25
	Gujarat Makai-4	40.20
	Gujarat Makai-6	8.28
	Narmada Moti	15.00
	Amber Popcorn	3.53
	WOSC	0.95
	HQPM-1(SCH)	13.58
	GAYMH-1(SCH)	29.35
	GAWMH-2(SCH)	17.34
2003-16	**Total**	**1244.46**

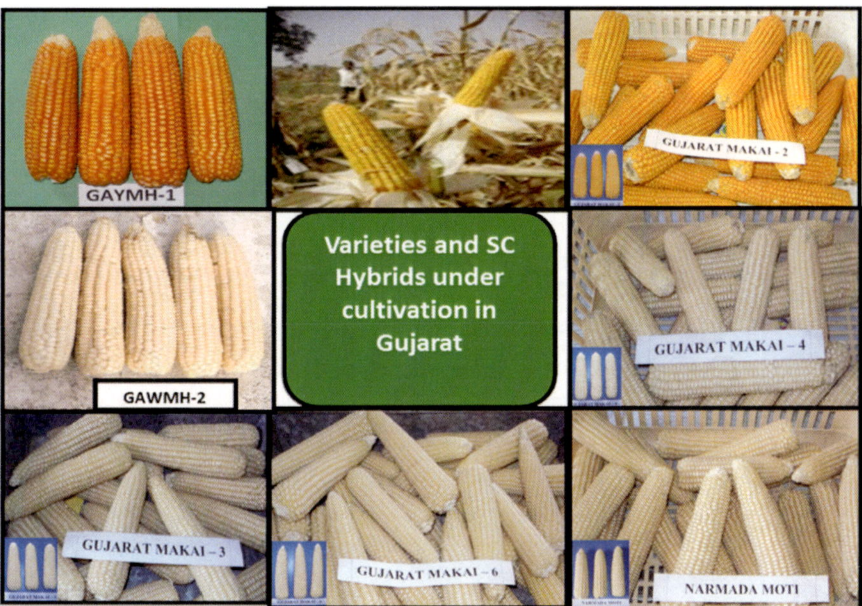

Fig. 4. Varieties and hybrids at farmers field developed by Main Maize Research Station, A.A.U., Godhra (Gujarat).

Changes in cropping system

Based on profitability, the farmers preferred maize-wheat, maize-chickpea, maize-green gram cropping systems. In addition, paddy-maize cropping system is also adopted as and when *Kharif* rainfall is more than 900 mm. There is no remarkable change in cropping system; however, the recently introduced soybean crop, which replaces maize area, is used as sole as intercrop in Dahod district adjoining to Madhya Pradesh. The production technologies recommended by AICRP maize are Summarised in Table 5.

Table: 5. Production technologies recommended and package of practices

Year	AICRP Recommendations
2008-09	Fertilizer dose of 150:65:65 NPK kg/ha was found better for early and Extra early
2009-10	maize hybrids in *Kharif* season. Fertilizer dose of 200:80:80 NPK kg/ha was found better for medium & late maturing maize hybrids in *Kharif* season
2010-11	Fertilizer dose of 200:80:80 NPK kg/ha was found better for medium maturing maize hybrids in *Rabi* season
2012-13	Sowing of early & extra early maturing maize hybrids are advisable in delayed monsoon conditions while under normal rainfall – planting of medium and late maturing maize hybrids are recommended. For enhancing production under rainfed conditions paired row maize planting and residue retention @ 5t/ha maize recommended.

Intercropping in Maize

1991 The farmers of Panchmahal district of Middle Gujarat Agro-climatic Zone growing maize recommended to take Tur variety BDN-2 as an intercrop in the ratio of 1:1 at a distance of 45 cm between rows with fertilizer application of 80 kg N and 40 kg P_2O_5/ha.

1997 The farmer of Agro climatic Zone– III (AES-II)recommended to apply fertilizer i.e. zero application to main crop pigeon pea and 100 % fertilizer (100 kg Nitrogen + 50 kg P_2O_5 /ha.) to inter crop maize for intercropping system.

2001 The farmers of North Gujarat Agro climatic Zone-IV (AES-I) recommended for growing groundnut, pigeon pea, castor, green gram, sesame and black gram as an intercrop in *Kharif* maize in the ratio of 2:1 and advised to grow Maize + castor intercrop maize + pigeon pea for higher profit.

1992 **Sprinkler irrigation on maize:** The farmers of Middle Gujarat Zone growing winter maize are advised to irrigate their crop with sprinkler to get extra income and to save about 40 % irrigation water with sprinkler spaced at 12 m grid. They should operate the system for 2.5 hrs. to apply 40 mm depth of water.

1992 **Sowing dates of *Rabi* maize:** The farmers of Middle Gujarat Agro-climatic Zone advised to plant recommended hybrid varieties of *Rabi* maize during first week of November to get the maximum grain yield and highest net returns.

1993 **Spacing and nitrogen in *Rabi* maize:** Farmers of Middle Gujarat Agro- Climatic Zone advised to sow hybrid maize with 60 cm spacing between two rows and fertilized the crop with 100 kg N/ha for getting higher net return.

1995 **Time of application of fertilizer & levels:** The maize growing farmers of Middle Gujarat Zone advised to apply 60 kg N in three splits (30 kg basal +15 kg at knee height +15 kg at tasseling) and 30 kg P_2O_5 (basal) per ha in early maize varieties.

1998 **Irrigation in *Rabi* maize:** The *Rabi* maize growing farmers of Middle Gujarat are advised to give 11 irrigations in shallow depth sandy loam soil. The first irrigation should be at sowing, followed by the second one at 20 days later with subsequently 7 irrigations at 10 to 12 days interval and last two irrigation at weekly interval.

1998 **Weed management in *Rabi* maize:** The farmers of Middle Gujarat Zone growing *Rabi* maize are advised to apply Atrazine @ 1.0 kg a.i/ha as pre-emergence along with 1 hand weeding at 45 DAS in order to get the maximum profit and effective control of weed.

1998 **Integrated nutrient management in *Kharif* maize:** The farmers of Middle Gujarat Agro-climatic Zone growing early maturing maize during *Kharif* season are advised to apply 5 tonne FYM in furrow in addition to recommended dose of fertilizer .

1998 **Phosphorus management in hybrid maize:** The farmers of Middle Gujarat Zone growing hybrid maize in *Kharif* season are advised to apply 40 kg P_2O_5/ha at time of sowing for realizing 44 % higher profit. The crop should be fertilized with recommended dose of N.

2001 **Use of Biofertilizer:** Farmers of Middle Gujarat Agro-climatic Zone (AES-II) growing *Kharif* maize are advised to inoculate the seeds with Azotobacter (ABA-1) bio-fertilizer @ 25 g/ kg seed containing 10 CFU/gram carriers along with 80 kg N/ha to get maximum yield and net returns.

2001 **Date of Sowing:** Farmers of Middle Gujarat Agro-climatic Zone-III (AES-II) growing *Kharif* maize are advised to sow the maize crop on 15th of June (on set of monsoon) to minimize the *Maydis* leaf blight (2.38%). Brown stripe downy mildew (1.81 %)

disease incidence and stem borer infestation (3.9 %) for getting significantly highest grain and fodder yield.

2001	**Maize Based Cropping System:** The farmers of Middle Gujarat Agro-climatic Zone – III (AES-II) growing *Rabi* crops after *Kharif* maize are advised to adopt Maize- Onion cropping system to get maximum net income where sufficient irrigation water is available. However, where the irrigation water is limited they should adopt maize – gram cropping system.
2002	**Effect of Spacing and nitrogen on *Rabi* sweet corn:** The farmers of Middle Gujarat Agro-climatic Zone-III (AES-II) growing sweet corn (Cv. Madhuri) for green cobs in *Rabi* season are advised to sow the crop at 45 cm row spacing and to fertilize it with 120 kg N/ha in order to obtain higher yield and net return.
2003	**Efficiency of multi-micronutrient formulation in improving crop production:** The farmers of Middle Gujarat Agro-climatic Zone-III (AES-II) growing *Kharif* maize (Var. GM -4) are advised to use soil application of micronutrient mixture grade (20 kg/ha) having Fe- 2%, Mn- 0.5%, Zn – 5 %, Cu- 0.5 % and B – 0.5 % equivalent to government notified general grade of soil application at the time of sowing.
2004	**Phosphorus management in hybrid maize in rainfed condition:** The farmers in *Kharif* are advised to give 60 kg P_2O_5/ha with recommended dose of nitrogen (100 kg/ha) to get maximum net return .
2004	The farmers of AES-I and AES-II growing *Kharif* maize for grain as well as fodder purpose are advised to grow the crop at the 30 cm row spacing with 90 kg N/ha and pull out the plants at 10 days interval commencing from 20 DAS at the rate of 10% upto 50 DAS (40% plant) to get higher net realization.
2007	**Irrigation scheduling in *Rabi* maize:** The maize growing farmers of AES-II are advised to give total 7 irrigations of 60 mm depth in *Rabi* season for securing higher yield and net return. The first, second and third irrigation should be given at the time of sowing, 6 DAS, and 30 DAS, respectively. The remaining four irrigations should be applied at 15 days interval to get higher yield and net return (CBR-1:2.15). Withholding irrigation at silking stage is detrimental to maize yield.
2010	The farmers of AES II growing rainfed maize CV. GM-4 are advise to make ridging at 2m across the furrow after sowing of maize for securing higher yield with net return.
2014	The farmers of AES II growing sweet corn (Madhuri variety) in *Rabi* season are advise to fertilizer the crop with 2.0 tonnes vermicompost + 1.2 tonne castor cake/ha with 90 kg nitrogen and 45 kg phosphorus for high green cob yield with high net return.
2016	The farmers of AES II growing *Rabi* maize recommended to remove tassel after 15 days of anthesis in alternate rows for securing higher grain yield and net return.

Changes in disease scenario over the years

Maydis leaf blight (MLB), *Turcicum* leaf blight (TLB) , banded leaf and sheath blight (BLSB) and late wilt are major prevalent foliar diseases in *Kharif* and *Rabi* seasons. However, the intensity observed low in *Rabi* season. Earlier, medium to high appearance of Brown Stripe Downey Mildew and Sorghum Downey Mildew observed in the *Kharif* season during 1950-80.

Changes in pest scenario over the years

The major insect pests appeared are stem borers (*C. partellus* Swinhoe and *S. inferens* Walker) in *Kharif* and *Rabi* seasons in the state. However, the nematode infestation is recently emerged in *Rabi* season due to continuous cultivation of maize in both seasons and maize after paddy with delay sowing. Another sucking pest is now emerged i.e. Aphids (*Rhopalosiphum maidis*) in both *Kharif* and *Rabi* seasons when maize plants are succulent under favorable environment.

Insect pests management

The *Trichogramma chilonis* @ 8 cards/ha should be released at 12 and 22 days after germination to control stem borer. Alternatively, an application of Carbofuran 3G into whorls of infested plant @ a pinch of granule per plant could be done. The use of seed of tolerant variety/ hybrid is best way to control the pests. Integrated insect pests management is also best approach to effectively control the pests. The appropriate crop rotation and field sanitation may also another measure to control the insect pests.

Disease management

The use of resistant varieties /hybrids and application of integrated disease management modules are the best approach to control the maize diseases during *Kharif* and *Rabi* seasons. The seed treatment with 2-3 g/kg seed with Thiram fungicide and soil application of 1.0 ton/ha neem cake and 2.0 tons/ha castor cake before 15 days of sowing are the alternative recommendations to control MLB, TLB and BLSB including nematodes. The appropriate crop rotation and field sanitation also effectively control diseases.

Post harvest management

The symptoms of cob husks turn dry and brown in colour with appearance of dark scars on abscission layer of grain and 25-30 per cent grain moisture are the indications of physiological maturity in maize. However, farmers prefer to harvest at complete dry of plants. To avoid infestation of storage insect pests and diseases, the shelled grain should be dried up to 10-12 per cent moisture under bright sunlight, cleaned and processed to grade appropriately. The grain should be packed in clean and dry gunny bags. The packed bags should be piled on either wooden or metal pallets or store in dry cool storage. The metal beans are also used to store the farm produce. The use of 3 tablets of Aluminium phosphide (phosphine (PH_3)/ tonne grain advised to be done to fumigate with tight shielding.

PPVFRA protection granted cultivars

The GM-2, GM-3, GM-4, GM-6 and Narmada Moti open pollinated varieties have been registered under PPV & FR Act, 2004 (Table 3).

Extension activities

The State Level Training Programmes (SLTP) of two days duration are regularly organized to train district agriculture officers and gram sevaks on latest improved maize production technology since inception of scheme. In addition, more than 200 FLDs and adoptive trials of GAYMH-1, GAWMH-2, HQPM-1 and Co 6 single cross hybrids in cluster approach with production technologies allotted in Annual Maize Workshops and Social Science Extension group meetings of SAUs were regularly conducted to demonstrate latest maize technologies on farmer's field in Panchmahal, Dahod, Mahisagar and Vadoadara in middle Gujarat zone and Bhiloda and Khedbrahma in north Gujarat zone. The yield gap observed to be 28-40 per cent and 22- 31 % over the local technology in *Rabi* and *Kharif* seasons, respectively. Since 2004, the Government of Gujarat launched Krishi Mahotsav which have great impact on making awareness of latest improved production technologies to the farmers through exhibition, interaction, Farmer Fairs, Sibirs and Field Days etc. with aim of achieving improvement in targeted maize production and yield.

Rainfall pattern: month-wise *Kharif* and *Rabi*

Based on decadal rainfall, the data revealed that the rainfall ranged from 795 mm (1965-70) to 1006 mm (2011-2016) with mean of 52 years was 892.2 mm (Fig. 5). More than 95 per cent rainfall precipitated in *Kharif* season and received unpredictable in *Rabi* season in the state since 1965. The maximum rain precipitation recorded in July and August months of the year across the 5 decades excluding some of years created drought and recently found changes in intensity and distribution in last 10 years (2006-16). In *Rabi* season, the farmers are dependent on residual soil moisture of previous season and lift irrigation for maize cultivation in the state.

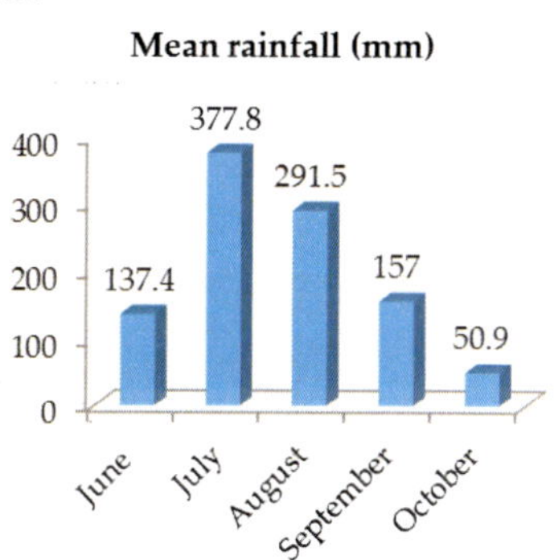

Fig. 5. Mean rainfall of *Kharif* season months during 1990-16.

Key research and development challenges and their probable solutions in maize for the state

The state is being experienced acute shortage of maize grain for starch and its derivatives, food and feed based industries and depends upon other states of India. The specialty corns like, sweet corn and popcorns cultivation do not have appropriate variety/hybrid for agro- ecological situations and dependent on private hybrids. The seed cost is very high which is not affordable to the farmers.

The crop experienced 20-25 days dry period in July-August at V8, flowering and grain filling stages which drastically lowered yield in *Kharif* season. The low to medium N-fertility and sandy loam with minimum clay particles type soil create another constraint to hamper in maize production specifically tribal belts in the state. The probable solutions are to develop drought avoidance/tolerant variety/hybrids for dual purposes as food and fodder, quality protein maize, sweet corn and popcorn to meet future demand of consumers. The precision crop production technology suitable to agro ecological situations with contingency plan is another urgent need of the farmers. The public private partnership should further strengthen to produce sufficient quality seed of notified variety/hybrids for the farmers. The eco-friendly integrated nutrient and insect pests disease managements are to be intensified with low cost technology. The extension education plays key role to educate farmers, need to be further accelerated and get feedback to rectify the production problems.

Factor important for increase/decrease of maize area/production/productivity in the state

The sufficient availability of quality seed of climate resilient notified variety/hybrids at sowing time, adaption of integrated weed, fertilizer and insect pests including disease managements, post harvest and storage handling are key factors for increase/decrease maize production and productivity under climate change. The crop damage due to wild pigs and blue cows are another factor to discourage maize cultivation and shifted to castor, tobacco, cotton and *eucalyptus* plantation in the state.

Future research and development strategy for maize in Gujarat

1. Development of early to medium maturing, high yielding, climate resilient normal/QPM single cross hybrids and open pollinated varieties suitable for rainfed *Kharif* and *Rabi* cultivation under varied agro-ecological situations of the state.

2. Development of specialty corns variety/hybrid for sweet corn and popcorn cultivations and late maturing high starch hybrids for starch industries.

3. Development of low cost and eco-friendly precision crop production technology to get optimum grain and fodder yield for *Kharif* and *Rabi* seasons.

4. Development of integrated and efficient soil moisture and nutrient management strategies with appropriate micro- irrigation scheduling to arrest climatic variation in maize yield.

5. Development of integrated eco friendly insect-pests and disease management in maize.
6. Dissemination of latest technologies to farmers through effective extension education.

Summary

Maize is an predominant crop in eastern part of the state adjoining to Madhya Pradesh and Rajasthan. It cultivates mainly in *Kharif* and winter/*Rabi* seasons as sole and intercrops, with either pigeon pea/black gram/soybean/ castor/cotton. It is staple food and feed of tribal and also used as raw material for starch and poultry feed industries in the state. The consumers prefer both white and yellow orange flint grain maize for food preparations. The Main Maize Research Station, Anand Agricultural University located at Godhra played key role to represent Gujarat state for maize research and development in coordination with Government of Gujarat. The station has developed composite varieties (OPVs) *viz.,* Gujarat Makai-1, Gujarat Makai-2, Gujarat Makai-3, Gujarat Makai-4, Gujarat Makai-6, Narmada Moti and single cross hybrids, GAYMH-1, GAWMH-2 with production technologies for *Kharif* and *Rabi* cultivation in the state. The 14 starch based industries need 9-10 tons/day maize grain to manufacture starch and its derivatives including food and poultry feed these industries are dependent upon maize import from Andhra Pradesh, Karnataka, Maharashtra and Madhya Pradesh. Hence, enhancing maize production and productivity are need of the state.

19

Maize Research and Development in Rajasthan

Dilip Singh, S.S. Sharma, P. Rokadia, V. Nepalia, D. Chouhan B.L. Baheti, M. Mahala, M. Vyas and Hargilas*

Maharana Pratap University of Agriculture and Technology, Udaipur-313001
**Corresponding Author's Email: dilipagron@gmail.com*

Maize (*Zea mays* L.), the "Queen of Cereals" is one of the most important highly evolved coarse cereal crop originated at Mexico. This is most widely distributed crop of the world cultivated predominantly in tropics and sub-tropics right from sea level to 4000 m altitude as irrigated or rainfed crop. The demand for maize as an animal feed will continue to grow faster than the demand for its use as a human food, particularly in Asia, where, doubling of production is expected from the present level of 165 m t to almost 400 mt.

Maize (*Zea mays* L.) is the third most important crop of country after rice and wheat and is cultivated round the year. Its grain is used as feed, food and industrial raw material. Enormous progress has been made during last six decades to enhance yield potential through genetic improvement and alleviate effects due to various biotic and abiotic-stresses. In India, it is an important crop not only in terms of acreage but also in context to its versatility for adoption under wide range of agro-climatic conditions. Of the different forms used for human consumption, 45% is consumed as staple food in various forms *viz.*, bread, biscuits, cookies or transformed into corn flakes, soups, fresh-roasted sweets, boiled cobs and vegetables *etc*. Maize grain is main ration for poultry birds while forage maize is used as fresh or dry fodder for dairy industries. In Rajasthan maize occupies 0.89 m ha area with an annual production of 1.55 mt and productivity of 1.74 t/ha (Govt of Rajasthan, 2015). Being major *Kharif* crop in Rajasthan, the productivity levels of maize are still below the national average and farmers are not getting better economic returns. However, profitability can be accelerated further by growing recently popular speciality corns.

The maize varieties recommended for Sub-Humid Southern Plain and Aravali Hills of Rajasthan have low protein content with unbalanced composition of essential amino acids. In these varieties, the content of desirable amino acids *viz.,* lysine and tryptophan are low while that of undesirable leucine and isoleucine are high (Jat *et al.,* 2009). The low protein and unbalanced amino acids content in maize cause protein deficiency diseases like kwashiorkor and malnutrition in the poor peoples who have maize as principle dietary cereal. The high yielding single cross hybrid of quality protein maize popularly known as 'QPM' assumes a great significance in overcoming problem of malnutrition in tribal population of southern Rajasthan where, maize is a staple food crop. There is enormous scope for cultivation of QPM further, due to increasing global demand, value addition potential and better prices in market compared to traditional varieties of maize. The QPM is a hybridized variety of maize specially bred by addition of *Opaque-*2 mutant gene, which improve lysine and tryptophan and reduces leucine and isoleucine contents and produce quality protein with balanced composition of amino acids (Prasanna *et al.,* 2001). The most important goal of QPM research in state is to reduce malnutrition as direct human consumption (Sofi *et al.,* 2009). Presently full maturing variety "HQPM-1", "HQPM-5" released for commercial cultivation in maize growing areas of Rajasthan. In recent past one medium maturing variety "Pratap QPM Hybrid Maize-1" released by MPUAT for commercial cultivation in Rajasthan. However, due to dull yellow colour and small grain size of these varieties, their cultivation is the last choice of farmers. Hence, there is need to have medium maturing QPM having bold grain and good attractive colour suitable under prevailing agro-climatic conditions of southern Rajasthan.

Amongst other specialty corn, use of "sweet corn" at immature stage is a popular practice as the kernels are sweet. Due to its sweet taste and tenderness, cultivation and demand of sweet corn is increasing in tourist state like Rajasthan. After harvesting green cobs, plant of sweet corn is used as green fresh fodder. In addition to these, "pop corn" is upcoming speciality corn which is nutritionally rich food (Kumar *et al.,* 2009). Due to its immense nutritionally rich value and higher remuneration, cultivation of pop corn is another choice now days amongst farmers of Rajasthan. Similarly "baby corn" is another speciality corn wherein immature cob used for table purpose as a salad or cooked as vegetable is also gaining popularity in state. For baby corn cultivation early maturing prolific corn cultivars *i.e.* those with two of more ears are suitable. After harvesting cobs, plants of baby corn used as green/dry fodder. Therefore, increased cultivation of specialty corn can assist in resolving malnutrition besides high economic returns for maize growers of southern Rajasthan which can further add to improve livelihood security.

Decadal seasonal trend of maize in Rajasthan

In Rajasthan, maize can be grown in all the seasons, however, because of favorable climatic conditions; it is manly grown as rainy season crop. Recently area under winter maize is increasing day by day in Banswara and adjoining Southern areas of Rajasthan because of higher temperature stress and assured irrigation facilities through "Mahi Dam". From 1970 to 2010 the area, under maize cultivation increased marginally and varied from 0.8 to 1.0 lakh ha. However, from 2010 onwards the area under maize cultivation is showing decreasing trend (Fig. 1). The decrease in area of maize crop is because of increasing soybean areas in maize growing area during rainy season.

During same period production and productivity of maize is showing fast increasing trend which is largely associated with significant genetic gain enhancements from the era of OPVs and composite to double cross and three ways cross hybrid and recently to single cross hybrid during this period (Fig. 2 and 3).

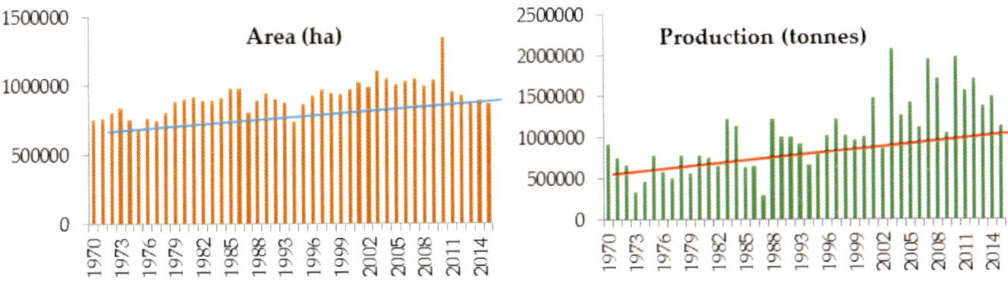

Fig. 1. Area under maize cultivation in state from 1970 to 2015.

Fig. 2. Production of maize in Rajasthan from 1970 to 2015.

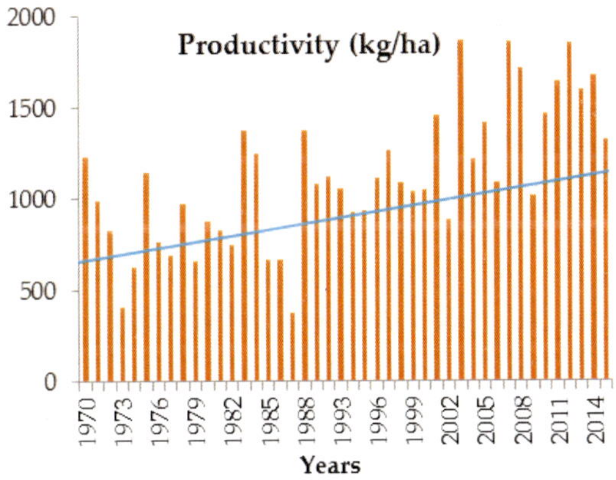

Fig. 3. Productivity of maize in Rajasthan from 1970 to 2015.

Major maize districts in Rajasthan

In Rajasthan, during *Kharif* maize is predominantly grown in Bhilwara, Udaipur, Chittorgarh, Banswara, Rajsamand, Kota, Baran, Bundi, Dungarpur, Pratapgarh, Jhalawar and Sirohi districts. Whereas *Rabi* maize cultivation is increasing in Banswara and adjoining irrigated areas with Mahi Dam.

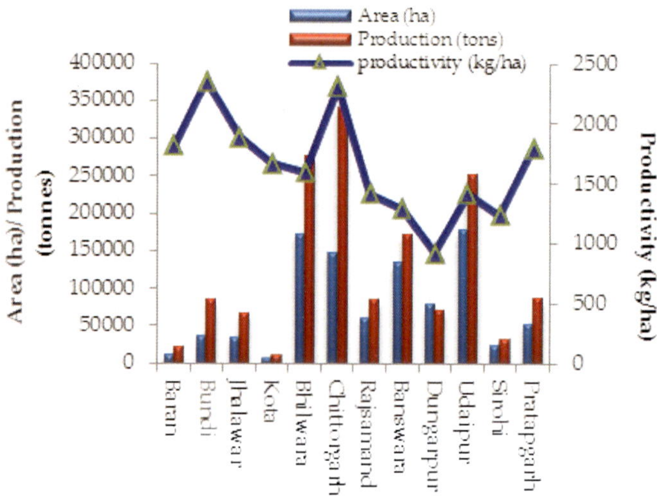

Fig. 4. Area, production and productivity of maize in major district of Rajasthan (Mean 2006-07 to 2015-16).

The mean average of last 10 years shows that maximum area occupied by Udaipur (178379 ha) followed by Bhilwara (171900 ha) and Chittorgarh (148421 ha) whereas, Chittorgarh ranks first in production (340926 t) followed by Bhilwara (276629 t) and Udaipur (178379 t). However highest yield recorded from Bundi (2351 kg/ha) followed by Chittorgarh (2303 kg/ ha) and Jhalawar (1884 kg/ha) amongst 12 major maize growing districts of Rajasthan.

State map showing maize growing districts

The south eastern districts of state comprising Baran, Bundi, Jhalawar, Kota, Bhilwara, Chittorgarh, Rajsamand, Banswara, Dungarpur, Udaipur, Sirohi and Pratapgarh are major districts where maize is predominantly grown in *Kharif* season. From recent past because of higher productivity (5-10 t/ha) *Rabi* maize cultivation becoming popular in Banswara and adjoining areas.

Industrial utilization of maize in Rajasthan

The cob grain when ground to powder used as fillers for explosives in the manufacture of plastics, glues, adhesives, reyon, resin, vinegar and artificial leather and as diluents and carrier in the formulation of insecticides and pesticides. Based on the chemical properties the processed grain find its use in the manufacture of furfural, fermentable sugars, solvents, liquid fuels, charcoal gas and other chemicals by destructive distillation, and also in the manufacture of pulp, paper and hard boards. The water in which the maize grains are soaked during the synthesis of glucose is used for growing penicillin moulds.

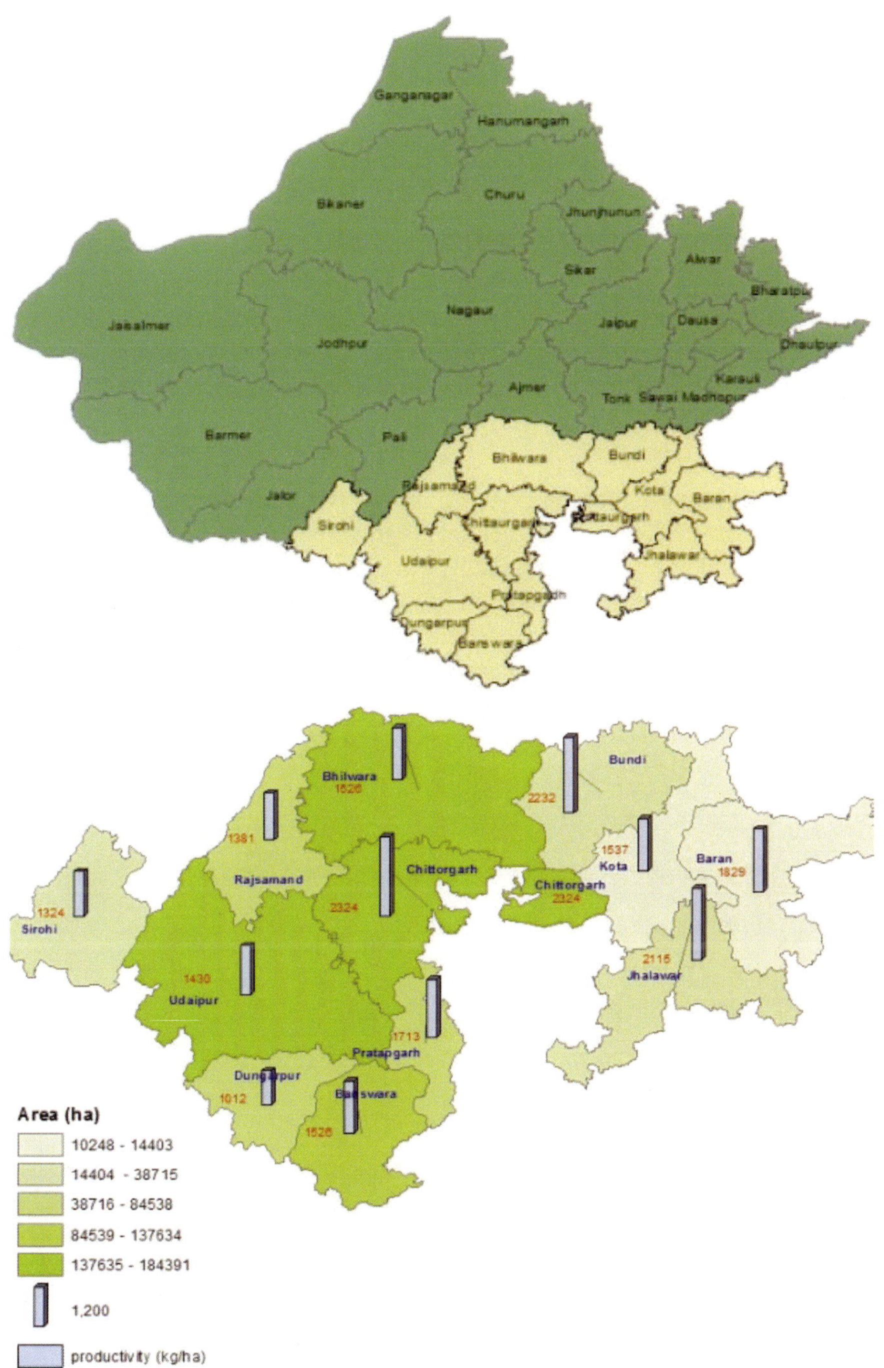

Fig. 5. Major maize districts in Rajasthan.

Corn flour is used as a thickening agent in the preparation of many edibles like soups, sauces and custard powder. Corn gel on account of its moisture retention character is used as a bonding agent for ice-cream cones. The huge industrial potential of maize increased maize based industry in many part of India but their establishment is still awaited in Rajasthan.

Major cropping system in Rajasthan

Maize has wide adaptability and compatibility under diverse soil and climatic conditions. Therefore, it is cultivated in sequence or in association with different crops under varied agro-ecologies of the state. The crop has compatibility with several crops of different growth habit which led to development of various intercropping systems. The crop is often intercropped with soybean, black gram, green gram, and groundnut.

Among maize based copping systems, maize-wheat ranks first in rainfed as well as irrigated ecologies. The other double cropping systems are maize-mustard and maize-chickpea. In some irrigated pockets maize-wheat-green gram is also becoming popular. With availability of full maturing single cross hybrids (SCHs) and irrigation through Mahi Dam, *Kharif* maize-*Rabi* maize cropping system is rapidly spreading in entire irrigated belt of Banswara owing to the very high yield level of *Rabi* maize (8-10 t/ha). The major maize based cropping systems in irrigated and rainfed conditions of state are given in Table 1.

Table 1. Major maize based cropping systems in Rajasthan.

Irrigated	Rainfed
Maize-wheatMaize-mustardMaize-maize	Maize-mustardMaize-gram

AICRP Centre: genesis and mandate

The ICAR has played a pioneering role in ushering Green Revolution and subsequent developments of agriculture in Rajasthan through its research and technology development that has **enabled the country to increase the production of maize grains for** the national food and nutritional security. Because of vast area under maize cultivation in state, ICAR has established AICRP on maize centre in 1969 with four major disciplines *viz.,* Breeding, Agronomy, Pathology and Entomology at Udaipur and another at Banswara with discipline of Breeding and Agronomy in 1980. Subsequently, discipline of nematology was added at Udaipur in 1993.

The mandates of the project are as follows:

- Development and identification of high yielding hybrids/composites for different farming situations with special emphasis on the development of early maturing white-seeded maize hybrids.

- Production of breeder's and foundation seed of suitable maize varieties.
- Development and refinement of production and protection technology of maize for different farming situation.

In recent past the varietal improvement and production technologies developed under AICRP on Maize other sector have resulted in an overall increase in area, production and productivity of maize in the Rajasthan.

Productivity increased from 10-11 q/ha 1950's to about 16-18 q/ha during 2015 16. In the same period, production of maize in southern Rajasthan increased tremendously.

Key production problems in the state for maize

In recent past it was proved that use of single cross hybrids developed and released for Rajasthan are better compared to composite. However, in major maize growing areas of high altitude and tribal region specially where hybrid seed availability cannot be ensured and farmers are still using composite varieties of maize. Although several composites of different maturity group have been developed and released for such areas of state, but for enhancing maize productivity it is very essential to ensure availability of hybrid seed for these areas.

About 80% of maize is cultivated during the monsoon season, mostly under rainfed conditions. During this period no similar trend for dry spell was noted however, majority of maize growing area in the state is exposed to 15 to 20% probability of the occurrence of moderate drought may be at initial stage, knee high stage or flowering and grain filling stage causes severe reduction in grain yield of maize or sometimes almost crop failure. The state thus needs drought resistant early to medium maturing maize cultivars for this 80 % rainfed area.

Weeds are a serious problem in maize due to wide row spacing, and warm and humid weather during rainy season, and slow initial growth in the winter season. In Rajasthan, in general the size of holding is small and the weeds are controlled mechanically by manual hoeing and animal drawn implements. Hoeing in standing crop at 20-25 days of growth followed by ploughing at 30-35 days after sowing with bullock drawn implement is common practice. This help in uprooting of weeds as well as earthing-up of plants along with formation of furrows , which stores rain water to sustain the crop during drought period. Pre-emergence application of atrazine @ 0.5 to 1.0 kg/ha is quite effective. With these integrated practices the crop remains weed free for first 25-30 days. However, for control of second flush of weeds a post-emergence herbicide is essentially required. Some of post emergence herbicides are under testing and hopefully they may prove best as post emergence herbicide.

There are several diseases of maize that occurs during different *Kharif* and *Rabi* seasons in various part of state that causes yield loss of about 10-15 %. Likewise, maize borer (*Chilo partellus*) and many other insects as well as cyst nematode (*Heterodera zeae)* are causing severe damage to the maize crop.

Popular land races with their peculiar characteristic for their popularities

The diversity of maize not only reflects physical differences in seed colour, shape and texture, but also the physiological differences that make some varieties suited to grow in specific environment. These seed holds the gene that has accumulated through natural and human selection over hundreds or even thousands of years. In Rajasthan, "Malan", "Bassi Local", ''Tageri" and "Negri" etc. are some of the oldest continuously inhabited maize races. Some important characters of these races are given in Table 2.

Table 2. Popular maize land races in Rajasthan.

Name	Maturity duration	Area
Malan	Late	Kumbhalgarh and some tribal belt of state
Bassi Local	Early to medium	Bassi and Chhittorgarh area of state
Tageri	Early to medium	In border part of MP and Rajasthan. Presently out of cultivation
Negdi	Early to medium	In order part of MP and Rajasthan. Presently out of cultivation .
Satthi	Early to medium	All Maize growing area of Rajasthan.
Taliya	-	In tribal belt of Rajasthan having high oil content. Presently out of cultivation.
Moti	Late maturity	Variety with small round grains. Presently out of cultivation.
Safed Makai	Early maturity	Variety with small round white grains. Presently out of cultivation.
Gurez Local Maize Mewa	Early maturity	Variety with small round white grains. Presently out of cultivation.

In most of the primitive land races, kernels are sweet, creamy, tender and crispy and therefore used for staple food purpose. Due to its sweet taste and tenderness, cultivation of some of races is still popular in tribal belt and interior part of maize growing area. After harvesting crop plant of these races is used as dry fodder.

Popular hybrids

The climate of the southern Rajasthan is very favourable for maize crop. In this zone it is mainly grown as rainfed crop with composites of different maturity group during monsoon season. In recent years, the maize production has shown a remarkable increase which is mainly associated with significant genetic enhancement from the area of open pollinated varieties and composite breeding to double and three way hybrids and recent development in single cross hybrids. The following hybrids are most popular so far in the state (Table 3).

Table 3. Hybrid so far popular in the Rajasthan.

Name	Grain colour	Maturity Duration/ type	Yield (q/ha)
Ganaga Safed-2	White	Full	45-50
Ganga-11	Orange	Full	45-55
K.H. 510	Yellow	Medium	35-40
Bio-9637	Yellow	Medium	34-40
Bio-9681	Yellow	Full	45-50
PEHM-2	Yellow	Medium	33-40
Pratap Hybrid Maize-1	White	Medium	33-40
HQPM-1	Yellow	Medium to full	45-55
HQPM-5	Yellow	Medium to full	45-55
HM-4	Yellow	Baby corn	-
Sugar-75	Yellow	Sweet corn	-
Daccan 103	Yellow	Full	40-45
K.H.528	Yellow	Full	40-45
Pro-311	Yellow	Full	40-45
Trishulta	Orange	-	44-48
Daccan 107	Orange	Medium	40-45
Him129	Yellow	Medium	30-33

Coverage of cultivars

The climate of the southern Rajasthan is very favourable for maize crop. In this zone it is mainly grown as rainfed crop during monsoon season. In recent years, the maize production has shown a remarkable increase which is mainly associated with significant genetic enhancement from the area of open pollinated varieties and composite to double and three way hybrids and recently single cross hybrids (SCH). The SCHs replaced older varieties to the extent of 40-60 %. On the basis of last two years mean data , 100 % seed replacement rate was recorded in Bundi and Kota districts. However, amongst other major districts, the seed replacement rate of Baran, Banswara and Dungarpur was 70-75 %, Bhilwara, Udaipur and Chittorgarh was 45-65% only (Fig. 5).

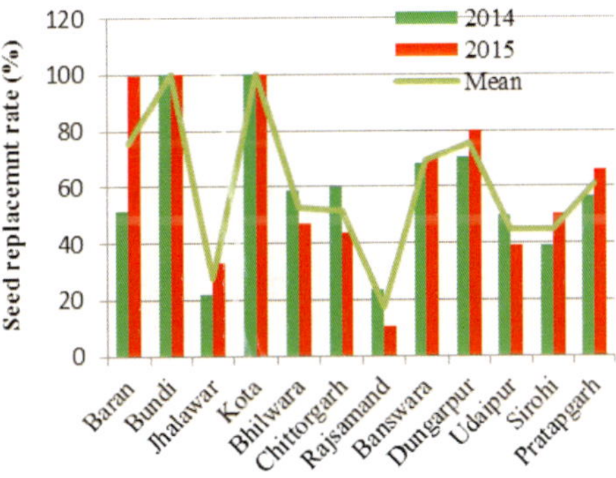

Fig. 5. Seed replacement rate of major maize growing districts of Rajasthan (Source: Govt. of Raj. 2015).

Cultivars released by the AICRP Maize centres for state

The climate of the southern Rajasthan is very favourable for maize crop. Where it is mainly grown as rainfed crop. A good number of composite and hybrid cultivars released by the AICRP on Maize centres (Table 4).

Table 4. Composite and hybrid released for state by AICRP centres (*Banswara and **Udaipur).

Cultivar	Year	By	Characteristics
Mahi Kanchan*Composite	1992	SVRC	Yellow; Maturity: 80-85 days; Height: 150-190 cm; Yield: 30-35 q/ha
Pratap Kanchan-2*	2009	SVRC	Yellow; Maturity: 80-85 days; Height: 150-190 cm; Yield: 30-35 q/ ha
Pratap Hybrid Maize 3**	2014	CVRC	Yellow; Maturity: 75-80 days; Height: 190-210 cm; Yield: 55-60 q/ ha
Pratap Makka-9** Composite	2014	CVRC	Yellow; Maturity: 81-87 days; Height: 177-227 cm; Yield: 45-50 q/ ha
Mahi Dhawal*Composite	1996	SVRC	White; Maturity: 90-100 days; Height: 180-200 cm; Yield: 40-45 q /ha
Arawali Makka-1**Composite	2001	SVRC	White; Maturity: 80-85 days; Height: 151-173 cm; Yield: 30-40 q/ ha
Pratap Hybrid Maize-1**Hybrid	2003	SVRC	White; Maturity: 80-82 days; Height: 170-180 cm; Yield: 35-38 q /ha
Pratap Makka-3**Composite	2004	SVRC	White; Maturity: 75-78 days; Height: 170-180 cm; Yield: 40-45 q /ha

In recent past Government of Rajasthan has produced huge quantity of seed of some QPM varieties and distributed to the tribal, farmers of Rajasthan under mini kits distribution programme. It shows that Banswara and adjoining areas may be developed or/used as hub of seed production of maize hybrid seed. Maize hybrid seed amounting 35 thousand quintal was produced at farmers' field by department of Agriculture, Banswara during 2012-13 to 2014-15 under technical support provided by MPUAT and RSSC (Table 5).

Table 5. Maize hybrid seed produced at farm.

Year	Maize Hybrid	Area (ha)	Production (q)
2012-13	HQPM-1PEHM-2	380.4	3600
2013-14	HQPM-1 &HQPM-5	892.0	11377
2014-15	HQPM-1 &HQPM-5PQPMH-1	1120.0	20000
Total		**2392.4**	**34977**

Change in cropping system over years

Rajasthan is a land of multiplicity of cropping systems. Since food crops are predominantly grown in the state, the various cropping systems are named as per main crop. Amongst them maize-wheat and maize-gram, cropping system are common in maize growing area.

In recent past soybean was introduced as a *Kharif* crop in several parts of Rajasthan. With availability of some post emergence herbicide in soybean, its area is increasing day by day and soybean-wheat, soybean-green gram cropping system emerged out another new important cropping system in maize growing area of state in recent past.

Production technologies for maize in Rajasthan

Operation	Recommendations
Field and its preparation	Well drained sandy loam soil having neutral reaction. For good soil tilth one ploughing followed by a harrowing and cross cultivator planking is required.
Seed treatment	Recommended fungicide, insecticide and *Azotobacter* + *PSB* bacteria in sequence.
Sowing time	Onset of monsoon
Seed rate	20-25 kg/ha
Spacing	60 x 25 cm
Fertilizer dose and its time of application	90-175.60:40 kg/ha N: P_2O_5 : K_2O. Full dose of P, K an 1/3 N at sowing. Remaining N in two equal split at knee high and tasseling stage.
Weed control	Atrazine 0.5 kg/ha as pre-emergence followed by one hoeing at 25-30 days after sowing
Diseases and pest control	• Application of 10 kg phorate/ha for control of termite, white grub and maize stem borer. • Release of *Trichgramma* egg parasitoids @ 1.5 lakh/ha for biological control of borers. • Dusting of methyl parathion dust @ 25 kg/ha for control of grass hopper. • Spray of Methyl dematon 25 EC @ 1 l/ha for control of Aphids.
Irrigation	*Kharif* crop is rainfed, however, on stress at knee high and tasseling stage irrigation should be provided .
Harvesting	As per maturity period at yellowing and drying of leaves.
Expected yield	30-50 q/ha

Changes in disease scenario in Rajasthan

In 1961-70, there was severe incidence of RDM and from 1968-onwards PFSR dominated which continued to be serving till date. CLS also occurred after 1967 which is now taking a big toll. This decade documented onset of new diseases and new introduction was observed. In Downy mildew, new alternate host reported as *Heteropogon melanocarpus* in addition to *H. contortus.* MLB was also observed with moderate incidence in 1957-58.

In 1971-80, alongwith all the major diseases of maize for the first time Head smut and Brown stripe downy mildew were observed and those found in pockets in later years in serve form. However, BSDM was found only on few fields in a village but this did not spread in large area.

Likewise, from 1981-90, it was observed a great shift in the disease severity from moderate to high. However, during low rainfall years, there were fewer incidences of foliar and other pathogens. Foliar pathogens were increasing that became alarming for future. From 1991-2000, post flowering stalk not became the top most devastating disease and management experiments were laid out to give a package of practices. For downy mildew seed treatment with Melalaxyl while Azoxystrobin found very effective against BLSB.

In recent past one and half decade from 2001-2017, the maize area is increasing but somewhere, it has been replaced by soybean due to less requirement of water and labour. However, in Zone IV b of Rajasthan, which covers tribal areas of Banswara, Dungarpur recorded maximum area of approx 40,000 ha in *Rabi* season. In these areas including *Kharif* in Udaipur region PFSR is one of the most serious pathogen followed by BLSB and CLS. MLB is also gaining importance and somewhere, soft rot and virus symptoms were also observed.

Future projections of disease

In coming year's erratic rainfall pattern and temperature aberrations will cause more PFSR, foliar diseases like MLB, CLS and also common rust. Soil borne pathogen like PSFR and BLSB will also be severe, but these have to be taken care while preparing the field for sowing . More use of organic means will keep the soil healthy. Foliar diseases can only be controlled by using fungicides, but we have to accept the challenge of global warming and climate change. Abiotic stresses will be important aspect to be addressed in view of change in disease scenario.

Change in pest scenario over the years

In Rajasthan, in spite of taking due care of the production components, the insect pests take a heavy toll to the crop thus bringing crop yield very low. The damage caused by insect pests is an important limiting factor of profitable cultivation. High yields in maize cannot be realized under different agro-climatic regions due to economic damage inflicted by large number of insect pests from sowing to harvest. Among the insect pests, so far recorded, stem borer [*Chilo partellus* (Swinhoe)], cob worm [*Helicoverpa armigera* (Hubner)], armyworm [*Mythimna separata* (Walker)], maize aphid [*Rhopalosipum maidis* (Fitch)], grey weevil [*Myllocerus discolour* (Bohemann)], and *Kharif* grasshopper [*Hieroglyphus nigrorepletus* (Bolivar)] are causing major economic damage to the maize crop. The maize stem borer (*C. partellus*) and the pink stem borer (*Semamia inferens* Walker*)* are prevalent throughout the maize growing area. Of these, *C. partellus* is widely distributed and is considered as a pest of *Kharif* maize, while *S. inferens* occurs occasionally during the rainy season however; it is reported to remain active during *Rabi* season in maize growing area. The injury caused by maize stem borer in maize includes leaf feeding, tunnelling within stalk, disruption of the flow of nutrients to the ear, and subsequent development of "dead hearts" by damage to the central growing shoot of young plant. The first symptom of damage caused by *C. partellus* is the appearance of "shoot hole" injury to whorl leaves. Plants that survive the initial attack show reduced inter-nodal length resulting in shoots 'rosetting'. Yield loss is attributed the physiological effects on final ear size, lodging or the complete loss of ears and formation of "dead hearts". The stem borer, *C. partellus* usually appears 10-15 days after germination and cause serious infestation, sometimes leading to complete failure of the crop. The development of newer and resistant maize varieties can play an important role in reduction of yield losses which are developed for their resistance through artificial release technique under field conditions against maize stem borer.

The knowledge of incidence pattern plays a great role in management of these pests. Farmers of the tribal belt of southern Rajasthan are poor economy and having small holdings. Further they do not use costly insecticides for the control of the insect pests. It has been established that cultural and biological control of various insect pests would be more beneficial as compared to application of pesticides. Under such circumstances, knowledge of natural bio-agent complex of vital importance in protecting the crop losses and avoiding damage by these insect pests. The efficient bio-agent like *Cotesia flavipes* (Cameron), *Chrysoperla carnea* (Stephen) and cocinellids could be utilized in the management of the insect pests in maize.

At the reproductive stage , *H. armigera* and *Euproctis* spp. were found feeding on tender tassels and immature cobs. Larvae fed on the soft shank and milky seeds of the immature cobs. Though cob damage due to *H. armigera* has been reported, decrease in grain yield is only a minor one. Female moths prefer silk, since neonates feed first on silk and then penetrate the cob through apical side, where at beginning eat tip of the cob and then grains. Around 2-5 numbers of just hatched larvae are seen on silk and they prefer milk grains. Irrespective of initial egg density on silks, population of final instar larvae per infected cob rarely exceeded one, because of the protection afforded by tight husks around cob, migration to other cobs/plants and intra-specific competition (cannibalism) among the younger instar larvae. Loss of corn due to *H. armigera* has been estimated at 262 kg/ha and if larvae damage early silks, pollination will be reduced resulting in even greater yield reductions. Control has generally not been practiced because of high cost associated with repeated insecticide application required during silking.

Nematode in maize

Maize cyst nematode (*Heterodera zeae*) has been reported to cause significant losses in Rajasthan due to monocropping of maize and favorable soil and environmental conditions. Nematodes apart from causing losses by themselves interact with other disease causing agents and adversely affect the quality and quantity of maize production. It was first reported from Chhapli village of Rajsamand district of Rajasthan and widely distributed in maize growing areas of Rajasthan. *H. zeae* is a sedentary semi-endoparasite which feeds on roots. Affected plants are stunted, pale in colour, with narrow leaves. In the field, stunting frequently occurs in irregular patches. The development of maize tassels may be noticeably delayed and the maize plants bear smaller cobs with relatively fewer grains. The root system is poorly developed with a bushy appearance, and the presence of cysts on the root surface can be observed.

As with most cyst nematodes, dissemination is largely ensured by passive transport with soil, water, and plant material. The mobile stages (juveniles, males) can only move over very short distances. It is likely that *H. zeae* cysts remain viable in the soil for long periods. Two-three deep summer ploughings during May-June at 10-15 days interval effectively manage the nematode population. Crop rotation / intercropping (2:2) with non-host crops (sesamum, clusterbean, soybean, black gram etc.) for 3-4 years and Application of organic material *i.e.* no edible oil-cakes (neem, karanj, mustard and castor cakes) or lantana/ neem/ karanj leaves @ 2-5 q/ha at the time of sowing significantly reduced the infection of nematode and enhanced the yield of maize in nematode prone areas. However, data on economic losses and integrated nematode management is generally lacking.

Commercialization of maize products and potential

Centre of Excellence on Maize Processing & Value Addition, MPUAT, Udaipur, Rajasthan, India has worked in this direction. Utilization of 100 % maize to different level of incorporation of maize in wide variety of products belonging to baked (cake, biscuit, cookies, muffins, bread etc.), extruded (puffballs, pastine), traditional food products (*Khaman, ladoo, mathri, bhujiya, shakkarpara* etc.) ready to eat and instant mixes (Cake mix, khaman mix, chapatti mix etc.) have been done using standard protocols. The developed maize products have acceptable sensory scores and improved nutritional profile to the consumers in terms of its protein, fiber, fat and mineral content. Products have been launched in local market and are found to be in most demand by the local population of the area. In developing countries like India where half of the population still malnurished, improving maize utilization with increased human consumption will help in ensuring nutritional security (Murdia *et al.,* 2016).

Rainfall characteristics of Rajasthan

The mean annual rainfall of Rajasthan is 592.7±217 mm with a coefficient of variation of 38.6%. Banswara district receive the maximum annual rain (1081 mm) followed by Jhalawar (921 mm). The southwest monsoon which has its beginning in the last week of June in the eastern parts, may last till mid-September. The mean southwest monsoon rainfall of Rajasthan is 535±213 mm with a coefficient of variation of 42.0%. Pre-monsoon showers begin towards the middle of June and post-monsoon rains occasionally occur in October and November. In the winter season also, there is sometimes, a little rainfall (12 mm) associated with the passing western disturbances over the region. The winter season rainfall ranges from 3.6 mm in Sirohi to 26.2 mm in Alwar. At most places in the state, the highest normal monthly rainfall is during July and August.

The mean annual rainfall of 592.7 mm is received over 29 rainy days with a variability of 29 %. Variability in rainy days is relatively high during monsoon season compared to annual rainfall. This high variability reflects the frequency of intermittent dry spells during continuous wet spells. Most of the rain events in Rajasthan occur during the southwest monsoon season. Number of annual rainy days are high in Banswara districts (41 days) followed by Pratapgarh (40 days), Baran, Jhalawar (38 days) and lowest in Jaisalmer (11 days) districts. The events are highly variable during the north eastern monsoon compared to southwest monsoon period.

Start of growing season, length of growing season, choice of cropping systems, allocation of resources and inputs depend significantly on the weekly distribution of rainfall. Considerable rain (> 20 mm/week) over the state occurs in the period from 26 Standard Meteorological Week (SMW) (25 June-1 July) to 36 SMW (3-9 Sep). This indicates a total growing period of 22 weeks (around 150 days). Less than 40% rainfall is exhibited in only one week (22 SMW). The maximum CV (54 per cent) was observed during 35 SMW.

In majority of the districts in eastern region, the rainy season commences in the 26 SMW and ends by 37 SMW. Banswara district recorded the highest weekly rainfall of 102 mm in 32 SMW, however, this rainfall situation is associated with high coefficient of variation (>100%) due to large inter-annual variability and thus has a low dependability. Isolated rainfall events are noted over the entire state during 40 to 47 SMW and the rains get momentum with the onset of monsoon from 24 SMW onwards.

Key research and development challenges

1. Lack of funds for R & D restricts maize scientific interventions.
2. Lack of public private partnership for overall development of ambience for phenomenal growth of maize crop.
3. Inability of the farmers to spend on good quality seeds, fertilizers, pesticides and fungicides etc and persistence of traditional cultivation practices in the

core and remote areas affect the yield as well as quality of maize produced in the state.

4. Maize suffers heavy post-harvest losses estimated at 20-30 %. The main underlying factor is the lack of farmers' education coupled with poor infrastructure and handling for transportation, improper storage and drying facilities, resulting in wastage and pilferage.
5. Lack of hybrids specific to prevailing agroclimates of the Rajasthan state.

Factors important for increase /decrease of maize in the state

1. Fragmented land holdings of traditional maize belt restricts wide scale adoption of best management practices by resource poor farmers.
2. Non-availability of varieties especially hybrids to match crop pendency and rainfall period is the most important divergence in bumper harvest.
3. Farmers do not differentiate between seeds and grains and hence they are in a habit of using stocked grains as seeds for sowing crops in the following season which is one of the major causes of low yield.
4. Bulk of maize is cultivated in rainy season which is prone to vagaries of nature in terms of monsoon rainfall quantum and duration.
5. Onset of monsoon has a great impact on sowing time of maize. It was estimated long back that late onset of rains adversely affects the productivity of this crop.
6. Withdrawal of monsoon is another factor that influences the reproductive stage of the crop.
7. Rainy season is congenial for weed growth which has a silent adverse effect on maize yield. Untimely and unpredicted rainfall does not allow farmers to plan weeding during critical stage. Moreover, herbicides are not available for wide spectrum of weed flora.
8. Blanket recommendations and imbalance in the use of fertilizers has caused deficiencies of secondary and micronutrients resulting in decreasing yield.
9. Inappropriate methods of fertilizer application like broadcasting are still prevalent which does not allow utilization of applied nutrient for maize growth and yield.
10. Lack of credit facilities and shortage of inputs at critical times are the managerial bottlenecks in maize cultivation.
11. Crop failure due to unpredicted insect pest problems like grasshoppers in not uncommon in bumper harvest.
12. Lack of contract farming opportunities due to the absence of industrial support in the state which could otherwise promote production for specific product/produce manufacturing or processing units.

13. Yield gap between conventional practices and improved production technologies is very high calling for urgent stronger extension services for educating the cultivators in the implementation of improved production technology.

Future research and development needs

i. An enormous increase in area under hybrids could transform maize scenario. Single cross hybrids are the preferred seeds of maize for achieving high yield. Uses of hybrid seed can alone double the maize yield. Despite best efforts presently hybrids constitute only 40 % .

ii. The private sector involvement in Indian agriculture is a recent development. The success like Bt cotton needs attention to boost maize yield.

iii. *Rabi* maize as an area of seed production and wide scale hybrid seed production needs to be focused.

iv. In order to decrease post harvest losses and increase the life of produce proper infrastructure for storage is required. Silos form an integral part of bulk handling of commodities and have many advantages over traditional warehouses in the Indian context such as requirement of lesser area for installation, larger storage life span, inbuilt system to protect grain from bacteria, reduce wastage of grain etc.

v. Farmers should be educated on handling post-harvest cleaning, grading and switching to standardized packaging of produce to meet export market requirement.

vi. Set-up of small-scale enterprise in rural areas generating employment and livelihood security.

vii. Economic empowerment of maize farmers. Innovative farm to agribusiness linkages to enable hassle free procurement of produce. These linkages help in increasing the bargaining power of small farmers and improve their income from the marketplace, potentially increasing agricultural viability.

viii. Farmers are unaware of other uses of maize and its products. They are also not familiar with specialty maize types and products such as baby corn, sweet corn, popcorn, corn oil, and corn syrup. If these alternative maize types and products could be introduced in conjunction with assured markets and agro-processing industries, this would go a long way towards improving livelihoods of maize producers.

ix. There is a need to increase contacts between farmers and scientists to disseminate new information and allow higher returns to investment from agricultural research. Farmers can call the help-line during a given time period and get the solution to their problems directly from experts at the university.

Based on major constraint of low yield in state, the future research priorities of the maize shall be as follow

1. Suitable early/medium/full season high yielding hybrid for rainfed conditions.
2. Suitable full season hybrid for *Rabi* cultivation in Banswara and adjoining areas.
3. Development of cultivars possessing inherent potentiality to survive in drought conditions especially flowering and grain filling stage.
4. Testing and development of suitable speciality corn hybrid and their production technologies.
5. Technology of integrated nutrient management for maize based cropping system along with SSNM based nutrient management.
6. Integrated weed management for maize based cropping system and selection of appropriate post-emergence herbicide.
7. Technologies for moisture conservation.
8. Maize intercropping and its technology.
9. Fodder maize and its utilization (hay, silage) technologies.
10. Technology for biological control of soil-borne diseases.
11. Post harvest management of toxigenic seed borne pathogens.
12. Information about distribution of major maize diseases and their endemic and epidemic area.
13. Information of population biology/virulence diversity of maize pathogens.
14. Evaluation of eco friendly plant products/bio agents/pesticides for their field evaluation in pest management maize nematode crop.
15. Development of location specific IPM/IDM modules, for insect /diseases/ nematode.
16. Technology of easy and fast rearing of bio agents and effective technology for its release in the field for screening genotype.
17. Development and improvement of safe storage technique for minimization of post harvest losses.
18. Studies on interaction of cyst nematode with other insect and pathogens.

Summary

In Rajasthan, *Kharif* maize is predominantly grown in Bhilwara, Udaipur, Chittorgarh, Banswara, Rajsamand, Kota, Baran, Bundi, Dungarpur, Ptratapgarh, Jhalawar and Sirohi districts whereas *Rabi* maize cultivation is increasing in southern district of Banswara and adjoining area. The average of last 10 years data show that maximum area was occupied by Udaipur (178379 ha) followed by Bhilwara (171900 ha) and Chittorgarh (148421 ha) whereas Chittorgarh ranked I[st] in production (340926 tons) followed by Bhilwara (276629 tons) and Udaipur (178379 tons). On the basis of last 10 years mean data, highest average yield was record from (Bundi 2351 kg/ha) followed by Chittorgarh (2303 kg/ ha) and Jhalawar (1884 kg/ha) in 12 major maize growing districts of Rajasthan. The maize productivity is seriously hampered due to non-availability of quality seed/hybrids, dependence on rains and vagaries of monsoon, incidence of insect pests, diseases and weeds, non adoption of best management practices due to resource poor farmers and lack of credit facilities. Lack of scientific and agribusiness attitude do not allow the economic empowerment of the farmers of maize growing belt. Therefore, the role of maize scientists to overcome the constraints and farmers friendly extension education for their technological empowerment is the need of the hour. Development of location specific technological modules can play a vital role in propelling maize growing farmers towards prosperity. Not only productivity, issues related to processing also needs to be addressed for economic empowerment of maize farmers in Rajasthan.

References

Govt. of Rajasthan. (2015). Vital Agriculture Statistics, Statistical Cell, Directorate of Agriculture, Pant Krishi Bhawan, Jaipur. Pp: 47.

Jat, M.L., Dass, S., Yadav, V.K., Sekhar, J.C. and Singh, D.K. (2009). Quality protein maize for food and security in India. Directorate of Maize Research, Pusa campus, New Delhi .

Kumar, A. (2009). Production potential and nitrogen use efficiency of pop corn (*Zea mays* L.) influenced by different planting density and nitrogen levels. *Ind J Agric Sci.* 79(5): 351-355.

Murdia, L.K., Wadhwani, R., Wadhawan N., Bajpai, P. and Shekhawat, S. (2016). Maize Utilization in India: An Overview. *American J Food Nutri.* 4(6): 169-176.

Prassanna, B.M., Vasal, S.K., Kassahun, B. and Singh, N.V. (2001). Quality protein maize. *Curr Sci.* 81: 1308-1319.

Sofi, P.A., Wani, S.A., Rathore, A.G. and Wani, S.H. (2009). Quality protein maize (QPM): Genetic manipulation for the nutritional fortification of maize. *J Plant Breed Crop Sci.* 1: 244-253.

20

Maize Research in India : Historical Perspective, Current Scenario and Future Challenges

Vinay Mahajan and Bhupender Kumar*

ICAR-Indian Institute of Maize Research, PAU Campus, Ludhiana, Punjab
**Corresponding Author's Email: vinmaha9@gmail.com; pdmaize@gmail.com*

Maize (*Zea mays* L., 2n=20) is a cereal crop that is grown widely in more than 165 countries in a range of agro-ecological environments. Even in Asia, the rapid increase in maize acreage in the last decade is an indication that it is winning the battle as compared with rice. The Asian region will see itself as a major global player in the near term. In the past it was mainly confined to food in India and many other countries however, now it has become the industrial crop. Maize has several industrial applications and more than 1000 products are being developed from maize in India and around 3500 products in USA and other countries. It is a raw material for many of the important industries *viz.* pharma, textile, paper, film, tyre food, processing, packing and biofuel, etc. The growth rate of poultry, livestock, fish, wet and dry milling industries are very high therefore, maize demand will continue to increase in year to come.

Global Maize Scenario

Globally, maize is primarily used as feed livestock and has many industrial uses; however in many developing countries it is still used as a food crop. Small-scale farmers in both sub-Saharan Africa and Mesoamerica generally grow maize as a food crop for household consumption and for saline urban markets. Maize is particularly important in the diets of the rural and urban poor in sub-Saharan Africa and Latin America. Globally, maize is grown in 165 countries in a total of 185.1 million hectares area with 1018.1 million tones of production and 5.52 tonnes/ha productivity (Anonymous 2015-16). Almost 70% of the total maize production in the developing world is coming from low and lower middle income countries. In Asia, maize has recorded the fastest annual growth (around 4%), as compared to other cereals. Maximum maize production was in USA (353.7 million tonnes) followed by China (218.6 million tonnes) and Brazil (80.3 million tonnes) (Fig. 1), while the area under maize cultivation was in China (36.3 million hectare) followed by USA (35.5 million hectare) and Brazil (15.3 million hectare) (Fig. 2). The average world maize productivity is 5500 kg/ha, but the highest productivity of maize was achieved in USA (9970 kg/ha) followed by Canada (9591 kg/ha) and France (8137 kg/ha) (Fig. 3).

India Maize Scenario

In India, maize production has been consistently growing in recent years. The pressure of population, a severe natural resource crunch and an impending ecological decline may trigger a major agricultural crisis, unless we act now and bring in paradigm scientific interventions to deal with these issues. As Asia's agribusiness and food processing industries and economies continue to grow, the opportunities for the use of maize as food, feed, fodder and in industry will also increase significantly. Besides, the growing needs of the poultry sector, the expansion of maize seed sector and increasing interest by consumers in nutritionally enriched and specialty maize products indicates the growing potential. The maize productivity in India, has reached more than 2452kg/ha with a total production of 24.1 million tons from an area of 9.2 million ha. By 2050, 3.25 times increase in production, 2.2 times increase in productivity and 1.4 times increase in acreage is anticipated (Dass *et al.*, 2009).

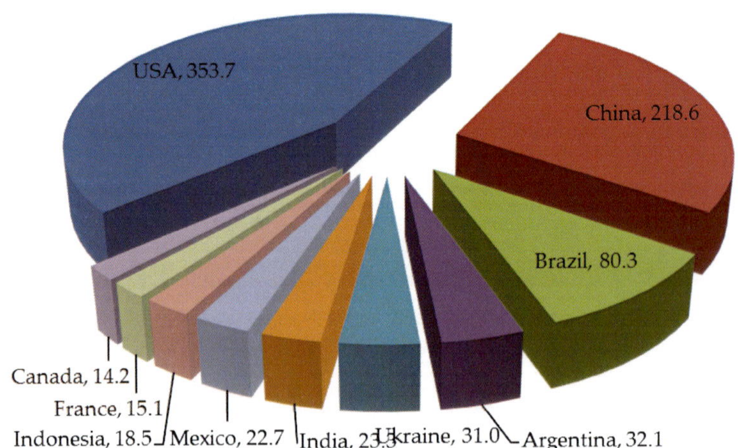

Fig. 1. Maize production (million tonnes) in different countries (Anonymous, 2015).

Phases of maize research in India

A high level committee constituted by GoI in 1953, chaired by Dr E.J. Wellhausen along with member Dr U. J. Grant, submitted a report in 1954 and recommended maize crop for its highest productivity, wider adaptability and multi-facet uses. In 1957, All Indian Coordinated Maize Improvement Programme (AICRP) was established with multi-location and multi-disciplinary testing programme. The winter nursery was established in 1962 giving more emphasis to germplasm development and advancement. It was followed by establishment of hotspots for important diseases in 1963 and today AICRP has 17 centers including 10 hotspot locations for screening of important maize diseases under artificial inculcation condition. Maize has the earliest coordinated program and has involved very strongly since then. The first trials started in 1957 and the first maize workshop was held in 1958. The first phase involved improvement of landraces during 1950-60. The principal breeding efforts were mainly focused towards improvement of local material through mass selection and hence the productivity remained very low (547 kg/ha).

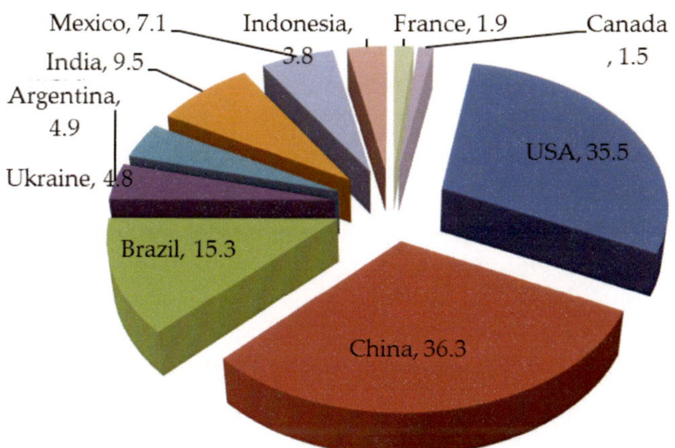

Fig. 2. Area (million ha) under maize cultivation in different countries (Anonymous, 2015).

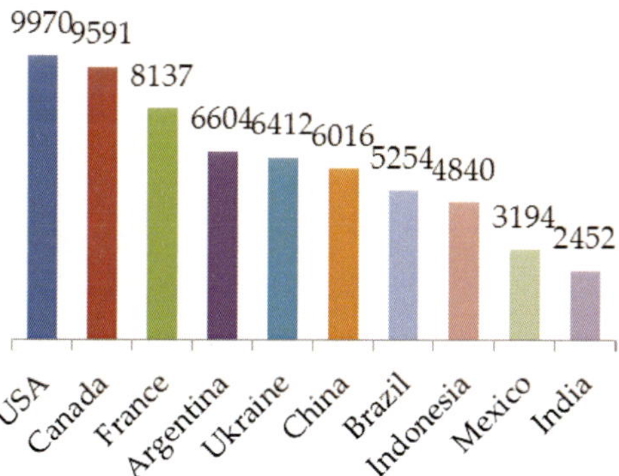

Fig. 3. Maize productivity (kg/ha) in different countries (Agriculture Statistics at a Glance, 2015).

The Indian Council of Agriculture Research (ICAR) launched the first All India Coordinated Research Project on Maize in 1957 to develop and evaluate improved cultivars of maize (Fig, 4). The program started with an emphasis on hybrids development. However, due to none availability of productive inbreds leading to the uneconomical hybrid seed production the emphasis, in sixties and seventies, was shifted to composite breeding in public sector, while the private sector shifted towards mutli-parent hybrids approach to address the problem of economic seed production. In 1961, the first set of four double-cross hybrids, *viz.* Ganga 1, Ganga 101, Ranjit and Deccan were released for commercial cultivation in India. These were followed by series of double-cross hybrids like VL 54, Him 123, Deccan 101, etc. These hybrids showed distinct superiority (30-40%) over the landraces. Ganga Safed 2, Hi-Starch and Ganga 5 were prominent hybrids released between 1963 and 1968. Further, the first OPV was notified in 1967 (Jawahar) and the first TWC was notified in 1969 (Hybrid Ganga 5). In these years maize breeders had emphasized on the development of productive maize inbreds with a target of economic seed production. However, with the development of productive inbred lines the SCH took the emphasis in nineties onwards and the first single cross hybrid (SCH) was notified in 1993 (COH 2). Simultaneously, the emphasis was on breeding for quality parameters and the first OPV Quality Protein Maize (QPM) was notified in 1973 (Shakti) and SCH QPM in 2004 (Shaktiman 2) (Fig. 4).The first QPM hybrid developed through MAS was notified in 2008 (Vivek QPM 9) by VPKAS, Almora. In addition, first popcorn (OPV) was notified in 1982 (VL Ambar popcorn), baby corn (OPV) in 1999 (COBC-1) and SCH in 2005 (HM 4), On the other hand, Madhuri was the first OPV sweet corn which was notified in 1990. The first SCH sweet corn was released and notified in 2010 (HSC 1).

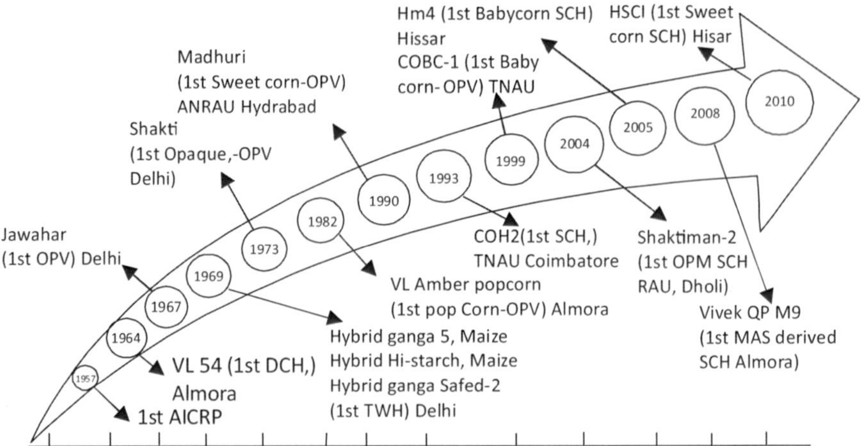

Fig. 4. Timeline of AICRP Maize Achievements First Hybrids/OPV notified in Maize AICRP

Dawn of single cross hybrid development

Late eighties witnessed the launching of single cross hybrid (SCH) breeding programme and adoption of New Seed Policy. SCH breeding activities witnessed many positive changes and accomplishments in generating vital scientific information as well as commercial products. Research efforts were focused on the development of vigorous genetically diverse inbred lines that have good performance per se as well as in cross combinations. This resulted in development of high yielding SCHs for different agro-ecological regions of the country (Fig. 5). The major strategy therefore, became to evolve and disseminate inbred-based hybrid technology.

Fig. 5. An excellent single-cross hybrids (Vivek Maize Hybrid 45) developed under AICMIP

A total of 212 hybrids and 119 composites of maize have been released till date. These represent a wide range of maturity to cater to the need of farmers in different production ecologies of various states. The improved cultivars have been widely adopted by Indian farmers. The adoption of improved cultivars and production technology had a synergistic effect on crop productivity and provided encouraging results. Even though public sector has release more hybrids especially single cross hybrids, however in last 10 years private sector has been neck-to-neck in release and notification of hybrids (Fig. 6 and 7).

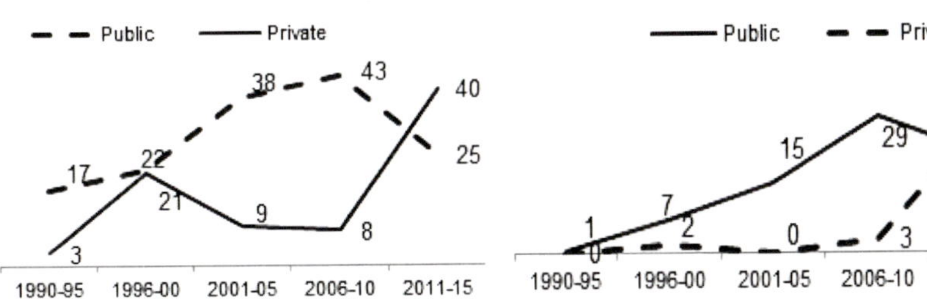

Fig. 6. Total Hybrids/composites Notified (226)

Fig. 7. SCH Hybrids Notified (Public: 73 and Private: 33)

This lead to increase and doubled maize production regularly in past (Table 1). Today there is need to double our maize production to 50 million tones till 2022 with achieving the *Kharif* productivity up to 10 tonnes/ha.

Table 1. Maize production scenario till date

Year	Production (million tonnes)	Percent Increase Area	Yield	Status
1950-51	1.73	-	-	-
1959-16	3.46	35	48	Doubled
1969-70	7.49	37	58	Doubled
2003-04	14.98	25	60	Doubled
2015-16	25	26	66	Added 10 million tones
2025-26	50*	19	70	To be Doubled

Different strategic issues have also been addressed in maize improvement. Heterotic pools were developed as a long-term strategy to derive inbred lines for use in hybrid development. Sources of resistance to major insect-pests and diseases have been identified and utilized in resistance breeding programmes. The application of molecular tools and techniques has also been used to understand the natural genetic diversity existing in the Indian maize germplasm and its utility in maize breeding.

Development of nutritional enriched maize cultivars

Recently, parental lines of 'Vivek hybrid 9' were introgressed with '*opaque2*' (o2) gene through marker assisted selection (MAS) and the hybrid between introgressed line was released as 'Vivek QPM 9'. '*opaque2*' gene alters the maize kernel endosperm composition of amino acid profile in desired direction with enhanced levels of essential amino acids like lysine and tryptophan. The success of 'Vivek QPM 9' has given an impetus for large scale conversion of elite normal maize inbred lines into Quality Protein Maize (QPM) lines. A number of maize hybrids/ composites/ synthetics are developed and released by the public sector. In addition, the efforts are on development of process-able maize hybrids especially for corn-flakes, high oil etc.

Indian maize breeding has a very strong linkage with the International Maize and Wheat Improvement Centre (CIMMYT), Mexico. CIMMYT has supplied diverse kind of germplasm (yellow maize, white maize, quality protein maize, waxy corn, high oil maize etc.) to Indian maize breeding programme. The QPM breeding is one of the most benefitted programmes. QPM possesses improved nutritional value as compared to normal maize due to enhanced levels of amino acids namely lysine and tryptophan. In 1970s, three QPM composites *viz.*, Shakti, Protina and Rattan were released. In the last decade India has developed nine quality protein hybrids, *viz.* Shaktiman 1, Shaktiman 2, Shaktiman 3, Shaktiman 4, HQPM 1, HQPM 4, HQPM 5, HQPM 7 and Vivek QPM 9.

Research on natural resource management

Resource management research has focused on development of agro-techniques for resource conservation in maize-based production systems, development of technologies for better management of abiotic stresses like drought, excess moisture, cold, heat etc. Reduced or conservation tillage system is gaining more attention in recent years with the rising concern over natural resource degradation. In India, conservation agriculture is being practiced under rice-maize cropping system in coastal Andhra Pradesh and north eastern Karnataka. Farmers are taking up sowing of maize in residual soil moisture after rice harvest and saving about 20-30 days of land preparation time. In rice-maize cropping system, farmers are harvesting >10,000 kg/ha of maize grain.

Impact on maize export

Average exports of maize was 3.6 million tones and gain income Rs. 5127 Cr (2010-11 to 2014-15) (Fig.8) and are exported to number of adjacent countries (Fig.9).

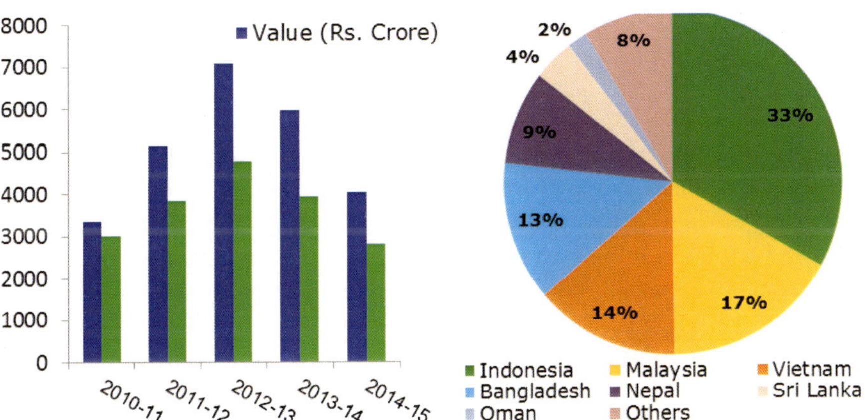

Fig. 8. Maize exports in value and quantity.

Fig.9. Maize exports from India.

The effects of climate change: Challenges

Climate change, the most serious environmental challenge facing humanity, is expected to have far-reaching impacts on maize, rice and wheat. Historical data on climatic parameters as well as yield data on maize are important to look into past setbacks and achievements. At a global level, it is estimated that higher temperatures and precipitation trends since 1980 have lowered yields of wheat by 5.5 percent and of maize by 3.8 percent below what they would have been when the climate remained stable.

On studying the effect of climate change in rainfall, observed a shift in peak of rainfall, reduction in peak of total rainfall during the rainiest months and low rainfall in the initial months as well as in September i.e. during grain filling period of the maize (Mahajan *et al.*, 2012) (Fig. 10). This change has been beneficial for maize cultivation because hybrid 'Seed Tech 2324', days to anthesis and silk have decreased by 0.31 and 0.11 days per year, respectively, while in 'Bio 9681' days to anthesis and silk decreased by 0.27 and 0.07 days per year, respectively. Concomitantly, there is an increase in yield by 0.29 per cent and 0.10 per cent per year.

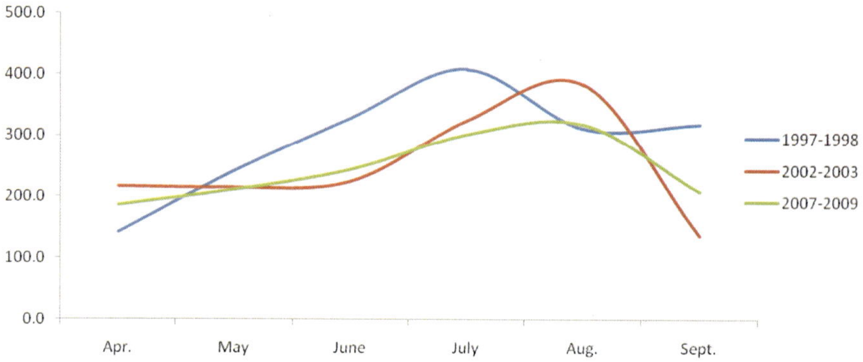

Fig.10. Effect of climate change since 1996: Jorhat, Assam, India – a high rainfall area.

On analyzing the grain yield of the best entry, the best check and trial mean of full season maturity group grown during the monsoon (*Kharif*) season in All India Coordinated Maize Improvement Project (AICMIP) trials from 1991 to 2012 in four agro-climatic zones of the country, Mahajan *et al.* (2015) found that the monthly fluctuations in rainfall during maize cropping season are high in all the zones during maize crop season and the trend was significantly increasing in the month of August in Zone V (Fig.11). The seasonal rainfall decreased during the month of October in all the Zones. The yield performance of best test entry was *at par* or higher than the best check with significant and strong correlation with each other and with trial mean.

There was increasing trend for maize yields from 1991 to 2012 in all the zones. The new maize hybrids from AICMIP are superior even under the changing climatic conditions in different agro-climatic zones, however, there was a stabilization of the zonal yields (pooled over states within agro-climatic zones) of maize around year 2002 for 7 to 12 years in different zones of the country. In spite of high fluctuations in rainfall, there was continuous improvement in the new genotypes developed at AICMIP especially in past two decades. Thus, AICMIP had an important contribution in overall increase in zonal yields of the country even though the differential impact in different zones. The impact of new hybrids in zonal or state productivity and production took less time for Zone II and Zone III and this impact lasted for more years in Zone II than Zone III.

Abiotic and Biotic Constraints

Improving the genetic potential of Indian maize would continue to be a major challenge. Today, hybrids with yield potential of up to 14 t/ha are available. However, it becomes difficult to achieve even half of this potential because of high incidence of biotic and abiotic stresses in farms. It is clear that the major challenge in germplasm enhancement lies in introducing stress tolerance traits.

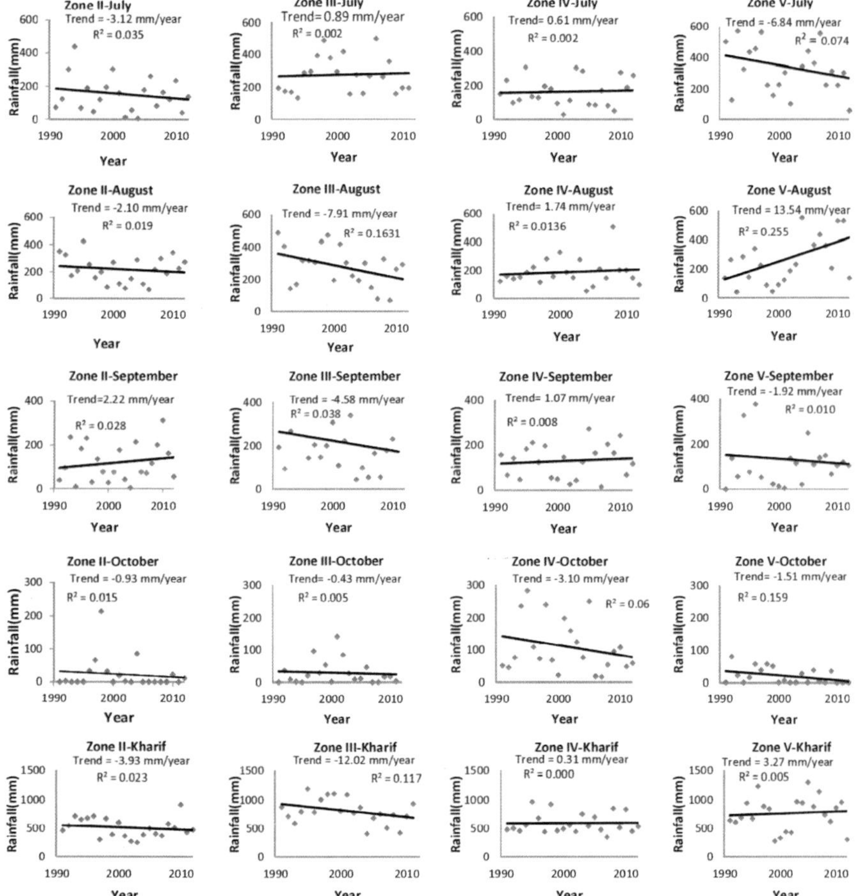

Fig. 11. Rainfall pattern in different months of *Kharif* season and different agro-climatic zones.

Development of high yielding cultivars with built-in higher level of resistance against stresses is a daunting task. This becomes all the more challenging due to unpredictability of plant-pest-natural enemies interaction in the context of changing climate. Maize seed, grain and processed maize product are highly vulnerable to stored grain pests. With the changing climatic conditions preparedness against changing spectrum of disease and pest in different agro-ecological zones is need to relook. BLSB, TLB and MLB are important disease throughout the India. In addition, in northern India, PFSR, BSDM and RDM are important and in southern part, Sorghum DM, *Polysora* Rust, are important. In addition, in eastern India, *Pythium* Stalk rot, *Fusarium* SR are also important while in western India, Common Rust is important. The screenings for these diseases were taken at 17 locations including 10 hot spot locations in different zones of the country. Deployment of new sources of resistance in diverse background, along with information in pathogenic races prevalent in the zones are need to be carried out.

Natural Resources Degradation

The natural resources degradation, fading organic carbon from soil, declining factor productivity, decreasing farm land due to more land under non-agricultural uses in future and profitability due to escalating input prices in agricultural production will further aggravate the problem of sustaining maize production systems.

Enhanced Feed and Fodder Requirements

Fodder is an important issue as the country is presently facing a net deficit of 61% for green fodder. In the absence of nutritious fodder, the farmers are feeding their cattle with low quality roughage or rice straw, thus adversely affecting the milk production potential of the animals. The competition for land and meeting the feed and fodder need for the support of livestock and poultry production will be another challenge in this scenario. A number of babycorn hybrids like 'HM4' and 'Vivek Maize Hybrid 27' hybrids are released by public sector, which also provides 70-80 qt of green fodder after 60 days.

Malnutrition

Micronutrient and protein malnutrition is a serious concern particularly in the rural masses as they solely depend upon cereals for food. Millions of people suffer from protein, Vitamin-A, iron and zinc deficiency. It is estimated that about 123 million people in Southeast Asia are at risk of vitamin A deficiency and 1.7 million are affected by exophthalmia (Fig. 12). Iron, zinc and lysine and tryptophan amino acids that determine protein quality are usually found deficient in maize consuming population. In times to come, systematic research in maize needs to be undertaken in India to address this issue. Till date, 13 single cross QPM hybrids are released by the public sector and among them the most outstanding was 'Vivek QPM 9' – an early maturing, attractive, orange semi-flint grain, developed through marker assisted selection (MAS) (Fig.13). The details of their nutrients are given below (Table 2).

Table 2. Quality of 'Vivek QPM 9' developed through MAS.

Hybrid	Protein (%)	Tryptophan (%)	Lysine (%)	Fe (ppm)	Zn (ppm)
Vivek QPM 9	8.46	0.83	4.19	37.0	29.0
Vivek 9	9.45	0.59	3.25	30.0	25.0
Increase	-	40.7	28.9	23.3	16.0

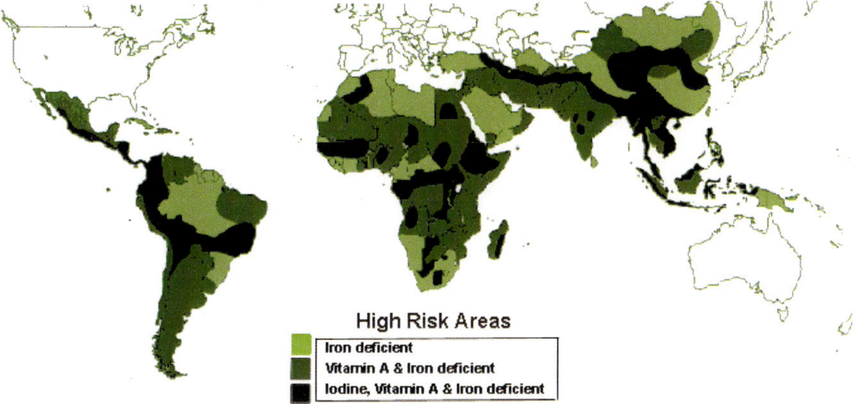

Fig. 12. Micro-nutrients deficit areas (Source: Harvest Plus, Breeding crops for better Nutrition).

Peri-urban Agriculture

Fig.13. Vivek QPM 9 – A QPM hybrid developed through MAS.

In context of peri-urban agriculture and urbanization, the profitability will be an issue for enhancing maize area and utilization, as majority of the population in India will be urbanized by 2050. The higher income, increasing urbanization and changing food habits will likely to shift the food preference in the country and there is possibility for increasing demand of protective healthy foods based on milk and meat. A number of hybrids and composites/synthetics specialty corns *viz.* popcorn, baby corn and sweet corn are released by Indian Council of Agricultural Research (ICAR) to meet the future upcoming needs of per-urban area. In addition, efforts are going on to develop process-able hybrids especially for corn-flakes, high oil, waxy corn etc.

which may add to value addition and may boost maize industries in the country. Thus, the futuristic production system involving specialty and nutritious maize are needed for sustaining agriculture in peri-urban areas.

Demanding Seed Supply

Low cost and quality seed production will be a major issue for enhancing adoption rate of hybrid maize cultivation. Farmer, private, cooperative and public sector participatory seed production programme needs to be evolved in future for assuring quality seed availability.

Should maize be diverted for ethanol production?

As of now, United States is already using 30 per cent of its maize for biofuel production. Biofuel from maize will help in energy security along with high procurement prices. The expanding spectrum of demand would necessitate focused research tailored for specific segments of the myriad value chains. In India, 65 per cent of total production of maize is used as feed while only 23 per cent used as food. From net return comparisons, it is clear that use of maize in ethanol production fetches better returns to the farmer. This has the potential of impacting food security only partially, if current level of supply demand position is taken into account. With increasing energy consumption both in India and world over, tapping of alternate sources of energy assumes significant importance. In this regard, biofuel from maize holds distinct potential. Maize is the most important natural multiplier and an economical source of starch. Starch comprises about 68-74 per cent of maize Kernel weight, which can easily be converted into glucose and subsequently fermented into ethanol.

Achieving targeted production Challenges by 2025

The increased production of maize has helped India to become one of the top exporters of maize in the South East Asian region. India being near to South East Asian countries it is supplying maize to neighboring countries as India also enjoys the comparative advantage over the American competitors because of low cost of transportation. The higher prices of U.S., Brazil and Argentina maize as well as their longer transport time gives Indian maize an edge in the South East Asian region. The following research interventions may help in achieving the future targeted maize production (vision 2050):

1. **Screen germplasm and discover trait specific novel genes from maize genomes:** Germplasm screening has been the core competency of maize research. Over several years, thousands of maize lines for traits like insect resistance, disease resistance, drought, heat and water-logging resistance,

etc. have been evaluated and sources of resistance/ tolerance have been identified. However, very few such sources have been effectively used in breeding programme or genetic engineering because the genes for the resistance are yet to be isolated. Once these genes are isolated, characterized and validated, they can be effectively used by marker assisted selection (MAS) in breeding schemes or over-expressed by genetic engineering.

2. **Deliver superior single cross hybrids with diverse genetic base for various segments:** The single cross hybrids have been the bedrock of the Indian maize programme for the last one decade, which saw quantum jump in maize productivity. However, narrowing of the genetic base of the elite germplasm is a matter of concern. There is a need to broaden the genetic base, develop new heterotic pools and continue to breed newer single cross hybrids with diverse base. Availability of doubled haploid technology is expected to enhance the speed of inbred line development. The inbred lines would be used to test various hybrid combinations. The stability and yield of hybrids would be ascertained in multi-location trials before they are commercialized.

3. **Thoroughly integrate marker assisted selection and doubled haploids in breeding schemes:** Speed and precision of maize breeding will be enhanced by using well established techniques of marker-assisted and whole genome selections. The genome sequence of maize is now available and re-sequencing costs have come down dramatically. Maize is one of the crops with richest genomic resources available in terms of SNP and conventional molecular markers.

4. **Spearhead development of public sector events of transgenic maize:** As indicated in previous sections advent of even more number of transgenic traits is a reality of global operating environment with respect to maize and we have to adequately respond to this aspect. The two most important pre-requisites for any transgenic development programme are: availability of few good genes and a robust transformation protocol. While we have developed remarkable competency in maize transformation, we lack key genes of our own. Our initial focus would be use of native genes and promoters and use of system independent synthetic genes, like artificial miRNAs and other RNAi cassettes. Non-native constructs would be developed in various partnerships. The prioritized traits are: insect resistance, herbicide tolerance, disease resistance and abiotic stress tolerance in short-term; mycotoxin management, low phytate and nutrient use efficiency in medium term and nutritional enrichment, modified cellulose and molecular pharming in long term research agenda.

5. **Invent a new generation of eco-friendly and biosafe technologies for maize value chain:** In the times to come, biosaftey concerns would dominate public discourse on food and agriculture. Biological interventions would be deemed more desirable than chemical interventions at all the stages of value chain- from farm to fork. Bio-fertilizers, bio-pesticides and pest control though cropping system, habitat management and Integrated Pest Management approaches would be in demand. The Institute is committed to address these concerns by delivering innovative biological solutions.
6. **Develop and popularize high yielding, profitable and ecologically sustainable maize-based farming systems:** The newly developed genotypes and associated technologies have to be constantly contextualized in an on-farm agronomic system. With large scale urbanization, in future, production technologies would be heavily mechanized. The technologies available from other sources have to be appropriately integrated to work out efficient, profitable and sustainable maize-based farming systems.
7. **Precision input management for higher productivity, profitability and environmental sustainability:** Under this initiative, the focus would be to develop cutting-edge precision agriculture technologies for maize and bringing out modules of Good Agricultural Practices based on Low External Input Sustainable Agriculture (LEISA) concept.
8. **Popularize resource conservation technologies in maize systems:** We would like to expand our activities in the domain of conservation agriculture that "strives to achieve acceptable profits together with high and sustained production levels while concurrently conserving the environment". The availability of herbicide tolerance trait would give a massive fillip to conservation tillage in next few years, which would necessitate higher research intensity in this critical area.

Future strategy

Keeping in mind the high productivity of *Rabi* maize and major area under cultivation during *Kharif* season, the present strategy shall be to increase area under *Rabi* maize such as Banswara area of Rajasthan wherein *Rabi* maize is more profitable due to its high productivity. In addition, new area like that of upland rice in West Bengal and around during *Kharif* season. The low productivity of *Kharif* maize need to be address and our targets shall be single cross hybrids with at least 10 tonnes/ha, so that we can achieve our targets of 50 million tonnes by 2025. Only hybrids of 10 tonnes/ha during *Kharif* season help in achieve our targets of 2025. New areas in Central Western Zone (CWZ) need to be explored using drought tolerant hybrids during *Kharif* season. The growth in maize area is highest in Peninsular Zone (PZ) followed by North-Eastern

Zone (NEPZ). The highly productive land of NEPZ and large area of CWZ are the area of future expansion to increase the productivity and production in both *Kharif* and *Rabi* season. In spite of 2009 and 2015 as drought years the maize productivity has maintained almost at the same level indicating spread of climate resilient maize hybrid released and spread in last one decade.

It is essential to use the new technology like MAS and DH, on large scale develop new hybrids which shall be high yield and nutritionally rich. As an outcome of the farmer's training programme, it was observed that most of the farmers are not aware of the latest technology to achieve highest profitable productivity and production. Hence, farmers should be educated on large scale in mission-mode approach coupled with timely and sufficient good quality seed should be supplied even in the remotest area. Presently, nearly 60 percent area is under hybrid cultivation and 25 percent area under single cross hybrids. In order to achieve our targets of 50 million tonnes of maize production there is urgent need to increase the area under hybrids to 90 percent and single cross hybrids up to at least 50 percent.

We need much more than improved varieties and development of climate resilient cultivars to sustainably and cultivation of new choice of hybrids in niche areas to enhance maize production and improve farmers' livelihoods and income. This will facilitate targeted maize improvement through a conventional breeding program coupled with molecular techniques.

In order to double the profitability of the farmers in next five years, it is essential to concentrate on process-able maize (like corn flakes, biscuits, chips etc.) hybrids, which could fetch better price to the farmers. The maize programme should emphasize on diversification of maize germplasm, sharing of germplasm, developing germplasm for abiotic and biotic stress under changing climatic conditions, emphasizing on data mining to understand the future behavior of the maize crop in changing climatic conditions, low cost of cultivation, zero-tillage etc. Stabilized market prices throughout the year coupled with attractive 'Minimum Support Price' will help in achieving, not only the targeted maize production but also doubling the farmer's income.

References

Anonymous. (2015). Agriculture Statistics at a Glance, Govt. of India.

Anonymous. (2015-16). State of Indian Agriculture 2015-16. Pp. 252. Ministry of Agriculture and Farmers Welfare Department of Agriculture, Cooperation and Farmers Welfare Directorate of Economics and Statistics, New Delhi.

Dass, S., Kaul, J., Manivannan, A., Singode, A. and Chikkappa, G.K. (2009). Single cross hybrid maize – a viable solution in the changing climate scenario. *Ind J Genet Plant Breed.* 69: 331-334.

Mahajan, V., Singh, K.P., Bansal, P., Kumar, V. and Sai Kumar, R. (2015). Rainfall trends and maize productivity in diverse agro-climatic regions of India. *Ind J Genet Plant Breed.* 75(4): 468-477.

Mahajan, V., Singh, K.P., Rajenderan, R.A. and Kanya. (2012). Response of maize to changing climatic conditions in Himalayan region. *Ind J Genet Plant Breed.* 72(2): 183-188.

Vision 2050. (2015). ICAR-Indian Institute of Maize Research, New Delhi. pp. 32.

21

Maize Disease Research in India Progress and Future Thrust

Meena Shekhar[1] *and Pravin Kumar Bagaria*[2]
ICAR-Indian Institute of Maize Research, Ludhiana
Corresponding Authors' Email: shekhar.meena@gmail.com

Maize (*Zea mays* L.) occupies an important place in world agriculture due to its high yield potential and greater demand. Because of its divergent character, maize is grown over a wide range of climatic conditions, ranging from near sea-level to several thousand meters above sea-level (2,700 *msl*). It can also be grown in all types of soils ranging from sandy to heavy clay. Among the factors adversely affecting productivity, incidence of diseases is prominent. In India Maize is prone to a number of biotic stresses like foliar diseases, downy mildews, banded leaf and sheath blight, stalk rots and ear rot caused by fungi and bacteria, incidence of which are variable in time, space, and genotype. Disease may be minimized by the reduction of the pathogen's inoculum, inhibition of its virulence mechanisms, and promotion of genetic diversity in the crop (Richard *et al.*, 2005). Conventional plant breeding for resistance has an important role to play that can now be facilitated by marker-assisted selection.

Chemical control measures for some of the diseases are helpful in reducing losses, however for stabilizing production it is necessary to build level of genetic resistance. Germplasm improvement programme undertaken in sixties and early seventies by All India Co-ordinated Maize Improvement Project provided an opportunity for combining resistance against important diseases with other desirable agronomic traits. Thus, breeding for resistance had been an important objective of the overall crop improvement programme and it had been a co-operative venture of the plant breeders and plant pathologists. Consequently, it was recognized that there should be an efficient technique for the mass multiplication of inoculums, artificial inoculation technique and scoring/rating of plant disease as a pre-requisite for the successful identification of resistant sources in such improvement programme. With these objectives in to consideration, a six days seminar on the "Techniques of scoring for resistance to diseases of maize in India" was held at Andhra Pradesh Agriculture University, Hyderabad during 15-20 March, 1982 (Payak and Sharma, 1983).

Based on research efforts for the last six decades, under the aegis of Indian Institute of Maize Research, erstwhiled Directorate of Maize Research, 16 out of 62 diseases adversely affecting this crop had been identified, *viz.*, turcicum leaf blight, Maydis leaf blight, Banded leaf and sheath blight, Post flowering stalk rots, Common rust, Polysora rust, Downy mildews Pythium stalk rot and Bacterial stalk rot are the major threat to the potential yield of maize (Table 1). Globally, losses due to maize diseases have been estimated to the tune of 9.4%, annually, for the countries of Asia the figure is 12%, while for the African countries the estimate is as high as 14% (Cramer, 1967; James, 1981). Even for the developed countries like USA, 12% of the produce is lost due to diseases, annually. In India the total loss in economic products of the crop due to diseases has been estimated to the tune of Rs. 17, 83, 320 (Table 1) and in terms of percent losses is 13.2% (Payak and Sharma, 1985).

Table 1. Major disease affecting the maize crop at a glance.

Disease	Affected part	Pathogen	Losses
Seed and Seedling Blights	Seedlings	Species of *Fusarium, Rhizoctonia, Pythium Penicillium Aspergillus, Acremonium, Cephalosporium* etc.	By reducing plant stand
Turcicum Leaf Blight (TLB)	Foliar part at knee high stage	*Exserohilum turcicum* (Pass) Leon. & Sugs	Severe infection causes up to 70% yield reductions
Maydis Leaf Blight (MLB)	Foliar part at knee high stage	*Drechslera maydis* Niskado Syn. *H. maydis*	Severe infection causes a premature death and up to 83 % yield reductions
Common Rust (C. rust)	Foliar part	*Puccinia sorghi* Schw	Yield losses as high as 50% have been recorded
Polysora Rust (P. Rust)	Foliar part	*Puccinia polysora* Underw.	Yield losses in excess of 45% have been recorded
Brown Stripe Downy Mildew (BSDM)	Foliar part	*Sclerophthora rayssiae* var. *zeae* Payak and Renfro	Losses ranging from 20-100% depending on the species and host cultivars in maize.
Sorghum Downy Mildew (SDM)	Foliar part & in severe case tassels.	*Peronosclerospora sorghi* (Weston & Uppal) Shaw	Severe outbreaks have occurred in all over world including India, The yield loss as high as 90% has been reported.
Rajasthan Downy Mildew (RDM)	Foliar part	*Peronosclerospora hetropogoni* Siradhana *et. al.*	Very serious in Rajasthan & Yield losses in excess of 70% have been recorded.
Brown Spot	Leaf & mid rib	*Physoderma maydis* Miyake	It is not so important in India, however, in favorable condition, disease can develop in severity and cause alarm situations.
Banded Leaf and Sheath Blight (BLSB)	Started from lower leaf & whole plant	*Rhizoctonia solani* f. sp. *Sasakii* Exner	yield losses close to 40% have been attributed to BLSB
Curvularia leaf spot	Leaf area	*Curvularia lunata*	

Pythium Stalk Rot (PSR)	Stalk at pre flowering stage	*Pythium aphanidermatum* (Eds) Fitz	The yield reduction in susceptible genotypes has been reported to the tune of 100%
Bacterial/Erwinia stalk rot (ESR)	Stalk at pre flowering stage	*Erwinia chrysanthemi* p. v. *zeae* (Sabet) Victoria, Arboleda & Munoz	Early infection led to complete death of plant, at flowering stage resulted in 92.2% yield loss and late infection in 57.0 to 36.3 % yield losses
Post Flowering Stalk Rots(PFSR)	Stalk at post flowering stages	*Macrophomina phaseolina*	Disease influenced by climatic conditions, crop density *etc.* losses in Karnataka recorded10-42% (Desai *et al.*, 1991).
Charcoal Rots (C. rot)	Stalk at post flowering stages	*Fusarium verticillaides* Sheld	Experimental loss ranges from 36-38% at Udaipur
Late Wilt	Stalk at post flowering stages	*Cephalosporium maydis* Samara, Sabeti & Hingorani	In India where it occurs endemically, with incidence as high as 70% and economic losses up to 51% (Johal *et al.*, 2004).

Avoidable yield losses in experimental field

The experiment conducted on avoidable yield losses against major diseases were conducted in different regular locations of the country during 2010–16. Genotypic variation on the extent of the avoidable yield losses was observed due to BLSB, PFSR and TLB, MLB, SDM and RDM (Table 2). In Vivek QPM 9 and Pant Sankul Makka-3, the losses due to BLSB was from 13.21 to 36.23 at Delhi & Pant Nagar, respectively.

Table 2. Avoidable yield losses in experimental field.

Genotype	Disease	Location	Yield loss (%)	Range of losses
Pant Sankul Makka 3	BLSB	Pant Nagar	29.02	BLSB –
Pant sankul makka 3	BLSB	Pant Nagar	36.23	29.02 – 36.23
Vivek QPM 9	BLSB	Delhi	13.21	
30V92	PFSR(*M. phaseolina*)	Hyderabad	20.74	
30V92	PFSR (*M. phaseolina*)	Hyderabad	19.26%	PFSR–
30V92	PFSR (*M. phaseolina*)	Hyderabad	20.11	C. Rot
HM 9	PFSR(*M. phaseolina*)	Delhi	31.4	19- 42%
Pro 311	PFSR(*M. phaseolina*)	Delhi	42%	
PMH1	PFSR(*M. phaseolina*)	Ludhiana	11.10%	
PMH2 (mean of 2 year)	PFSR(*M. phaseolina*)	Ludhiana	15.70%	
Mahi Dhawal	PFSR(*F. verticillaides*)	Udaipur	36.23	PFSR (F. Rot)
Mahi Dhawal	PFSR(*F. verticillaides*)	Udaipur	38.93%	36.23-38.93%
DHM – 2	TLB	Arabhavi	14.26	
EH 434042(Arjun)	TLB	Arabhavi	11.97	TLB –
Bio 9681	TLB	Arabhavi	29.84	11.00 – 29.84%
DHM – 2	TLB	Arabhavi	20.84%	
Bio 9681	TLB	Arabhavi	16.55%	
Vivek Hybrid5	TLB	Almora	17.21	
HM 8	MLB	Delhi	14.03	MLB –
Bio 9681	MLB	Delhi	10.58	10.58-14.03%
COH(M)-5	SDM	Coimbatore	54.95	SDM - 54.95%
Pratap Makka	RDM	Udaipur	55.49%	RDM 48.22-70.50

Economically important diseases of Maize in India

Seed and Seedling Blights

Causal organism: Species of *Fusarium, Rhizoctonia, Penicillium, Aspergillus, Acremonium, Pythium,* and *Cephalosporium* etc.

Distribution

The disease is prevalent in compacted, poorly drained and wet soils when temperature exists between 10-15°C at planting time. This disease is more severe in wet soil, in low lying areas & in soil that remained wet for an extended period of time. Disease severity is also affected by planting depth, soil type, seed quality, mechanical injury to seed etc.

Symptoms

Symptom appears as brown sunken lesions on mesocotyl, rotting at collar region leading to wilting & toppling of seedlings (Fig. 1). The disease poses a serious problem in temperate areas by reducing plant stand. However, it is not a serious threat in the major tropical environments of India because of rapid emergence of seedlings. The problem becomes severe with the use of old and poor quality seeds stored under high temperature and sown during winter (Dey *et al.*, 1992). Disease can be controlled by using good quality seed and treated with a protective fungicide. A variety of pathogens are associated with seed rots and seedling blights including species of *Pythium, Fusarium, Acremonium, Penicillum, Rhizoctonia, Macrophomina* and *Sclerotium* etc.

Fig 1. Symptoms of seedling blight.

Extent of Losses

Germinating maize seedlings are attacked by number of soil borne or seed borne fungi that causes seed rot & seedling blight consequently plant stand is reduced.

Foliar Diseases

Turcicum Leaf Blight

Causal organism *Exserohilum turcicum* (Pass) Leon. & Sugs (Teleomorph: *Setopharia turcica*).

Distribution

The disease is prevalent in areas where cooler condition with high humidity prevails and maize is planted in high lands. It is common in winter planting in the plains as moderate humid conditions (18-27°C) favours disease. It is common in Jammu & Kashmir, Himachal Pradesh, Sikkim, West Bengal, Meghalaya, Tripura, Assam, Rajasthan, Uttar Pradesh, Uttarakhand, Bihar, Madhya Pradesh, Gujrat, Maharashtra, Andhra Pradesh, Karnataka and Tamil Nadu.

In India, for the first time this disease was reported from Bihar by Butler during 1907, later from many parts of the country, *viz.*, Lalmardi, Srinagar (Kaul, 1957), Punjab, Himachal Pradesh (Chenulu and Hora, 1962) and Kashmir valley (Payak and Renfro, 1968). Disease intensity depends on cultivar susceptibility weather factors and location (Gowda *et al*., 1989).

Symptoms

Slightly oval water-soaked, small spots are produced on leaves, turned in to long, elliptical (2.5-15cm) grayish green or tan (Fig. 2) develop first on the lower leaves and later spread on upward. Complete blighting of leave, when disease incidence occurs after anthesis. In damp weather, large numbers of grayish black spores are produced on lesions, which may appear on the outer husks. Severe infection causes a premature death and resembles frost or drought injury (Fig 2B). Pant *et al.* (2001) reported about 91% reduction in the rate of photosynthesis when severity of TLB in maize exceeded 50%.

Fig. 2. Symptoms of TLB.

Genetics of resistance

TLB resistance in maize genotypes is quantitative and gene conferring resistance is designated as Ht, which is either identical or very closely linked if related in two sources (Hooker, 1961, 1962, 1963a, 1963b). Homozygous plants with Ht gene have highest level of resistance however it confer comparatively less protection to the host than in the case of polygenic resistance under heavy disease pressure. Lim *et al.* (1974) studied and documented that virulence in the pathogen is monogenic in nature. However, Gevers (1975) reported the major gene HtN is reasonably stable. Soon after resistance of HtN gene was broken down by reporting a new race of *E. turcicum,* was virulent to this gene (Thakur *et al*., 1989). Later physiological races of *E. turcicum viz.* race 0, 23, 23N and race 2N were reported from Uganda (Adipala *et al*., 1993). Gowda *et al.* (1994) studied and established four maize differentials *viz.*, H-4460Ht, H-4460Ht2, H-4460Ht3 and A-503HtN, carrying different Ht genes and he reported, isolates from Almora, Bajaura and Nagenahalli belong to Race 4. Recently, Nagenahalli isolates was found virulent in terms of disease intensity as compared to Almora isolates (Kumar *et al.,* 2011). Further, it was studied that isolates from Hyderabad, Coimbatore, Udaipur and Ludhiana belong to Race1, whereas isolates from Godhra, Dholi, Chindwara and Kolhapur belong to Race 2. And isolates from Jashipur and Jorhat belong to Race 3 by Anonymous, 2003.

Spread of disease

Pathogen overwinters as mycelium and chlamydo spores in infected crop debris and at the onset of the favourable condition, begin to sporulate. Levy (1984) reported that *E. turcicum* overwintered on sorghum and maize plant debris, spores are disseminated by wind and rain splash to freshly planted maize.

Extent of losses

- If disease establishes before silking and spreads to upper leaves during grain filling, severe yield losses can occur. Crop lodging is a particular concern where maize is mechanically harvested.
- The local susceptible cultivars recorded 66% reduction in yield due to TLB (Payak and Sharma, 1985). Yield loss up to 83% (Babu *et al.*, 2004) in susceptible varieties.

Maydis Leaf Blight

Causal organism: *Drechslera maydis* Niskado Syn. *H. Maydis* (Teleomorph: *Cochliobolus heterostrophus*).

Distribution

Disease is prevalent in warm humid, temperate to tropical climate (20-30^{0}C). In 1970, the disease was first reported from southern Florida. In India it is distributed in Jammu & Kashmir, Himachal Pradesh, Sikkim, Meghalaya, Punjab, Haryana, Rajasthan, Delhi, Uttar Pradesh, Uttarakhand, Bihar, Madhya Pradesh, Gujrat, Maharashtra, Andhra Pradesh, Karnataka and Tamil Nadu.

Symptoms

Plant affected at knee height stage young lesions on leaves are small and diamond shaped, become elongate (2-3 cm). At maturity lesions may coalesce and produce a complete burning of large areas (Fig. 3). Lesions caused by Race O are restricted to leaves. However lesion produced by race 'T' is tan, (0.6-12 x 0.6-2.7cm) with reddish brown borders and may occur on the leaves,

Fig. 3. Symptoms of MLB.

stalks, leaf sheath, ear husk, and cobs. Seedling from infected kernels (Race 'T') may wilt and die within three to four weeks after planting. Race T and Race O are morphologically similar, although Race T is pathogenic to maize where the Texas male sterility has been incorporated. Additionally, Race T produces larger lesions than Race O.

Host range

In addition to maize, *D. maydis* is also known to infect sorghum.

Spread of disease

Pathogen overwinters as mycelium on infected crop debris that remains on the soil. They start sporulating and produce conidia when condition is favorable and disseminated through wind and rain splash, for secondary infection.

Extent of losses

- Photosynthetic area reduced, resulting in lower in photosynthate production. Race T infects stems, sheaths and ears, which is cause of ear drop, and lodging.
- Damage is critical if infection occurs prior to silking under favourable condition. If cultivating varieties with the Texas source of male sterility, infection by Race T can lead to extensive stalk and ear rots.
- Increased application of nitrogen fertilizer and increased crop density are associated with increased disease severity.

Common Rust

Causal organism is *Puccinia sorghi* Schw of this disease.

Distribution

In India it was recorded by Barclay (1891) from Shimla, further from Poona (Butler and Bisby, 1931). It is prevalent worldwide in a cool (16-20°C), subtropical, temperate and high land environments, coupled with high RH. It is common in Jammu and Kashmir, Himachal Pradesh, Sikkim, Meghalaya, West Bengal, Punjab (*Rabi*), Haryana (*Rabi*), Rajasthan, Uttar Pradesh, Bihar (*Rabi)*, Madhya Pradesh, Maharashtra, Andhra Pradesh, Karnataka and Tamil Nadu.

Symptoms

Disease is common at tasseling stage, lesions are circular to elongate (0.2-2 mm), dark brown pustule (uredinia) scattered over both the surfaces of leaf giving a rusty appearance (Fig. 4). Pustules may emerge in circular bands that occurred in the whorl and break through the leaf epidermis to release reddish-brown spores (uredinios-pores). Under severe disease pressure, leaves may turn chlorotic and senesce prematurely. At maturity they release brownish-black spores (teliospores).

Fig. 4. Symptoms of common rust.

Spread of Disease

Two hosts (maize and *Oxalis* species) and five spore stages (teliospores, basidiospores, spermatia, aeciospores and urediniospores) are involve in completing the life cycle of *P. sorghi*. Urediniospores are disseminated by wind spread from tropical/subtropical regions to temperate regions in spring and summer when maize is cultivated. Teliospores are source of secondary inoculum.

Extent of losses

- Reduction in photosynthetic leaf area, chlorosis, and premature leaf senescence, leading to incomplete grain filling and poor yields.
- Damage is most critical at conducive (16-25°C, RH above 95%, and over 6 hours of leaf wetness) in early stage and in susceptible hybrids.
- Yield losses from Diara region in Bihar ranges from 11.2-33.6% (Gupta, 1981), and 32.18% reported by Sharma *et al.* (1982).

Polysora rust

Causal organism is *Puccinia polysora* Underw, also known as southern corn rust.

Distribution

In India the disease was first noticed from Mysore district (Payak, 1992). In early eighties it was considered as disease of minor importance and has been emerging as a threat in coastal areas of A.P. and Karnataka. Incidence of P. rust has taken a heavy toll in majority of the cultivars grown in southern districts of Karnataka. It is favored by mild temperature (27°C) and high relative humidity.

Symptoms

It is very similar to common rust (*P. sorghi*) and can be distinguished, as the pustules of Polysora rust (uredinia) are golden-orangish in color, circular to oval (0.2-2.0 mm long), raised, and most prominent on the upper leaf surfaces, sometimes on stems and sheaths

Fig. 5. Symptoms of Polysora rust.

(Fig. 5). When pustules rupture, uredinios-pores are released, are source of secondary inoculum. On severity, leaves turned in to chlorotic and plant senesce prematurely.

Genetics of resistance

A single dominant gene for resistance to Southern rust was found in a recovered inbred line B-1139 TRpp. The original source of resistance (B-1138T) comes from South Africa and has resistance to *Bipolaris turcicum* (Pass.). Hulbert *et al.* (1990) observed that chromosome 10 of maize carries a cluster of genes which controls resistance to the rust species. Recombination frequencies between Rp genes depend heavily on the percentages used in the crosses and results from crosses involving reciprocal translocation stocks, which suggest that chromosomes 4, 8 and 9 of B73 and 2, 4 and 7 of MO-17 were responsible for general resistance among the 33 inbreds and 23 single cross hybrids tested.

Host range

An alternate host for *P. polysora* has not been identified urediniospores of *P. polysora* are known to infect maize, and various grasses.

Spread of Disease

Urediniospores are source of both primary and secondary inoculum. They infect maize plants through the stomata and spread rapidly under favorable condition.

Extent of Losses

- Chlorosis and leaf senescence leading to incomplete grain filling and poor yields.
- Damage is most serious in early infected plants and disease spreads to leaves above the ear. Late planted maize is particularly vulnerable.
- Yield losses up to 45% especially on late planted maize (Rodriguez *et. al.*, 1980) and 60% losses in grain yield recorded by Frederekson *et. al.* (1990).

Downy Mildews

Downy mildews are important and destructive diseases losses recorded up to 70% caused by different species of Oomycetes fungi *i.e.* genus *Peronosclerospora, Scleropthora* and *Sclerospora.* They constitutes one of the most important factors limiting maize production in India. The important downy mildews in maize in India are the Sorghum downy mildew, Brown stripe downy mildew and Rajasthan downy mildew. Maize is most vulnerable to downy mildew infection at seedling stage (15-20 days after planting).

Brown Stripe Downy Mildew

It is caused by *Sclerophthora rayssiae* var. *zeae* Payak and Renfro.

Distribution

It is prevalent in Himalayan areas and limited upto 1500 *masl* with cool, wet and high humid conditions. This disease was first reported from *tarai* area of Uttarakhand (Payak and Renfro, 1967). It is commonly distributed in Himachal Pradesh, Sikkim, West Bengal, Meghalaya, Punjab, Haryana, Rajasthan, Uttarakhand, Bihar, Madhya Pradesh and Gujarat.

Symptoms

Initially, lesions develop on the leaves as narrow, chlorotic or yellowish stripes, 3-7 mm wide with well-defined margins and delimited by the veins. The stripes later become reddish to purple (Fig. 6). Lateral development of lesions causes severe striping and blotching prior to flowering. Downy or wooly cottony whitish growth occurs in early morning hours on lower surfaces of the lesions.

Fig. 6. Symptoms of brown stripe downy mildew.

Sorghum Downy Mildew

Peronosclerospora sorghi (Weston and Uppal) Shaw

Distribution

The disease prevails in cool, wet and high humid conditions. It is commonly distributed in Gujarat, Maharashtra, Andhra Pradesh, Karnataka, and Tamil Nadu.

Symptoms

Infected plants are chlorotic that includes the base of the blade with transverse margin more erect as compared to healthy and easily can be defined between diseased and healthy tissue (Fig. 7A). White downy growth may appear on lower surfaces of infected leaves (Fig. 7B). Tassels of diseased plants turned in

Fig. 7A. Field view of resistant and susceptible plants of sorhum downy mildew.

Fig.7B. White downy growth on lower surface of leaf.

to green leafy structure called phyllody, with no seed settings when disease is severe. In tolerant varieties, the plant show disease symptoms but have normal seed setting.

Spread of disease

Conidia and oospores are sources of inoculums, can survive in the soil for several seasons. The germ tube infects the underground part of seedlings leading to characteristic systemic symptoms *i.e.* extensive chlorosis and stunted growth. Conidia are produced during the growing season are disseminated by the wind, providing secondary inoculum, which can induce systemic infection in plants up to four weeks. Pathogen can also be seed borne and seed-transmitted in cases where seeds are fresh and have high moisture content.

Rajasthan Downy Mildew

Caused by *Peronosclerospora hetropogoni* Siradhana.

Distribution

Disease is common in hot and humid conditions of Rajasthan and surrounding areas. A total of 11 districts of major maize growing belt of Rajasthan are considered as major yield constraint. This downy mildew was first reported by Siradhana *et al.* (1980). Several outbreaks due to this disease have been recorded in 3 districts *viz.,* Udaipur, Rajsamand, and Bhilwara during 1973-1980 (Rathore *et al.*, 2002).

Symptoms

Systemic symptoms in early seedling stage are characterized by pale appearance at the bases of second & third leaves giving complete chlorotic strips (Fig. 8A). Severely infected plants give yellowish appearance and plants die at about knee-high stage. The secondary symptoms starts at 2-3 leaf stage up to tassels and silks are formed. Whitish cottony growth can be observed on the lower and upper leaf surfaces under hot and humid conditions.

Fig. 8A. Symptoms of RDM.

Fig. 8B. *Heteropogon melonocarpus* showing Primary infection.

Symptoms of this disease are similar to *P. Sorghi*, the only difference is that *P. Heteropogoni* can infects the grass, *Heteropogon spp* serve as an alternate host during offseason. Conidia produced on *H. contortus* or *H. melonocarpus*

(Fig. 8B). Tassels may be malformed producing less pollen while ears may be aborted resulting in partial or complete sterility. In early symptoms plants are stunted and may die.

Spread of disease

Conidia and oospores are sources of inoculum. Oospores are produced abundantly on alternate host, a grass *H. contorts* but not on maize. Conidia produced on *H. contortus* or *H. melonocarpus* are the primary source of inoculum to infect maize.

Extent of losses

Rajasthan downy mildew causes heavy losses under favourable condition. The disease severity increased with the increase of inoculum density that resulted in 88.7% reduction of grain yield of maize (Rathore *et al.*, 2002; Rathore and Siradhana, 1987).

Brown Spot

Causal organism of this disease is *Physoderma maydis* (syn. *P. zeae-maydis*), which is closely related to the oomycete or water mold fungi.

Distribution

Disease is favored by moderate temperature, high humidity. It is prevalent in subtropical areas where abundant rainfall with moderate temperature prevails. It common in Jammu and Kashmir, Himachal Pradesh, Sikkim, West Bengal, Punjab, Rajasthan, Madhya Pradesh and Karnataka.

Fig. 9. Symptoms of brown spot.

Symptoms

Lesions are small and round to oblong, yellowish to brown on the leaf blade develop in distinct bands on leaf, stalk, sheath, and husks. Young lesions on can resemble those caused by early southern rusts. However, white lesions on the lamina continue as chlorotic spots (Fig. 9). Such brown lesions appear on nodes and internodes also. As the disease progresses, the lesions coalesce, become larger in size and turned from chocolate to reddish brown or purple in colour.

Spread of inoculum

The pathogen overwinters in infested plant debris or soil where reduced tillage practices are employed and produces brown sporangia that are packed inside infected cells which releases motile zoospores that require both light and water to germinate to infect the healthy plant. Infection most commonly occurs in the whorl where water tends to accumulate during periods of rain and irrigation.

Extent of losses

This disease is not considered to be important in India, however, when conditions are more conducive, the disease can develop and greater incidence and cause alarm because its symptoms are similar to some other more serious diseases.

Banded Leaf and Sheath Blight

Causal organism is *Rhizoctonia solani* f.sp. *sasakii* Exner (Teleomorph: *Corticium sasakii*). It has a significant impediment to maize production in many hot and humid environments in the tropics and subtropics. In particular, BLSB is recognized as a serious constraint to maize production in China, South Asia and Southeast Asia.

Distribution

It was first recorded from Sri Lanka (Bertus, 1927) and in India it was recorded from *Tarai* region of Uttarakhand for the first time in 1960 (Payak and Renfro, 1966). In early sixties it was considered as a disease of minor importance, now it is gaining importance due to its severity. Hot and humid conditions and irrigated fields are highly favorable. Disease is widespread in tropical and subtropical regions of the world and once established in a field it remains indefinitely. It is common in Jammu & Kashmir, Himachal Pradesh, Uttarakhand, Sikkim, Meghalaya, Assam, Nagaland Punjab, Haryana, Rajasthan, Madhya Pradesh, Delhi, Uttar Pradesh and Bihar.

Symptoms

Initially symptom appears on lower leaves and sheaths at 40-50 days old plant in the form of concentric bands and rings (Fig. 10A). On severity, it spread to the husk characterized by light brown, cottony mycelium with small, round, black sclerotia compact mass of hyphae that can survive in unfavorable conditions). The developing ear is completely damaged and dried up prematurely with cracking of the husk leaves.

Fig. 10. Symptoms of BLSB.

Host range

R. solani is pathogenic to a wide range of cultivated crops. In addition to maize, anastomosis group AG1-IA is also pathogenic to rice, wheat, sorghum, bean (*Phaseolus* species), and soybean.

Spread of disease

R. solani survives in the soil and on infected crop debris as sclerotia or mycelium for several years. The fungi spread by water, irrigation, movement of contaminated soil, and plant debris. At the onset of the growing season, in response to high humidity and temperatures (15-35°C), fungal growth is attracted to freshly planted host crops by chemical stimulants released by growing plant cells.

Extent of losses

- Maximum damage is caused when ears are infected kernels are wrinkled, dry, chaffy and light weight.
- In India, estimated yield losses recorded from 23.9 to 31.9% (Lal *et al.*, 1980). Singh and Sharma (1979) estimated 40.5% yield loss with 71% disease index.
- The disease caused yield loss up to 97% (Butchaiah, 1977) and exhibited a direct correlation with other yield parameters (Barua, 1979). In India, yield losses as high as 40% have been recorded.

Stalk Rots

Pre-flowering Stalk Rots

Pythium Stalk Rot

Disease caused by *Pythium aphanidermatum* (Eds) Fitz, fungi from Oomycete.

Distribution

In India, the disease was recorded for the first time by Srivastava and Rao (1964) from Delhi, it is prevalent in hot, humid, tropical and subtropical zones and sometime in temperate areas where soil is poorly drained with temperature (25-35°C), with RH 80-100%. High crop density, and high levels of nitrogen fertilizer favor disease severity, it is common in Sikkim, Himachal Pradesh, West Bengal, Punjab, Haryana, Rajasthan, Delhi, Uttar Pradesh and Bihar.

Symptoms

Pythium stalk rot occur prior to flowering and confined to a single inter-node just above the soil line. Infected stalk is brown, water-soaked, and soft, twisted and finally collapsed (Fig. 11). However infected plant does not break off completely and remain green, turgid up to several weeks because the vascular bundles remain intact.

Fig. 11. Symptoms of *Pythium* stalk rot.

Host range

Pythium aphanidermatum is known to infect a wide range of cultivated crops, including cereals and grasses, cucurbits, horticultural crops, and cotton. *P. aphanidermatum* is also an important greenhouse pathogen.

Spread of disease

P. aphanidermatum survives in the soil as mycelium or oospores for several years. They produce sporangia which release motile zoospores to produce characteristic symptoms. The pathogen is dispersed with either the movement of infected crop debris or with flooding or excess wetness. However the pathogen can survive on the turf grass (*Agrotis palustris*) throughout the year (Saladini *et al.,* 1983).

Extent of losses

- Damage is usually localized in a field, valley areas and river bottom fields are susceptible to infection, results in lodging of the plant.
- The yield reduction in susceptible genotypes has been reported to the tune of 100% (Payak & Sharma, 1978).

Bacterial stalk rot

Disease caused by bacteria *Erwinia chrysanthemi* p.v. *zeae* (Sabet) Victoria, Arboleda and Munoz, it is a motile, gram-negative, rod shaped bacterium.

Distribution

Disease is prevalent in high temperature (32-35^{0}C) with high RH and flooding. It is commonly distributed in Himachal Pradesh, Sikkim, West Bengal, Punjab, Haryana, Rajasthan, Delhi, Uttar Pradesh, Uttarakhand, Bihar, Madhya Pradesh and Andhra Pradesh.

Symptoms

Primary symptoms of discoloration generally appear when plant suddenly falls over and are seen scattered in the field (Fig. 12). Splitting of stalk exposes internal discoloration and soft slimy rot at the nodes

Fig. 12. Symptoms of bacterial stalk rot.

with foul odor macerated tissues. Disease builds up in the stalk and rapidly spreads up the stalk and into the leaves leading to collapse of the plant affecting normal tasseling and pollination. Affected plants may remain green for several days.

Host range

The bacterium has a wide host range. It can attack tubers of potato and sweet potato, onion bulbs, bean pods, roots of carrot, turnip, radish and sugarbeet, fruit of tomato, brinjal, tobacco, cabbage (Thind, 1970; Rangarajan and Chakravarty, 1971; Hingorani *et al.*, 1959).

Spread of disease

These bacteria survive in infected maize and sorghum stalks and residue. The bacteria enter the plants through natural openings; wounds from hail, high winds, or insect feeding (*e.g.* stalk borers) can provide additional entry sites into the plant. Saxena and Lal (1984) reported that the pathogen survives for 120 days in a field soil containing infected debries (Prasad and Sinha, 1977).

Extent of losses

Early infection led to complete death of plant however infection at flowering stage resulted in 92.2%, yield loss and late infection in 57.0 and 36.3% yield losses (Saxena and Lal, 1980).

Post-flowering stalk rots

Post-flowering stalk rots are one of the most serious, destructive and widespread group of diseases in maize. It caused internal decay and discoloration of stalk tissue, directly reduced yield by blocking translocation of water and nutrients, and can result in death and lodging of the plants. Potential grain yields are lost because kernels on rotted plants are lightweight and ears are missed during harvest. Premature drying is a critical sign of this stalk rot. Fungi *viz.*, *Macrophomina phaseolina and Fusarium verticilloides* are associated with stalk rots disease, are ubiquitous (Kaiser and Mukherjee, 1979; White, 1999). Severity of stalk rots is greatly influenced by temperature, rainfall, soil drainage, and soil type, available nutrient. Mechanical injuries and insect damage also tend to increase the severity of the stalk rots. Evidence of the carbohydrate sink size of the grain also influencing the occurrence of stalk rot supported the photosynthetic stress-translocation balance concept of stalk rot (Dodd, 1980).

Fusaium Stalk Rot

It is caused by fungus *Fusarium verticillioides* Sheld

Distribution

Fusarium stalk rot is distributed worldwide and most common in dry, warm regions and cause extensive crop damage by premature plant death. The disease is most common in Rajasthan, Uttar Pradesh, Bihar and Andhra Pradesh.

Symptoms

The infected plants typically wilted, leaves turn to dull green and the lower stalk becomes yellowed/straw-colored finally whole plant wilted (Fig. 13A). The internal pith of the lower nodes gets disintegrated and softened which can be easily pressed between thumb and finger. Fungal mycelium can often be seen at such nodes. The stalks show pink-purplish discoloration on splitting (Fig. 13B). Pathogen commonly affects the crown regions of roots and lower internodes.

Fig. 13A. Field view of *Fusarium* stalk rot infested crop

Fig. 13B. Symptoms of *Fusarium* stalk rot.

Hosts Range

The pathogen has wide host range: infects many crops, including maize, sorghum, sugarcane, wheat, cotton, banana, pineapple, and tomato *etc.*

Spread of disease

Primary source of inoculums is mycelium in infected crop debris remain in the soil and produces macro and micro conidia, which are disseminated through wind and rain splash to freshly planted maize. Disease is more severe when crop density and nitrogen fertilizer rate is increased. Stressed crop/plants are more prone to stalk rots. Stalks are already vulnerable at the time of grain filling stages as more sugars are diverted to ear leaf for grain filling.

Extent of losses

- Damage is caused by premature death of plant, lodging, and interference with translocation of water and nutrients during grain filling when plants infected at early stages leading to poor yields.
- Precise yield loss data for most stalk rots is difficult to ascertain. However, extensive losses are possible under severe disease infestation.

Charcoal Rot

This disease is caused by *Macrophomina phaseolina* (Tassi) Goid.

Distribution

Charcoal stalk rot is an important disease of maize and prevalent in hot, dry environments caused by the fungi *Macrophominaphaseolina.* Drought and high temperature (30-42°C) favors disease severity. In India it is commonly distributed in Jammu & Kashmir, West Bengal, Punjab, Haryana, Rajasthan, Delhi, Uttar Pradesh, Madhya Pradesh, Andhra Pradesh, Karnataka and Tamil Nadu.

Host Ranges

M. phaseolina infects an extremely wide range of hosts, including sorghum, soybean, cucurbits and various weed species. Over 500 hosts have been documented for *M. phaseolina* including peanut, Cabbage, pepper, chick pea, citrus, cucumis, soybean, sunflower, sweet potato, alfalfa, Pinus spp., Prunus spp., sesame, potato, sorghum, and bean etc., therefore rotation with these crop can increase inoculums levels and enable the pathogen to penetrate in the fields. High crop density and poor soil nutrient status also favors disease severity.

Symptoms

Symptoms observed once the crop/plant approaches maturity, affected plants dry prematurely (Fig. 14A) as internodes become disintegrated with black discoloration. Presence of numerous, minute black sclerotia, charring and shredding of the pith tissue of the stalks is a distinguishing character (Fig. 14B, 14C). The disease is confined to first or second internode above soil level.

Fig.14A. Toppled plant susceptible to C. rot showing toothpick

Fig. 14B. Susceptible reaction to *M. phaseolina*

Fig. 14C. Micro sclerotia on the cortical region of inter-node

Spread of Disease

M. phaseolina overwinters as sclerotia in the soil and can remain viable for several years. In dry and hot conditions fungi infect the roots of plants and colonize the lower stalk, eventually giving rise to abundant, minute, black sclerotia which are source of inoculums. Isolates that are pathogenic to maize are not known to produce conidia. Water stress at or after flowering period is a predisposing factor to infection.

Extent of losses

- Infection of the stalk can lead to premature senescing of the crop and lodging, increased losses may be experienced where maize is machine harvested due to lodging.
- It is prevalent in drought and in arid regions occurring huge losses where maize is regularly cultivated in rotation with maize and other host crops.
- Yield losses was recorded in Karnataka ranged from 10 to 42% (Desai and Hegde, 1991), in experimental avoidable losses was 42% (Shekhar *et al.*, 2012).

Late Wilt

It is caused by *Cephalosporium maydis* Samara, Sabeti and Hingorani. Late wilt, or black bundle disease, is a vascular wilt disease of corn that is caused by the soilborne fungus, *Harpophora maydis* (Samra *et al.*, 1966) synonyms: *Cephalosporium maydis* Samra, Sabet, and Hingorani and *Acremonium maydis* (CABI, 1999; El Shafey and Claflin, 1999). It was first reported as a vascular wilt disease of corn in Egypt in 1960 and is now considered endemic throughout

Egypt. Late wilt occurs in Andhra Pradesh, Uttar Pradesh, Bihar, and Rajasthan provinces of India (Payak *et al.,* 1970; Jain *et al.*, 1975). It develops rapidly at 20-32°C, with optimum disease development at 21-27°C (Singh and Siradhana, 1987a), however pathogen can grow over a wide range of soil pH from 4.5-10, with an optimum at pH 6.5 (Singh and Siradhana, 1987b).

Distribution

Andhra Pradesh, Uttar Pradesh and Rajasthan.

Symptoms

Initial symptoms observed as moderately to rapid wilting of from lower to upper leaves at tasseling time. Vascular bundles in the stalk are discolored (Fig. 15A). As leaf wilting advanced, yellowish or reddish brown streaks appeared on the basal internodes of the stalk, which dried up and became shrunken (Fig. 15B) and hollow with or without wrinkling turn purple to dark brown which is more prominent on lower 1-3 internodes. On splitting open stalk, a brown discoloration extended along the internodes. Some secondary organisms also develop on stalk rots that cause wet rot with some typical sweetish smell.

Fig. 15A. Symptoms of late wilt

Fig. 15B. Resistant and susceptible plants to late wilt

Spread of Disease

It is primarily soil borne and may infect young maize plants more readily than other plants. Infected seed, biological culture, or an infected crop residue could provide pathways for introduction. Secondary dissemination could occur through contaminated seed and the movement of agricultural equipment. Disease symptoms appear late in the season. Late summer planting reduced disease severity in India (Singh and Siradhana, 1988).

Extent of losses

- A yield loss of up to 40% was reported (El-Shafey and Claflin, 1999; Jain *et al.*, 1975). Whereas up to 100% reported by Satyanarayana (1995).

- Late wilt, or black bundle disease, poses a moderate to severe threat to corn production in India where it occurs endemically, with incidence as high as 70% and economic losses up to 51% (Johal *et al.*, 2004).

Ear and cob rot

Number of field fungi that invade the ear and kernels before harvest while the maize is in the field which affects the quality and appearance of kernels. The common fungi responsible for the ear rot in India are *A. flavus, Fusarium verticilloides, Gibberella, Trichoderma, penicillium* etc. Among them *A. flavus,* and *F. verticilloides* are important.

Aspergillus ear rot

Genus *Aspergillus* may infect maize in field as well as in the storage causing ear and kernel rot. *A. flavus* produces yellow & green masses of spores (Fig. 16) while *A. niger* produces black powdery masses of spore on the that cover both cob and kernel. Disease is favoured by high temperature and dry weather. Spores are spread by wind or insects to maize silks where they initiate infection. Insect's damage and other stresses tend to increase the *Aspergillus* infection. These infected kernels are responsible for aflatoxin production, which is fungal metabolite that is harmful for animal and human health. Crop planted and harvested late and grown under nitrogen stress more commonly contains aflatoxins prior to harvest than corn grown under good management practices and supplied with adequate nitrogen.

Fig. 16. *Aspergillus* ear rot symptoms.

Fusarium ear rot

Fusarium ear rot caused by *F. verticillioides* and occurs when harvest is delayed. Symptom shows pink to reddish brown discoloration on the kernels and later on it spreads on whole ear (Fig. 17). As the disease progresses, infected kernel is covered with a powdery/cottony pink mold growth whereas whitish streaks on the grains observed when infection occur late

Fig. 17. *Fusarium* ear rot symptoms

in the season. Air born spores present on residues can land on corn silks when it turns dark brown, whereas green silks are relatively resistant. Infection follows some form of injury, bird damage, feeding of corn borers. Disease development is favored by dry warm weather after pollination.

As molds grow and become established their metabolic activities and create microenvironments with elevated temperature and moisture content. When moisture content increases, conditions become suitable for other, less xerophilic moulds in a process known as fungal succession *Aspergillus* sp. and *Penicillium* sp. are the common molds in some tropical countries like India which can grow in low water active value and deteriorates the quality of maize grains in addition to the formation of mycotoxins. Such contaminated seed/grain again spoiled at the time of poor storage conditions due to colonization of *Aspergillus* sp. and other microflora and consequently mycotoxin formation takes place on such contaminated grains that cause serious feeding problems in a wide range of animals, when used as animal feed.

Maize crop planted late and grown under nitrogen stress is predisposed to aflatoxins contamination. Maize grown in cooler areas usually contains low amount of mycotoxin (Doko *et al.*, 1995). *Aspergillus flavus* competes poorly under cool conditions and prevalent in warmer environments, above 25°C (Shearer *et al.*, 1992). However occurrence of fungi and moulds is governed by prevailing environmental factors and post-harvest operations like harvesting of cobs, drying, sorting, processing and storage where farmers can play a major role (Shekhar *et al.*, 2016).

Management options for major diseases in maize

I. General cultural practices for the management of maize diseases

- Deep summer ploughing down of crop debris may reduce early infection.
- Residue management through tillage and sanitation.
- Soil solarization to reduce inoculum of soil borne diseases.
- Crop rotation with non-host crops
- Maintain recommended plant. population and use crop production practices.
- Use a balanced fertility programme
- Timely irrigation to reduce drought stress.
- Proper field drainage to reduce water logging conditions.
- Adopt practices to minimize the insect and nematode damage.

Specific cultural practices for management of maize diseases

a. For sorghum downy mildew

- Avoid maize-sorghum crop rotation as well as avoid sowing of maize adjacent to a field of maize or sorghum to check the spread of secondary infection in field.

b. For Rajasthan Downy mildew

- Destroy alternate host *Heteropogon melonocarpus* .

c. For Banded Leaf and Sheath Blight

- Stripping of lower 2-3 leaves along with their sheath considerably lowers incidence and also does not affect grain yield.

d. For Pythium stalk rot

- Plant population not to exceed 50,000/ha with good field drainage in endemic areas.
- Soil fumigation and soil drenches with bioagents and fungicides to mitigate the soil inoculums.

e. For Erwinia stalk Rots

- Planting of the crop on ridges and avoid water logging condition.
- Avoid use of sewage water for irrigation.

f. For Post Flowering Stalk Rots

- In stalk rot affected field, balance soil fertility specially increases the potash level up to 80 kg/ha help in minimizing the disease.
- Use Trichoderma Formulation in furrows after mixing with FYM @ 10g/kg FYM (1kg/100 kg FYM/acre) at least 10 days before its use in the field in moist condition.

g. For ear and cob rots

- Early planting.
- Hybrid selection (Use high yielding hybrid, resistant for diseases)
- Harvest crop in time.
- Minimize mechanical damage.
- Dry and store the grains properly at 13% or less moisture.
- Segregate, blend or destroy contaminated grains.
- Keep storage facilities clean.

II. Adoption of resistant hybrids/sources for management of different maize diseases

Disease	Resistant varieties/Hybrids/Sources of resistance
Turcicum Leaf Blight (TLB)	PEMH-5, Vivek21, Vivek23, Vivek25, Pratap Kanchan2, Nithyashree for Karnataka & Andhra regions
Maydis Leaf Blight (MLB)	HM10, PAU352, Malviya Hybrid Makka2, PEMH1, HQPM7, HQPM5, HQPM1, PEMH5, HQPM4, and HSC1
Common rust (C. rust)	Pusa prakash, HHM 1, HHM 2 and HQPM 1, Nithyashree
Polysora rust (P. rust)	NAH 2049 resistant hybrid for Karnatka and resistant sources NAI112, SKV18, SKV21,
Brown stripe downy mildew	PAU352, Pratap Makka3, Gujarat Makka4, Shalimar KG1, Shalimar KG2, PEMH5, Bio9636, NECH- X1280
Sorghum downy mildew	DMH1, NAC6002, COH (M) 4, COH (M)5, Nithyashree
Brown spot	JH10655, FH-3113, DMR-1, DMR-5
Pythium stalk rot	Pusa early hybrid, and X1280.
Erwinia Stalk Rot	PAU352, PEMH5, DKI 9202, DKI 9304.
Post Flowering Stalk Rot	JH6805, Bio9636, Pusa early hybrid, X1280, JKMH-1701, JH-6805, Bio-9639, Bio-9636, and X-1280.

III. Management options for maize diseases through chemical means

Disease	Chemical Control
TLB and MLB;	• Foliar Spray of mancozeb (Dithane M 45, Indofil M 45) or zineb (Dithane Z 78) at first appearance of disease/pustules @ 2 -2.5g/litre of water followed by 2 to 4 applications at 10 days interval if needed.
C. rust and P. rust	• Two to three fungicidal applications are recommended at 15 days interval when weather conditions are favorable for disease development to reduce secondary infection.
Downy mildews	• Seed treatment with metalaxyl (Ridomil 25WP, Apron 35SD) @ 2.5g/kg seed • Foliar spray of systemic fungicide, such as metalaxyl (Apron 35FN) @ 2-2.5g/L at first appearance of diseases.
Brown spot	• Spraying with copper fungicides at the rate of 3g/L in the whorls of plant twice a week for four weeks before silking • Seed treatment with fungicide *viz.*, Benlate, Bavistin and Plantavax.
BLSB	• Seed treatment of peat based formulation @ 16 g/kg of Pseudomonas fluorescence or as soil application @ 7 g/L of water • Foliar spray of Rhizolex 50 WP@10 g/10L of water or Sheethmar (Validamycin) @2.7 ml/L water at first appearance of disease.
Pythium stalk rot	• Soil drenching with 75% captan @ 12 g/100 litre of water.
ESR	• Bleaching powder with 33% of chlorine@10 kg/ha as soil drenching or by mixing with irrigated water.

Present status and future thrust

Changing in disease spectrum scenario

Disease can fluctuate according to climate variation, widespread occurrence of particular disease in a particular time use to be in climax when environmental conditions are conducive, as result epidemics take place. If changes in climate and global climate continue in the future as predicted, there may be chances of relocation and occurrence new diseases. In India there is warming trend in the climate along the west coast, in central, interior peninsula and Northeast that creates favorable conditions for diseases like Polysora rust, Sorghum downy Mildew and TLB, while a trend of decreasing monsoon seasonal rainfall has been observed over eastern Madhya Pradesh, North India, Central India, some parts of Gujarat and Kerala such condition is suitable for Post flowering stalk which is also gaining importance (Gautam *et al.*, 2013)

Responses to climate change are already visible as there is a major shift in disease pattern during the past years as major diseases like Pythium stalk rot and Erwinia Stalk Rot are gradually becoming diseases of lesser economic importance. In early sixties the diseases like BLSB, Polysora Rust and Pre harvest cob rots were considered as a minor disease, but due to changing the weather condition and pattern of disease spectrum these diseases are considered as major diseases.

Curvularia leaf spot

Curvularia lunata is the causal organism of this disease, which is a facultative pathogen of many plant species and of the soil. It is a very common in tropical, subtropical as well as in temperate regions. In India early nineties it was not so common, recently as per the survey report IIMR erstwhile DMR earlier during the year 2004, this disease was recorded in traces from most of the places and severe in some pockets from Rajasthan but, now during the year 2010-11 and onwards it was recorded in severe from Hardwar, Dehradun and Kashipur of Uttarakhand state. However based on report of 2011-12, it has shown widespread occurrence in Rajasthan with moderate incidence and gaining importance by exhibiting sever incidence recorded from 22 places.

Some major epidemics due to maize diseases in India

- Bacterial stalk rot (1959, 1969).
- Brown stripe downy mildew (1962-1965).
- Sorghum downy mildew. (1968-1973).
- Common rust (1973 in *Rabi* maize).
- Turcicum leaf blight (1974).

After strengthening maize pathology programme, no such epidemic was occurred due to development of resistant cultivars and management strategies. Some economically important diseases of this crop along with their pathogen, symptoms, and their distribution along with their management practices are being described to achieve better management of these diseases.

Critical analysis for future thrust

Effect of CO_2 concentration

An increase in CO_2 levels may positively correlate with production of plant biomass, but regulated by the availability of water and nutrients, competition against weeds and damage by pests and diseases. Consequently, high concentration of carbohydrates in the host tissue promotes the development of biotrophic fungi such as rust (Chakraborty *et al.*, 2002). Hence, there may be chances of increase the severity of Polysora rust as well as common rust in the pockets where elevated CO_2 is prevailed.

Effect of temperature

A change in temperature may favors the development of dormant pathogens, warmer mean air temperatures in India in early springs especially during winter, favours Post flowering stalk rots in winter maize. However, warmer and drier summers may hinder spread of most fungal pathogens, and finally slow down or completely inhibit disease progress of foliar diseases like Turcicum leaf blight, Maydis leaf blight, Banded leaf and sheath blight and Downy mildews etc., resulting in the regional distribution patterns of these diseases is going to be modified and foliar diseases started to infect *Rabi*/winter maize due to warming effect.

On the other hand, pathogens also have the capacity to adopt warmer conditions (Zhan and McDonald, 2011; Mboup *et al.*, 2012) and temperature/ humidity dependent on disease resistance of crop cultivars may be altered in the future (Huang *et al.*, 2006; Juroszek and Tiedemann, 2011). That's why TLB which was prevalent cooler and temperate region, has now become common in tropical in winter maize. Stalk rot diseases are anticipated to increase in hot and dry areas in summer crops as well as in winter crops of Bihar, Rajasthan, Andhra Pradesh and Karnataka due to rise in temperature in early springs at the time of flowering. Increasing trend in cultivation of *Rabi*/winter maize in various states, as a result host is available throughout the year in major maize growing areas which leads to multiplication of inoculum of soil borne disease, which is an important factor for increasing the extent of crop losses. Therefore, the total number of diseases may not change, but there might be some changes in relocation and diversification of maize diseases in future in India. Boland *et al.* (2004) also considered stalk rots of maize, particularly related to sweet corn under temperate climatic conditions. The importance of *Fusarium*/*Gibberella* stalk rots may increase in sweet corn, whereas Pythium stalk rot might decrease.

Effects on ear/cob rots and related mycotoxin contamination

Mycotoxin contamination is truly an alarming issue, as ear rot is associated with mycotoxin contamination with negative human and animal health consequences.

The potential risk may be expected to increase in a future climate change scenario. If, temperature, drought, and insect injury in subtropical and tropical regions would increase, an increase of *Aspergillus flavus* and *Fusarium verticillioides* may occur. Aflatoxins, produced by *A. flavus* and *A. parasiticus* are hepato carcinogenic in humans. Fumonisins produced by *F. verticillioides* and *F. proliferatum* are classified as Group 2B 'possibly carcinogenic to humans' with a possible link to increased oesophageal cancer (Wild and Gong, 2010). DON and ZEA are also harmful to human and animal health, but appear to be less acute toxic than aflatoxin and fumonisin (Murphy *et al.*, 2006).

If hot humid weather prevailed at the time of critical stage of maize crop like in North east, some part of Western hills with limited sunshine hours at the maturity of the maize crop are predisposed to mycotoxin contamination. Sometimes due to unpredictable weather condition the *Rabi* and *Kharif* maize crop faces the same situation and ultimately spoiled due to mycotoxin contamination. Extreme dry condition at the time of flowering is also one of the predisposing factors to AFB_1 contamination. Although maize is grown mainly in wet, hot climates, it has been said to thrive in cold, hot, dry or wet conditions. Hence, in future there are more chances of mycotoxin contamination due to rise in temperature and unpredictable rainfall in Indian subcontinent.

Suggestions and new management strategies

In present changing climate and shift in seasons, choice of crop management practices based on the prevailing situation is important. In such scenarios, weather-based disease monitoring may play a significant role. In such a climate change scenario, there is an urgent need to predict which maize diseases are likely to increase in importance but this requires crop-disease-weather interaction models.

Breeding Climate-Resilient Cultivars

In order to minimize the impacts of climate and other environmental changes, it will be crucial to breed new varieties for improved resistance to abiotic and biotic stresses such as resistant to cold and heat stress, as well as for drought and water logging condition. Considering erratic monsoon, late onset and/or shorter duration of winter, there is chance of delaying and shortening the growing seasons for certain *Rabi*/cold season crops.

Rescheduling of Crop Planting

Due to increase in global temperature and altered rainfall patterns may result in shrinking of crop growing seasons with intense problems of diversification and relocation of maize diseases. Hence there is need to rescheduling crop planting dates as per suitability to maize crop in changing environment scenario. Maize farmers/growers have to change disease management strategies in accordance with the projected changes in disease incidence and extent of crop losses in view of the changing climate.

Sensitization of farmers/stakeholders

Considering the impacts of future climate change on sustainability and productivity of maize, there is an urgent need to sensitize the growers/farmers, stakeholders, extension workers, about major diseases at zonal and regional level and management strategies to cope with the situation. Sensitization of farmers, industries and exporters about the importance of mycotoxin and their management is also needed. This can be achieved through organization of awareness campaigns, training and capacity-building programmes, development of learning material and support guides for different risk scenarios of pest, etc.

Screening of Pesticides with Novel Mode of Actions

Screening of nano-pesticides against important maize diseases is needed. The application of zinc oxide nano particles against disease powdery mildews has antifungal activity (Lamsal *et al.*, 2011; He *et al.*, 2010). Salicylic acid which is associated with plant defense responses enhance plant vigour and abiotic stress tolerance (Gonias *et al.*, 2003; Theilert, 2006; Horii *et al.*, 2007; Chiriboga *et al.*, 2009; Ford, 2010). This gives an insight into role of new molecules in enhancing stress tolerance in plants.

Summary

Considering the declining production efficiency of agro-ecosystems due to depleting natural resource base, serious consequences of climate change on diversity and abundance of pests and the extent of crop losses, food security for 21st century is the major challenge for human kind. Being a tropical country, India is more challenged with impacts of looming changes. We need to urgently take steps to increase our adaptive ability. This would require increased support to adaptation research, developing regionally differentiated contingency plans for temperature and rainfall related risks, enhanced research on seasonal weather forecasts and their applications for reducing risks, and evolving new land use systems, including heat and drought tolerant varieties which are, adapted to climatic variability and climate change. Dealing with the climate change is a tedious task owing to its complexity, uncertainty, unpredictability and differential impacts over time and place. However impact of climate change on crop production mediated through changes in disease and pest dynamics need to be given careful attention for planning and devising adaptation and mitigation strategies for future pest management programmes.

References

Adipala, E., Lipps, P.E. and Madden, L.V. (1993). Occurrence of *Exserohilum turcicum* on maize in Uganda. *Plant Diseases.* 77: 202-205.

Anonymous. (2003). Forty-sixth Ann. Prog. Rep. of *Kharif.* All India Coordinated Maize Improvement Project, IARI, New Delhi. pp.1-62.

Babu, R., Mani, Pandey, A.K., Pant, S.K. and Singh, R. (2004). Maize Research at Vivekanand Parvatiya Krishi Anusandhan Sansthan, An Overview, *Techn. Bull*, Vivekanand Parvatiya Krishi Anusandhan Sansthan, Almora. 21: 31.

Barclay, A. (1891). Additional uredineae from the neighborhood of Simla. *J Asiatic Soc Bengal.* 60: 214.

Barua, P. (1979). *Studies on Some Aspect of Banded Sclerotial Disease of Maize Caused by Rhiznctoni solani* f. *sp. sasakii. M.Sc. Thesis*. G.B. Pant Univ. of Agric. and Tech., Pantnagar. pp. 56.

Bertus, L. (1927). A sclerotial disease of maize *(Zea mays* L.) due to *Rhizoctonia solani* Kuhn. *Year Book Deptt Agric Ceylon.* pp. 46-48.

Boland, G.J., Melzer, M.S., Hopkin, A., Higgins, V. and Nassuth, A. (2004). Climate change and plant diseases in Ontario. *Can J Plant Pathol.* 26: 335-350.

Butchaiah, K. (1977). *Studies on Banded Sclerotia! Disease of Maize. M.Sc. Thesis*. G.B. Pant Univ. of Agric. and Tech., Pantnagar. pp. 65.

Butler, E.J. and Bisby, G.R. (1931). The Fungi of India. Scientific Monograph No. 1. The Imperial Council of Agricultural Research. Government of India Central Publication Branch, Calcutta, India.

CABI. (1999). Late Wilt. Crop Protection Compendium. CAB International CD-Rom, Wallingford. UK.

Chakraborty, S., Murray, G. and White, N. (2002). Potential impact of climate change on plant diseases of economic significance to Australia. *Australas Plant Pathol.* 27: 15–35.

Chenulu, V.V. and Hora, T.S. (1962). Studies on losses due to *Helminthosporium* blight of maize. *Ind Phytopath.* 15: 235-237.

Chiriboga, A., Herms, D.A. and Royalty, R.N. (2009). Effects of imidacloprid (Merit®2F) on physiology of woody plants and performance of two spotted spider mite, fall webworm, and imported willow leaf beetle. Bayer Environmental Science. https://kb.osu.edu/dspace/bitstream/handle/1811/36552/ACHIRIBOGA %20OARDC% 20Poster% 20Apr09.pdf, Accessed on 06/04/2011.

Cramer, H.H. (1967). Plant Protection and World Crop Production. Bayer, Leverkusen. Germany.

Desai, S. and Hegde, R.K. (1991). A preliminary survey of incidence of stalk rot complex of maize in two districts of Karnataka. *Ind Phytopath.* 43(4): 575-576.

Dey, S.K., Dhillon, B.S. and Khera, A.S. (1992). Diseases of maize in Asia-Problem and Progress. *In Progress of Plant Pathological Res.* (Eds. S.S. Sokhi; *et. al.*) pp. 88-108.

Dodd, J.L. (1980). Grain sink size and predisposition of *Zea mays* to stalk rot. *Phytopath.* 70: 534-535.

Doko, M.B., Rapior, S., Visconti, A. and Schjoth, J.E. (1995). Incidence and levels of fumonisin contamination in maize genotypes grown in Europe and Africa. *J Agric Fd Chem.* 43: 429-434.

El-Shafey, H.A. and Claflin L.E. (1999). Late Wilt. *In*: D. G. White, (ed.), Compendium of Corn Diseases (3rd). St. Paul: APS Press. pp. 43-44.

Ford, K.A., Casida, J.E. and Chandranb, D. (2010). Neonicotinoid insecticides induce salicylate associated plant defense responses. *PNAS.* 107: 17527-17532.

Frederekson, R.A., Craig, J. and Smith, J.D. (1990). Characterization of slow rusting components in maize inbreds and single crosses. *Phytopathology*. 80: 1069.

Gautam, H.R., Bhardwaj, M.L. and Kumar, R. (2013). Climate change and its impact on plant diseases. *Current Sci.* 105(12): 25.

Gevers, H.O. (1975). A new major gene for resistance to *Helminthosporium turcicum* leaf blight of maize. *Plant Dis Rep.* 59: 296-300.

Gonias, E.D., Oosterhuis, D.M., Bibi, A.C. and Brown, R.S. (2003). Yield, growth and physiology of TrimaxTM Treated Cotton. *Summaries of Arkansas Cotton Res.* Univ. Arkansas, USA. pp. 139-144.

Gowda, P.K.T., Gowda, B.J. and Rajasekharaiah, S. (1989). Variability in the incidence of turcicum leaf blight of maize in southern Karnataka. *Current Res.* 18: 115-116.

Gowda, P.K.T., Lal, S., Shekhar, M., Mani, V.P. and Singh, W.W. (1994). Additional source of resistance in Maize to *Exserohilum turcicum*. *Ind J Agric Sci.* 64: 498–500.

Gupta, U. (1981). Assessment of losses in grain yield due to maize rust (*P. sorghi* Schw.) Seventh Annual Progress Report of Pathological Research on Diseases of maize presented at *Rabi* Workshop. New Delhi.

He, L., Liu, Y., Mustapha, A. and Lin, M. (2010). Antifungal activity of zinc oxide nanoparticles against *Botrytis cinerea* and *Penicillium expansum*. *Microbiological Res.* DOI: http://dx.doi.org/10.1016/j.micres. 2010.03.003.

Hingorani, M.K., Grant, U.J. and Singh, N. (1959). *Erwinia carotovora* f.sp. *zeae*, destructive pathogen of maize of India. *Ind Phytopath.* 12: 151-157.

Hooker, A.L. (1961). A new type of resistance in corn to *Helminthosporium turcicum* in seedling corn. *Plant Dis Rep.* 45: 780-781.

Hooker, A.L. (1962). Additional sources of resistance to *Puccinia sorghi* in the United States. *Plant Dis Reptr.* 46:14-16.

Hooker, A.L. (1963a). Inheritance of chlorotic-lesion resistance to *Helminthosporium turcicum* in seedling corn. *Phytopath.* 53: 660-662.

Hooker, A.L. (1963b). Monogenic resistance of *Zea mays* L. to *Helminthosporium turcicum*. *Crop Sci.* 3: 381-383.

Horii, A., McCue, P. and Shetty, K. (2007). Enhancement of seed vigour following insecticide and phenolic elicitor treatment. *Biores Tech.* 98: 623-632.

Huang, Y.J., Evans, N., Li, Z.Q., Eckert, M., Chevre, A.M., Renard, M. and Fitt, B.D.L. (2006). Temperature and leaf wetness duration affect phenotypic expression of Rlm6-mediated resistance to *Leptosphaeria maculans* in *Brassica napus*. *New Phytol.* 170: 129-141.

Hulbert, S.H., Richter, T.E., Axtell, J.D. and Bennetzen, J.L. (1990). Genetic mapping and characterization of sorghum and related crops by means of maize DNA probs. *Proc Natl Acad Asc USA*. 87: 4251.

Jain, K.L., Dange, R.A.S. and Kothari, K.L. (1975). Stalk rots of maize in Rajasthan. *Farmers Parliament*. 9: 21-22.

James, W.C. (1981). Estimated losses in crops from plant pathogens. In: *Handbook of Pest* Management in Agriculture. I (Ed. By D. Pimentel) pp. 79-84. *Agricultural Sciences – Vol. II - Pest Control in World Agriculture*. Boca Rato. Florida.

Johal, L., Huber, D.M. and Martyn, R. (2004). Late wilt of corn (maize) pathway analysis: Intentional introduction of *Cephalosporium maydis*. *In*: Pathways Analysis for the Introduction to the U.S. of Plant Pathogens of Economic Importance. United States Department of Agriculture, Animal and Plant Health Inspection Service. *Technical Report* No. 503025.

Juroszek, P. and Tiedemann, A. (2011). Potential strategies and future requirements for plant disease management under a changing climate. *Plant Pathol.* 60: 100-112.

Kaiser, S.A.K.M. and Mukherjee, N. (1979). Stalk rot complex of maize in West Bengal and their management. *Ind J Mycol Res.* 17(2): 77-83.

Kaul, T.N. (1957). Outbreaks and new record. *FAO Plant Prot Bull.* 5: 93-96.

Kumar, S., Gowda, P.K.T., Pant, S.K., Shekhar, M., Kumar, B., Kaur, B., Hettiarachchi, K., Singh, O.N. and Parsanna, B.M. (2011). "Sources of Resistance to *Exserohilum turcicum* (Pass) and *Puccinia polysora* (Underw) Incitant of Turcicum Leaf Blight and Polysora rust of maize in addition to variability among isolates of *E. turcicum*". *Archives Phytopath Plant Protection*. 44(6): 528–536.

Lal, S., Baruah, P. and Butchaiah, K. (1980). Assessment of yield losses in maize cultivars to banded sclerotial disease. *Ind Phytopath*. 33: 440-443.

Lamsal, K., Kim, S.W., Jung, J.H., Kim, Y.S., Kim, K.S. and Lee, Y.S. (2011). Inhibition effects of silver nanoparticles against powdery mildews on cucumber and pumpkin. Mycobiology. 39(1): 26–32.

Levy, Y. (1984). Overwintering of *Exserohilum turcicum* in Israel. *Phytoparasitica*. 12: 177-182.

Lim, S.M., Kinsey, J.G. and Hooker, A.L. (1974). Inheritance of virulence in *Helminthosporium turcicum* to monogenic resistant corn. *Phytopath*. 64: 1150–1151.

Mboup, M., Bahri, B., Leconte, M., De Vallavieille-Pope, C., Kaltz, O., and Enjalbert, J. (2012). Genetic structure and local adaptation of European wheat yellow rust populations: the role of temperature-specific adaptation. *Evol Appl.* 5: 341-352.

Murphy, P.A., Hendrich, S., Landgren, C. and Bryant, C.M. (2006). Food mycotoxins: an update. *J Food Sci* .71: R51-65.

Pant, S.K., Kumar, P. and Chauan, V.S. (2001). Effect of Turcicum leaf blight on photosynthesis in maize. *Ind Phytopath*. 54: 251-252.

Payak, M.M. (1992). Interception of *Puccina polysora* Southern rust of maize in India, NBPGR, IARI, New Delhi. pp.1-5.

Payak, M.M. and Renfro, B.L. (1966). Diseases of maize new to India. *Indian Phytopath Soc Bull.* 3: 14-18.

Payak, M.M. and Renfro, B.L. (1967). A new downy mildew disease of maize. *Phytopath.* 57:394-97

Payak, M.M. and Renfro, B.L. (1968). Combating maize diseases. *Ind Farmer Dig.* 1: 53-58.

Payak, M.M. and Sharma, R.C. (1978). Research on diseases of maize. Pl. 480 Project Final Technical Report (April 1969-March 1975), Indian Council of Agricultural Research, New Delhi. pp. 228.

Payak, M.M. and Sharma, R.C. (1983). Disease rating scales in maize in India. In: *Techniques of Scoring for Resistance to Important Diseases of Maize*. Indian Agricultural Research Institute, New Delhi. pp. 1-4.

Payak, M.M. and Sharma, R.C. (1985). Maize diseases and their approach to their management. *Trop Pest Manage.* 31: 302-310.

Payak, M.M., Lal, S., Lilaramani, J. and Renfro, B.L. (1970). *Cephalosporium maydis* -a new threat to maize in India. *Ind Phytopath.* 23(3): 562-569.

Prasad, M. and Sinha, S.K. (1977). Survival and retention of infectivity of bacterial stalk rot of maize and its perpetuation in varied cropping patterns. *Plant Soil.* 47: 245-248.

Rangarajan, M. and Chakravarti, B.P. (1971). *Erwinia carotovora* (Jones) Holland, the inciting agent of corn stalk rot in India. *Phytopathologia Mediterranea.* 10: 41-45.

Rathore, R.S. and Siradhana, B. (1987). Estimation of losses caused by *Peronosclerospora hetropogoni* on Ganga 5maize hybrid. *Phytoparasitica.*19: 119-120

Rathore, R.S., Trivedi, A. and Mathur, K. (2002). Rajasthan downy mildew of maize: the problem and management perspectives. In: Proceedings of the 8th Asian Regional Maize Workshop, Bangkok, Thailand: August 5e8, 366e378.

Richard, N., Strange and Peter, R.S. (2005). Plant Disease: A Threat to Global Food Security. *Annual Review Phytopath*. 43: 83-116.

Rodriguez–Ardon, R., Scott, G.E. and King, S.B. (1980). Maize yield losses caused by Southern Corn rust. *Crop Sci.* 20: 812–814.

Saladini, J.L., Schmitthenner, A.F. and Larsen, P.O. (1983). Prevalence of *Pythium* species associated with cottony-blighted and healthy turfgrasses in Ohio. *Plant Dis.* 67: 517-519.

Samra, A.S., Sabet, K.A., Abdel Rahim, M.F., El Shafey, H.A., Mensour, I.M., Foll, F.A., Dawood, N.A. and Jhalil, H.I. (1966). Investigation on stalk rots diseases of maize in UAR. *Maize and Sugarcane Dis. Cont Plant Prot Dept. UAR*. pp. 204.

Satyanarayana, E. (1995). Genetic studies of late wilt and turcicum leaf blight resistance in maize. *Madras Agri J.* 82: 608-609.

Saxena, S.C. and Lal, S. (1980). Assessment of yield losses due to bacterial stalk rot of maize (abstract). *J Mycol Plant Pathol.* 10.

Saxena, S.C. and Lal, S. (1984). Use of meterological factors in prediction of Erwinia stalk rots of maize. *Tropical Pest Manage*. 30: 82-85.

Sharma, R.C., Payak, M.M, Shankerlingam, S. and Laxminarayan (1982). A comparison of two methods of estimating yield losses in maize caused by common rust. *Ind Phytopath.* 35: 18-20.

Shearer, J.F., Sweets, L.E., Baker, N.K. and Tiffany, L.H. (1992). A study of *Aspergillus flavus*, *Aspergillus parasiticus* in Iowa crop fields 1988-1990. *Plant Disease*. 76: 19-22.

Shekhar, M., Kumar, S. and Sharma, S.S. (2012). Avoidable yield losses due to Post Flowering Stalk Rots in Maize Paper ID No. IAC2012/Sym-V Third International Agronomy Congress on Agriculture Diversification. Climate Change Management and Livelihoods .

Shekhar, M., Singh, N., Kumar, S. and Kiran, R. (2016). Role of mould occurrence in aflatoxin build-up and variability of *Aspergillus flavus* isolates from maize grains across India. Quality Assurance and Safety of Crops and Foods: DOI: http://dx.doi.org/10.3920/QAS2015.0720.

Singh, B.M. and Sharma, Y.R. (1979). Evaluation of germplasm to banded sclerotial disease and assment of yield losses. *Ind Phytopath.* 29: 129-132.

Singh, S.D. and Siradhana, B.S. (1987a). Influence of some environmental conditions on the development of late wilt of maize induced by *Cephalosporium maydis*. *Ind J Mycol Plant Patho.* 17: 1-5.

Singh, S.D. and Siradhana, B.S. (1987b). Survival of *Cephalosporium maydis*, incitant of late wilt of maize. *Indian J Myco Plant Patho.* 17: 83-85.

Singh, S.D. and Siradhana, B.S. (1988). Date of sowing in relation to late wilt disease of maize. *Ind Phytopath.* 41: 489-491.

Siradhana, B., Dange, S.R.S., Rathore, R.S. and Singh, S.D. (1980). A new downy mildew of maize in Rajasthan, India. *Current Sci.* 49: 316–317.

Srivastava, D.N. and Rao, V.R. (1964). Pythium stalk rot of corn in India. *Current Sci.* 33: 119-120.

Thakur, R.P., Leonard, K.J. and Jones, R.K. (1989). Characterization of a new race of *Exserohilum turcicum* virulent on corn with resistance gene *HtN*. *Plant Diseases.* 73: 151-155.

Theilert, W.A. (2006). Unique product: The story of the imidacloprid stress shield. *Pflanzenschutz-Nachrichten Sci Forum Bayer.*59: 73-86.

Thind, B.S. (1970). Investigations on bacterial stalk rot of maize (*Erwinia carotovora* var. *zeae* Sabet). *Ph.D thesis*, Indian Agricultural Research Institute, New Delhi. pp.113.

White, D.J. (1999). Fungal stalk rots. Compendium of Corn Diseases 3rd Edition. D.G. White ed. APS Press, St. Paul. M.N.

Wild, C.P. and Gong, Y.Y. (2010). Mycotoxins and human disease: a largely ignored global health issue. *Carcinogenesis*. 31: 71-82.

Zhan, J. and McDonald, B.A. (2011). Thermal adaptation in the fungal pathogen *Mycosphaerella graminicola. Mol Ecol.* 20: 1689-1701.

22

Recent Innovations in Maize Research for Development

B.M. Prasanna

Global Maize Grogram, CIMMYT & CGIAR Research Program MAIZE
ICRAF House, UN Avenue, Gigiri, Nairobi 00621, Kenya
Corresponding Author's E-mail: b.m.prasanna@cgiar.org

The ability to fast-track development and delivery of improved maize cultivars with capacity to provide climate-resilience (high and stable yields in stress-prone environments), coupled with improved nutritional quality, will be critical for sustainable intensification and diversification of maize-based cropping systems, and for providing enhanced income opportunities to the smallholders. This warrants coordinated multi-institutional efforts and implementation of innovative strategies to further increase genetic gains through breeding, and to translate these gains in the farmers' fields. This article will focus only on the novel tools/technologies/strategies that are being employed by CIMMYT for increasing genetic gains, especially for grain yield under stress-prone environments in the tropical/sub-tropical regions of sub-Saharan Africa, Latin America and Asia.

Novel tools/techniques/strategies for increasing genetic gains in the tropics

"Genetic gain" (also called as selection response) refers to the difference in the mean value of the selection criterion between original generation and next generation (formed from only the selected individuals), when they were compared in the same environment. The selection criterion is the trait(s) on which selection is based. While conventional breeding has been successful in improving maize yields over the decades, and especially in the last one decade in India through single-cross hybrids, the significantly rising demand for maize in the tropical regions of sub-Saharan Africa and Asia warrants sharp increases in genetic gains. Thus, the major challenge for the present generation of product development teams is to double the genetic gains for grain yield under stressed environments, i.e., from the present <1% to at least 1.5% within the next 5-6 years.

This requires maize breeding programs in Africa and Asia to quickly acquire capacity to adapt and integrate an array of novel tools/technologies/strategies (Figure 1) that enable rapid genetic gains and improvements in breeding efficiency. These include doubled haploid (DH) technology, high-throughput and precise field-based phenotyping, genomics-assisted breeding (including molecular marker-assisted selection, and rapid-cycle genomic selection with year-round nurseries), mechanization/automation of breeding operations, breeding management system and decision support tools, and transgenic technology for prioritized traits through public-private partnerships.

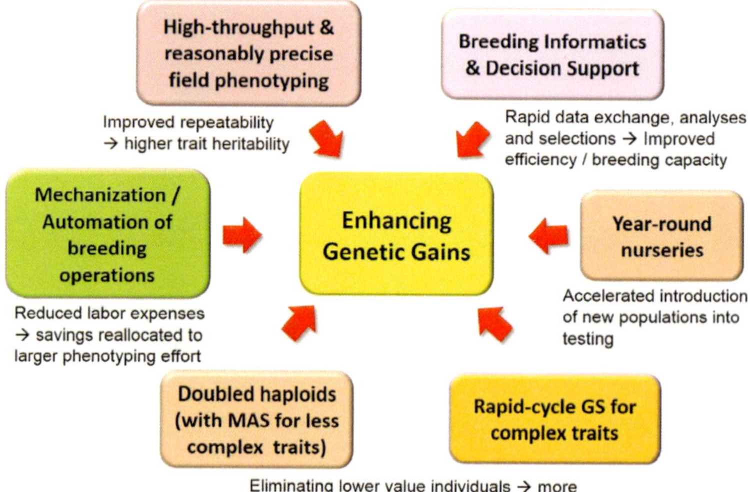

Fig. 1. Enhancing genetic gains in maize through integration of multiple components in breeding programs

a) Doubled haploid (DH) technology in maize breeding

Development of stable and productive inbred/parental lines to produce hybrid seed is the cornerstone of successful and affordable hybrid maize technology. The *in vivo* maternal haploid induction in desired source populations, haploid identification, chromosome doubling, and derivation of genetically homozygous DH lines is now an integral part of well-established large-scale maize breeding programs (Prasanna *et al.* 2012). The DH technology enhances "forward breeding" and provides an opportunity to have an earlier look at the potential of new lines, greater knowledge about their environmental adaptability before they are fully tested, and further used as parental lines for hybrid development and commercial cultivation. Use of DH technology can also enhance the efficiency of recurrent selection or genomic selection based schemes for traits with low heritability, particularly for breeding programmes without access to offseason nurseries (Bouchez and Gallais, 2000). Furthermore, the DH technology enables shifting of resources from the labour-intensive task of repeated inbreeding to generate inbred lines, and spending more time on evaluation of the DH lines for yield and other adaptive traits, and using the identified lines for producing high-yielding and climate-resilient hybrids.

In collaboration with the University of Hohenheim, Germany, since 2007, CIMMYT has developed first-generation tropicalized haploid inducers (with haploid induction rate 8-10%), by introgressing the haploid induction trait from the temperate haploid inducers into CIMMYT's improved maize lines (Prasanna *et al.* 2012). These tropicalized haploid inducers have been shared with several public and private sector institutions worldwide (including India), for research or commercial use under specific terms and conditions. CIMMYT team has also optimized the DH protocol, including various steps along the DH development process (Prasanna *et al.* 2012), and also fine-mapped a major genomic region (*qhir1*) that confers maternal haploid induction in maize (Nair *et al.* 2017). In 2017, CIMMYT Global Maize Program will announce the availability of second-generation tropicalized phaploid inducer lines with high haploid induction rate (10% or more), better plant vigour and pollen-producing capacity (as compared to first-generation inducers), besides disease resistance. These second-generation haploid inducers will further widen the application of DH technology in maize breeding programs in the tropics.

The *R1-nj* (Navajo) phenotypic marker, which is widely used in the haploid inducers worldwide for detection of haploids, is ineffective in some maize germplasm (especially landraces) with natural anthocyanin expression in pericarp tissue (masking the Navajo marker expression on the aleurone layer of the the endosperm). To overcome this limitation, CIMMYT team developed haploid inducer lines with triple anthocyanin color markers, including the expression of anthocyanin coloration in the seedling roots and leaf sheaths, in addition to the Navajo marker on the seed (Chaikam *et al.* 2016). We are at present improving the agronomic performance of these red root marker-based haploid inducers especially for tropical environments.

CIMMYT, in partnership with Kenya Agriculture and Livestock Research Organization (KALRO) has also established a centralized, state-of-the-art maize DH facility at Kiboko (Kenya) in 2013, for strengthening maize breeding programs, including those of national agricultural research systems (NARS) and small and medium enterprise (SME) seed companies in sub-Saharan Africa. The facility is now producing >60,000 DH lines every year. A similar centralized maize DH facility is proposed by CIMMYT for establishment in India (in partnership with the Indian Council of Agricultural Research (ICAR), to cater to the needs of the NARS, SME seed companies (with maize breeding programs) across Asia, besides CIMMYT. Such a centralized DH facility, established jointly by ICAR and CIMMYT at an appropriate location in India, and operated with technical guidance and professional management by CIMMYT, can help achieve economy of scale (similar to the maize DH platform at Kiboko-Kenya), as this is critical for cost-efficiency of maize DH operations and continuous process improvement.

Beyond DH line development, it is important for the maize breeding programs to develop and implement an effective selection strategy, especially using molecular markers for key traits. This is important to reduce the breeding and product development costs, and improve the efficiency of breeding programs. This is being implemented at CIMMYT, especially in sub-saharan Africa, with remarkable success in the past few years. The result: 36 of the 40 high-yielding, drought-tolerant and disease-resistant maize hybrids derived by CIMMYT and released through partners in Kenya, Tanzania and Uganda during 2015-2016, had DH lines among the parental lines of these hybrids.

b) High-throughput and precise field-based phenotyping

Field-based phenotyping is costly and remains a major bottleneck for breeding progress. High-throughput field-based phenotyping of selected traits, using low cost, easy-to-handle proximal and remote sensing tools, is now possible (Prasanna *et al.* 2013; Araus and Cairns, 2014; Zaman-Allah *et al.* 2015; Vergara-Diaz *et al.* 2016), and should become an integral and key component of the maize breeding programs. There is also a distinct need for the public and private institutions to come together and establish abiotic and biotic stress phenotyping networks for comprehensive and efficient characterization of genetic resources and breeding materials for prioritized target traits.

Under the CGIAR Research Program MAIZE, CIMMYT, in partnership with the University of Barcelona (Spain), AirElectronics (Spain), Consejo Superior de Investigaciones Científicas (Spain), and Crop Breeding Institute (Zimbabwe), developed low-cost remote-sensing platform for phenotyping an array of important traits that help analyse responses of maize genotypes to drought, heat and low soil nitrogen stresses. Zaman-Allah *et al.* (2015) and Vergara-Diaz *et al.* (2016) demonstrated the utility of multispectral imaging and RGB-based phenotyping, respectively, in prediction of grain yield of maize under different conditions of nitrogen fertilization.

Beyond such technological advances in high-throughput field-based phenotyping for key traits, there is also an immense need for measuring and reducing the effects of field variability, and thereby increasing the genetic signal-to-noise ratio to detect real differences between genotypes. CIMMYT is making intensive efforts for characterizing field variability at the key phenotyping sites worldwide, and for improving heritability of data generated through field-based phenotyping (Prasanna *et al.* 2013). Furthermore, CIMMYT physiology and abiotic stress phenotyping team has recently brought out technical manuals on phenotyping for drought (Zaman-Allah *et al.* 2016), heat (Zaidi *et al.* 2016a), and water-logging (Zaidi *et al.* 2016b) stresses.

c) Genomics-assisted breeding

Molecular marker-assisted or genomics-assisted breeding is the way forward in effectively meeting the greater challenge of developing cultivars with combinations of relevant adaptive traits, including biotic and abiotic stress tolerance, and nutritional quality. With the rapid reduction in genotyping costs, new genomic selection technologies have become available that allow the maize breeding cycle to be greatly reduced, facilitating inclusion of information on genetic effects for multiple stresses in selection decisions.

Molecular markers can be used to speed-up the development and deployment of improved germplasm in different ways, including marker-assisted backcrossing (MABC), marker-assisted recurrent selection (MARS), and genomic selection (GS). MABC is the simplest form of marker assisted selection (MAS) used to transfer one or few genes or major effect quantitative trait loci (QTL) that are fine-mapped and validated across different genetic backgrounds. MARS is a marker-based breeding method that seeks to accumulate favorable alleles from several genomic regions within a single population (Edwards and Johnson 1994). In contrast to MABC, MARS aims at accumulating relatively large number of medium-effect QTLs in a given population using a subset of markers that are significantly associated with target traits (Bernardo 2008). Every MARS population is handled independently so QTL information generated in one population is not often transferable into another population. Genomic selection or GS (Meuwissen *et al.* 2001) is another marker-based strategy that incorporates all available marker information simultaneously into a model to predict the genetic value of progenies for selection (Lorenz 2013). Each marker is considered a putative QTL, reducing the risk of missing small-effect QTLs (Guo *et al.* 2012).

Molecular marker-assisted backcrossing (MABC)

One of the successful examples of MABC for maize improvement is the utilization of *opaque2*-specific simple sequence repeat (SSR) markers in conversion of improved maize lines into Quality Protein Maize (QPM) lines with enhanced nutritional quality (Babu *et al.* 2005; Gupta *et al.* 2009; Prasanna *et al.* 2010). A MAS-derived QPM hybrid, 'Vivek QPM Hybrid 9' has been released by the Vivekananda Parvatiya Krishi Anusadhan Sansthan (VPKAS) at Almora, India.

This QPM hybrid was developed through marker-assisted transfer of the *o2* gene and phenotypic selection for endosperm modifiers in the parental lines (CM145 and CM212) of Vivek Hybrid 9. This strategy has been successfully used to develop QPM versions of several elite, early maturing inbred lines adapted to the hill regions of India (Gupta *et al.* 2009), as well as QPM versions of elite inbred lines that are the parents of three single-cross hybrids (Hossain *et al.* manuscript under review).

Another application of MAS in maize is improving the provitamin A content of maize lines. Quantifying the provitamin A content of maize samples is difficult, time-consuming and expensive, and breeding programs, therefore, benefit greatly from use of MAS for provitamin A trait that reduces the need for phenotypic assays during MABC. Collaborative research work under HarvestPlus led to the detection of allelic variation for key genes in carotenoid biosynthetic pathway, and development of breeder-ready markers for favourable alleles, especially for two key genes in the carotenoid biosynthetic pathway, namely *LycE* (*Lycopene epsilon cyclase*) and *CrtRB1* (*Carotene beta-hydroxylase 1*) (Harjes *et al.* 2008; Yan *et al.* 2010). Molecular marker-assisted selection for *CrtRB1* favorable allele is an integral part of CIMMYT's provitamin A breeding program (Babu *et al.* 2013). This has already led to deployment of 3 first-generation provitamin A-enriched maize varieties (with 6-8 ppm provitamin A) in Zambia in 2013, and 8 second-generation provitamin A-enriched maize varieties (with 9-14 ppm provitamin A) in Zambia (3), Malawi (4), and Zimbabwe (1) during 2015-2016. The same strategy was followed by the ICAR-IARI Maize Team in India, leading to the development of a provitamin A-enriched versions of seven inbred parents of four commercial maize hybrids released by the public sector; the trait donors used in MABC were from CIMMYT-HarvestPlus Program (Muthuswamy *et al.* 2014).

MAS for disease resistance

Besides the two above-mentioned nutritional quality traits, another recent and successful example with regard to MAS is for resistance to Maize Streak Virus (MSV), a major disease that affects maize productivity in several countries in sub-Saharan Africa. CIMMYT team has fine-mapped and identified SNP markers for a major QTL for MSV resistance (*Msv1*) and validated these markers on a set of DH lines that were phenotyped for responses to MSV in different locations in sub-Saharan Africa (Nair *et al.* 2015). Marker-assisted forward breeding as well as MABC are now being routinely implemented by CIMMYT maize in Africa, using a three-marker haplotype for *msv1* selection.

CIMMYT team in Africa also discovered important genomic regions associated with resistance to the deadly Maize Lethal Necrosis (MLN) disease in eastern

Africa (Gowda *et al.* 2015). Genome association study (GWAS) of 690 CIMMYT lines derived from four parents revealed that chr. 3, 6 and 7 harbor most of the genomic regions associated with MLN resistance. The results were further validated on an array of biparental mapping populations. Based on these studies, we have implemented massive work on MABC for conversion of 43 CIMMYT's Africa-adapted elite but MLN-susceptible parental lines of widely-used commercial hybrids in Africa, using 8 MLN-resistant donor lines. As many as 473 new versions of lines (BC_4F_3) are in final stages of testing. In addition, 22 MABC projects (with KS23-5 and KS23-6 as trait donors) are presently in BC_3F_1.

CIMMYT is also employing joint GWAS (genome-wide association study) and linkage mapping approach for identifying genomic regions and breeder-ready markers for resistance to an aray of major diseases affecting maize production and productivity in the tropics, including Turcicum leaf blight (TLB), Grey leaf spot (GLS), Maize lethal necrosis (MLN), Tar spot complex (TSC), Corn stunt complex (CSC), and Post-flowering stalk rots (PFSR).

MARS and rapid-cycle genomic selection (RCGS) for complex traits

Several of the abiotic stress tolerance traits in maize follow polygenic inheritance, and several haplotypes contributing to trait improvement. MARS and RCGS are powerful strategies for enhancing genetic gains in breeding for complex traits. These strategies rely on the availability of low-cost, high-throughput genotyping platforms. CIMMYT Global Maize Program teams in sub-Saharan Africa and Latin America, with highly encouraging results, have recently undertaken pilot projects on implementation of RCGS using appropriate marker densities.

Recent comparative studies on pedigree selection, MARS and GS undertaken in sub-saharan Africa demonstrated the advantage of both MARS and RCGS schemes over pedigree selection in most populations. As part of the Drought Tolerant Maize for Africa (DTMA) and Water Efficient Maize for Africa (WEMA) projects, CIMMYT has developed more than 34 bi-parental populations since 2009. Testcrosses of progenies derived from these populations were evaluated in 2-4 managed drought and 3-4 well-watered environments, followed by advancing the populations using either MARS or GS approaches. Eight of the 34 bi-parental populations were improved using GS. The results demonstrated the efficiency of GS in maize, with an average gain per cycle of 0.086 Mg ha^{-1} under drought stress without significant changes in maturity and plant height. Higher gains in grain yield (GY) under drought stress were achieved using GS compared to conventional pedigree selection. Moreover, GS offered the advantage of significant time-savings over conventional breeding methods, since up to three cycles of GS could be conducted within a year (Beyene *et al.*, 2015).

Beyene *et al.* (2016) estimated genetic gains in GY using MARS with selection for a set of <200 SNPs (tagging QTL influencing drought tolerance) in 10 biparental tropical maize populations. A genetic gain study was conducted using testcrosses of lines from the different cycles F_1 and founder parents. Test crosses, along with five commercial hybrid checks were evaluated under four water-stressed (WS) and four well-watered (WW) environments. The overall gain for GY using MARS across the 10 populations was 105 kg ha^{-1} yr^{-1} under WW and 51 kg ha^{-1} yr^{-1} under WS. Results from this study demonstrated the potential of MARS for increasing genetic gain under both drought and optimum environments in Africa.

One of the most important applications of GS in maize breeding is to predict and identify the best untested lines from biparental populations, when the training and validation sets are derived from the same cross. The CIMMYT maize team at Meico evaluated 19 tropical biparental populations in multienvironment trials to assess prediction accuracy of different quantitative traits using low-density (~200 markers) and genotyping-by-sequencing (GBS) SNPs, respectively (Zhang *et al.* 2015). The study showed that: a) low-density SNPs (~200 markers) were largely sufficient to get good prediction in biparental maize populations for simple traits with moderate-to-high heritability, but GBS outperformed low-density SNPs for complex traits and simple traits evaluated under stress conditions with low-to-moderate heritability; and b) heritability and genetic architecture of target traits affected prediction performance.

d) Mechanization of breeding operations

Recent trend in large-sized maize breeding programs has been to increase row numbers per breeder and research support staff, increase specialization, decrease staff costs through mechanization/automation, and increase the scale of operations. This has resulted in increased genetic gain through reduced cycle time, increased volume of test materials, reduced costs in some cases, and general operational efficiency.

CIMMYT has been focusing on mechanization/autiomation of maize breeding operations, especially at its breeding hubs in Mexico, Kenya, Zimbabwe and India; this includes: (i) planning field trials and breeding nurseries; (ii) automated seed preparation and seed treatment; (iii) mechanized planting; (iv) state-of the art phenotyping, data collection and trait scoring; (v) fully mechanized harvest; and (vi) modern post-harvest operations. Maize breeding programs in many countries, especially in the public sector as well as in seed companies, currently invest very little on mechanization of breeding operations. A good start is to improve mechanization in the trialling and nursery operations, incuding automated seed preparation, electronic data collection, planters, and high throughput ear shellers.

e) Breeding management system and decision support tools

A modern maize breeding program, with strong trait- and product-pipelines, requires an effective breeding management system with a set of decision support tools to optimize on time, cost and genetic gain. For coordinated workflows, standardized software tools are needed for processing large-volumes of high-throughput, field-based phenotypic data, genotypic analyses and prediction, molecular marker-assisted forward breeding, and rapid-cycle genomic selection. This will enable: (a) effectively integrating pedigree, genotypic, phenotypic and environmental data; (b) selecting desirable recombinants through an optimum combination of phenotypic and genotypic information; and (c) developing a breeding program that minimize population sizes, number of generations and overall costs, while maximizing genetic gain for traditional and novel target traits (Prasanna *et al.* 2014).

Transgenic technology and public-private partnerships

Transgenic technology is now a major part of maize cultivars in the developed world, especially USA. Weeds and insect-pests are two serious factors impacting maize production and productivity in the tropics, and can be effectively tackled through the transgenic technology. Effective transgenic technologies for weed management and stem borer control in maize are already available and are being used by maize farmers in other countries on a large scale, including countries like Philippines and Vietnam. South Africa is the only country in Africa that is presently commercializing insect-resistant (*Bt*) maize. Adaptive trials of herbicide-tolerant and *Bt* maize cultivars were also going on for the past 10 years in India. India, however, needs a practical and futuristic policy on GM crops, with a timely and objective assessment of the GM technologies, so as to provide appropriate technological options to the Indian farmers (Yadav *et al.* 2016).

Public-private partnerships are important for effective development and deployment of improved transgenic maize cultivars for prioritized traits, such as control of weeds, and resistance to stem borers. Under the Water Efficient Maize for Africa (WEMA) project, with partners including African Agricultural Technology Foundation (AATF), Monsanto, CIMMYT and NARS partners from Kenya, Uganda, Tanzania, Mozambique and Soyth Africa, significant progress has been made in terms of demonstrating the efficacy of Bt and Bt + Drought Tolerant (DT) transgenic maize under confined field trials (CFTs). In this project, the Bt and Bt+DT transgenes have been provided by Monsanto under humanitarian license for the target countries in sub-Saharan Africa, and the transgenes have been introgressed in CIMMYT's Africa-adapted maize genetic backgrounds.

References

Araus, J.L. and Cairns, J. (2014). Field high-throughput phenotyping—the new crop breeding frontier. *Trends Plant Sci.* 19: 52–61.

Babu, R., Nair S.K., Kumar, A. *et al.* (2005). Two-generation marker-aided backcrossing for rapid conversion of normal maize lines to quality protein maize (QPM). *Theor Appl Genet.* 111: 888–897.

Babu, R., Palacios, N.P., Gao, S. *et al.* (2013). Validation of the effects of molecular marker polymorphisms in *LcyE* and *CrtRB1* on provitamin A concentrations for 26 tropical maize populations. *Theor Appl Genet.*. 126: 389–399.

Bernardo, R. (2008). Molecular markers and selection for complex traits in plants: Learning from the last 20 years. *Crop Sci.* 48: 1649–1664.

Beyene, Y., Semagn, K., Crossa, J., Mugo, S., Atlin, GN., Tarekegne, A., Meisel, B., Sehabiague, P., Alvarado, G., Machida, L., Olsen, M., Prasanna, B.M. and Banziger, M. (2016). Improving maize grain yield under drought stress and non-stress conditions in sub-Saharan Africa using molecular marker-assisted recurrent selection. *Crop Sci.* 56: 344-353.

Beyene, Y., Semagn, K., Mugo, S., Tarekegne A., Babu, R., Meisel, B., Sehabiague, P., Makumbi D., Olsen, M. *et al.* (2014). Genetic gains in grain yield through genomic selection in eight biparental maize populations under drought stress. *Crop Sci.* 55: 154-163.

Bouchez, A. and Gallais, A. (2000). Efficiency of the use of doubled haploids in recurrent selection for combining ability. *Crop Sci.* 40: 23–29.

Chaikam, V., Martinez, L., Melchinger, A., Schipprack, W. and Prasanna, B.M. (2016). Development and validation of red root marker-based haploid inducers in maize. *Crop Sci.* 56: 1678-1688.

Edwards, M. and Johnson, L. (1994). RFLPs for rapid recurrent selection: Analysis of molecular marker data, American Society for Horticultural Science, CSSA, Madison, WI. pp. 33–40.

Gowda, M., Das, B., Makumbi, D., Babu, R., Semagn, K.F., Mahuku, G., Olsen, M.S., Bright, J.M., Beyene, Y. and Prasanna, B.M. (2015). Genome-wide association and genomic prediction of resistance to maize lethal necrosis disease in tropical maize germplasm. *Theor Appl Genet.* 128: 1957–1968.

Guo, Z,G., Tucker, D.M., Lu, J.W., Kishore, V. and Gay, G. (2012). Evaluation of genome-wide selection efficiency in maize nested association mapping populations. *Theor Appl Genet.* 124: 261–275.

Gupta, H.S., Agrawal, P.K., Mahajan, V. *et al.* (2009). Quality protein maize for nutritional security: rapid development of short duration hybrids through molecular marker assisted breeding. *Curr Sci.* 96: 230-237.

Harjes, C.E., Rocheford, T.R., Bai, L. *et al.* (2008). Natural genetic variation in Lycopene epsilon cyclase tapped for maize biofortification. *Sci.* 319: 330-333.

Lorenz, A.J. (2013). Resource allocation for maximizing prediction accuracy and genetic gain of genomic selection in plant breeding: A simulation experiment. *Genes Genomes Genet.* 3: 481–491.

Meuwissen, T., Hayes, B.J. and Goddard, M.E. (2001). Prediction of total genetic value using genome-wide dense marker maps. *Genet.* 157: 1819–1829.

Muthusamy, V., Hossain, F., Nepolean, T., Choudhary, M., Saha, S., Bhat, J.S., Prasanna, B.M. and Gupta, H.S. (2014). Development of β-carotene rich maize hybrids through marker-assisted introgression of β-carotene hydroxylase allele. *PLOS One.* 9(12): e113583.

Nair, S., Babu, R., Magorokosho, C., Mahuku, G., Semagn, K., Beyene, Y., Das, B., Makumbi, D., Kumar, L., Olsen, M. and Prasanna, B.M. (2015). Fine mapping of *Msv1*, a major QTL for resistance to Maize Streak Virus leads to development of production markers for breeding pipelines. *Theor Appl Genet.* 128: 839-1854.

Nair, S.K., Molenaar, W., Melchinger, A.E., Prasanna, B.M. *et al.* (2017). Dissection of a major QTL *qhir1* conferring maternal haploid induction ability in maize. *Theor Appl Genet.* [In Press].

Prasanna, B.M, Babu, R., Nair, S., Semagn, K. *et al.* (2014). Molecular marker-assisted breeding for tropical maize improvement. In: Genetics, Genomics and Breeding of Maize (eds. R Wusirika, M Bohn, J Lai, C Kole), Science Publishers/CRC Press, Boca Raton, USA. pp. 89-118.

Prasanna, B.M., Araus, J.L., Crossa, J., Cairns, J.E., Palacios, N., Das, B. and Magorokosho, C. (2013). High-throughput and precision phenotyping for cereal breeding programs. In: Cereal Genomics-II (eds. PK Gupta, RK Varshney). Springer-Verlag, Dordrecht. pp. 341-374.

Prasanna, B.M., Chaikam, V. and Mahuku, G. (2012). Doubled haploid technology in maize breeding: theory and practice. Mexico, CIMMYT. pp. 50.

Prasanna, B.M., Pixley, K.V., Warburton, M. and Xie, C. (2010). Molecular marker-assisted breeding for maize improvement in Asia. *Mol Breed.* 26: 339-356.

Vergara-Díaz, O., Zaman-Allah, M.A., Masuka, B. *et al.* (2016). A novel remote sensing approach for prediction of maize yield under different conditions of nitrogen fertilization.*Frontiers in Plant Science* http://dx.doi.org/10.3389/fpls.2016.00666.

Yadav, O.P., Prasanna, B.M., Yadava, P., Jat, S.L., Kumar, D., Dhillon, B.S., Solanki, I.S. and Sandhu, J.S. (2016). Doubling maize production and productivity of India by 2025 – challenges and opportunities. *Indian J Agric Sci.* 86: 427–434.

Yan, J., Kandianis, C.B., Harjes, C.E. *et al.* (2010). Rare genetic variation at *Zea mays crtRB1* increases â-carotene in maize grain. *Nature Genet.* 42: 322-327.

Zaidi, P.H., Zaman-Allah, Z., Trachsel, S., Seetharam, K., Cairns, J.E. and Vinayan, M.T. (2016a). Phenotyping for abiotic stress tolerance in maize – Heat stress. A field manual. CIMMYT: Hyderabad, India.

Zaidi, P.H., Vinayan, M.T. and Seetharam, K. (2016b). Phenotyping for abiotic stress tolerance in maize: Waterlogging stress. A field manual. CIMMYT: Hyderabad, India.

Zaman-Allah, M., Vergara, O., Araus, J.L., Tarekegne, A. *et al.* (2015). Unmanned aerial platform-based multispectral imaging for field phenotyping of maize. *Plant Methods.* 11:35.

Zaman-Allah, M., Zaidi, P.H., Trachsel, S., Cairns, J.E., Vinayan, M.T. and Seetharam, K. (2016). Phenotyping for abiotic stress tolerance in maize – Drought stress. A field manual. CIMMYT, Mexico.

Zhang, X., Perez-Rodriguez, P., Semagn, K. *et al.* (2015). Genomic prediction in bi-parental tropical maize populations in water-stressed and well-watered environments using low density and GBS SNPs. *Heredity.* 114: 291-299.

23

Nutritional Quality Improvement of Maize in India

Firoz Hossain, Vignesh Muthusamy, Rajkumar U. Zunjare, Abhijit K. Das, Konsam Sarika, Sweta Dosad, Sunil K. Jaiswal, Rashmi Chhabra, Shankar L. Jat#, Jayant S. Bhat, Bhupender Kumar# Mukesh Choudhary, Dharam Paul Chaudhary#, Satish K. Guleria$, Narendra K. Singh@, Javaji C. Sekhar#, Vinay Mahajan#, Om P. Yadav# and Hari S. Gupta*

ICAR-Indian Agricultural Research Institute (IARI), New Delhi

 # ICAR-Indian Institute of Maize Research (IIMR), Ludhiana

$ CSK-Himachal Pradesh Krishi Vishvavidyalaya (CSKHPKV), Bajaura

@ GB Pant University of Agriculture and Technology (GBPUA&T), Pantnagar

**Corresponding Author's Email: fh_gpb@yahoo.com*

Maize occupies an important position in the world economy, and serves as an important source of food and feed (Gupta *et al.,* 2015a). Together with rice and wheat, it provides at least 30% of the food calories to more than 4.5 billion people in 94 developing countries (Shiferaw *et al.,* 2011). In India, maize is an important cereal too, and is grown on an area of 9.2 million hectares with production of 24.2 million tonnes (www.indiastat.com). The demand for cereals will continue to increase as a consequence of the expanding human population. The world will have around 7.7 billion people by 2020, and it will reach up to 9.3 billion by 2050, and the demand for maize between now and 2050 will double in the developing world (Rosegrant *et al.,* 2009). By 2025, India too would require to double the production (50 million tonnes) of maize grain to meet the domestic demand (Yadav *et al.,* 2016).

Malnutrition caused by inadequate consumption of balanced diet has emerged as one of the alarming problems especially in the under-developed and developing world. Mostly the resource-poor suffer from 'hidden hunger', a term often used to describe malnutrition caused due to micronutrient deficiencies in staple food diet (Bouis and Welch, 2010).

An estimated two billion people suffer from micronutrient deficiency, while 795 million people are undernourished (IFPRI, 2016a). Of the 667 million children under the age of five worldwide, 159 million are stunted, while 50 million do not weigh enough for their height (IFPRI, 2016b). An estimated 45% of deaths of children under age five are linked to malnutrition (Black *et al.*, 2013). Malnutrition contributes to global burden of disease, and loss in annual gross domestic product (GDP) in Asia and Africa to extent of 11%. South Asia is home to more than 35% of the world's poor, and 21.9% of the population of India lives in poverty, and thus vulnerable to various health problems (IFPRI, 2016a). 15.2% of the people in India are undernourished, and more than one third of the world's underweight malnourished children live in India, posing severe socio-economic loss to the country.

At the beginning of the new millennium, world leaders set eight 'Millennium Development Goals' (MDGs) at the United Nations, of which (i) eradication of extreme poverty and hunger (Goal 1), (ii) reduction of child mortality (Goal 4) and (iii) improvement of maternal health (Goal 5), pertain to providing healthy and nutritious food to the people worldwide. During 2015, global community further set new goals known as "Sustainable Development Goals" (SDGs) to chart a path toward meeting current human needs without compromising the ability of future generations to meet their needs. Of the 17 SDGs, 12 contain indicators that are highly relevant for nutrition, reflecting nutrition's central role in sustainable development, and improved nutrition being the platform for progress in health, education, employment, female empowerment, and poverty. It has been estimated that alleviating malnutrition is one of the most cost-effective steps with every $1 invested in proven nutrition programme offers benefits worth $16 (IFPRI, 2016b). Thus, efforts directed towards providing the balanced and nutritious food assumes great significance (Gupta *et al.,* 2015b; Yadav *et al.,* 2015).

Agricultural systems have traditionally focussed mostly on increasing productivity, but research must now formulate new policies that not only provide enough calories to meet the energy needs of the poor, but also deliver all the essential nutrients needed for adequate nutritional health (Graham *et al.,* 2007). A number of methods such as 'food fortification', 'supplementation' and 'dietary diversification' are practised all over the world for alleviating the micronutrient deficiencies. However, these measures in general have not been found viable in the long run due to various reasons. Lack of purchasing power of the poor due to poverty, restricts the access to the fortified foods, thereby reducing their efficiency and application. Poor infrastructure in developing countries has limited the widespread use of direct vitamin supplementation. Diet diversification is often limited by crop seasonality, expense, low bioavailability of green leafy plant carotenoids and lack of purchasing power by the poor (Lieshout and Pee

2005). 'Biofortification', a process in which micronutrient density in crops is increased and proposed as a sustainable and cost-effective mean for providing the required levels of micronutrients in pure form to alleviate malnutrition in humans (Bouis *et al.*, 2011). Among various micronutrients, lysine, tryptophan, vitamin A, vitamin E, iron (Fe) and zinc (Zn) found to be deficient in kernel among traditional maize varieties. Availability of natural variants of key genes provides great opportunity to develop micronutrient rich maize.

A. High lysine and tryptophan

Role in human health

Human requires 0.66 g protein/kg body weight/day to meet the requirement for proper growth and development (WHO/FAO/UNU 2007). Of the twenty amino acids that are required for protein synthesis, nine of them *viz.*, lysine, threonine, methionine, phenylalanine, tryptophan, isoleucine, leucine, valine, and histidine must be acquired through diet (Galili, 2002). Our body uses amino acids in a specific ratio to each other, and lower levels of lysine or tryptophan affect the body's ability to use other amino acids (Tome and Bos, 2007). From the human nutrition point of view, lysine followed by tryptophan is the most important limiting amino acid in the maize endosperm protein. Maize protein contains 2.0% lysine, which is less than half of the recommended dose specified for human nutrition (Young *et al.*, 1998). The daily requirement of lysine is 30 mg/kg body weight/day for adults, while it is 35 mg/kg body weight/day for children of 3 to 10 years of age. Tryptophan is required at the rate of 4 mg/kg body weight/day and 4.8 mg/kg body weight/day in adults and children, respectively (WHO/FAO/UNU, 2007). Besides playing vital role in protein synthesis, lysine and tryptophan serve as precursors for several neuro-transmitters and metabolic regulators.

Deficiency symptoms

Intake of food deficient in essential amino acids results in increased susceptibility to diseases, decreased blood proteins and eventually retarded mental- and physical- development especially among young children. Malnutrition caused due to deficiency of proteins is observed most frequently in developing countries (Bain *et al.*, 2013; Temba *et al.*, 2016). The deficiency of lysine in protein leads to fatigue, dizziness, nausea, anemia, delayed growth, loss of appetite and reproductive tissue. While depression, anxiety, impatience weight loss, slow growth in children are the symptoms of tryptophan deficiency.

Seed storage proteins

The maize grain is composed of pericarp (6%), endosperm (82%) and germ (12%). The main structural component of the endosperm is starch; it constitutes on an average 71% of the grain. Maize grains contain 8-10% protein, of which 70% of protein is composed of prolamins. Prolamins are a group of plant storage proteins and known as 'zeins' in maize, and are inherently deficient in lysine and tryptophan. The remaining fractions of protein, collectively called non-zeins, comprised of globulins (3%), glutelins (34%), and albumins (3%), are balanced in essential amino acids (Vasal, 2001). Zeins composed of α- (19- and 22- kDa), β- (15-kDa), γ- (10- and 18- kDa) and δ- (16-, 27- and 50- kDa) fractions, are deficient in lysine and tryptophan (Olsen and Phillips, 2001; Wu and Messing, 2011). However, the expression of the genes varies across different maize backgrounds. The synthesis of zeins starts about 12 days after pollination (DAP) and is most active between 16 and 35 days, and can occur up to 50 DAP (Wall and Bietz 1987). In case of maize, protein bodies (PB) are retained within the endoplasmic reticulum lumen, and a typical PB at 18-20 DAP is spherical, discrete, and has a highly ordered architecture of α- and δ-zeins which are deposited in the centre, while γ- and β-zeins are located in the peripheral layer (Lending and Larkins, 1989).

Genetic loci for higher lysine and tryptophan

Until 1960s, efforts were limited only to screening elite maize germplasm and the improvement strategy such as recurrent selection could not be easily implemented in the absence of specific genes enhancing lysine and tryptophan. The search for mutants that could alter the amino acid profile and increase the concentration of lysine and tryptophan of maize endosperm led to the discovery of several genetic mutations *viz.*, *floury1* (*fl1*), *floury2* (*fl2*), *floury3* (*fl3*), *opaque5* (*o5*), *opaque6* (*o6*), *opaque7* (*o7*), *Defective endosperm* (*Def-B30*), *Mucronate* (*Mc*) and others (Balconi *et al.,* 2007). However, none of them could be successfully used in the breeding programme. The discovery of the *opaque2* (*o2*) mutant by Jones and Singleton in the 1920s provided a significant breakthrough, as it causes enhanced accumulation of lysine and tryptophan in the endosperm (Mertz *et al.,* 1964). The gene was mapped on the chromosome 7L, and it encodes a basic leucine zipper (bZIP) transcriptional factor that regulates α-zeins synthesis. The *o2* mutant generally reduces the accumulation of 22- and to lesser extent 19-kDa α-zeins, with an overall reduction of 50-70% in zeins (Ueda *et al.,* 1992). Reduction in the α-zeins results in PB with abnormal morphology, size and number; it leads to opaque grains with soft and starchy texture. The mutant *o2* is generally accompanied by an increase in non-zeins such as cytoskeleton-associated carbohydrate metabolizing enzymes and eEF1A, which are relatively rich in lysine and tryptophan (Habben *et al.,* 1993). Furthermore, *o2* also down-regulates the synthesis of lysine ketoglutarate reductase, the enzyme that converts lysine to further components, and the loss of enzyme activity thus results in increased levels of free lysine (Kemper *et al.,*

1999). Besides, *o2* is also involved in regulation of various metabolic pathways and causes enhanced synthesis of various lysine-rich proteins and enzymes (Jia *et al.,* 2013).

Endosperm modifiers

The elation over the discovery of *o2* and its utility in breeding programme was short-lived with the discovery of pleiotropic effects such as soft endosperm resulting in damaged grains, increased susceptibility to pests and fungal diseases, and inferior food processing (Shewry and Thatam, 1990). Since grain weight is reduced due to loosely packed starch granules with larger air spaces, there is a corresponding decline in the yield which could be almost to the tune of 10% or above (Singh and Venkatesh, 2006). This grain phenotype or appearance is also a major barrier to their acceptance in the developing countries where hard/ vitreous grain is preferred. In an effort to overcome the problems associated with original, soft *o2* materials, researchers at CIMMYT, Mexico, and University of Natal, South Africa, came out with viable strategies to overcome the pleiotropic effects (Geevers and Lake, 1992; Villegas *et al.,* 1992). The inferior quality of *o2* could be overcome with the accumulation of *o2*-specific endosperm modifiers without affecting the concentration of lysine and tryptophan, and led to the birth of hard endosperm based 'quality protein maize' (QPM). The mechanism of *o2* endosperm modification in QPM is not clear. The inheritance pattern of these modifier genes is complex and likely to involve several loci, which complicates the genetic analysis of QPM. A possible role of 27-kDa γ-zein in recovering the vitreous phenotype has been put forward.

The most prominent biochemical feature observed in QPM endosperm is the accumulation of the 27-kDa γ-zein at 2-3 fold higher levels than soft *o2* (Wallace *et al.,* 1990; Geetha *et al.,* 1991; Lopes and Larkins, 1995). Endosperm with increased amount of this protein contains more PB that are tightly packed due to cross linking by disulfide bonds between cysteine residues of 27-kD γ-zein (Wu *et al.,* 2010). Endosperm modifier genes have been used in breeding programmes worldwide for development of QPM genotypes that are vitreous/ semi-vitreous in grain phenotype (Hossain *et al.,* 2008a & b). QPM is essentially interchangeable with normal maize in cultivation and grain phenotype, and has become a popular choice among the farming community (Hossain *et al.,* 2007a & b; Gupta *et al.,* 2013; Pandey *et al.,* 2016).

Nutritional quality of QPM

QPM genotypes have 2-3 fold enhanced lysine and tryptophan in the endosperm proteins (Prasanna *et al.,* 2001; Vasal, 2001; Chaudhary *et al.,* 2012; Agrawal *et al.,* 2015). Average lysine concentration in *o2* maize is about 4% as compared to 2% in normal maize, while the same for average tryptophan is 0.8% as compared to 0.4% in the wild types (Vivek *et al.,* 2008). The decreased leucine content in QPM produces a favourable leucine-isoleucine ratio, which liberates more tryptophan for niacin biosynthesis and in turn reduces pellagra significantly (Vasal, 2001). The nutritional and biological superiority of QPM has been well established in infants, and small children as well as adults. In Guatemala, it was demonstrated that *o2* maize has 90% of the nutritive value of milk protein in young children. Children in Colombia suffering from kwashiorkar, a severe protein deficiency disease, were brought back to normalcy on a diet containing only *o2* (Bressani, 1994; Prasanna *et al.,* 2001).

Development of QPM genotypes

Large-scale efforts were employed to develop diverse QPM germplasm in tropical, subtropical and highland genetic background with different maturity groups, grain colour and texture (Bjarnason and Vasal, 1992). A number of advanced maize populations in International Maize and Wheat Improvement Centre (CIMMYT) were successfully converted to QPM populations, and led to the development of several QPM populations, pools, inbreds, and hybrids adapted to subtropical and tropical environments, which are widely used in the development of an array of QPM cultivars in several countries including Brazil, China, Ghana, India, and several Latin American countries (Vasal, 2001; Gupta *et al.,* 2009). In Ghana, the successful utilization of a QPM variety 'Obtanpa' (meaning 'good mother', released in 1992) is particularly noteworthy (Prasanna *et al.,* 2001). Several single- and three-way cross hybrids such as GH132-88, GH110-81 and GH2823-88 were developed with superior grain yields. South Africa pursued vigorously to develop several hybrids such as HL1, HL2 and HL8, possessing hard endosperm, good yield potential and tolerance to diseases. QPM hybrid breeding in Brazil has been greatly strengthened with the release of QPM cultivars, BR-451 and BR-473. Very few countries grew QPM in 1977, but now more than 23 countries have released QPM varieties for large-scale cultivation. Mexico accounts large area under QPM with the support and commitment of the Mexican government. In China, a number of high yielding QPM hybrids are under cultivation and expected that about 30% of maize area will be under QPM cultivars by 2020. 'Zhong Dan 206', the first high-lysine single-cross hybrid with soft endosperm, was released for commercial use in 1988. With the QPM intervention, in Guizhou Province, one of the poorest provinces in China, wellbeing of the farmers has been effectively improved. Several other countries in Asia, Africa and Latin America, are part of QPM research and development programme for the improvement and promotion of QPM in developing countries.

In India, Shakti, Rattan and Protina, the *o2*-based soft endosperm maize composites were released during 1971 by All India Coordinated Research Project (AICRP) on Maize (Table 1), and these are perhaps the first set of biofortified varieties developed through targeted breeding approaches across crops in the country. Later, a nutritionally superior *o2* composite with hard grain texture, 'Shakti-1', was released in 1997. Since 1998, intensive efforts resulted in the release of a series of QPM hybrids. The first white-grained QPM hybrid 'Shaktiman-1' (a three-way cross hybrid using CIMMYT-QPM lines) was released during 2001 by Rajendra Agricultural University (RAU), Pusa, Bihar. The same institute released 'Shaktiman-2', a single-cross hybrid with white grain in 2004. The first yellow-grained single-cross hybrid 'HQPM-1' was released by CCS-Haryana Agricultural University (CCSHAU), Uchani, during 2005, followed by the releases of 'Shaktiman-3', 'Shaktiman-4', 'HQPM-5', 'HQPM-7', HQPM-4, Pratap QPM Hybrid-1, and Shaktiman-5. Diverse QPM inbreds have been developed through conventional breeding methods, and are now available in the breeding programme (Gupta *et al.,* 2015a).

The cloning and characterization of the *O2* gene, followed by detection of gene specific three simple sequence repeats (SSRs) *viz.*, *phi057*, *phi112* and *umc1066*, offers advantages in molecular marker-assisted conversion of non-QPM lines into their QPM versions (Schmidt *et al.,* 1987; Motto *et al.,* 1988). In India, a marker-assisted selection (MAS) derived QPM hybrid, 'Vivek QPM 9', was released during 2008 by the ICAR-Vivekananda Parvatiya Krishi Anusandhan Sansthan (VPKAS), Almora (Gupta *et al.,* 2013). It utilized the *o2*-specific SSRs for foreground selection of the *o2* allele and phenotypic selection for endosperm modifiers in the parental lines. It possesses 41% more tryptophan and 30% more lysine over the original hybrid, with similar grain yield potential of Vivek Hybrid-9. Stringent selection of endosperm modification helped in having high degree of vitreous grains in the reconstituted version of parental inbreds and hybrids. Vivek QPM-9 earned the distinction of being the *'first MAS-based maize cultivar'* released for commercial cultivation in India. Vivek QPM-21, developed through marker-assisted introgression of *o2* allele into Vivek Hybrid-21 is yet another QPM hybrid released in 2012 for Uttarakhand state. Several institutions under the Indian Council of Agricultural Research (ICAR) and State Agricultural Universities (SAUs) have targeted enhancement of lysine and tryptophan in selected normal maize hybrids using accelerated breeding strategy. Research efforts at ICAR-Indian Agricultural Research Institute (IARI), New Delhi have led the development of QPM version of five commercial hybrids, *viz.*, HM-4, HM-8, HM-9, HM-10, and HM-11 (Hossain *et al.,* 2014). DHM-117 and Palam Sankar Makka-2 from Acharya NG Ranga Agricultural University (ANGRAU), Hyderabad, and CSK-Himachal Pradesh Krishi Vishvavidyalaya (CSK-HPKV), Bajaura, respectively have been improved for protein quality.

Further, single-cross hybrids, *viz.*, Buland and PMH-1 have been targeted for conversion to QPM using MAS at Punjab Agricultural University (PAU), Ludhiana. Several of the normal inbreds have also been converted to QPM at GB Pant University of Agriculture and Technology (GBPUA&T), Pantnagar using MAS (Tufchi *et al.*, 2015). Various institutions *viz.*, ICAR-Indian Institute of Maize Research (IIMR), Ludhiana and Tamil Nadu agricultural University (TNAU), Coimbatore have initiated MAS programme to convert some of the promising normal maize hybrids to QPM.

Further enhancement of lysine and tryptophan in QPM

In 2005, a recessive *opaque16* (*o16*) mutant, located on chromosome 8 was discovered and isolated from Robertson's Mutator (Mu) stock (Yang *et al.*, 2005). The recessive *o16o16* possessed nearly two-fold more lysine (0.247%) and tryptophan (0.072%) in mutants, than wild type (0.125% lysine and 0.035% tryptophan (Sarika *et al.*, 2017a). Additionally, *o16* does not influence the endosperm attributes such as grain hardness and vitreousness, and its mode of action is not yet known. The microscopic organisation of starch granules and proteinaceous matrix depicted by scanning electron microscopy is similar with the normal maize, thus forming a compact packaging, explaining the hard vitreous endosperm of *o16* lines (Sarika *et al.*, 2017b). Synthesis of zein fractions is not affected in the mutant. The mechanism of *o16* on nutritional improvement seems to be completely different from the *o2*. Since the *o16o16* genotypes possessed vitreous endosperm with similar grain hardness to normal line, the mutant provides tremendous advantage to the breeders as accumulation of endosperm modifiers in the genetic background need not to be looked into while breeding for high lysine and tryptophan, which otherwise is mandatory for *o2*-based breeding (Sarika *et al.*, 2017b). Further, several *o16o16* based inbreds have also been developed and can be readily utilized in the Indian maize breeding programme (Sarika *et al.*, 2017a).

Further, combination of *o2* and *o16* offers possibility of enhancement of lysine by 40-80% over *o2* genotype alone (Yang *et al.*, 2005). In India, four parental inbreds *viz.*, HKI161, HKI163, HKI193-1 and HKI193-2 of popular Indian QPM hybrids (HQPM-1, HQPM-4, HQPM-5 and HQPM-7) have been targeted for introgression of *o16* using MAS at IARI (Sarika *et al.*, 2015; Sarika 2016). The linked SSRs *viz.*, *umc1141* and *umc1149* have been successfully used to pyramid *o16* in *o2* genetic background, and MAS-derived inbreds possess higher lysine and tryptophan as compared to their original parents. Several researchers especially in China have also successfully pyramided *o2* and *o16* in diverse genetic background using MAS, and reported higher concentration of lysine and tryptophan (Yang *et al.*, 2013; Zhang *et al.*, 2013).

Table 1. Details of the QPM maize cultivars released in India (Adopted from Gupta *et al.*, 2015a)

S. No.	Name	Type	Parentage	Kernel colour	Year	Centre	Area of adaptation
Soft endosperm							
1.	Shakti	Composite	JLo2, Cuba 1J o2, Antigua 2D o2	Yellow	1971	Delhi	MP, Rajasthan and Tarai belt of UP
2.	Rattan	Composite	J1o2	Yellow	1971	Ludhiana	Punjab and Rajasthan
3.	Protina	Composite	(Jowatigua × Antigua car II) o2 × (Doeto × G.C.C.) o2	Yellow	1971	Pantnagar	Punjab, Rajasthan, Mysore and Tarai belt of UP
Hard endosperm/QPM							
4.	Shakti-1	Composite	Antigua, Ver 181 HEo2, Amarillo crstallino HEo2, Ant Rep Dom, HEo2,temperate HEo2	Yellow	1997	Delhi	Across the country
5.	Shaktiman-1	Three way cross	(CML142 × CML150) × CML186	White	2001	Dholi	Bihar
6.	Shaktiman-2	Single cross	CML176 × CML186	White	2004	Dholi	Bihar
7.	HQPM-1	Single cross	HKI193-1 × HKI163	Yellow	2005	Uchani	Across the country
8.	Shaktiman-3	Single cross	CML161 × CML163	Yellow	2006	Dholi	Bihar
9.	Shaktiman-4	Single cross	CML161 × CML169	Yellow	2006	Dholi	Bihar
10.	HQPM-5	Single cross	HKI163 × HKI 161	Yellow	2007	Uchani	Across the country
11.	HQPM-7	Single cross	HKI193-1 × HKI161	Yellow	2008	Uchani	Karnataka, AP, TN & MH
12.	Vivek QPM-9*	Single cross	VQL1 × VQL2	Yellow	2008	Almora	J & K, Uttarakhand, HP, AP, TN, Karnataka and MH
13.	HQPM-4	Single cross	HKI-193-2 × HKI 161	Yellow	2010	Uchani	Across the country
14.	Vivek QPM-21*	Single cross	VQL1 × VQL17	Yellow	2012	Almora	Uttarakhand
15.	Pratap QPM Hybrid-1	Single cross	DMRQPM-106 × HKI-193-1	Yellow	2013	Udaipur	Gujarat, Rajasthan, MP, Chattisgarh
16.	Shaktiman-5	Single cross	CML161 × CML165	Yellow	2013	Dholi	Orissa, Bihar, Jharkhand, West Bengal and Eastern UP

*Hybrids developed through MAS

B. Provitamin A

Role in human health

Vitamin A is required for normal functioning of the visual system, maintenance of cell function for growth, epithelial integrity, red blood cell production, immunity and reproduction (Sommer and West, 1996). An estimated average requirement (EAR) of 250 and 500 retinol equivalents (RE) per day for children and adults, respectively has been recommended by World Health Organisation (WHO) for normal growth and development (Bouis and Welch, 2010). Based on the available information, 200 and 400 g of daily maize consumption is required to meet daily requirement of children and women, respectively considering only 50% retention after post-harvesting and cooking, and a 12:1 bioconversion rate of provitamin A to retinol. Thus, 15 ìg/g of provitamin A in kernel was set as the target level in maize biofortification (Li *et al.*, 2007; Ortiz-Monasterio *et al.*, 2007). However, vitamin A needs to be supplied through dietary means as it cannot be synthesized by human body.

Deficiency symptoms

Vitamin A deficiency (VAD) has been defined as a liver retinol reserve of <0.1 ìmol g^{-1} liver (Tanumihardjo, 2011). About 4.4 million preschool-age children have suffered from visible eye damage due to VAD. Close to 20 million pregnant women are also vitamin A deficient, of which about one-third are clinically night-blind. Nearly one-half of these cases occur in India with severe form of VAD (www.harvestplus.org). Xerophthalmia is the world's leading cause of blindness, and a cardinal indicator of VAD (Sommer and West, 1996). Of the world's children with xerophthalmia, nearly half reside in South or South-East Asia, with more than 85% of these living in India (West, 2002). The VAD also results in night blindness, keratomalacia, and subsequent inflammation and infection causing irreversible blindness (Mayer *et al.*, 2008). Moreover, VAD affected people are easily exposed to measles, diarrhoea and respiratory diseases (Sommer and Davidson, 2002).

Carotenoids in maize kernel

Maize exhibits considerable natural variation of kernel carotenoids, with some genotypes accumulating as much as 66 ìg/g of total carotenoids in grains (Harjes *et al.*, 2008). Sufficient variation for total carotenoids has also been reported in Indian maize germplasm (Mishra and Singh, 2010; Das and Singh, 2012; Tiwari *et al.*, 2012; Sivaranjani *et al.*, 2013 & 2014; Rashmi *et al.*, 2014). Provitamin A carotenoids (β-cryptoxanthin, α- and β-carotene) are the precursors of vitamin A, and essential for vitamin A associated functions. Lutein and zeaxanthin have been associated with lowering the risk of cataract, age-related mascular degeneration, and other degenerative diseases (Burt *et al.,* 2011). However, the amount of provitamin-A carotenoids is typically only 10-20%, whereas zeaxanthin and lutein, each commonly represents 30-50% of total carotenoids in maize (Ortiz-Monasterio *et al.,* 2007). Most yellow maize cultivated and eaten worldwide, however, has less of provitamin-A carotenoids (Pixley *et al.,* 2013). In India as well, large genetic variation for non-provitamin A carotenoids was observed, while the concentration of provitamin A in Indian inbreds was reported to be very low (<2 ìg/g) to meet the target level (Vignesh *et al.,* 2012; Muthusamy *et al.,* 2015a & 2015b; Choudhary *et al.,* 2014 and 2015). The pre-dominant carotenoids in maize kernels, in decreasing order of concentration, are lutein, zeaxanthin, β-carotene, β-cryptoxanthin, and α-carotene (Suwarno *et al.,* 2014).

Stability assessment studies revealed that carotenoids are quite stable over locations, and environments play minor role in causing variation (Vignesh *et al.,* 2012; Muthusamy *et al.,* 2015a). Suwarno *et al.* (2014) reported that provitamin-A expression is more influenced by genotype and environment than by G × E effects. Egesel *et al.* (2003) reported that general combining ability (GCA) effects, or additive gene action, accounted for 72-87% of the variation for β-carotene, β-cryptoxanthin, and total carotenoids. However, in some of the crosses, non-additive gene action was also reported in their study. Senete *et al.* (2011) and Muthusamy *et al.* (2016a) also reported preponderance of additive gene action. Suwarno *et al.* (2014) while evaluating 156 maize hybrids at four locations reported additive gene action, which was mostly responsible for determining β-carotene concentration. *Yellow1* (*phytoene synthase1*) is dominant over recessive *y1* (which in homozygous conditions produces white colour with no carotenoids), and also exhibits dosage effects with *Y1Y1Y1* having intense yellow/orange colour as compared to *Y1Y1y1* and *Y1y1y1*. *Lycopene epsilon cyclase* (*lcyE*) and *β-carotene hydroxylase 1* (*crtRB1*), the two key genes associated with total provitamin-A concentration are partially dominant and partially recessive, respectively (Babu *et al.,* 2013).

Carotenoid biosynthesis pathway

The carotenoid metabolic pathway has been well researched in model species and key genes governing critical steps have been identified (Vallabhaneni *et al.,* 2009). The key regulatory step of the pathway involves the condensation of two geranylgeranyl pyrophosphate (GGPP) to form 15-cis-phytoene that is further converted to all-translycopene (a red pigment) by four desaturation and an isomerization reaction.

The carotenoid biosynthesis pathway has two major branches that occur after the biosynthesis of the linear carotenoid, all-trans-lycopene (Della Penna and Pogson, 2006). Lycopene may be cyclized to form two β rings, as found in β-carotene and its derivatives, β-cryptoxanthin and zeaxanthin. Alternatively, lycopene may be cyclized to form one β ring and one å ring, as found in α-carotene and its derivatives, zeinoxanthin and lutein. The two predominant provitamin-A carotenoids in maize, β-carotene and β-cryptoxanthin are produced by the β, β branch of the biosynthesis pathway, whereas the third common provitamin A carotenoid, α-carotene, is produced by the β, å branch.

Genetic loci for higher provitamin-A

Wong *et al.* (2004) and Chander *et al.* (2008a) identified quantitative trait loci (QTL) governing accumulation of carotenoids in maize grains, and one of which explained 6.6–27.2% phenotypic variation. In maize, three genes have been proposed to play crucial roles in the final accumulation of provitamin A carotenoids in the grain. *Phytoene synthase1* (*Y1 or Psy1*) catalyses the first committed step in the pathway leading to formation of phytoene from GGPP, and is primarily responsible for the shift from white to yellow maize (Li *et al.,* 2010). Two genes, *lcyE* and *crtRB1* have been shown to regulate the accumulation of provitamin A compounds. Natural *lcyE* converts lycopene into æ-carotene and eventually to α-carotene through the action of other associated genes. Favourable *lcyE* allele forces pathway flux towards β-carotene branch (Harjes *et al.*, 2008). Though the favourable *lcyE* allele increases the proportion of β-carotene in the pathway, a large amount is hydroxylated to produce β-cryptoxanthin (with 50% provitamin A activity) and zeaxanthin (0% provitamin A activity).

crtRB1 is a hydroxylase gene that converts β-carotene into β-cryptoxanthin. However, naturally available favourable *crtRB1* allele blocks the process of hydroxylation of β-carotene into further components, thereby increasing concentration of β-carotene in the kernel (Yan *et al.,* 2010). Babu *et al.* (2013) validated the effects of *crtRB1* gene in 26 diverse tropical maize genetic backgrounds. The favourable allele is rare and occurs with low frequency in the maize germplasm, and the same observation was also reported in Indian maize inbreds by Muthusamy *et al.* (2015a).

Estimation of β-carotene through high performance liquid chromatography (HPLC) showed that the CIMMYT genotypes with *crtRB1* had high kernel β-carotene, whereas the Indian inbreds with the same allele had low β-carotene (Vignesh *et al.,* 2013). Among the several single nucleotide polymorphisms (SNPs) and InDels, SNP4, SNP13, InDel6 and InDel7 identified in the 32 -UTR region clearly differentiated the high and the low β-carotene genotypes.

Bioconversion of provitamin-A in human body

The amount of 1 μg retinol is referred as 1 RE in the FAO/WHO recommendations, which is equivalent to 12 ìg of provitamin A carotenoids.

Although there is no consensus about the exact bioconversion ratio for β-carotene to retinol, recent estimates of 2.8:1 (Howe and Tanumihardjo, 2006), 3.2:1 (Muzhingi *et al.*, 2011), and 6.5:1 (Li *et al.*, 2010) indicate that bioconversion of β-carotene from biofortified maize is considerably more efficient than the standard ratio of 12:1 proposed by the United States Institute of Medicine (IOM 2001). However, HarvestPlus a global initiative under Consultative Group for International Agricultural Research (CGIAR) has adopted 12:1 ratio for β-carotene to retinol and 24:1 for β-cryptoxanthin to retinol (www.harvestplus.org).

Two studies have indicated that the provitamin A carotenoids in biofortified maize is efficiently converted to vitamin A in human body. A study by Li *et al.* (2010) measuring the triacylglycerol rich lipoprotein fraction of human blood estimated biofortified maize porridge (containing 8 g of fat) to have a vitamin A equivalence of 6.5:1 in North American women. In another study conducted by Muzhingi *et al.* (2011), the vitamin A equivalence of intrinsically labeled high β- carotene yellow maize porridge (containing 21 g of fat) was estimated as 3.2:1 in Zimbabwean men. Both of these studies indicate that consumption of biofortified maize may supply sufficient vitamin A to meet daily requirements. An animal study testing bioavailability and an *in vitro* simulated digestion/Caco-2 cell study also testing bioaccessibility of biofortified maize have supported the findings of human studies in terms of efficient absorption (Howe and Tanumihardjo, 2006; Thakkar and Failla, 2008). In addition, none of the studies observed a negative influence of the high concentration of the large variety of carotenoids in maize on absorption. Furthermore, Gannon *et al.* (2014) in Zambia working on 140 rural Zambian children has concluded that β-carotene from maize was efficacious when consumed as a staple food in this population and could avoid the potential for hyper-vitaminosis A that was observed with the use of preformed vitamin A from supplementation and fortification. A Mongolian gerbil feeding study also indicated that biofortified maize containing high concentrations of β-cryptoxanthin resulted in a more efficient bioconversion (2.4:1) than a β-carotene supplement (4.6:1) (Davis *et al.*, 2008). Liu *et al.* (2012) reported that provitamin A equivalents increased in eggs from hens fed high-β-cryptoxanthin rich maize, and this could lead to human health benefit if widely adopted.

Development of provitamin A rich maize cultivar

With the development and access to reliable PCR-based gene specific markers for *lcyE* and *crtRB1* genes, MAS has become an attractive option for provitamin A enrichment in maize (Harjes *et al.*, 2008; Yan *et al.*, 2010; Zhang *et al.*, 2012; Babu *et al.*, 2013). Quantifying the provitamin A carotenoid of maize samples using HPLC is difficult, time-consuming and expensive, and breeding programme thus would benefit greatly using MAS to reduce the need for phenotypic assays. By selecting the favourable alleles of the two key genes *viz.*, *lcyE* and *crtRB1*, provitamin A concentration can be increased in the maize endosperm. Using these markers, CIMMYT breeders have already conducted preliminary evaluations of tropical maize genotypes (Pixley *et al.,* 2011). Large scale validation of the effects of molecular marker polymorphisms in these two genes was carried out at CIMMYT, and the results suggest that by doing MAS for *lcyE* and *crtRB1*, it would be possible to breed tropical maize varieties with 15 μg/g of provitamin A, the target level as set by HarvestPlus for alleviating the widespread VAD in humans (Babu *et al.,* 2013). Liu *et al.* (2015) have improved elite QPM inbreds *viz.*, CML161 and CML 171; showed improvement of 5.25 μg/g from 1.60 μg/g of provitamin A in the parent of CML161 and 8.14 μg/g from 1.80 μg/g in the parent of CML171. Globally, three maize hybrids from Zambia (GV662A, GV664A, and GV665A), two hybrids (Ife maize hyb-3, and Ife maize hyb-4) and two synthetics (Sammaz 38, and Sammaz 39) from Nigeria and one synthetic from Ghana (CSIR-CRI Honampa) were released that contain 6-8 ìg/g of provitamin A (www.harvestplus.org).

Due to lower level of provitamin A in Indian maize germplasm, CIMMYT-HarvestPlus genotypes having favourable allele of *lcyE* and *crtRB1* with high β-carotene have been used as donors in the Indian maize biofortification programme. Maize breeders at IARI, New Delhi, have introgressed the favourable allele of *crtRB1* gene in seven elite genetic background (VQL1, VQL2, V335, V345, HKI161, HKI323 and HKI1105) using gene-based MAS approach (Muthusamy *et al.,* 2014). These inbreds are the parents of two early (Vivek QPM-9 and Vivek Hybrid-27) and two medium maturing (HM-4 and HM-8) single cross hybrids adapted to diverse ecological regions of India. The improved inbreds contained kernel β-carotene ranging from 8.6 to 17.5 ìg/g. The reconstituted hybrids developed from improved parental inbreds also showed enhanced kernel β-carotene as high as 21.7 ìg/g, compared to 2.6 ìg/g in the original hybrid (Muthusamy *et al.,* 2014; Gupta *et al.,* 2015b). These improved hybrids possessed similar grain yield potential as compared to original hybrids. This is the 'first-ever demonstration of conversion of elite maize hybrids into β-carotene-rich version'.

In another study at IARI, four popular QPM hybrids (HQPM-1, HQPM-4, HQPM-5 and HQPM-7) were enriched with provitamin-A by introgressing the favourable alleles of both *crtRB1* and *lcyE* genes using MAS. The CIMMYT-HarvestPlus genotypes, HP704-22 and HP704-23 were used as donors, while four elite QPM parents *viz.* HKI161, HKI163, HKI193-1 and HKI193-2 were the recipients. Introgressed versions of the selected QPM inbreds having favourable alleles of *opaque2*, *crtRB1* and *lcyE* possess provitamin A as high as >20 ìg/g (Zunjare *et al.,* 2015 & 2016). Also, HKI1128, elite inbred line was improved for kernel β-carotene at IARI, New Delhi (Goswami *et al.,* 2015). Further, at IARI, favourable allele of *crtRB1* has been combined with *shrunken2* (*sh2*) to develop vitamin A rich sweet corn hybrids (Baveja *et al.,* 2016; Chauhan *et al.,* 2016). Currently, maize breeding programme at IARI, New Delhi; IIMR, Ludhiana; VPKAS, Almora; CSK-HPKV, Bajuara; Professor Jayashankar Telangana State Agricultural University (PJTSAU), Hyderabad; PAU, Ludhiana; GBPUA&T, Pantnagar and TNAU, Coimbatore are involved in generating/ selecting diverse inbreds with high β-carotene. Enrichment of carotenoids in maize has also been attempted using transgenic approach. Over expression of *crtB* and *crtI* from *Erwinia herbicola*, had resulted in accumulation of 10 µg/g of β-carotene in Hi-II maize genotype (Aluru *et al.,* 2008). Zhu *et al.* (2008) and Naqvi *et al.* (2009) further developed transgenic maize lines (with ~60 µg/ g β-carotene) using combination of five genes (*psy1*, *crtI*, *lycb*, *bch* and *crtW*). Though development of β-carotene rich maize through transgenic approach is a rapid method, its successful adoption depends upon (i) regulatory mechanisms and (ii) political as well as socio-economic factors.

C. Vitamin E

Role in human health

Vitamin E (Tocopherol) performs role in human body which includes scavenging of various reactive oxygen species (ROS) and free radicals, quenching of singlet oxygen (high energy oxygen), and membrane stability by protecting polyunsaturated fatty acids (PUFA) from lipid peroxidation (Bramley *et al.,* 2000). Essential role of vitamin E for the development of tissues and organs such as brain and nerves, muscle and bones, skin, bone marrow and blood has been well documented. It has emerged as an immuno-homeostatic factor, and response of immune cells to vitamin E supplementation has also been recognized. It has been proposed that vitamin E positively affects age-related pathophysiology of the immune system (Wu and Meydani, 2014). Further, modulation of lung anaphylaxis and a defensive role in the pathogenesis of allergic disease have been recently proposed (Abdala-Valencia *et al.,* 2013). α- tocopherol is also known to act as a cytoprotective factor with advocated involvement in inhibiting inflammatory and degenerative processes in the liver (Mitcheva *et al.,* 1993; Hashem *et al.,* 2016).

Recommended dietary allowance (RDA) for vitamin E is 4 mg/day (d) for 0-6 months old child, while the same for e"14 years is 15 mg/d for both males and females. However, for the lactating mother the RDA is 19 mg/d (Institute of Medicine 2000). The intake of vitamin E around the globe is generally lower than optimum. Peter *et al.* (2015) reported that intake of α-tocopherol and other vitamin E forms are below the RDA for a major part of the population both in industrialized and developing countries. In a recent study in South Korea, 2/3 of adults have been reported to have suboptimal vitamin E level and almost 25% are deficient (Kim and Cho, 2015). About 1/3 of the pregnant women in rural Nepal are severely affected by vitamin E deficiency (Jiang *et al.,* 2005). In Bangladesh, the vitamin E deficiency is more critical and worse as about 2/3 of the women in early pregnancy are severely vitamin E deficient. Even in United States, vitamin E intake has been reported below the optimum in majority of the population (Dwyer *et al.,* 2013).

Deficiency symptoms

Vitamin E deficiency symptoms are similar to Freidriech's ataxia which causes progressive damage to nervous system. Inadequate level of α-tocopherol can lead to sensory neuropathy that can affect sensory nerves (Traber *et al.,* 2008). In children, areflexia (absence of reflex) is first manifested and neurologic abnormalities become more and more apparent after 3-5 years. Initial mild loss of sensations in adults with vitamin E deficiency can be easily overlooked, but neurologic abnormalities once progressed to clinically significant level are often irreversible. Vitamin E deficiency is also associated with haemolytic anaemia in the premature infant.

Tocopherols in maize kernel

Vitamin E is a generic term for eight structurally related lipophilic compounds which include four tocopherols (α, β, δ, γ) and four tocotrienols (α, β, δ, γ) (Abbasi *et al.,* 2007). All eight forms of vitamin E are absorbed to a comparable degree from the gastrointestinal tract and transported to the liver, which sorts out α-tocopherol and preferentially transports it into the bloodstream. As a result, in human serum, α-tocopherol is observed at highest concentration, followed by γ-tocopherol, whereas tocotrienols are generally not detected (Frank *et al.,* 2012). Involvement of α-tocopherol transfer protein (α-TTP), a hepatic 32 kDa cytosolic protein was initially suggested to be involved in selective transport of α-tocopherol from liver (Hosomi *et al.,* 1997; Meier *et al.,* 2003). This was partly attributed to relative binding affinities of α-TTP for different tocopherols (natural α-tocopherol, 100%; β-tocopherol, 38%; γ-tocopherol, 9%; δ-tocopherol, 2%) (Hosomi *et al.,* 1997). An alternate explanation of favoured retention of α-tocopherol is based on metabolic pathway of vitamin E. In the pathway, vitamin E is degraded by CYP-mediated ù-hydroxylation and subsequent rounds of β-oxidation to side chain-shortened carboxy ethyl hydroxyl chromanol (CEHC) metabolites, and non-α-tocopherols are preferentially metabolized than α-tocopherol (Birringer *et al.,* 2002; Sontag and Parker, 2002; Sontag and Parker, 2007).

The maize germ contains 70-86% of the total tocopherols, with endosperm having 11-27% of the fractions. Most of the α-tocopherol (94-96%) and γ-tocopherol (93-96%) are found within the germ, whereas δ-tocopherol is restricted in the germ only. Among the constituents, γ-tocopherol constitutes ~80% of the total tocopherol, while α-tocopherol accounts ~20% of the total pool (Grams *et al.,* 1970). However, γ-tocopherol is less absorbed in the body due to lack of affinity of receptors. On the contrary, α-tocopherol is most favoured fraction and well absorbed in the body (Bramley *et al.,* 2000). Due to favourable interaction with the receptor, α-tocopherol is present 10 times more than γ-tocopherol in plasma. The role of tocopherols in human system is generally measured in terms of the levels of α-tocopherol. In 15 maize inbreds, range of γ-tocotrienol was reported from 2-28% with a mean of 9.35 μg/g, and 4-21% for α-tocotrienol with a mean of 5.63 μg/g. Concentration of δ-tocopherol in corn grain is much less than α-tocopherol and γ-tocopherol. In one mapping population reported by Rocheford *et al.* (2002), δ-tocopherol accounted for only 2-10% with a mean 1.6 μg/g and 1.0-5.33% with a mean 4.28 μg/g in another population.

The very first attempt to estimate the GCA and SCA effects of maize inbreds and hybrids for tocopherol was accomplished by Egesel *et al.* (2003). Ten inbreds and resulted 45 hybrids, were used to evaluate α- and γ-tocopherol. Several cross combinations had high SCA effect, inbred R84 revealed high GCA for γ-tocopherol and total tocopherol which shows the possibility to develop vitamin E enriched maize hybrid. Feng *et al.* (2015a) evaluated 10 sweet corn inbreds at 15, 18, 21, 24, 27 and 30 DAP for their kernel tocopherol. There was no significant difference among tocopherols at 21, 27 and 30 DAP.

Vitamin E biosynthesis pathway

Precursor for tocochromanol synthesis in plants comes from two separate pathways. Common head group, i.e., homogentistic acid (HGA) is delivered by shikkimate pathway, whereas tail phytyldiphosphate (PDP) and GGPP is synthesized via plastidicisoprenoid pathway. Conversion of p-hydroxy phenylpyruvate (HPA) from shikimic acid pathway to HGA occurs through cytosolic enzyme p-hydroxyphenylpyruvic acid dioxygenase (HPPD); the first committed step to synthesize the aromatic head group common to all tocochromanols. Thereafter in plastid, HGA is either condensed with phytoldiphosphate (PDP) by homogentisate phytyl transferase (*VTE2*) to produce 2-methyl-6-phytyl-1,4-benzoquinol (MPBQ), particular to tocopherols or geranyl geranyl pyrophosphate (GGPP) from isoprenoid pathway by homogentisate geranyl geranyl transferase (HGGT) to yield 2-methyl-6-geranylgeranyl benzoquinol (MGGBQ), committed precursors to tocotrienols, respectively (Lipka *et al.,* 2013). MPBQ and MGGBQ are either converted to γ-tocopherol and γ-tocotrienol, respectively by the action of one methyl transferase (*VTE3*) and one tocopherol cyclase (*VTE1*) or to δ-tocopherol and δ-tocotrienol, respectively by ocopherol cyclase (*VTE1*). Finally, γ-tocopherol methyl transferase (*VTE4*) converts the δ- and γ-tocopherols (and tocotrienols) to β- and α-tocopherols (and tocotrienols), respectively (DellaPenna and Pogson, 2006).

Genetic loci for high vitamin E

Several researchers (Wong *et al.,* 2003; Chander *et al.,* 2008b; Shutu *et al.,* 2012; Feng *et al.,* 2013) have reported a number of quantitative trait loci (QTL) for kernel α-tocopherol, γ-tocopherol, total tocopherol and ratio of α/γ tocopherol. Li *et al.* (2012) has reported two insertion/deletions (*InDel7* and *InDel118*) within the gene *ZmVTE4* involved in tocopherol biosynthesis pathway and an SNP at 85 kb upstream of *ZmVTE4* significantly affecting level of α-tocopherol. *ZmVTE4* was suggested as a major gene to control phenotypic variation of α-tocopherol as both the haplotypes together could explain 33% of α-tocopherol. *InDel118*, located 9-bp upstream of the putative transcription start site, controls α-tocopherol by regulating *ZmVTE4* transcript level, whereas *InDel7* by affecting translation efficiency. Association of a new polymorphism at *ZmVTE1* (tocopherol cyclase) and tocotrienol composition was reported in addition to earlier reported effect of *ZmVTE4* haplotypes and α-tocopherol (Lipka *et al.,* 2013). Further pathway level analysis using sequence similarity search of maize B73 RefGen_v2 genome for trocochromanol precursor (HGA, GGPP and PDP) and core pathway (*ZmVTE*) genes from a*Rabi*dopsis and HGGT in maize, identified genetic contribution of 60 a priori candidate genes. Association studies of 122 SNPs within ± 250 kb of these 60 loci identified two additional genes *viz., ZmHGGT1* and one prephenate dehydratease parolog that also moderately contributes to tocotrienol variation. Altogether in addition to most favourable *ZmVTE4* haplotype, three new gene targets have been reported that could be used to increase vitamin E and antioxidant levels in maize.

Bioconversion of vitamin E in human body

Though all four tocopherols (α, β, δ, γ) and tocotreinols (α, β, δ, γ) are naturally occurring forms of vitamin E, only the α-tocopherol form of the vitamin is maintained in human plasma (Brigelius-Flohe and Traber, 1999). Chemically synthesised α-tocopherol (all-rac α-tocopherol or dl-α-tocopherol) is a racemic mixture of eight different isomers of α-tocopherol in equal proportion (12.5%) of which four stereoisomers are of 2R configuration (RRR, RRS, RSS and RSR) and rest four possess a 2S configuration at position 2' of the phytyl tail (SRR, SSR, SRS, and SSS). Because of more stability than its free form, in feed additives acetate and succinate esters of α-tocopherol (RRR α-tocopheryl acetate, RRR α-tocopheryl succinate, all-rac α-tocopheryl acetate, all-rac α-tocopheryl succinate) are frequently used. Based on the rat foetal resorption bioassay, United States Pharmacopeia (USP) had reported one International Unit (IU) of vitamin E is equivalent to 1 mg all-rac α-tocopheryl acetate, 0.74 mg of RRR α-tocopheryl acetate and 0.67 mg of RRR α-tocopherol (EFSA 2010). However, for human, USP conversion factor has been redefined by Institute of Medicine (IOM) because 2S stereoisomers of all-rac α-tocopherol were not maintained in human plasma (Burton, 1993; Acuff *et al.*, 1994; Kiyose *et al.*, 1997) which indicates that active form is limited to the 2R stereoisomeric forms of α-tocopherol, representing 50% of all-rac α-tocopheryl acetate and 100% of RRR α-tocopheryl acetate formulations.

Development of vitamin E rich maize cultivar

Researchers had reported wide genetic variability for α-tocopherol, γ-tocopherol, δ-tocopherol and total tocopherol (Wong *et al.*, 2003; Chander *et al.*, 2008b; Li *et al.*, 2012; Shutu *et al.*, 2012; Lipka *et al.*, 2013; Feng *et al.*, 2015a). Range for α-tocopherol was reported to vary from a minimum 0.11 μg/g (Shutu *et al.*, 2012) to the maximum 88.80 μg/g (Chander *et al.*, 2008b). Lowest γ-tocopherol (0.71 μg/g) and total tocopherol (4.11 μg/g) was reported by Shutu *et al.* (2012) while the highest (244.58 μg/g and 296.10 μg/g, respectively) was reported by Wong *et al.* (2003).

Favourable alleles of *InDel7* and *InDel118* at *ZmVTE4* locus have been recently transferred from a suitable donor parent (SY999) to four *shrunken2* (*sh2*) based sweet corn recipient lines (M01, M14, K140 and K185) through MAS (Feng *et al.*, 2015b). Except inbred M14, average α-tocopherol of three backcross progenies increased significantly; average increment in α-tocopherol of K140 and K185 were 7.73 and 5.33 μg/g, respectively. Converted recipient lines also showed improved γ-tocopherol and total tocopherol. Average γ-tocopherol of four recipient lines before and after conversion was 29.11μg/g and 51.45 μg/g, respectively, whereas total tocopherol was 36.63 μg/g and 63.34 μg/g respectively.

In India, an effort to enhance vitamin E level in maize has already been initiated at IARI, New Delhi. To identify inbreds having favourable alleles of *ZmVTE4*, a large set of indigenous and exotic inbreds were screened using gene-based markers. Only fifteen genotypes had favourable allele of *VTE4* (Das *et al.*, 2016). Four Indian elite QPM inbreds (HKI161, HKI163, HKI193-1 and HKI193-2, parental inbreds of HQPM-1, HQPM-4, HQPM-5 and HQPM-7), already improved for provitamin A, were targeted for enhanced vitamin E to develop multi-nutrient maize. One of the exotic inbred with favourable allele for both the InDels and higher α-tocopherol, was used as donor for enhancement of α-tocopherol. Introgression of favourable alleles of *ZmVTE4* from donor to the recipient was successfully carried out by MAS. *sh2*-based sweet corn hybrids are gaining popularity worldwide including India (Khanduri *et al.,* 2010 & 2011, Hossain *et al.,* 2015, Chhabra *et al.,* 2016). Favourable allele of *ZmVTE4* has been introgressed into the background of parental inbreds of two promising sweet corn hybrids (ASKH-1 and ASKH-2) for enrichment of vitamin E (Chauhan *et al.,* 2016).

D. Iron (Fe) and Zinc (Zn)

Role in human health

Among the various mineral elements, Fe and Zn are the most common micronutrients that have been found deficient predominantly in cereal-based human diet (Pfeiffer and McClafferty, 2007; Bouis *et al.,* 2011; Agrawal *et al.,* 2012; Gupta *et al.,* 2015c). Humans require Fe for basic cellular functions and proper functioning of the muscle, brain and red blood cells (Roeser, 1986). Most of the Fe in the human body is present in erythrocytes as haemoglobin, where its main function is to carry oxygen from the lungs to tissues. Iron is also an important component of various enzyme systems, such as cytochromes, involved in oxidative metabolism. Fe is generally stored in the liver as ferritin and as haemosiderin.

Zn is an essential mineral for humans, animals and plants for many biological functions. It plays a crucial role for more than 300 enzymes in the human body for the synthesis and degradation of carbohydrates, lipids, proteins and nucleic acids (Sandstorm, 1997; Gupta *et al.,* 2015c). As per estimated average requirement, human requires 1460 µg/d of Fe, while it is 1860 µg/d for Zn. Considering 200g and 400g consumption daily by children and adults, respectively, bioavailability (5% for Fe; 25% for Zn) and 90% retention after processing, 60 µg/g of Fe and 38 µg/g of Zn (on dry weight basis) have been fixed as the target level in maize (Bouis and Welch, 2010).

Deficiency symptoms

It has been estimated that over 60% of the world's population is Fe deficient, while it is 30% for Zn (White and Broadley, 2009; Gupta *et al.,* 2015c). Since, human body cannot synthesize micronutrients, they must be made available through diet (Welch and Graham, 2002 and 2005; Poletti *et al.,* 2004; White and Broadley, 2005; Gibson, 2006; Graham *et al.,* 2007). Anemia is considered as one of the symptoms of Fe deficiency in humans (DeMaeyer and Adiels-Tegman 1985). Fe-deficiency anaemia is present in all age groups, and has emerged as a public health problem in most regions of the world. The highest prevalence of Fe deficiency is found in infants, children, adolescents and especially pregnant women (Lozoff *et al.,* 1991). The clinical symptoms of severe Zn deficiency in humans are growth retardation, delayed sexual and bone maturation, skin lesion, diarrhoea, impaired appetite, increased susceptibility to infections mediated via defects in the immune system and the appearance of behavioural changes (Prasad, 1996). A reduced growth rate and impairments of immune defence are so far the only clearly demonstrated signs of mild form of Zn deficiency in humans. Severe form of Zn deficiency is characterized by short stature, hypogonadism, impaired immune function, skin disorders, cognitive dysfunction and anorexia (Goldenberg *et al.,* 1995; Brown *et al.,* 1998).

Fe and Zn in maize kernel

The concentration of Fe and Zn in maize is reported to be highest in the seed coat and scutellum as compared to endosperm (Bityutskii *et al.,* 2002). However, since endosperm occupies greater proportion of the maize kernel, the total content of these micronutrients is much higher in the endosperm (60-80%), followed by scutellum (15-35%) and seed coat (8-12%). Hence, endosperm and scutellum are the major micronutrient reserves in the maize kernel as compared to other cereal grains (Bityutskii *et al.,* 2001 & 2002, Gupta *et al.,* 2015c). Major portion of Fe and Zn is located in the aleurone (Banziger and Long, 2000; Ortiz-Monasterio *et al.,* 2007). Multiple aleurone layers (MAL) with an average of 2.0-3.7 layers per kernel and maximum of six layers in some of the kernels have been reported in Coroico maize landrace found in Bolivia, Peru, Ecuador and Brazil. Since majority of Fe and Zn is located in the aleurone layers, increase in proportion of aleurone layers in relation to starchy endosperm is advantageous in increasing the micronutrient concentration (Wolf *et al.,* 1972).

Preponderance of additive gene action was observed for both Fe and Zn in maize kernel (Brkic *et al.,* 2003; Long *et al.,* 2004; Pixley *et al.,* 2011; Simic *et al.,* 2011). In contrast, Qin *et al.* (2012) reported that genetic effects for kernel Fe and Zn were predominantly of partially dominant and overdominant nature. On the other hand, Chakraborti *et al.* (2010) observed that additive gene action was of higher magnitude as compared to dominance for kernel Fe, while dominance was relatively higher in case of kernel Zn. Besides, additive × dominance component was significant for kernel Fe, whereas additive × additive component was predominant for kernel Zn. Significant positive correlation between kernel Fe and Zn has been reported in several studies (Brkic *et al.,* 2003; Oikeh *et al.,* 2003; Guimaraes *et al.,* 2004; Menkir, 2008; Chakraborti *et al.,* 2009; Lungaho *et al.,* 2011; Baxter *et al.,* 2013). This positive correlation can be due to tight linkage between the genes affecting the accumulations or pleiotropic effects of the genes governing the accumulation of micronutrients. Qin *et al.* (2012) reported the positive correlation between kernel Fe and Zn in maize due to co-localization of QTLs for both kernel -Fe and -Zn on the same chromosome thereby suggesting the feasibility of their simultaneous improvement. In contrast, Simic *et al.* (2009) reported weak association between kernel Fe and Zn, while no association was observed by Prasanna *et al.* (2011), Agrawal *et al.* (2012) and Pandey *et al.* (2015).

Accumulation of Fe and Zn in plants

The acquisition of mineral elements such as Fe and Zn by plant roots depends on the forms of mineral elements supplied to them and phytoavailability of mineral elements in the rhizosphere solution (White and Broadley, 2009). Mineral elements are present in the soil as free ions or adsorbed onto mineral or organic surfaces as dissolved compounds or precipitates as part of lattice structures or contained within the soil. The important properties of soil that govern mineral availability are soil pH, redox conditions, cation exchange capacity, activity of microbes, soil structure, organic matter and water content (Frossard *et al.,* 2000; White and Broadley, 2009).

In case of graminaceous monocots, plant synthesizes and releases non-proteinaceous amino acids known as phytosiderophores (structural derivatives of mugineic acid). Phytosiderophores once released in the soil chelate with the Fe^{3+} and form stable Fe in soil (Roberts *et al.,* 2004). A highly specific Fe transport system then transports Fe^{3+}- phytosiderophores across the plasma membrane of the root cell (Graham and Stangoulis 2003). Zn prefers to enter the graminaceous system as divalent cation (Zn^{2+}) rather than phytosiderophore-mediated chelated complex (Bell *et al.,* 1991; Norvell and Welch, 1993). However, Von-Wiren *et al.* (1996) demonstrated that roots of Fe-efficient maize do absorb Zn in the form of Zn-phytosiderophores. Higher uptake rates of free Zn compared to its chelated species imply that free Zn^{2+} remains the preferential form for Zn uptake even in the presence of Zn^{2+}-phytosiderophores (Frossard *et al.,* 2000).

After absorption of Fe and Zn within the root cells, they are transferred to xylem vessels for transportation to shoot portions. Several transporters have been identified that mediate translocation of Fe and Zn in plants (Maser *et al.,* 2001; Clemens *et al.,* 2002; Hall and Williams, 2003; Mallikarjuna *et al.,* 2016). However, the entire amount of Fe and Zn absorbed by the roots is not transported to the shoot portion. Some amount of the minerals is stored in the root itself, probably in the vacuoles, and once the demand from the shoot increases, the stored Fe and Zn are then again transported to the shoot portion (Ghandilyan *et al.,* 2006). Minerals are distributed over all the tissues but certain plant parts accumulate more minerals than the others. Fe and Zn are loaded into the phloem along with the carbohydrates for their transportation to different parts. Fe is found to form complex with nicotianamine and moves as a chelate during loading into the phloem vessels (Von-Wiren *et al.,* 1999). However, the loading of Zn on the Zn-chelator is yet to be established, although both Fe and Zn are easily transported through phloem. During deficiency, remobilization of Fe and Zn from the rich older leaves takes time till the senescence is initiated in the older tissue, and thus, the deficiency symptoms become more prominent in the younger leaves rather than the older leaves. In general, mineral concentration in the leaf blades on a dry matter basis is higher than in other tissues (Frossard *et al.,* 2000).

Genetic loci for high Fe and Zn

QTL mapping studies by Lungaho *et al.* (2011), Simic *et al.* (2011), Qin *et al.* (2012), Baxter *et al.* (2013) and Jin *et al.* (2013) reported that accumulation of kernel Fe and Zn in maize is controlled by numerous genetic loci having minor effects. This suggests that there are different sets of genes governing the accumulation of micronutrients, and genetic improvement could be undertaken independent of each of the traits. Using genome sequence data, candidate genes for Fe and Zn transporters were predicted in maize (Chauhan, 2006; Sharma and Chauhan, 2008; Jin *et al.,* 2015). Further, over-expression of nicotianamine synthase (NAS) and nicotianamine aminotransferase (NAAT) leads to increased phytosiderophore synthesis (Zheng *et al.,* 2010). Expression of soybean ferritin transgene in maize endosperm altered the expression of native Fe homeostasis genes and accumulated significantly higher concentrations of calcium and magnesium in addition to Fe (Kanobe *et al.,* 2013).

Development of high Fe and Zn rich maize genotype

Sufficient genetic variation for kernel Fe and Zn in maize has been reported worldwide (Banziger and Long 2000; Pixley *et al.,* 2011). In India, Chakraborti *et al.* (2009) [Fe: 13.23-40.09, Zn: 13.44-46.39], Chakraborti *et al.* (2011a) [Fe: 12.02-38.46, Zn: 17.57-49.14], Chakraborti *et al.* (2011b) [Fe: 13.95-39.31, Zn: 21.85-40.91], Prasanna *et al.* (2011) [Fe: 11.28-60.11, Zn: 15.14-52.95], Agrawal *et al.* (2012) [Fe: 20.38-54.29, Zn: 7.01-29.88], Guleria *et al.* (2013) [Zn: 3.81-35.83], Mallikarjuna *et al.* (2014) [Fe: 16.6-83.4, Zn: 16.4-53.2], Mallikarjuna *et al.* (2015) [Fe: 18.88-47.65, Zn: 5.41-30.85] and Pandey *et al.* (2015) [Fe: 23.8-42.7, Zn: 12.6-39.4] reported wide genetic variation for kernel Fe and Zn in maize germplasm. In QPM germplasm, *o2* was found to have influence on higher accumulation of Fe and Zn in maize (Arnold *et al.,* 1977; Welch *et al.,* 1993). Chakraborti *et al.* (2009) reported significant difference between normal and QPM inbreds for kernel Zn concentration, although the same could not be found for kernel Fe. However, significant influence of G × E interaction for Fe and Zn has been observed in maize, and especially soil profile plays pivotal role in accumulation (Long *et al.,* 2004; Oikeh *et al.,* 2004; Prasanna *et al.,* 2011;

Agrawal *et al.,* 2012). In India, hybrids with higher Fe (~40 μg/g) or/and Zn (~35 μg/g) have been identified, but the levels are below the target level set by HarvestPlus. Integration of modern genomic approaches with breeding efforts is the way forward to develop hybrid maize with the required level of Fe and Zn (Gupta *et al.,* 2015c; Jha *et al.,* 2015).

Enhancement of Fe and Zn in maize grains through fertilizer application has also been attempted as a short term complementary approach to genetic biofortification (Velu *et al.,* 2014). For deficient soils, the agronomic approach for enhancement of Zn has been very effective, however it is ineffective to enhance Fe (Aciksoz *et al.,* 2011). Fe fortification can be done more effectively by adding nitrogen (N) rather than Fe fertilization. The N fertilization also improves Zn due to enhanced mobilization and transport, a result of increased root growth and phloem retranslocation (Cakmak *et al.,* 2010). Soil Zn applications are, however, less effective in increasing grain Zn, compared to foliar Zn applications that results in remarkable increase in grain Zn concentration. Though foliar application is effective, soil-applied nutrients may also contribute to the micronutrient supplementation. Further, agronomic biofortification reduces the accumulation of seed phytic acid, as Zn in poor soils reduces the absorption of phosphorus and therefore synthesis of phytic acid (Lyons and Cakmak, 2012). Such a feature improves Zn bioavailaibility in the digestive system (Cakmak 2008). Although not much information is available on agronomic biofortification in corn, some studies indicated that it possesses good potential in the South Asian, South Africa, Zimbabwe, and other African countries (Prasad *et al.,* 2014). The combined soil and foliar application recorded 27% higher Zn concentration in maize grain as well as in stover, with the treatments falling in the following order of combined Γfoliar Γsoil through Zn-coated urea Γsoil (Shivay and Prasad, 2014). A study in Zimbabwe showed that the application of cattle manure (supplying 113 g Zn/ha) NPK and leaf litter (supplying 430 g Zn/ ha) NPK significantly increased Zn concentration in corn grain over NPK alone (Manzeke *et al.,* 2012).

Bioconversion of Fe and Zn in human

Only 5% of the Fe and 25% of the Zn are bioavailable in human gut (Bouis *et al.,* 2011). The bioavailability can be increased by reducing the concentration of 'antinutrients' such as phytate, oxalate, polyphenolics (tannins) which interfere with the absorption of Fe and Zn by gut, and increasing the concentrations of 'promoter' substances, such as ascorbate (vitamin C), β-carotene, cysteine-rich polypeptides and certain organic compounds and amino acids, which stimulate the absorption of essential mineral elements by the gut (White and Broadley, 2009). Several researchers have reported the synergistic effect of *o2* allele in enhancing the concentration of micronutrients especially Zn (Arnold *et al.,* 1977; Welch *et al.,* 1993; Chakraborti *et al.,* 2009; Pandey *et al.,* 2015). Thus, QPM genotypes which possess higher lysine have the potential to enhance the kernel Fe and Zn concentration and its absorption in digestive system. The presence of β-carotene in maize increased the absorption of Fe up to 1.8-fold (Garcia-Casal *et al.,* 2000). Besides, the addition of lutein in maize-based diet increased the bioavailability of Fe by two folds (Garcia-Casal, 2006). Among various factors, enhancement of bioavailability of Fe and Zn by lowering the levels of phytic acid has emerged as viable and easy option.

Phytic acid in maize kernel

Phytic acid (PA) (*myo*-inositol-1, 2, 3, 4, 5, 6-hexakisphosphate or InsP6) is a ubiquitous and the most abundant inositol phosphate found in all the eukaryotic cells. Phosphate groups impart PA a strong negative charge at cellular pH, as a result it tightly binds positively charged monovalent or bivalent mineral cations to form mixed salts referred to as phytate or phytin. In maize seeds, 88% of phytate accumulates in the embryo, 10% in the outer layers and only 2% in endosperm (O'Dell *et al.,* 1972). In the developing seeds, approximately 80-90% of total phosphorus (P) is converted into PA, a non-available form of P. The remaining 10-20% of total seed P i.e. non-phytic acid P or available P which is found as soluble inorganic phosphate and cellular P (in all the P containing compounds such as nucleic acids, proteins, lipids, carbohydrates etc.). Generally, in traditional maize, matured seeds contain 10-15 mg/g (seed dry weight) of phytic acid while in mutant it is reduced to 5-8 mg/g. At maturity, inorganic P in seed is about 2-7% of seed total P or about 0.2 mg/g (Raboy, 2009). PA serves as a chief reservoir of not only phosphorus and myo-inositol but also of mineral cations which are utilized during seed germination and seedling growth. PA is synthesized in the cytoplasm and then imported inside the protein storage vacuole (Greenwood and Bewley, 1984; Otegui *et al.,* 2002). Evidence accumulated over the past years indicates that plants synthesize PA by two alternative routes *viz.,* (i) lipid independent and (ii) lipid dependent pathway. The first one operates predominantly in seed, whereas in all the other eukaryotic cells of plant system, PA biosynthesis predominantly proceeds via lipid dependent pathway (Raboy, 2001; Sparvoli and Cominelli, 2015).

Phytic acid as anti-nutrient

Owing to the high density of negative charge, PA or $InsP_6$ and lower isomers of inositol phosphate ($InsP_5$, $InsP_4$ and $InsP_3$) are potent chelators of nutritionally important positively charged mineral ions *viz.*, Fe and Zn. When consumed, dietary PA and its isomers continue to bind not only seed derived minerals but also other minerals they encounter in the upper gastrointestinal tract from other food items, rendering them unavailable for absorption in human gut (Moretti *et al.*, 2014). Undigested P from animal wastes reaches water bodies and ultimately contributes to deterioration of water quality.

Genetic loci for low phytic acid

Various methods such as processing, food diversification, food and feed fortification and, transgenic plants and animals, have been adopted over time to avert the undesirable effects of PA thereby providing required amount of minerals and phosphate to animals, hence increasing absorption and mitigating P management problem. However, the simplest, consumer friendly and the most effective strategy to assuage both the environmental and nutritional issues is to breed plants that produce a minimal amount of PA necessary for their survival. In low PA mutants (*lpa*), total phosphorus content remains the same as in wild type grains, however PA content is greatly reduced coupled with proportional increase in free inorganic P (Raboy, 2000; Pilu *et al.*, 2003). Extensive research in seed PA content has led to the isolation of three *lpa* mutations in maize *viz.*, *lpa-1*, *lpa-2* and *lpa-3,* compared to the wild-type kernels, they contain 66%, 50% and 50% less PA, respectively (Raboy, 2000; Shi *et al.*, 2005). *lpa1* located on distal region of 1S chromosome, encodes a multidrug resistance-associated protein (MRP) designated as *ZmMRP4*, which belongs to subfamily of ATP-binding cassette transmembrane transporters (Shi *et al.*, 2007). In mutants, transporters are defective as a result PA cannot be stored inside the protein storage vacuoles, however, PA is synthesized and accumulates in the cytoplasm which leads to a negative feedback mechanism that suppresses the expression of genes involved in PA biosynthesis. So far, four mutant alleles of *lpa1* have been identified in maize namely *lpa1-1*, *lpa1-2*, *lpa1-7* and *lpa1-241*. *lpa2* maps on proximal end of 1S chromosome (Raboy 2000) and arises due to a mutation in inositol phosphate kinase (*ZmIPK*) encoding gene (Shi *et al.*, 2003). Two versions *viz.*, *lpa2-1* and *lpa2*-2 have been identified. Maize *lpa3* gene is located near *adh1* locus on chromosome 1, and it encodes *myo*-inositol kinase (Shi *et al.*, 2005).

Development of low phytate maize cultivar

Phytic acid reduction is generally coupled with negative pleiotropic effects on plant and seed performance, such as reduced germination and emergence rate, stunted vegetative growth, lower seed filling and susceptibility to various biotic and abiotic stresses (Raboy, 2000; Pilu *et al.*, 2005; Doria *et al.*, 2009; Badone *et al.*, 2010). Hence, the challenge now is to breed *lpa* mutants with good agronomic performance and yield stability. The first effort to breed low PA maize inbreds and hybrids was made by Ertl *et al.* (1998) via backcross methods. Fourteen near-isogenic maize hybrid pairs, each consisting of non-mutant and *lpa1-1* mutant were produced and tested. They observed little or no effect on germination potential, stand establishment, lodging, plant height, ear height, growth rate and grain moisture. However, eight *lpa1-1* hybrids exhibited an average yield loss of 5.5% as compared to the non-mutant hybrids. Recently, Beavers *et al.* (2015) developed low PA maize population without negatively affecting seed quality through three rounds of selections in broad based synthetic populations. The *lpa* population thus developed had only 18% more PA than *lpa* mutants. *lpa* population was found to outperform *lpa* mutants in field emergence and agronomic characteristics. The success of *lpa* populations developed with recurrent selection suggests that further cycles of breeding may reduce phytate concentration while preserving seed quality. Naidoo *et al.* (2013) crossed two inbred parental lines *viz.*, CM32 (temperate LPA line) and P16 (tropical wild type line) to produce F1 heterozygotes, and designed SNP primers for *lpa1-1* and used high resolution melt analysis to successfully distinguish among homozygous dominant (wild type), homozygous recessive (*lpa*) and heterozygous genotypes. In India, *lpa2-2* was successfully introgressed into regionally well adapted and productive elite inbred lines *viz.*, UMI 395 and UMI 285 through MAS at TNAU, Coimbatore (Suresh kumar *et al.,* 2014a; Tamilkumar *et al.,* 2014). Further, CM145 and V334 have been introgressed with *lpa1* and *lpa2*, respectively at VPKAS, Almora. These lines can be used for the development of hybrids with low PA. Efforts are also underway at IARI and IIMR to develop low PA maize hybrids (Dosad *et al.,* 2016 and 2017; Muthusamy *et al.,* 2016b).

Drakakaki *et al.* (2005) simultaneously expressed *Aspergillus phytase* (*phyA2*) and soybean *ferritin* genes in the seed endosperm which resulted in upto 95% decrease in phytate and a 3-fold more iron was bioavailable to Caco-2 intestinal epithelial cells compared to non-transgenic grains. Similar results were obtained by Aluru *et al.* (2011) when they over-expressed endosperm specific soybean *ferritin* gene in *lpa1-1* maize using a modified and highly active *γ-zein* promoter. Expression of *Aspergillus phytase* (*phyA2*) gene under the control of maize embryo-specific *globulin-1* promoter resulted in a 50-fold increase in phytase activity without affecting seed germination and other agronomic performance (Chen *et al.,* 2008).

Challenges and opportunities

Maize cultivars with high lysine, tryptophan and provitamin A have been developed; and hybrids for higher vitamin E, Fe, Zn and low phytate are under developmental stages in India. These biofortified maize hybrids hold great promise in providing the balanced food. It is noteworthy to mention here that despite well documented health benefits, QPM cultivars account for only 1% or less of 90 million hectares grown in Mexico, Latin America, Sub-Saharan Africa and Asia

(CIMMYT 2012). India is also not an exception in this regard, despite the availability of dozens of diverse QPM hybrids. Successful adoption of biofortified maize cultivars depends on various factors related to (i) research and development, (ii) socio-economic issues, (iii) awareness generation, and (iii) policy interventions (Gupta *et al.,* 2015a; Anonymous, 2015).

Assurance of high yield

In general, it is perceived that nutritionally enriched crops possess low yielding potential. QPM, provitamin A, vitamin E, Fe and Zn do not have any yield penalty, and nutritionally enriched maize for these traits can provide grain yield comparable to normal maize (Vivek *et al.,* 2008; Gupta *et al.,* 2015). In India, MAS-derived QPM version of Vivek Hybrid-9, HM-4, HM-8 and HM-9 have been tested under the AICRP trials, and were found to be *at par* with the original version for grain yield potential (Gupta *et al.,* 2013; Gupta *et al.,* 2015a). Similarly, the MAS-derived provitamin-A version of Vivek QPM-9, Vivek Hybrid-27, HM-4 and HM-8 also possessed similar grain potential to their original versions (Muthusamy *et al.,* 2014). As a short term goal, improvement of already existing high yielding normal maize hybrids for nutritional quality traits would be an effective strategy to provide diverse choices to the farmers.

Broadening of genetic base

Germplasm base of nutritionally enriched maize are extremely narrow as compared to normal maize, primarily due to the fact that very few breeding centres have active quality breeding programmes (Hossain *et al.,* 2016). Thus, strengthening breeding programme through development of diverse heterotic pools for nutritional quality traits, and deriving promising inbreds with high *per se* productivity are the way forward. Integration of doubled haploid technology coupled with marker-assisted selection would accelerate the process of germplasm development. To further strengthen the quality breeding programme, ICAR has initiated two important Consortia Research Platform on 'biofortification' and 'molecular breeding', where several research institutes such as (i) IARI, New Delhi, (ii) IIMR, Ludhiana, (iii) VPKAS, Almora, (iv) CCSHAU, Uchani, (v) TNAU, Coimbatore are involved to develop diverse maize inbreds and hybrids for higher lysine, tryptophan, provitamin A, Fe, Zn and low phytate. Further, Department of Biotechnology (DBT), Government of India has also been funding various research projects on biofortification in maize. However, further strengthening of research collaborations among various national partners of the National Agricultural Research System (NARS) and international research organizations like CIMMYT and HarvestPlus would help in sharing novel germplasm and expertise for the development of biofortified maize.

Precise and rapid phenotyping

Establishment of 'nutritional quality service labs' at various places are essential for assessing large number of segregating progenies/inbreds for micronutrients in the breeding programme (Anonymous, 2015). Creation of trained human resources for precise estimation of the nutritional quality is also a key to the success of biofortification programme. Further the effects of micronutrients such as lysine, tryptophan, provitamin A (among yellow/orange maize), vitamin E, Fe and Zn are invisible, and farmers would face difficulty in convincing the trader regarding the extent of quality of his produce while selling in the market. Development of a portable device that rapidly determines the quality of the produce would be of great help to the farmers. Formulation of a separate 'biofortification' trial under the AICRP could aid in precise evaluation of micronutrient-rich entries, as additional precautions during pollination, harvesting and storage are required for biochemical estimation.

Maintenance of quality attributes

Maize hybrids with *o2, crtRB1, lcyE, VTE4* and *lpa* once pollinated by pollen from normal maize, show xenia effects leading to dilution of nutritional quality (Gupta *et al.,* 2015). The level of contamination is high in the border area, while it tends to reduce in the middle of the field. However, the extent of contamination depends on the proximity of the normal maize, synchrony of flowering with other cultivars, direction of wind during pollination; and competitiveness of pollen for fertilization (Vivek *et al.,* 2008). In case of QPM, xenia effects caused contamination to an extent of 11% of the total harvest (Twumasi-Afriyie *et al.,* 1996; Ahenkora *et al.,* 1999). Thus, growing of trait specific biofortified variety in larger area would reduce the loss in nutritional quality. Mechanical mixture of normal maize grains with biofortified grains during threshing and packaging may also lead to the reduction of nutritional benefits. Ahenkora *et al.* (1999) reported that QPM benefits are lost when 20% or more of normal grains are mixed with QPM grains. Thus, the requirement of separate post-harvest processing and storage arrangements for biofortified maize grains is an essentiality to avoid contamination from normal maize grains.

Combination of various micronutrients

Development of multi-nutrient maize instead of single trait is more attractive to the consumers. Combination of (i) QPM + provitamin A, (ii) QPM + vitamin E, (iii) QPM + provitamin A + vitamin E, (iv) QPM + Fe + Zn, and (v) QPM + low phytate, can be easily developed as donors with combination of various traits and robust marker system are now available for introgressing the target genes. Various such combinations are in the developmental stages in India (Das *et al.,*

2016; Zunjare *et al.,* 2016). In India, Vivek QPM-9 with high provitamin A, developed by IARI is first such example of multinutrient maize. 35-40% of the area, especially in the North-Eastern states and tribal belts are still dependent on low yielding local landraces, and incidentally maize is consumed as food in much higher proportion. Deployment of available biofortified maize hybrids with higher yield may prove beneficial for health and well being of the people.

Promotion of food processing- and value-addition- industries

Biofortified maize with higher micronutrients may find important place among health conscious urban population, as consumers are ready to pay 20-70% premium price for the biofortified foods (Steur *et al.,* 2015). Attractive labelling and suitable branding highlighting the health benefits on products made from biofortified maize would help the consumers to choose more nutritious foods over conventionally available ones. Various food products from QPM such as 'Pusa-Shakti', 'Kheer-mix/Dilkhush' Kadhi-mix/Proteino-H' have been developed in RAU, Pusa (Singh and Chandra, 2014). Corn flakes, snack items and maize based multi-grain *atta* also offer attractive options to the urban people. Nutritionally enriched sweet corn kernels with higher lysine, tryptophan, provitamin A and vitamin E would offer attractive options to the urban community over conventional sweet corn grains. To meet the industrial requirement, biofortified maize grains need to be systematically evaluated, and 'contract farming and buy-back policy' would ensure continuous supply of grains to the industry. Technologies such as 'Nixtamalization' enhance the shelf-life of grains and prevention of aflatoxin development on grains offers tremendous advantage, and should be used in India.

Sensitization of poultry industry

In Asian countries including India, 60-70% of the maize is used as animal feed, and biofortified maize is advantageous over normal maize. QPM in poultry diet improves the growth performance of broilers and results in higher weight gains when replaced with normal maize (Panda *et al.,* 2013). Recently, a study on effect of provitamin A biofortified maize diet inclusion on meat quality in Ovambo chickens has been conducted in Africa (Odunitan-Wayas *et al.,* 2016). The results revealed that the provitamin-A fed chickens had higher redness and yellowness, lower lightness in the meat and skin colour than white maize fed chickens. Liu *et al.* (2012) fed chickens with maize grains rich in high β-cryptoxanthin, that eventually increased the provitamin-A levels in egg, besides enhancing the colour of the yolks, a trait consumers prefer over light coloured yolk. Meat quality was also shown to be improved when livestock rations were supplemented with α-tocopherol. Poultry having high level of tissue tocopherol have been correlated with positive effect that has economic ramifications.

Supplementation of vitamin E in poultry enhanced the stability of poultry meat (Dewinne and Dirinck, 1996; Williams, 1997). India being the fifth largest poultry producer and third largest egg producer possesses enormous growth potential. So far, parameters such as body weight gain and number of eggs produced per unit time and feed efficacy have been traditionally used to compare the effects of QPM *vis-a-vis* normal maize on poultry. Since synthetic lysine is available at much cheaper rate, thus high lysine QPM in feed looks less appealing. However, sensitization of poultry growers on other benefits of QPM such as high tryptophan that helps in regulation of egg laying would further help in its popularization. Further, nutritive value of meat and eggs of chicken fed with biofortified maize needs to be analyzed in details. Liu *et al.* (2012) reported that chicken fed with higher β-cryptoxanthin accumulates more β-cryptoxanthin in egg yolk, thereby enhancing the nutritive value of the eggs. Awareness generation among the local poultry growers on advantages of biofortified maize on poultry meat and egg, would help them earn more profits.

Generation of public awareness

Altered phenotypes caused due to introgression of new trait may prove to be a deterrent to the easy acceptability of biofortified grains among consumers. In specific areas including India, white maize is still preferred, and provitamin-A rich maize having orange/yellow colour may not be easily accepted (De Groote and Kimenju, 2008; Muzhingi *et al.,* 2008; De Groote *et al.,* 2010). Strong extension activities may play a major role in the popularization of biofortified crops. Further, specific training of extension workers and volunteers, arrangement of community drama, radio broadcasts, and other activities such as field days, training for grandmothers and community leaders, and market promotion events would help in the promotion of biofortified maize. A study in Zimbabwe revealed that ~94% of the respondent agreed to consume yellow maize instead of traditional white maize, if educated on health benefits (Stevens and Winter-Nelson, 2008). Pillay *et al.* (2011) reported that the participants expressed willingness to consume biofortified maize, if it was cheaper than white maize, and was readily available in local grocery stores. Also, imparting education on the nutritional benefits of consuming yellow maize should be aimed at the adult caregivers, particularly women.

Support through policy intervention

Policy supports from the government are essential for the successful adoption of biofortified maize cultivars (Gupta *et al.,* 2015). Intensive awareness campaign supported by Government would help in popularization of biofortified maize for its nutritional value. The available biofortified maize can potentially contribute to the nutritional security especially in the North-Eastern states and tribal areas in

India. Inclusion of biofortified maize in the 'rural transformation' project of NITI AAYOG, and other Government sponsored programme like National Food Security Mission (NFSM), Rashtriya Krishi Vikas Yojna (RKVY) as well as nutrition intervention programme such as Integrated Child Development Scheme (ICDS) and 'Mid-day meal' would help in further popularization (Anonymous, 2015). In India, a pilot programme on nutri-farms for introducing micronutrient-rich maize (QPM), Fe-rich pearl millet and Zn-rich rice, has been implemented to improve the nutritional status of malnourished people of 100 districts of nine states (Gupta *et al.,* 2015). Assurance of remunerative price through minimum support price and/or premium price for biofortified maize grains in the market will encourage the farmers to grow more biofortified maize. Intervention such as creation of 'Seed village' especially at Tamil Nadu, Karnataka, Eastern Uttar Pradesh, Rajasthan, Bihar and West Bengal would strengthen the seed chain to produce and supply good quality seeds of biofortified maize to industry. Providing subsidized seeds and other inputs would further contribute to the rapid dissemination of nutritionally improved cultivars among the farmers. Value-added products provide excellent opportunities to develop small scale industry and in turn empower village women, providing loan at subsidized rate would help in popularizing biofortified maize in India.

References

Abbasi, A.R., Hajirezaei, M., Hofius, D., Sonnewald, U. and Voll, L.M. (2007). Specific roles of a and g-tocopherol in abiotic stress responses of transgenic tobacco. *Plant Physiol.* 143(4): 1720-1738.

Abdala-Valencia, H., Berdnikovs, S. and Cook-Mills, J.M. (2013). Vitamin E isoforms as modulators of lung inflammation. *Nutrients.* 5: 4347-4363.

Aciksoz, S.B., Yazici, A., Ozturk, L. and Cakmak, I. (2011). Biofortification of wheat with iron through soil and foliar application of nitrogen and iron fertilizers. *Plant Soil.* 349(1-2): 215-225.

Acuff, R.V., Thedford, S.S., Hidiroglou, N.N., Papas, A.M. and Odom, T.A. Jr. (1994). Relative bioavailability of RRR- and all-rac-alpha-tocopheryl acetate in humans: Studies using deuterated compounds. *Am J Clin Nutr.* 60(3): 397-402.

Agrawal, P.K., Jaiswal, S.K., Prasanna, B.M., Hossain, F., Saha, S., Guleria, S.K. and Gupta, H.S. (2012). Genetic variability and stability for kernel iron and zinc concentration in maize (*Zea mays* L.) genotypes. *Indian J Genet.* 72: 421-428.

Agrawal, V.K., Singh, R.M., Shahi, J.P., Agrawal, R.K. and Chaudhary, D.P. (2015). Inheritance of kernel quality attributes in quality protein maize (*Zea mays L.*). *Ind J Genet.* 75(2): 201-207.

Ahenkora, K., Twumasi-Afriyie, S., Sallah, P.Y.K. and Obeng-Antwi, K. (1999). Protein nutritional quality and consumer acceptability of tropical Ghanaian quality protein maize. *Food Nutr Bull.* 20: 354-360.

Aluru, M.R., Rodermel, S.R. and Reddy, M.B. (2011). Genetic modification of low phytic acid 1-1 maize to enhance iron content and bioavailability. *J Agric Food Chem*. 59: 12954-12962.

Aluru, M., Xu, Y., Guo, R., Wang, Z., Li, S., White, W., Wang, K. and Rodermel, S. (2008). Generation of transgenic maize with enhanced provitamin-A content. *J Exp Bot*. 59: 3551-3562.

Anonymous. (2015). Proceeding of brain storming workshop on up-scaling quality protein maize for nutritional security. May 20-21, NASC, New Delhi.

Arnold, J.M., Bauman, L.F. and Aycock, H.S. (1977). Inter relations among protein, lysine, oil, certain mineral element concentrations and physical kernel characteristics in two maize populations. *Crop Sci*. 17: 421-425.

Babu R., Rojas, N.P., Gao, S., Yan, J. and Pixley, K. (2013). Validation of the effects of molecular marker polymorphisms in *lcyE* and *crtRB1* on provitamin-A concentrations for 26 tropical maize populations. *Theor Appl Genet*. 126: 389-399.

Badone, C., Cassani, E., Landoni, M., Doria, E., Panzeri, D., Lago, C., Mesiti, F., Nielsen, E. and Pilu, R. (2010). The *low phytic acid1-241 (lpa1-241)* maize mutation alters the accumulation of anthocyanin pigment in the kernel. *Planta*. 231: 1189-1199.

Bain, L.E., Awah, P.K., Geraldine, N., Kindong, N.P, Sigal, Y. and Bernard, N. (2013). Malnutrition in Sub-Saharan Africa: Burden, causes and prospects. *Pan Afr Med J*. 15: 1-9.

Balconi, C., Hartings, H., Lauria, M., Pirona, R., Rossi, V. and Motto, M. (2007). Gene discovery to improve maize grain quality traits. *Maydica*. 52: 357-373.

Banziger, M. and Long, J. (2000). The potential for increasing the iron and zinc density of maize through plant breeding. *Food Nutr Bull*. 21: 397-400.

Baveja, A., Hossain, F., Panda, K.K., Muthusamy, V., Zunjare, R. *et al.* (2016). Development of diverse maize genetic resource possessing *crtRB1*, *opaque2* and *shrunken2* alleles through marker-assisted selection. In 1st International Agrobiodiversity Congress. November 6-9, 2016, New Delhi, India, Organized by Indian Society of Plant Genetic Resources & Bioversity International, Abstract Book. 189.

Baxter, I.R., Gustin, J.L., Settles, A.M. and Hoekenga, O.A. (2013). Ionomic characterization of maize kernels in the intermated B73 x Mo17 population. *Crop Sci*. 53: 208-220.

Beavers, A.W., Goggi, A.S., Reddy, M.B., Lauter, A.M. and Scott, M.P. (2015). Recurrent selection to alter grain phytic acid concentration and iron bioavailability. *Crop Sci*. 55: 2244-2251.

Bell, P.F., Chaney, R.L. and Angle, J.S. (1991). Determination of the copper Zt activity required by maize using chelator-buffered nutrient solutions. *Soil Sci Soc Am J*. 55: 1366-1374.

Birringer, M., Pfluger, P., Kluth, D., Landes, N. and Brigelius-Flohe, R. (2002). Identities and differences in the metabolism of tocotrienols and tocopherols in HepG2 cells. *J Nutr*. 132: 3113-3118.

Bityutskii, N., Magnitski, S., Lapshina, I., Lukina, E., Soloviova, A. and Patsevitch, V. (2001). Distribution of micronutrients in maize grains and their mobilization during germination. In: Horst, W.J. *et al.* (Eds.), Plant nutrition-food security and sustainability of agro-ecosystems. 218-219.

Bityutskii, N.P., Magnitskiy, S.V., Korobeynikova, L.P., Lukina, E.I., Soloviova, A.N., Patsevitch, V.G., Lapshina, I.N. and Matveeva, G.V. (2002). Distribution of iron, manganese and zinc in mature grain and their mobilization during germination and early seedling development in maize. *J Plant Nutr*. 25: 635-653.

Bjarnason, M. and Vasal, S.K. (1992). Breeding of quality protein maize. *Plant Breed Rev.* 9: 181-216.

Black, R.E., Victora, C.G., Walker, S.P., Bhutta, Z.A., Christian, P. *et al.* (2013). Maternal and child undernutrition and overweight in low-income and middle-income countries. *Lancet*. 382 (9890): 427-451.

Bouis, H.E. and Welch, R.M. (2010). Biofortification - a sustainable agricultural strategy for reducing micronutrient malnutrition in the global south. *Crop Sci*. 50: 20-32.

Bouis, H.E., Hotz, C., Mcclafferty, B., Meenakshi, J.V. and Pfeiffer, W.H. (2011). Biofortification: A new tool to reduce micronutrient malnutrition. *Food Nutr Bull.* 32(1): 31-40.

Bramley, P., Elmadfa, I., Kafatos, A., Kelly, F., Manios, Y., Roxborough, H., Schuch, W., Sheehy, P. and Wagner, K.H. (2000). Vitamin E. *J Sci Food Agric.* 80: 913-938.

Bressani, R. (1994). *Opaque-2* corn in human nutrition and utilization. In: Larkins, B.A. and Mertz, E.T. (Eds.), Proceedings of International Symposium on Quality Protein Maize. 41- 63.

Brigelius-Flohe, R. and Traber, M.G. (1999). Vitamin E: function and metabolism. *FASEB J.* 13: 1145-1155.

Brkic, I., Simic, D., Zdunic, Z., Jambrovic, A., Ledencan, T., Kovacevic, V. and Kadar, I. (2003). Combining abilities of corn-belt inbred lines of maize for mineral content in grain. *Maydica*. 48: 293-297.

Brown, K., Peerson, J.M. and Allen, L.H. (1998). Effects of zinc supplementation on children's growth. In: Sandstrom, B. and Walter, P. (Eds.), Role of Trace Elements for Health Promotion and Disease Prevention, Bibliotheca Nutritioet Dieta. 54: 76- 83.

Burt, A.J., Grainger, C.M., Smid, M.P., Shelp, B.J. and Lee, E.A. (2011). Allele mining of exotic maize germplasm to enhance macular carotenoids. *Crop Sci*. 51: 991-1004.

Burton, G. (1993). Vitamin E: Molecular and biological function. In: Proceedings of the Nutrition Society, Nottingham. 12-15: 251-262.

Cakmak, I. (2008). Enrichment of cereal grains with zinc: agronomic or genetic biofortification. *Plant Soil.* 302: 1-17.

Cakmak, I., Pfeiffer, W.H. and Mcclafferty, B. (2010). Biofortification of durum wheat with zinc and iron. *Cereal Chem*. 87: 10-20.

Chakraborti, M., Prasanna, B.M., Hossain, F. and Singh, A. (2011a). Evaluation of single cross Quality Protein Maize (QPM) hybrids for kernel iron and zinc concentrations. *Indian J Genet.* 71(4): 312-319.

Chakraborti, M., Prasanna, B.M., Hossain, F., Mazumdar, S., Singh, A.M., Guleria, S.K. and Gupta, H.S. (2011b). Identification of kernel iron- and zinc-rich maize inbreds and analysis of genetic diversity using microsatellite markers. *J Plant Biochem Biotech.* 20(2): 224-233.

Chakraborti, M., Prasanna, B.M., Hossain, F., Singh, A.M. and Guleria, S.K. (2009). Genetic evaluation of kernel Fe and Zn concentrations and yield performance of selected Maize (*Zea mays* L.) genotypes. *Range Manag Agrofor.* 30: 109-114.

Chakraborti, M., Prasanna, B.M., Singh, A. and Hossain, F. (2010). Generation mean analysis of kernel iron and zinc concentrations in maize (*Zea mays* L.). *Indian J Agric Sci.* 80(11): 956-959.

Chander, S., Guo, Y.Q., Yang, X. H. Zhang, J. and Lu, X.Q. (2008a). Using molecular markers to identify two major loci controlling carotenoid contents in maize grain. *Theor Appl Genet.* 116: 223-233.

Chander, S., Guo, Y.Q., Yang, X.H., Yan, J.B., Zhang, Y.R., Song, T.M. and Li, J.S. (2008b). Genetic dissection of tocopherol content and composition in maize grain using quantitative trait loci analysis and the candidate gene approach. *Mol Breed.* 22(3): 353-365.

Chaudhary, D.P., Mandhania, S.S. and Kumar, R. (2012). Inter-relationship among nutritional quality parameters of maize (*Zea mays*) genotypes. *Ind J Agric Sci.* 82: 681-686.

Chauhan, H.S., Hossain, F., Muthusamy, V., Zunjare, R., Das, A.K., Saha, S. and Gupta, H.S. (2016). Utilizing the exotic germplasm for development of multi-vitamin rich sweet corn genotypes through marker-assisted breeding. In: 1st International Agrobiodiversity Congress. November 6-9, 2016, New Delhi, India. Organized by Indian Society of Plant Genetic Resources & Bioversity International, Abstract Book. 217.

Chauhan, R.S. (2006). Bioinformatics approach towards identification of candidate gene for zinc and iron transporters in maize. *Curr Sci.* 91: 510-515.

Chen, R., Xue, G., Chen, P., Yao, B., Yang, W., Ma, Q., Fan, Y., Zhao, Z., Tarczynski M.C. and Shi, J. (2008). Transgenic maize plants expressing a fungal *phytase* gene. *Transgenic Res.* 17: 633-643.

Chhabra, R., Hossain, F., Muthusamy, V. and Zunjare, R. (2016). Molecular characterization of *shrunken2* gene encoding large subunit of ADP-glucose pyrophosphorylase for its utilization in sweet corn breeding. In: 1st International Agrobiodiversity Congress. November 6-9, 2016, New Delhi, India, Organized by Indian Society of Plant Genetic Resources & Bioversity International, Abstract Book. 194.

Choudhary, M., Hossain, F., Muthusamy, V., Thirunavukkarasu, N., Saha, S., Pandey, N., Jha, S.K. and Gupta, H.S. (2015). Microsatellite marker-based genetic diversity analyses of novel maize inbreds possessing rare allele of *b-carotene hydroxylase (crtRB1)* for their utilization in b-carotene enrichment. *J Plant Biochem Biotechnol.* DOI: 10.1007/s13562-015-0300-3.

Choudhary, M., Muthusamy, V., Hossain, F., Thirunavukkarasu, N., Pandey, N., Jha, S.K. and Gupta, H.S. (2014). Characterization of b-carotene rich MAS-derived maize inbreds possessing rare genetic variation in *b-carotene hydroxylase* gene. *Ind J Genet.* 74(4): 620-623.

CIMMYT. (2012). http://www.cimmyt.org/en/what-we-do/maize-research/item/kernels-with-a-kick-quality-protein-maize-improves-child-nutrition.

Clemens, S., Palmgren, M.G. and Kramer, U. (2002). A long way ahead: understanding and engineering plant metal accumulation. *Trends Plant Sci.* 7: 309-315.

Das, A.K. and Singh, N.K. (2012). Carotenoid and SSR marker- based diversity assessment among short duration maize (*Zea mays* L.) genotypes. *Maydica* 57: 106-113.

Das, A.K., Hossain, F., Muthusamy, V., Zunjare, R., Chauhan, H.S., Baveja, A., Saha, S. and Gupta, H.S. (2016). Characterization of Indian maize germplasm for kernel a-tocopherol using functional markers. In: 1st International Agrobiodiversity Congress. November 6-9, 2016, New Delhi, India. Organized by Indian Society of Plant Genetic Resources & Bioversity International, Abstract Book. 216.

Davis, C., Jing, H., Howe, J.A., Rocheford, T. and Tanumihardjo, S.A. (2008). b-Cryptoxanthin from supplements or carotenoid-enhanced maize maintains liver vitamin A in Mongolian gerbils (*Meriones unguiculatus*) better than or equal to b-carotene supplements. *Br J Nutr.* 100: 786-793.

De Groote, H. and Kimenju, S.C. (2008). Comparing consumer preferences for color and nutritional quality in maize: application of a semi-double bound logistic model on urban consumers in Kenya. *Food Policy*. 33: 362-270.

De Groote, H., Kimenju, S.C. and Morawetz, U.B. (2010). Estimating consumer willingness to pay for food quality with experimental auctions: the case of yellow versus fortified maize meal in Kenya. *Agric Econ.* 41: 362-370.

DellaPenna, D. and Pogson, B.J. (2006). Vitamin synthesis in plants: Tocopherols and carotenoids. *Annu Rev Plant Biol.* 57: 711-738.

DeMaeyer, E. and Adiels-Tegman, M. (1985). The prevalence of anaemia in the world. *World Health Stat Q.* 38: 302-316.

Dewinne, A. and Dirinck, P. (1996). Studies on vitamin E and meat quality. 2. Effect of feeding high vitamin E levels on chicken meat quality. *J Agric Food Chem.* 44: 1691-1696.

Doria, E., Galleschi, L., Calucci, L., Pinzino, C., Pilu, R., Cassani, E. and Nielsen, E. (2009). Phytic acid prevents oxidative stress in seeds: evidence from a maize (*Zea mays* L.) low phytic acid mutant. *J Exp Bot.* 60: 967-978.

Dosad, S., Muthusamy, V., Bellundagi, A., Jaiswal, S.K., Zunjare, R. and Hossain, F. (2016). Characterization of exotic- and indigenous maize genotypes for phytic acid and *lpa* genes. In: 1st International Agrobiodiversity Congress. November 6-9, 2016, New Delhi, India, Organized by Indian Society of Plant Genetic Resources & Bioversity International, Abstract Book. 190.

Dosad, S., Muthusamy, V., Zunjare, R.U., Jaiswal, S.K., Bhatt, V., Bellundagi, A. and Hossain, F. (2017). Influence of low phytic acid (*lpa*) mutants on seed quality traits in maize. In: XIV National Seed Seminar "Food Security Through Augmented Seed Supply Under Climate Uncertainties", January 28-30, 2017, Dr. B.P. Pal Auditorium, ICAR-IARI, New Delhi, India. 268- 269.

Drakakaki, G., Marcel, S., Glahn, R.P., Lund, E.K., Pariagh, S., Fisher, R., Christou, P. and Stoger, E. (2005). Endosperm specific coexpression of recombinant soybean ferritin and *Aspergillus* phytase in maize results in significant increases in the levels of bioavailable iron. *Plant Mol Biol.* 59: 869-880.

Dwyer, J., Nahin, R.L., Rogers, G.T., Barnes, P.M., Jacques, P.M., Sempos, C.T. and Bailey, R. (2013). Prevalence and predictors of children's dietary supplement use: the 2007 National Health Interview Survey. *J Am Clin Nutr.* 97: 1331-1337.

EFSA Panel on Additives and Products or Substances Used in Animal Feed (FEEDAP). (2010). Scientific Opinion on the Safety and Efficacy of Vitamin E as a Feed Additive for All Animal Species. *EFSA J.* 8(6): 1635.

Egesel, C.O., Wong, J.C., Lambert, R.J. and Rocheford, T.R. (2003). Combining ability of maize inbreds for carotenoid and tocopherols. *Crop Sci.* 43: 818-823.

Ertl, D. S., Young, K.A. and Raboy, V. (1998). Plant genetic approaches to phosphorus management in agricultural production. *J Environ Qual.* 27: 299-304.

Feng, F., Deng, F., Zhou, P., Yan, J., Wang, Q., Yang, R. and Li, X. (2013). QTL mapping for the tocopherols at milk stage of kernel development in sweet corn. *Euphytica.* 193(3): 409-417.

Feng, F., Wang, Q., Liang, C., Yang, R. and Li, X. (2015b). Enhancement of tocopherols in sweet corn by marker-assisted back-crossing of *ZmVTE4*. *Euphytica.* 206(2): 513-521. DOI:10.1007/s10681-015-1519-8

Feng, F., Wang, Q., Zhang, J., Yang, R. and Li, X. (2015a). Assessment of carotenoid and tocopherol level in sweet corn inbred lines during kernel development stages. 75(2): 196-200.

Frank, J., Chin, X.W., Schrader, C., Eckert, G.P. and Rimbach, G. (2012). Do tocotrienols have potential as neuroprotective dietary factors? *Ageing Res Rev.* 11: 163-180.

Frossard, E., Bucher, M., Machler, F., Mozafar, A. and Hurrell, R. (2000). Potential for increasing the content and bioavailability of Fe, Zn and Ca in plants for human nutrition. *J Sci Food Agric.* 80: 861-879.

Galili, G. (2002). New insights into the regulation and functional significance of lysine metabolism in plants. *Annu Rev Plant Biol.* 7: 153-156.

Gannon, B., Kaliwile, C., Arscott, S.A., Schmaelzle, S., Chileshe, J., Kalungwana, N., Mosonda, M., Pixley, K., Masi, C. and Tanumihardjo, S.A. (2014). Biofortified orange maize is as efficacious as a vitamin A supplement in Zambian children even in the presence of high liver reserves of vitamin A: a community based, randomized placebo-controlled trial. *Am J Clin Nutr.* DOI:10.3945/ajcn.114.087379.

Garcia-Casal, M. (2006). Carotenoids increase iron absorption from cereal based food in the human. *Nutr Res.* 26: 340-344.

Garcia-Casal, M.N., Leets, I. and Layrisse, M. (2000). b-carotene and inhibitors of iron absorption modify iron uptake by Caco-2 cells. *J Nutr.* 130: 5-9.

Geetha, K.B., Lending, C.R., Lopes, M.A., Wallace, J.C. and Larkins, B.A. (1991). *Opaque-2* modifiers increase gamma-zein synthesis and alter its spatial distribution in maize endosperm. *Plant Cell.* 3: 1207-1219.

Geevers, H.O. and Lake, J.K. (1992). Development of modified *opaque2* maize in South Africa. In: Mertz, E.T. (Ed.), Quality Protein Maize. American Association of Cereal Chemists, St. Paul. 49-78.

Ghandilyan, A., Vreugdenhil, D. and Aarts, M.G.M. (2006). Progress in the genetic understanding of plant iron and zinc nutrition. *Physiol Plant*. 126: 407-417.

Gibson, R.S. (2006). Zinc: the missing link in combating micronutrient malnutrition in developing countries. *Proc Nutr Soc*. 65: 51-60.

Goldenberg, R.L., Tamura, T., Neggers, Y., Copper, R.L., Johnston, K.E., DuBard, M.B. and Hauth, J.C. (1995). The effect of zinc supplementation on pregnancy outcome. *JAMA*. 274: 463-468.

Goswami, R., Hossain, F., Khan, S., Muthusamy, V., Baveja, A. *et al.* (2015). Marker-assisted introgression of *crtRB1*-favourable allele into QPM version of an elite maize inbred-HKI1128 for enrichment of kernel b-carotene. In: National symposium on Germplasm to genes: Harnessing Biotechnology for Food Security and Health, 9-11, August 2015, ICAR-NRCPB, New Delhi. 86.

Graham, R.D. and Stangoulis, J.C.R. (2003). Trace element uptake and distribution in plants. *J Nutr*. 133: S1502-S1505.

Graham, R.D., Welch, R.M., Saunders, D.A., Ortiz-Monasterio, I., Bouis, H.E., Bonierbale, M., de Haan, S., Burgos, G., Thiele, G. and Liria R. (2007). Nutritious subsistence food systems. *Adv Agron*. 92: 1-74.

Grams, G.W., Blessin, C.W. and Inglett, G.E. (1970). Distribution of tocopherols within the corn kernel. *J Am Oil Chem Soc*. 47: 337-339.

Greenwood, J.S. and Bewley, J.D. (1984). Subcellular distribution of phytin in the endosperm of developing castor bean: a possibility for its synthesis in the cytoplasm prior to deposition within protein bodies. *Planta* . 160: 113-120.

Guimaraes, P.E.O., Schaffert, R.E., Ribeiro, P.E.A., Sena, M.R., Costa, L.P. *et al.* (2004). Mineral grain content and association among Zn and other mineral in yellow QPM and normal endosperm maize lines, Harvest Plus. 2004. MILHO biofortificado (Brochure), Washington.

Guleria, S.K., Chahota, R.K., Kumar, P., Kumar, A., Prasanna, B.M., Hossain, F., Agrawal, P.K. and Gupta, H.S. (2013). Analysis of genetic variability and genotype × year interactions on kernel zinc concentration in selected Indian and exotic maize (*Zea mays* L.) genotypes. *Indian J Agric Sci.* 83(8): 836-841.

Gupta, H.S., Agrawal, P.K., Mahajan, V., Bisht, G.S., Kumar, A. *et al.* (2009). Quality protein maize for nutritional security: rapid development of short duration hybrids through molecular marker assisted breeding. *Curr Sci*. 96: 230-237.

Gupta, H.S., Hossain, F. and Muthusamy, V. (2015a). Biofortification of maize: An Indian perspective. *Indian J Genet*. 75(1): 1-22.

Gupta, H.S., Hossain, F. and Muthusamy, V. (2015b). Development of biofortified maize through molecular breeding. ISB News Report. Virginia Polytechnic Institute and State University. 1-5.

Gupta, H.S., Hossain, F., Nepolean, T., Vignesh, M. and Mallikarjuna, M.G. (2015c). Understanding genetic and molecular bases of Fe and Zn accumulation towards development of micronutrient enriched maize. In: Rakshit, A., Singh, H.B. and Sen, A. (Eds.), Nutrient Use Efficiency: From Basics to Advances, Springer Publications: 255-282.

Gupta, H.S., Raman, B., Agrawal, P.K., Mahajan, V., Hossain, F. and Nepolean, T. (2013). Accelerated development of quality protein maize hybrid through marker-assisted introgression of *opaque*-2 allele. *Plant Breed.* 132: 77-82.

Habben, I.E., Kirleis, A.W. and Larkins, B.A. (1993). The origin of lysine-containing proteins in *opaque*-2 maize endosperm. *Plant Mol Biol.* 23: 825-838.

Hall, J.L. and Williams, L.E. (2003). Transitional metal transporters in plants. *J Exp Bot.* 54:2601-2613.

Harjes, C.E., Rocheford, T.R., Bai, L., Brutnell, T.P., Kandianis, C.B. *et al.* (2008). Natural genetic variation in *lycopene epsilon cyclase* tapped for maize biofortification. *Science.* 319:330-333.

Hashem, R.M., Hassanin, K.M., Rashed, L.A., Mahmoud, M.O. and Hassan, M.G. (2016). Effect of silibinin and vitamin E on the ASK1-p38 MAPK pathway in d-galactosamine/lipopolysaccharide induced hepatotoxicity. *Exp Biol Med.* 241(11): 1250-1257.

Hosomi, A., Arita, M., Sato, Y., Kiyose, C., Ueda, T., Igarashi, O. and Inoue, K. (1997). Affinity for a-tocopherol transfer protein as a determinant of the biological activities of vitamin E analogs. *FEBS Letters.* 409(1): 105–108. DOI:10.1016/S0014-5793(97)00499-7

Hossain, F., Muthusamy, V., Bhat, J.S., Jha, S.K., Zunjare R., Das, A., Sarika, K. and Kumar, R. (2016). Maize: Utilization of genetic resources in maize improvement. In: Broadening the Genetic Base of Grain Cereals. Springer Publication. 10.1007/978-81-322-3613-9_4.

Hossain, F., Nepolean, T., Pandey, N., Vishwakarma, A.K., Saha, S. *et al.* (2014). Development of quality protein maize hybrids through marker-assisted breeding strategy. In: Prasanna, B.M. *et al.* (Eds.), 12th Asian Maize Conference and Expert Consultation on Maize for Food, Feed and Nutritional Security, Book of Abstracts. 30 October - 01 November, Bangkok, Thailand: 65.

Hossain, F., Nepolean, T., Vishwakarma, A.K., Pandey, N., Prasanna, B.M. and Gupta, H.S. (2015). Mapping and validation of microsatellite markers linked to *sugary1* and *shrunken2* genes in maize (*Zea mays* L.). *J Plant Biochem Biotechnol.* 24(2): 135-142.

Hossain, F., Prasanna, B.M., Kumar, R. and Singh, B.B. (2008a). Genetic analysis of kernel modification in Quality Protein Maize (QPM) genotypes. *Indian J Genet.* 68(1): 1-9.

Hossain, F., Prasanna, B.M., Kumar, R. and Singh, B.B. (2008b). The genotype x pollination mode interaction affects kernel modification in Quality Protein Maize (QPM) genotypes. *Indian J Genet.* 68(2): 132-138.

Hossain, F., Prasanna, B.M., Kumar, R., Singh, S.B., Singh, R., Prakash, O. and Warsi, M.Z.K. (2007a). Genetic analysis of grain yield and endosperm protein quality in Quality Protein Maize (QPM) lines. *Indian J Genet.* 67(4): 315-322.

Hossain, F., Prasanna, B.M., Sharma, R.K., Kumar, P. and Singh, B.B. (2007b). Evaluation of Quality Protein Maize (QPM) genotypes for resistance to stored grain weevil, *Sitophilus oryzae*. *Int J Trop Insect Sci.* 27(2): 114-121.

Howe, J.A. and Tanumihardjo, S.A. (2006). Evaluation of analytical methods for carotenoid extraction from biofortified maize (*Zea mays* L.). *J Agric Food Chem.* 54(21): 7992-7997.

IFPRI. (2016a). Global food policy report. Washington, DC: International Food Policy Research Institute.

IFPRI. (2016b). Global nutrition report: From promise to impact, ending malnutrition by 2030. Washington, DC.

Institute of Medicine. (2000). Dietary reference intakes for vitamin C, vitamin E, selenium and carotenoids. *Food and Nutrition Board.* National Academy Press, Washington.

Institute of Medicine. (2001). Dietary reference intakes for vitamin A, vitamin K, arsenic, boron, chromium, copper, iodine, iron, manganese, molybdenum, nickel, silicon, vanadium and zinc. National Academy Press, Washington.

Jha, U.C., Bhat, J.S., Patil, B.S., Hossain, F. and Barh, D. (2015). Functional Genomics: Applications in Plant Science. In: PlantOmics: The Omics of Plant Science, Springer India. 65-111.

Jia, M., Wu, H., Clay, K.L., Jung, R., Larkins, B.A. and Gibbon, B.C. (2013). Identification and characterization of lysine-rich proteins and starch biosynthesis genes in the *opaque2* mutant by transcriptional and proteomic analysis. *BMC Plant Biol.* 13:60 DOI: 10.1186/1471-2229-13-60.

Jiang, T., Christian, P., Khatry, S.K., Wu, L. and West Jr., K.P. (2005). Micronutrient deficiencies in early pregnancy are common, concurrent, and vary by season among rural Nepali pregnant women. *J Nutr.* 135: 1106-1112.

Jin T., Zhou J., Chen J., Zhu L., Zhao, Y. and Huang Y. (2013). The genetic architecture of zinc and iron content in maize grains as revealed by QTL mapping and meta-analysis. *Breed Sci.* 63: 317-324.

Jin, T., Chen, J., Zhu, L., Zhao, Y., Guo, J. and Huang, Y. (2015). Comparative mapping combined with homology based cloning of the rice genome reveals candidate genes for grain zinc and iron concentration in maize. *BMC Genet.* 16: 1-15.

Kanobe, M.N., Rodermel, S.R., Bailey, T. and Scott, M.P. (2013). Changes in endogenous gene transcript and protein levels in maize plants expressing the soybean ferritin transgene. *Front Plant Sci.* 4: 196.

Kemper, E.L., Neto, G.C., Papes, F., Moraes, K.C.M., Leite, A. and Arruda, P. (1999). The role of *Opaque2* in the control of lysine-degrading activities in developing maize endosperm. *Plant Cell.* 11: 1981-1993.

Khanduri, A., Hossain, F., Lakhera, P.C. and Prasanna, B.M. (2011). Effect of harvest time on kernel sugar concentration in sweet corn. *Indian J Genet.* 71(3): 231-234.

Khanduri, A., Prasanna, B.M., Hossain, F. and Lakhera, P.C. (2010). Genetic analyses and association studies of yield components and kernel sugar concentration in sweet corn. *Indian J Genet.* 70(3): 257-263.

Kim, Y.N. and Cho, Y.O. (2015). Vitamin E status of 20- to 59-year-old adults living in the Seoul metropolitan area of South Korea. *Nutr Res Pract.* 9: 192-198.

Kiyose, C., Muramatsu, R., Kameyama, Y., Ueda, T. and Igarashi, O. (1997). Biodiscrimination of alpha-tocopherol stereoisomers in humans after oral administration. *Am J Clin Nutr.* 65(3): 785-789.

Lending, C.R. and Larkins, B.A. (1989). Changes in the zein composition of protein bodies during maize endosperm development. *Plant Cell.* 1: 1011-1023.

Li, Q., Yang, X., Xu, S., Cai, Y., Zhang, D., Han, Y. and Yan, J. (2012). Genome-wide association studies identified three independent polymorphisms associated with/ a-tocopherol content in maize kernels. *PLoS ONE.* DOI:10.1371/ journal.pone.0036807.

Li, S., Nugroho, A., Rocheford, T. and White, W.S. (2010). Vitamin A equivalence of the b-carotene in b-carotene-biofortified maize porridge consumed by women. *Am J Clin Nutr.* 92: 1105-1112.

Li, S., Tayie, F.A.K., Young, M.F., Rocheford, T. and White, W.S. (2007). Retention of provitamin A carotenoids in high b-carotene maize (*Zea mays*) during traditional African household processing. *J Agric Food Chem.* 55: 10744-10750.

Lieshout, M.V. and Pee, S.D. (2005). Vitamin A equivalency estimates: understanding apparent differences. *J Am Clin Nutr.* 81: 943-945.

Lipka, A.E., Gore, M.A., Magallanes-Lundback, M., Mesberg, A., Lin, H., Tiede, T. and DellaPenna, D. (2013). Genome-wide association study and pathway level analysis of tocochromanol levels in maize grain. *G3 Genes|Genomes|Genetics.* 3(8): 1287-1299.

Liu, L., Jeffers, D., Zhang, Y., Ding, M., Chen, W., Kang, M.S. and Fan, X. (2015). Introgression of the *crtRB1* gene into quality protein maize inbred lines using molecular markers. *Mol Breed.* 35(8): 154. DOI: 10.1007/s11032-015-0349-7.

Liu, Y.Q., Davis, C.R., Schmaelzle, S.T., Rocheford, T., Cook, M.E. and Tanumihardjo, S.A. (2012). b-Cryptoxanthin biofortified maize (*Zea mays*) increases b-cryptoxanthin concentration and enhances the color of chicken egg yolk. *Poult Sci.* 91(2): 432-438.

Long, J.K., Banziger, M. and Smith, M.E. (2004). Diallel analysis of grain iron and zinc density in Southern African-adapted maize inbreds. *Crop Sci.* 44: 2019-2026.

Lopes, M.A. and Larkins, B.A. (1995). Genetic analysis of *opaque2* modifier gene activity in maize endosperm. *Theor Appl Genet.* 91: 274-281.

Lozoff, B., Jimenez, E. and Xolf, A.W. (1991). Long term development outcome of infants with iron deficiency. *N Engl J Med.* 325: 687-694.

Lungaho, M.G., Mwaniki, A.M., Szalma, S.J., Hart, J., Rutzke, M.A., Kochian, L.V., Glahn, R. and Hoekenga, O.A. (2011). Genetic and physiological analysis of iron biofortification in maize kernels. *PLoS ONE*. 6: e20429.

Lyons, G. and Cakmak, I. (2012). Agronomic biofortification of food crops with micronutrients. Fertilizing crops to improve human health: *A Scientific Review*. 1: 97-122.

Mallikarjuna, M.G., Nepolean, T., Hossain, F., Manjaiah, K.M., Singh, A.M. and Gupta, H.S. (2014). Genetic variability and correlation of kernel micronutrient among exotic quality protein maize inbreds and their utility in breeding programme. *Indian J Genet*. 74(2): 166-173.

Mallikarjuna, M.G., Nepolean, T., Mittal, S., Hossain, F., Bhat, J.S., Manjaiah, K.M., Marla, S.S., Mithra, A.C.R., Agrawal, P.K., Rao, A.R. and Gupta, H.S. (2016). In-silico characterisation and comparative mapping of yellow stripe like transporters in five grass species. *Indian J Agric Sci*. 86(5): 621-627.

Mallikarjuna, M.G., Thirunavukkarasu, N., Hossain, F., Bhat, J.S., Jha, S.K. *et al.* (2015). Stability performance of inductively coupled plasma mass spectrometry-phenotyped kernel minerals concentration and grain yield in maize in different agro-climatic zones. *PLoS One*. 10(10): e0140947.

Manzeke, G.M., Mapfumo, P., Mtambanengwe, F., Chikovo, R., Tendayi, T.A. and Cakmak, I. (2012). Soil fertility management effects on maize productivity and grain Zn content in small holder farming system of Zimbabwe. *Plant and Soil*. 361: 57-69.

Maser, P., Thomine, S., Schroeder, J.I., Ward, J.M., Hirschi, K. *et al.* (2001). Phylogenetic relationships within cation transporter families of *ARabidopsis*. *Plant Physiol*. 126: 1646-1667.

Mayer, J.E., Pfeiffer, W.H. and Beyer, P. (2008). Biofortified crops to alleviate micronutrient malnutrition. *Curr Opinion Plant Biol.* 11: 166-170.

Meier, R., Tomizaki, T., Schulze-Briese, C., Baumann, U. and Stocker, A. (2003). The molecular basis of vitamin E retention: structure of human alpha-tocopherol transfer protein. *J Mol Biol*. 331: 725-734.

Menkir, A. (2008). Genetic variation for grain mineral content in tropical adapted maize inbred lines. *Food Chem*. 110: 454-464.

Mertz, E.T., Bates, L.S. and Nelson, O.E. (1964). Mutant gene that changes protein composition and increases lysine content of maize endosperm. *Science*. 145: 279-280.

Mishra, P. and Singh, N.K. (2010). Spectrophotometric and TLC based characterization of kernel carotenoids in short duration maize. *Maydica* 55: 95-100.

Mitcheva, M., Astroug, H., Drenska, D., Popov, A. and Kassarova, M. (1993). Biochemical and morphological studies on the effects of anthocyans and vitamin E on carbon tetrachloride induced liver injury. *Cell Mol Biol*. 39: 443-448.

Moretti, D., Biebinger, R., Bruins, M.J., Hoeft, B. and Kraemer, K. (2014). Bioavailability of iron, zinc, folic acid, and vitamin A from fortified maize. *Ann N Y Acad Sci*. 1312: 54-65.

Motto, M., Maddolini, M., Panziani, G., Brembilla, M., Marrota, R., di Fonzo, N., Soave, C., Thompson, R. and Salamani, F. (1988). Molecular cloning of the *o2-m5* allele of *Zea mays*, using transposon tagging. *Mol Gen Genet*. 121: 488-494.

Muthusamy, V., Hossain F., Thirunavukkarasu, N., Saha S. and Gupta, H.S. (2015c). Allelic variations for lycopene å-cyclase and β-carotene hydroxylase genes in maize inbreds and their utilization in b-carotene enrichment programme. *Cogent Food Agric*. DOI: 10.1080/23311932.2015.1033141.

Muthusamy, M., Zunjare, R.U., Dosad, S., Bhatt, V., Jaiswal, S.K. and Hossain, F. (2016b). Breeding for low phytate maize to achieve nutritional quality. In: Souvenir and Conference Book, National Conference on Innovative and Current Advances in Agriculture and Allied Sciences, held at PJTSAU, Rajendranagar, Hyderabad (Telangana) from 10-11 December 2016. 140.

Muthusamy, V., Hossain, F., Nepolean, T., Choudhary, M., Saha, S., Bhat, J.S., Prasanna, B.M. and Gupta, H.S. (2014). Development of *b-carotene* rich maize hybrids through marker-assisted introgression of *b-carotene hydroxylase* allele. *PLoS ONE*. 9(12): e113583.

Muthusamy, V., Hossain, F., Thirunavukkarasu, N., Pandey, N., Vishwakarma, A.K., Saha, S. and Gupta, H.S. (2015b). Molecular characterization of exotic and indigenous maize inbreds for biofortification with kernel carotenoids. *Food Biotech*. 29: 276-295.

Muthusamy, V., Hossain, F., Thirunavukkarasu, N., Saha, S., Agrawal, P.K. and Gupta, H.S. (2015a). Genetic variability and inter-relationship of kernel carotenoids among indigenous and exotic maize (*Zea mays* L.) inbreds. *Cereal Res Commun*. 43: 567-578.

Muthusamy, V., Hossain, F., Thirunavukkarasu, N., Saha, S., Agrawal, P.K. and Gupta, H.S. (2016a). Genetic analyses of kernel carotenoids in novel maize genotypes possessing rare allele of *b-carotene hydroxylase* gene. *Cereal Res Commun*. 44(4): 669-680.

Muzhingi, T., Gadaga, T.H., Siwela, A.H., Grusak, M.A., Russell, R.M. and Tang, G. (2011). Yellow maize with high b-carotene is an effective source of vitamin A in healthy Zimbabwean men. *Am J Clin Nutr*. 94: 510-519.

Muzhingi, T., Yeum, K.J., Russell, R.M., Johnson, E.J., Qin, J. and Tang, G. (2008). Determination of carotenoids in yellow maize, the effects of saponification and food preparations. *Int J Vitam Nutr Res*.78: 112-120.

Naidoo, R., Watson, G.M.F., Tongoona, P., Derera, J. and Laing, M.D. (2013). Development of a single nucleotide polymorphism (SNP) marker for detection of the low phytic acid (*lpa1-1*) gene used during maize breeding. *Afr J Biotechnol*. 12: 892-900.

Naqvi, S., Zhu, C., Farre, G., Ramessar, K., Bassie, L., Breitenbach, J., Conesa, D., Ros, G., Sandmann, G., Capell, T. and Christou, P. (2009). Transgenic multivitamin corn through biofortification of endosperm with three vitamins representing three distinct metabolic pathways. *Proc Natl Acad Sci USA*. 106: 7762-7767.

Norvell, W.A. and Welch, R.M. (1993). Growth and nutrient uptake by barley (*Hordeum vulgare* L. cv Herta). Studies using an N-(2-hydroxyethyl) ethylenedinitrilotriacetic acid-buffered nutrient solution technique. I. Zinc ion requirements. *Plant Physiol.* 101:619-625.

O'Dell, B.L., de Boland, A.R. and Koirtyohann, S.T. (1972). Distribution of phytate and nutritionally important elements among the morphological components of cereal grains. *J Agric Food Chem.* 20: 718-721.

Odunitan-Wayas, F.A. Kolanisi, U. Chimonyo, M. and Siwela, M. (2016). Effect of provitamin A biofortified maize inclusion on quality of meat from indigenous chickens. *J Appl Poult Res.* 25: 581-590.

Oikeh S.O., Menkir, A., Maziya-Dixon, B., Welch, R. and Glahn, R.P. (2003). Genotypic differences in concentration and bioavailability of kernel iron in tropical maize varieties grown under field conditions. *J Plant Nutr.* 26: 2307-2319.

Oikeh, S.O., Menkir, A., Maziya-Dixon, B., Welch, R.M., Glahn, R.P. and Gauch, G. (2004). Environmental stability of iron and zinc concentrations in grain of elite early-maturing tropical maize genotypes grown under ûeld conditions. *J Agric Sci.* 142: 543-551.

Olsen, M.S. and Phillips, R.L. (2001). Molecular genetic improvement of protein quality in maize. In: Impacts of Agriculture on human health and nutrition. Encyclopaedia of Life Support Systems-UNESCO.

Ortiz-Monasterio, J.I., Palacios-Rojas, N., Meng, E., Pixley, K., Trethowan, R. and Pena, R.J. (2007). Enhancing the mineral and vitamin content of wheat and maize through plant breeding. *J Cereal Sci.* 46: 293-307.

Otegui, M.S., Capp, R. and Staehelin, L.A. (2002). Developing seeds of *ARabidopsis* store different minerals in two types of vacuoles and in the endoplasmic reticulum. *Plant Cell.* 14: 1311-1327.

Panda, A.K., Prakash, S.V., Rao, R., Raju, M.V.L.N. and Shyam Sundar, G. (2013). Utilisation of high quality protein maize in poultry. *World's Poult Sci.* 69: 877-888.

Pandey, N., Hossain, F., Kumar, K., Vishwakarma, A.K, Muthusamy, V. *et al.* (2016). Molecular characterization of endosperm- and amino acids- modifications among quality protein maize inbreds. *Plant Breed.* 135(1): 47-54.

Pandey, N., Hossain, F., Kumar, K., Vishwakarma, A.K., Nepolean, T. *et al.* (2015). Microsatellite marker-based genetic diversity among quality protein maize (QPM) inbred lines differing for kernel iron and zinc. *Mol Plant Breed.* 6(3): 1-10.

Peter, S., Holguin, F., Wood, L.G., Clougherty, J.E., Raederstorff, D. *et al.* (2015). Nutritional solutions to reduce risks of negative health impacts of air pollution. *Nutrients.* 7(12): 10398-10416.

Pfeiffer, W.H. and Mc Clafferty, B. (2007). HarvestPlus: Breeding crops for better nutrition. *Crop Sci.* 47: S88-S105.

Pillay, K., Siwela, M., Derera, J. and Veldman, F. (2011). Provitamin A carotenoids in biofortified maize and their retention during processing and preparation of South African maize foods. *J Food Sci Technol.* 48: 1-11.

Pilu, R., Landoni, M., Cassani, E., Doria, E. and Nielsen, E. (2005). The maize *lpa-241* cause a remarkable variability of expression and some pleiotropic effects. *Crop Sci.* 45:2096-2105.

Pilu, R., Panzeri, D., Gavazzi, G., Rasmussen, S., Consonni, G. and Nielsen, E. (2003). Phenotypic, genetic and molecular characterization of a maize low phytic acid mutant (*lpa*241). *Theor Appl Genet.* 107: 980-987.

Pixley, K., Palacios, N.R. and Glahn, R.P. (2011). The usefulness of iron bioavailability as a target trait for breeding maize (*Zea mays* L.) with enhanced nutritional value. *Field Crop Res.* 123: 153-160.

Pixley, K., Palacios-Rojas, N., Babu, R., Mutale, R., Surles, R. and Simpungwe, E. (2013) In: Tanumihardjo, S. (Ed.), Carotenoids and Human Health, Biofortification of Maize with Provitamin A Carotenoids, Springer Science+ Business Media New York 2013. 271-292. DOI: 10.1007/978-1-62703-203-2_17.

Poletti, S., Gruissem, W. and Sautter, C. (2004). The nutritional fortification of cereals. *Curr Opinion Biotech.* 15: 162-165.

Prasad, A.S. (1996). Zinc deficiency in women, infants and children. *J Am Coll Nutr.* 15: 113-120.

Prasad, R., Shivay, Y.S. and Kumar, D. (2014). Agronomic Biofortification of cereal grains with iron and zinc. In: Donald L.S. (Ed.), Advances in Agronomy. 125: 55-91.

Prasanna, B.M., Mazumdar, S., Chakraborti, M., Hossain, F., Manjaiah, K.M., Agrawal, P.K., Guleria, S.K. and Gupta, H.S. (2011). Genetic variability and genotype x year interactions for kernel iron and zinc concentration in maize (*Zea mays* L.). *Indian J Agric Sci.* 81(8): 704-711.

Prasanna, B.M., Vasal, S.K., Kassahun, B. and Singh, N.N. (2001). Quality protein maize. *Curr Sci.* 81: 1308-1319.

Qin, H.N., Cai, Y.L., Liu, Z.Z., Wang, G.Q., Wang, J.G., Guo, Y. and Wang, H. (2012). Identification of QTL for zinc and iron concentration in maize kernel and cob. *Euphytica.* 187: 345-358.

Raboy, V. (2000). Low phytic acid grains. *Food Nutr Bull.* 4: 423-427.

Raboy, V. (2001). Genetics and breeding of seed phosphorus and phytic acid. *J Plant Physiol.* 158: 489-497.

Raboy, V. (2009). Approaches and challenges to engineering seed phytate and total phosphorus. *Plant Sci.* 177: 281-296.

Rashmi, Tufchi, M., Shastry, M. and Singh, N.K. (2014). Phenotyping of maize inbred lines for beta-carotene and determining relationship with total carotenoids and kernel colour in maize. *Indian J Genet.* 74: 631-37.

Roberts, L.A., Pierson, A.J., Panaviene, Z. and Walker, E.L. (2004). Yellow Stripe1 expanded roles for the maize iron phytosiderophore transporter. *Plant Physiol.* 135: 115-120.

Rocheford, T.R., Wong, J.C., Egesel, C.O. and Lambert, R.J. (2002). Enhancement of vitamin E levels in corn. *J Am Coll Nutr.* 21: 191S-198S. DOI:10.1080/07315724.2002.10719265.

Roeser, H.P. (1986). Iron. *J Food Nutr*. 42: 82-92.

Rosegrant, M.R., Ringler, C., Sulser, T.B., Ewing, M., Palazzo, A. and Zhu, T. (2009). Agriculture and food security under global change: Prospects for 2025/2050. IFPRI, Washington, D.C.

Sandstorm, B. (1997). Bioavailability of zinc. *Eur J Clin Nutr*. 51: S17-S19.

Sarika, K. (2016). Marker-aided pyramiding of *opaque2* and *opaque16* genes for enhancing lysine and tryptophan in maize endosperm. Ph.D. Thesis. IARI, New Delhi.

Sarika, K., Hossain, F., Muthusamy, V., Baveja, A., Zunjare, R., Goswami, R., Thirunavukkarasu, N., Saha, S. and Gupta, H.S. (2017a). Exploration of novel *opaque16* mutation as a source for high lysine and tryptophan in maize endosperm. *Indian J Genet*. 77: 59-64.

Sarika, K., Hossain, F., Muthusamy, V., Zunjare, R., Baveja, A., Thirunavukkarasu, N., Goswami, R. and Gupta, H.S. (2015). Marker-assisted pyramiding of *opaque2* and *opaque16* alleles for the enhancement of lysine and tryptophan in maize endosperm. In: National Symposium on Germplasm to Genes: Harnessing Biotechnology for Food Security and Health, 9-11, August 2015, ICAR-NRCPB, New Delhi. 87.

Sarika, K., Hossain, F., Muthusamy, V., Zunjare, R.U., Baveja, A. *et al.* (2017b). Effects of novel *opaque16* mutant on physical and biochemical characteristics of maize endosperm. In: XIV National Seed Seminar "Food Security Through Augmented Seed Supply Under Climate Uncertainties", January 28-30, 2017, Dr. B.P. Pal Auditorium, ICAR-IARI, New Delhi, India. 270-271.

Schmidt, R.J., Burr, F.A. and Burr, B. (1987). Transposon tagging and molecular analysis of the maize regulatory locus *opaque-2*. *Science*. 238(4829): 960-963.

Senete, C.T., Guimaraes, P.E.D., Paes, M.C.D. and Souza, J.C. (2011). Diallel analysis of maize inbred lines for carotenoids and grain yield. *Euphytica*. 182: 395-404.

Sharma, A. and Chauhan, R.S. (2008). Identification of candidate gene-based markers (SNPs and SSR) in the zinc and iron transporter sequence of maize (*Zea mays* L.). *Curr Sci*. 95: 1051-1059.

Shewry, P.R. and Thatam, A.S. (1990). The prolamin storage proteins of cereal seeds: structure and evolution. *Biochem J*. 267: 1-12.

Shi, J., Wang, H., Schellin, K., Li, B., Faller, M. *et al.* (2007). Embryo-specific silencing of a transporter reduces phytic acid content of maize and soybean seeds. *Nat Biotech*. 25: 930-937.

Shi, J., Wang, H., Wu, Y., Hazebroek, J., Meeley, R.B. and Ertl, D.S. (2003). The maize low-Phytic acid mutant *lpa2* is caused by mutation in an *inositol phosphate kinase* gene. *Plant Physiol.* 131: 507-515.

Shi, J.H., Hazebroek, W.J., Ertl, D.S. and Harp, T. (2005). The maize low-phytic acid 3 encodes a myo- inositol kinase that plays a role in phytic acid biosynthesis in developing seeds. *Plant J*. 42: 708-719.

Shiferaw, B., Prasanna, B.M., Hellin, J. and Banziger, M. (2011). Crops that feed the world, 6. Past successes and future challenges to the role played by maize in global food security. *Food Security*. 3: 307-327.

Shivay, Y.S. and Prasad, R. (2014). Effect of source and methods of zinc application on corn productivity, nitrogen and zinc concentrations and uptake by high quality protein corn (*Zea mays*). *Egypt J Biol.* 16: 72-78.

Shutu, X., Dalong, Z., Ye, C., Yi, Z., Shah, T. *et al.* (2012). Dissecting tocopherols content in maize (*Zea mays* L.), using two segregating populations and high-density single nucleotide polymorphism markers. *BMC Plant Biol.* 12(1): 201.

Simic, D., Mladenovic, D.S., Zdunic, Z., Jambrovic, A., Ledencan, T., Brkic, J., Brkic, A. and Brkic, I. (2011). Quantitative trait loci for biofortification traits in maize grain. *J Hered.* 103: 47-54.

Simic, D., Sudar, R., Jambrovic, A., Ledencan, T., Zdunic, Z., Kovacevic, V. and Brkic, I. (2009). Genetic variation of bioavailable iron and zinc in grain of a maize population. *J Cereal Sci.* 50: 392-397.

Singh, N.N. and Venkatesh, S. (2006). Development of quality protein maize inbred lines. In: Kaloo, G., Rai, M., Singh, M. and Kumar, S. (Eds.). Heterosis in Crop Plants. Res. Book Center, New Delhi. 102-113.

Singh, U. and Chandra, S. (2014). Development and commercialization of value-added food products of quality protein maize for nutrition security. In: Prasanna *et al.* (Eds.), 12th Asian Maize Conference and Expert Consultation on Maize for Food, Feed and Nutritional Security, Book of Extended Summaries, 30 October-01 November, Bangkok, Thailand. 94-97.

Sivaranjani, R., Prasanna, B.M., Hossain, F. and Santha, I.M. (2013). Genetic variability for total carotenoid concentration in selected maize inbred lines. *Indian J Agric Sci.* 83(4): 431-436.

Sivaranjani, R., Santha, I.M., Pandey, N., Vishwakarma, A.K., Nepolean, T. and Hossain, F. (2014). Microsatellite-based genetic diversity in selected exotic and indigenous maize (*Zea mays* L.) inbred lines differing in total kernel carotenoids. *Indian J Genet.* 74(1): 34-41.

Sommer, A. and Davidson, F.R. (2002). Assessment and control of vitamin A deficiency: the annecy accords. *J Nutr.* 132: S2845-S2850.

Sommer, A. and West, K.P. (1996). Vitamin A deficiency: Health survival and vision. New York, Oxford University Press.

Sontag, T.J. and Parker, R.S. (2002). Cytochrome P450 omega-hydroxylase pathway of tocopherol catabolism. Novel mechanism of regulation of vitamin E status. *J Biol Chem.* 277: 25290-25296.

Sontag, T.J. and Parker, R.S. (2007). Influence of major structural features of tocopherols and tocotrienols on their ù-oxidation by tocopherol-ù-hydroxylase. *J Lipid Res.* 48: 1090-1098.

Sparvoli, F. and Cominelli, E. (2015). Seed biofortification and phytic acid reduction: A conflict of interest for the plant? *Plants.* 4: 728-755.

Steur, H.D., Blancquaert, D., Strobbe, S., Lambert, W., Gellynck, X. and Straeten, D.V.D. (2015). Status and market potential of transgenic biofortified crops. *Nat Biotech.* 33: 25-29.

Stevens, R. and Winter-Nelson, A. (2008). Consumer acceptance of provitamin A-biofortified maize in Maputo, Mozambique. *Food Policy*. 33: 341-351.

Sureshkumar, S., Tamilkumar, P., Senthil, N., Nagarajan, P., Thangavelu, A.U. *et al.* (2014a). Marker assisted selection of low phytic acid trait in maize (*Zea mays* L.). *Hereditas*. 151:20-27.

Suwarno, W.B., Pixley, K.V., Palacios-Rojas, N., Kaeppler, S.M. and Babu, R. (2014). Formation of heterotic groups and understanding genetic effects in a provitamin A biofortified maize breeding program. *Crop Sci.* 54: 14-24.

Tamilkumar, P., Senthil, N., Sureshkumar, S., Thangavelu, A.U., Nagarajan, P. *et al.* (2014). Introgression of low phytic acid locus (*lpa2*-2) into an elite Maize (*Zea mays* L.) inbred through marker assisted backcross breeding. *Austr J Crop Sci*. 8: 1224-1231.

Tanumihardjo, S. (2011). Vitamin A: Biomarkers of nutrition for development. *Am J Clin Nutr.* 94: 658S-665S.

Temba, M.C., Njobeh, P.B., Adebo, O.A., Olugbile, A.O. and Kayitesi, E. (2016). The role of compositing cereals with legumes to alleviate protein energy malnutrition in Africa. *J Food Sci Tech*. DOI: 10.1111/ijfs.1303.

Thakkar, S.K. and Failla, M.L. (2008). Micellarization of beta-carotene during in vitro digestion of maize and uptake by Caco-2 intestinal cells is minimally affected by xanthophylls. *The FASEB J.* 22: 1105.6.

Tiwari, A., Prasanna, B.M., Hossain, F. and Guruprasad, K.N. (2012). Analysis of genetic variability for kernel carotenoid concentration in selected maize inbred lines. *Indian J Genet*. 72(1): 1-6.

Tome, D. and Bos, C. (2007). Lysine requirement through the human life cycle. *J Nutr*. 137:1642-1645.

Traber, M.G., Frei, B. and Beckman, J.S. (2008). Vitamin E revisited: Do new data validate benefits for chronic disease prevention? *Curr Opinion Lipidol.* 19(1): 30-38. DOI:10.1097/MOL.0b013e3282f2dab6.

Tufchi, M., Rashmi, Jha, S.K. and Singh, N.K. (2015). Effect of *opaque-2* allele on accumulation of tryptophan among backcross-derived introgressed progenies of maize. *Indian J Genet*. 75(4): 453-458.

Twumasi-Afriyie, S., Ahenkora, K., Sallah, P.Y.K., Frempong, M. and Agyemang, A. (1996). Effect of extraneous pollen from normal maize in adjoining fields on the nutritional quality of quality protein maize. In: 12th South African Maize Breeding Symposium, Pietermaritzburg, South Africa.

Ueda, T., Waverczak, W., Ward, K., Sher, M., Ketudat, M., Schmidt, R.J. and Messing, J. (1992). Mutation of 22 and 27-kD zein promoters affect transactivation by the *opaque-2* protein. *Plant Cell.* 4: 701-709.

Vallabhaneni, R., Gallagher, C.E., Licciardello, N., Cuttriss, A.J., Quinlan, R.F. and Wurtzel, E.T. (2009). Metabolite sorting of a germplasm collection reveals the hydroxylase3 locus as a new target for maize provitamin A biofortification. *Plant Physiol.* 151:

1635-1645.

Vasal, S.K. (2001). Quality protein maize development: An exciting experience. In: Seventh Eastern and South Africa Regional Maize Conference. 3-6.

Velu, G., Ortiz-Monasterio, I., Cakmak, I., Hao, Y. and Singh, R.P. (2014). Biofortification strategies to increase grain zinc and iron concentrations in wheat. *J Cereal Sci.* 59(3): 365-372.

Vignesh, M., Hossain, F., Nepolean, T, Saha, S., Agrawal, P.K., Guleria, S.K., Prasanna, B.M. and Gupta, H.S. (2012). Genetic variability for kernel b-carotene and utilization of *crtRB1* 3'TE gene for biofortification in maize (*Zea mays* L.). *Indian J Genet Plant Breed.* 72:189-194.

Vignesh, M., Nepolean, T., Hossain, F., Singh, A.K. and Gupta, H.S. (2013). Sequence variation in 32 UTR region of *crtRB1* gene and its effect on b-carotene accumulation in maize kernel. *J Plant Biochem Biotech.* DOI 10.1007/s13562-012-0168-4.

Villegas, E., Vasal, S.K. and Bjarnason, M. (1992). Quality protein maize – what is it and how was it developed? In: Mertz, E.T. (Ed.), Quality Protein Maize. American Association of Cereal Chemists, Minnesota, USA. 27-48.

Vivek, B.S., Krivanek, A.F., Palacios-Rojas, N., Twumasi-Afiriye, S. and Diallo, A.O. (2008). Breeding quality protein maize (QPM) cultivars: protocols for developing QPM cultivars. CIMMYT, Mexico.

Von-Wiren, N., Klair, S., Bansal, S., Briat, J.F., Khodr, H., Shioiri, T., Leigh, R.A. and Hider, R. (1999). Nicotianamine chelates both FeIII and FeII-Implications for metal transport in plants. *Plant Physiol.* 119: 1107-1114.

Von-Wiren, N., Marschner, H. and Romheld, V. (1996). Roots of iron-efficient maize also absorb phytosiderophore chelated zinc. *Plant Physiol.* 111: 1119-1125.

Wall, J.S. and Bietz, J.A. (1987). Differences in corn endosperm protein in developing seeds of normal and *opaque-2* corn. *Cereal Chem.* 64: 275-280.

Wallace, J.C., Lopes, M.A., Paiva, E. and Larkins, B.A. (1990). New methods for extraction and quantitation of zeins reveal a high content of g-Zein in modified *opaque-2* maize. *Plant Physiol.* 92: 191-196.

Welch, R.M. and Graham, R.D. (2002). Breeding crops for enhanced micronutrient content. *Plant and Soil.* 245: 205-214.

Welch, R.M. and Graham, R.D. (2005). Agriculture: the real nexus for enhancing bioavailable micronutrients in food crops. *J Trace Elem Med Biol.* 18: 299-307.

Welch, R.M., Smith, M.E., Van-Campen, D.R. and Schaeffer, S.C. (1993). Improving the mineral reserves and protein quality of maize (*Zea mays* L.) kernels unique genes. *Plant Soil.* 155(156): 215-218.

West, K.P. (2002). Extent of vitamin A deficiency among preschool children and women of reproductive age. *J Nutr.* 132: 2857-2866.

White, P.J. and Broadley, M.R. (2005). Biofortifying crops with essential mineral elements. *Trends Plant Sci.* 10: 586-593.

White, P.J. and Broadley, M.R. (2009). Biofortification of crops with seven mineral elements often lacking in human diets- iron, zinc, copper, calcium, magnesium, selenium and iodine. *New Phytol.* 182: 49-84.

WHO/FAO/UNU. (2007). Protein and amino acid requirements in human nutrition. 1764 Report of a Joint WHO/FAO/UNU Expert Consultation, WHO Technical Report Series, No 935. Geneva.

Williams, P.E.V. (1997). Poultry production and science: future directions in nutrition. *World Poult Sci J.* 53: 33-48.

Wolf, M.J., Cutler, H.C., Zuber, M.S. and Khoo, U. (1972). Maize with multilayer aleurone of high protein content. *Crop Sci.* 12: 440-442.

Wong, J.C., Lambert, R.J., Tadmor, Y. and Rocheford, T.R. (2003). QTL associated with accumulation of tocopherols in maize. *Crop Sci.* 43(6): 2257-2266. DOI:10.2135/cropsci2003.2257.

Wong, J.C., Lambert, R.J., Wurtzel, E.T. and Rocheford, T.R. (2004). QTL and candidate genes phytoene synthase and æ-carotene desaturase associated with the accumulation of carotenoids in maize. *Theor Appl Genet.* 108: 349-359.

Wu, D. and Meydani, S.N. (2014). Age-associated changes in immune function: impact of vitamin E intervention and the underlying mechanisms. *Endocr Metab Immune Disord Drug Targets.* 14: 283-289.

Wu, Y. and Messing, J. (2011). Novel genetic selection system for quantitative trait loci of quality protein maize. *Genet.* 188: 1019-1022.

Wu, Y., Holding, D.R. and Messing, J. (2010). g-Zeins are essential for endosperm modification in quality protein maize. *Proc Natl Acad Sci USA.* 107: 12810-12815.

Yadav, O.P., Hossain, F., Karjagi, C.G., Kumar, B., Zaidi, P.H. *et al.* (2015). Genetic improvement of maize in India: retrospect and prospects. *Agric Res.* 4(4): 325-338. DOI 10.1007/s40003-015-0180-8.

Yadav, O.P., Prasanna, B.M., Yadava, P., Jat, S.L., Kumar, D. *et al.* (2016). Doubling maize (*Zea mays*) production of India by 2025 - Challenges and opportunities. *Indian J Agric Sci.* 86(4): 427-434.

Yan, J., Kandianis, B.C., Harjes, E.C., Bai, L., Kim, H.E. *et al.* (2010). Rare genetic variation at *Zea mays crtRB1* increases beta carotene in maize grain. *Nat Genet.* 42: 322-327.

Yang, L., Wang, W., Yang, W. and Wang, M. (2013). Marker-assisted selection for pyramiding the *waxy* and *opaque16* genes in maize using cross and backcross schemes. *Mol Breed.* 31: 767-775.

Yang, W., Zheng, Y., Zheng, W. and Feng, R. (2005). Molecular genetic mapping of a high-lysine mutant gene (*opaque-16*) and the double recessive effect with *opaque-2* in maize. *Mol Breed.* 15: 257-269.

Young, V.R., Scrimshaw, N.S. and Pellet, P.L. (1998). Significance of dietary protein source in human nutrition: animal and/or plant proteins. In: Waterlow, J.C., Armstrong, D.G., Fowden, L. and Riley, R. (Eds.), Feeding a world population of more than eight billion people. Oxford University Press in associate with Rank Prize Funds, New

York. 205-221.

Zhang, W., Yang, W., Wang, M., Wang, W., Zeng, G., Chen, Z. and Cai, Y. (2013). Increasing lysine content of waxy maize through introgression of *opaque2* and *opaque16* genes using molecular assisted and biochemical development. *PLoS ONE*. 8: 1-10.

Zhang, X., Pfeiffer, W.H., Palacios-Rojas, N., Babu, R., Bouis, H. and Wang, J. (2012). Probability of success of breeding strategies for improving provitamin-A content in maize. *Theor Appl Genet*. 125: 235-246.

Zheng, L., Chengz, A.C., Jiang, X., Bei, X., Zheng, Y., Glahn, R.P., Welch, R.P., Miller, D.D., Lei, X.G. and Shou, H. (2010). Nicotianamine, a novel enhancer of rice iron bioavailability to humans. *PLoS ONE*. 5: e10190.

Zhu, C., Naqvi, S., Breitenbach, J., Sandmann, G., Christou, P. and Capell, T. (2008). Combinatorial genetic transformation generates a library of metabolic phenotypes for the carotenoid pathway in maize. *Proc Natl Acad Sci USA*. 105: 18232-7.

Zunjare, R., Hossain, F., Muthusamy, V., Baveja, A., Chauhan, H.S., Thirunavakkarasu, N., Saha, S. and Gupta, H.S. (2016). Development of vitamin A enriched quality protein maize germplasm through introgression of *b-carotene hydroxylase* and *lycopene-å-cyclase*. In: 1st International Agrobiodiversity Congress. November 6-9, 2016, New Delhi, India, Organized by Indian Society of Plant Genetic Resources and Bioversity International, Abstract Book. 192.

Zunjare, R., Hossain, F., Muthusamy, V., Baveja, A., Thirunavukkarasu, N., Saha, S. and Gupta, H.S. (2015). Marker-aided pyramiding of *b-carotene hydroxylase* (*crtRB1*) and *lycopene-å-cyclase* (*lcyE*) genes for provitamin A enrichment in quality protein maize (QPM) inbreds. In: National Symposium on Germplasm to Genes: Harnessing Biotechnology for Food Security and Health, 9-11, August 2015, ICAR-NRCPB, New Delhi. 74.

24

Conventional and Molecular Breeding Approaches for Abiotic Stresses Tolerance in Maize

Bhupender Kumar, Mukesh Choudhary, Vishal Singh, Harpreet Kaur Shraddha Srivastava, Tanu Tiwari and Vinod Kumar*

ICAR-Indian Institute of Maize Research, Ludhiana-141004, Punjab

**Corresponding Author's Email: bhupender.iari@gmail.com*

The ever mounting population in 21st century is a great challenge for the researchers to provide ample food and that has to be achieved within the context of existing challenges such as deteriorated arable land along with increasing adverse affects of changing climate. There is demand to increase food production to the tune of 70% by 2050 (Wani and Sah, 2014). Although major crops have achieved good progress, vulne*Rabi*lity to climatic variability emerged as a serious concern in crop improvement which could be attributed to higher sowing densities that in turn lead to more competition for water and nutrients. Abiotic stresses can be considered as major yield restraining factors among different crops. The projected estimations for changes in abiotic stresses such as drought, water logging and heat reveals far serious concerns for crop productivity (Qin *et al.*, 2011; Bailey-Serres *et al.*, 2012). Therefore, the concern of global food security in the era of climate change has compelled plant scientists in different crops to shift to breeding for climate resilient genotypes. Maize, being a hardy crop has an inherent potential to perform well under sub-optimal environments with minimum crop management (Kole *et al.*, 2015). Abiotic stresses such as drought, heat and water logging affects maize yield adversely.

Although, conventional breeding approaches have exploited its fullest potential for production of elite maize hybrids but still little success attained for breeding climate resilient genotypes or varieties owing to the fact that stress tolerance is multi-genic or quantitative in nature. Therefore, the best option is to utilize the molecular breeding along with the conventional approaches for development of climate smart varieties. In the context of molecular mapping, several QTLs (Quantitative Trait Loci) have been mapped for the abiotic stresses such as drought, water-logging and heat in maize. Besides molecular mapping, advanced breeding strategies like genomic selection (GS) has also boosted maize stress breeding. In the era of molecular markers, "omics" has proved its potential through facilitating the identification of abiotic stress associated QTLs in maize along with functional genomics aided identification and characterization of abiotic stress related genes at transcriptional level (Cushman and Bohnert, 2000; Tuberosa and Salvi, 2004; Coraggio and Tuberosa, 2004). High throughput sequencing tools has further added new dimensions to functional genomics for understanding the molecular basis of tolerance related to abiotic stresses in maize. Proteomics has also been exploited in maize to understand the expression of proteins under stress conditions and regulation of stress machinery through transcription factors. However, all these biotechnological approaches can only be utilized as supplement to the conventional breeding approaches and alone they may not be able to discover the potential genotypes which can survive under abiotic stresses.

Abiotic stresses in Maize

Abiotic stresses considered in the form of adversarial effects on cellular homeostasis and growth, owing to the poor climatic or soil conditions. There is a need to combat the abiotic stresses through manifold mechanisms of yield maintenance because it has been noticed that abiotic stresses generally occur together like heat and drought at same time or flooding trailed by drought.

Maize being majorly a rainfed crop is prone to drought stress at critical stages and its effect can reflect in form of variable yields. With most maize production dependent on rainfall, especially in the developing world, maize is particularly vulnerable to drought and its yields fluctuate more widely from year to year. There has been a continuous decline in the irrigated area with parallel increase in the rainfed area and this decline can be attributed to the shrinking ground water table (Cooper *et al.*, 2014). Therefore breeding for drought tolerant cultivars is crucial for successful cultivation of maize in drought sensitive domains. In the high rainfall prone agro-ecologies of maize cultivation, waterlogging is a serious concern and poor soil drainage capacity makes it even worse. It has been estimated that water-logging affects more than 18 % of the total maize production area in South and South-east Asia leading to annual losses of 25-30% of maize production (Cairns *et al.*, 2012). This problem is going to get more intensified in major maize-growing regions due to predicted changing patterns of rainfall distribution in South Asia (IPCC, 2007). Hence, in order to get the optimum yield there is need to breed for water-logging adapted cultivars too. Heat stress is another type of stress which severely affects maize production. Reproductive stage is most sensitive to heat stress. It has been found that high temperature augments yield losses occurring due to drought stress. According to the projections of climate change, significant crop losses is going to occur in near future due to high temperatures especially in the drought-prone rainfed areas of South Asia (Cairns *et al*., 2013).

Physiological, biochemical and molecular responses to abiotic stresses

Abiotic stresses results into activation of various fundamental biochemical and physiological reactions of cells in plants (Wang *et al*., 2003; Hasanuzzaman *et al*., 2013). The general mechanism of response to various abiotic stresses and tolerance development in maize has been depicted in Figure 1. The response of maize to moisture stress is not straight forward as it affects diverse fundamental processes of plant growth and development such as osmotic regulation, antioxidant machinery, photosynthetic rate decline, and abscisic acid (ABA) buildup (Cramer *et al*., 2011). The complexity of response can be cited from the fact that different proteins with differential expression in various stress tolerant species control these processes by their involvement in developmental stage specific manner. Different proteins are involved in various biochemical pathway and thereby regulating the moisture stress responses in a complex manner. Drought affects the maize germination by hindering the basic processes of water absorption, imbibition and activation of enzymes involved in metabolism. The parameters of seedling growth-root and shoot elongation are also affected by drought stress. There is reduction in both parameters under drought stress but shoot elongation show higher reduction (Khodarahmpour, 2011).

Water-logging affects the plants by making it devoid of oxygen owing to the hypoxia/anoxia conditions created due to excess moisture. This in turn affects the water and nutrient uptake capacity of plants which ultimately results in wilting of plants. The oxygen deficiency forces the plant to shift from aerobic to anaerobic mode of respiration. Aerenchyma formation, improved availability of soluble sugars, better activity of glycolytic pathway and fermentation enzymes and antioxidant defense mechanism activation are peculiar characters exhibited by the water-logging tolerant cultivars under hypoxia/anoxia oxidative stress. Ethylene has been found to activate the enzymes governing the expression of genes related with glycolysis, fermentation and aerenchyma formation.

Heat stress negatively affects maize crop by shortening the life cycle and high sterility (Cairns *et al*., 2012). Heat stress affects the plants by excess production of reactive oxygen species (ROS) which ultimately results in creation of oxidative stress. Heat stress triggers the plants to change their metabolism through production of compatible solutes which helps to organize basic structure of proteins and cell. Besides metabolism alteration also occurs by maintaining cell turgor via osmotic adjustment, and modification of the antioxidant machinery to retain the cellular homeostasis.

The plant has the capacity to maintain its integrity in adverse environments through crucial mechanism of cellular respiration functionality and photosynthesis like rapid alterations in gene expression, enhanced heat shock protein (HSP) levels, membrane alterations, etc. Heat stress interrupts these basic cellular mechanisms and thereby plant loses the ability to cope with high temperature. Recently a new trait known as reduced kernel abortion recognised for high temperature tolerance in tropical maize cultivars (Rattalino and Otegui, 2013). Rubisco enzyme activity is also adversely affected due to high temperature and thereby resulting in low production owing to reduced photosynthesis and grain size (Steven *et al.*, 2002).

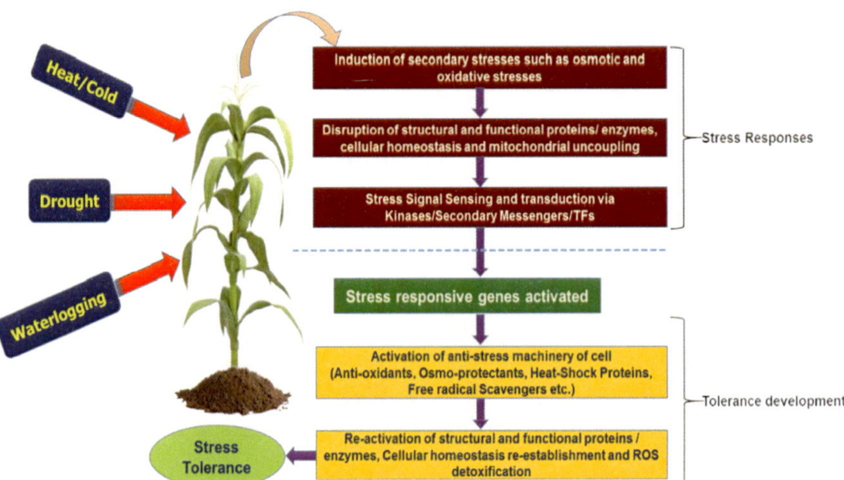

Fig. 1. General mechanism of response to abiotic stresses and tolerance development in maize.

Breeding for abiotic stresses tolerance in maize

Conventional breeding focuses on the extensive screening or characterization of the germplasm for stress tolerance related traits. Besides that it also emphasizes on the gene action and heritability studies of the identified traits for their effective utilization in stress breeding. Thus, it is very important to understand the inheritance of abiotic stress tolerance related morphological and physiological traits. Little information is available on the gene action of heat stress tolerance related traits in maize. Leaf temperature (Hussain *et al.*, 2009), cell membrane thermo-stability (Saleem *et al.*, 2015), leaf firing (Kaur *et al.*, 2010), ear length (Ahmed *et al.*, 2000), kernels per ear (Muraya *et al.*, 2006), 100-grain weight (Wu, 1987) and grain yield per plant (Iqbal *et al.*, 2007) are governed by both additive and non-additive gene action. The most important trait observed under drought stress is the widened interval of anthesis and silking commonly known as Anthesis-Silking Interval (ASI). The cause behind high ASI is slow rate of ear growth relative to tassel and therefore delayed silk emergence. Low ASI is the preferred trait for drought as well as water-logging as it helps in better synchronization of male and female flowering plants and therefore ensuring good seed setting. The fundamental strategies for conventional breeding always start with the characterization of drought/water-logging and heat-prone environments followed by the identification of donors with stress tolerant related traits. Then the focus should be given on optimization of procedures for conducting stress phenotyping trials followed by extensive multi-environmental trials for developing stable stress-tolerant maize germplasm.

Field screening for abiotic stresses

Choosing of appropriate selection environments for abiotic stress screening is the most critical step in breeding. Genotype-by-environment (G×E) interactions are common under abiotic stress and make breeding progress difficult (Kumar *et al.*, 2016a). G×E interactions may originate from environmental variation in the timing and severity of water deficits, genetic variation in flowering time, and nutrient deficiencies and toxicities whose occurrence and severity interact with water deficits. Therefore, the proper care must be taken while selecting the target environments. The seedling, flowering and grain filling stages are the most sensitive stages to majority of the abiotic stresses in maize. The managed abiotic stress screening at critical stage *viz.*, low moisture stress at flowering stage may be induced by restricting the irrigation around 12-15 days prior to flowering in maize trials may be helpful in effective screening for low ASI. Similarly the stress at early grain filling stage can be induced by restricting the irrigation in trials immediately when the pollination over. The same may be sustained at least for 20 to 22 days. Special attention is required to constitute the trials maturity wise for all types of stresses, this may help to coincide the same stage of all the genotypes in a trial to given stress. Further, the most important point is the validation of identified tolerant genotypes is highly recommended by screening under controlled conditions as well as through using genomic tools and techniques (Kumar *et al.*, 2016b). All these points may help to reduce the experimental error while field evaluation.

Breeding trials for stress tolerance are generally conducted in an environment mimicking to the target population environments (TPE) where ultimately the crop will be grown. This helps the plant breeders in passive selection of stress tolerance related traits (Yamamoto *et al.*, 2011; Kugblenu *et al.*, 2013). Although in comparison to controlled conditions, field experiments offer a more suitable phenotyping environment, but utmost care has to be taken to reduce experimental error in order to replicate the success in TPE. Therefore, maintenance of uniformity in stress management trials is an important point to be considered.

Uniform stand establishment helps to ensure evenness of water availability per plant. Besides, there is a need to opt for a reliable and realistic screening technique for effective selection of tolerant germplasm. Nowadays novel phenotyping tools based on remote sensing which have an advantage that plant water status can be estimated on the basis spectral reflectance and infrared thermometry and thereby non-destructive in nature. The conventional breeding has achieved a lot in development of stress tolerant cultivars however its efficiency can be further increase using molecular tools and techniques. The important point to be kept in mind is that conventional and molecular breeding are complementary under most breeding schemes and should not be considered mutually exclusive.

Use of molecular tools in abiotic stresses tolerance breeding

Molecular breeding can be defined as the molecular marker based selection as well as transfer of traits of interest for crop improvement (Jiang, 2013). Molecular breeding helps in quick accumulation of favorable alleles at target loci as increased genetic gain per cycle reduces the crop selection cycles (Delannay *et al.*, 2012). It includes various breeding approaches such as marker-assisted selection (MAS), marker-assisted backcross breeding (MABB), marker-assisted recurrent selection (MARS) and genomic selection (GS). Identification of several quantitative trait loci (QTLs) for abiotic stresses in maize has augmented the progress of the conventional breeding for stress tolerant cultivar development in short time (Collard and Mackill, 2008). MAS help in the indirect selection of traits with low heritability during gene transfer in lines, cultivars, populations. MABB helps in the precise transfer of major QTLs into widely adapted cultivars. MABB has been carried out in maize for the introgression of favorable alleles related to yield and flowering traits related to drought stress (Ribaut and Ragot, 2007). Marker-assisted gene pyramiding is also an effective approach employed to transfer of two or more resistance genes and thereby providing multiple stress tolerant cultivars. MARS and GS are specifically targeted for complex traits such as drought, waterlogging and heat stress as these techniques help in precise selection and thereby reducing the phenotyping cost. Thus, it can be noticed that plant breeders can utilize the various molecular-breeding approaches for developing stress-resilient high-yielding cultivars. A major QTL, *root-ABA1* has been identified in backcross derived lines (Tuberosa *et al.*, 1998; Sanguineti *et al.*, 1999) and further studies have validated the strong effect of this QTL on leaf ABA concentration across different water regimes (Giuliani *et al.*, 2005; Landi *et al.*, 2005). Success can be achieved by the identification of large effect constitutive QTLS for abiotic stress tolerance related traits and their validation in a related mapping population. Meta-QTL (mQTL) analysis based on the identification of genomic regions responsible for grain yield under multiple environments such as water-stress as well as well-watered conditions across a series of germplasm (Goffinet and Gerber, 2000; Swamy *et al.*, 2011). Three populations and multiple environments based mQTL analysis led to identification of seven genomic regions for grain yield and one for ASI, of which six mQTL for grain yield were constitutively expressed across water stress and well watered environments (Almeida *et al.*, 2013). A list of major QTLS identified for various traits related to drought, water-logging and heat stress has been provided in table 1. Association mapping studies led to identification of SNPs related to genes supposed to be involved in carbohydrate and ABA metabolite accumulation during moisture stress (Setter *et al.*, 2010). SNP from aldehyde oxidase gene was found to be associated with silk ABA concentration under drought stress.

MARS has emerged as the novel technique for handling the complex traits especially where the trait is governed by multiple QTLs. MARS focus on the accumulation of favorable QTLs in a population by inter-mating of selected individuals in every selection cycle (Bernardo, 2008). MARS can be considered superior to other molecular breeding strategies for complex trait improvement because there is enhanced frequency of the superior alleles owing to selection through MAS in each cycle (Eathington *et al.*, 2007). MARS studies for grain yield improvement under drought-stress as well as well watered conditions in10 biparental tropical maize populations resulted into higher grain yield than traditional breeding and the differences were much higher under drought stress. This provides an evidence for successful application of MARS for breeding of complex traits and that too within a short time span (Beyene *et al.*, 2016).

GS differs from MARS in the fact that it utilizes genome wide markers rather than limiting to selected markers. Genomic selection is based on genotyping the genome wide markers followed by incorporation of genotyping data into a model so that the genome estimated breeding values of the progenies can be predicted and thereby skipping the phenotyping frequency (Rutkoski *et al.*, 2010). Availability of representative training population with subsequent addition of lines over the time followed by accurate phenotyping are the key factors to be considered to achieve success in GS. Superiority of GS over traditional breeding can be cited from the comparatively higher yield gains (two to four folds) in drought stress environments (Beyene *et al.*, 2015).

The success of transgenic development for abiotic stress tolerance relies upon the identification of the true candidate genes followed by the choice of most appropriate promoters. Large number of events should be generated for getting the desirable transgenics. Stress responsive genes play an important role in alleviating the stress effects by adjusting cellular homeostasis and thereby developing stress tolerance (Munns and Tester, 2008). Therefore, stress tolerance

Table 1. List of cloned/fine mapped important QTLs for abiotic stresses tolerance in maize.

Trait	QTL/ Loci	Mapping Population	Cross(s)	Genotyping Markers	Environment	Chromosome	References
Drought	17 QTLS (Leaf chlorophyll, Plant senescence, Electric root capacitance)	RILs	CML444 × SC-Malawi	SSRs	Field	1, 2, 4, 5, 6 and 10	Messmer *et al.*, 2011
	22 QTLs(sugar concentration, root density, root dry weight, total biomass, relative water content, and ABA	$F_{2:3}$	DTP79 × B73	RFLP	Green House	1, 3, 5, 6, 7 and 9	Rahman *et al.*, 2011
	203 QTLs (ASI, ears per plant, stay-green and plant-to-ear height ratio)	RILs $F_{2:3}$ $F_{2:3}$	CML444 × MALAWI CML440 × CML504 CML444 × CML441	SNPs and SSRs	Field	1, 3, 4, 5, 7 and 10	Almeida *et al.*, 2014
	6 QTLs(Plant height, root length, root dry weight, shoot dry weight)	$F_{2:3}$	HZ32 × K12	SSRs	Glass House	1, 4, 6, 7 and 9	Osman *et al.*, 2013
	18 QTLs (Yield, brace roots, chlorophyll content, % stem and root lodging)	RILs	CML311-2-1-3 × CAWL-46-3-1	SNP markers	Field	1, 2, 3, 4, 5, 7, 8 and 10	Zaidi *et al.*, 2015
Water-logging	5 QTLs (Adventitious root formation)	F_2	B64 × *Z. mays ssp. huehuetenangensis*	SSRs and AFLPs	Green House	4 and 8	Mano *et al.*, 2005
	5 QTLs (Aerenchyma formation)	F_2	B73 × teosinte *Zea luxurians*	SSR	Green House	2,3,5,9 and 10	Mano *et al.*, 2008
	25 QTLs (Total brace root tier number and effective brace root tier number)	RILs	Huangzao 4 × CML288	SSRs	Field	1, 2, 3, 5, 6, 7, 8, 9, and 10	Ku *et al.*, 2012
	15 QTLs (seedling height, root length, shoot fresh weight, root fresh weight, shoot dry weight and root dry weight)	BC_2F_2	K12 × HZ32	SSRs and SNPs	Green House	5, 6 and 9	Zhang *et al.*, 2013
	169 QTLs (Grain yield per plant, ear length, kernel number per row, ear weight and hundred kernel weight)	NAM	11 bi-parental families(2000 RILs)	SNPs	Field	1, 3 and 10	Li *et al.*, 2016
Heat	6 QTLs (Cellular membrane stability)	RILs	T232 × CM37	RFLPs	Green House	1, 2, 4, 8, 9 and 10	Ottaviano *et al.*, 1991

can be developed in crops through engineering metabolic and stress-signalling pathways. Drought and heat tolerance has been achieved in maize transgenics through enhanced synthesis of glycine betaine (GB) synthesis (Chen and Murata, 2008). Similarly, transgenic plants with enhanced synthesis of trehalose have been found to be tolerant to different abiotic stresses due to its role as osmo-protectant (Almeida *et al.*, 2007).

Engineering the oxidative-stress-related genes is current are of interest for development of stress tolerant cultivars (Hussain *et al.*, 2009). Constitutive expression of the *Nicotiana PK1* gene in transgenic maize led to development of tolerance to heat stress (Shou *et al.*, 2004). Besides this Shou *et al.* (2004) developed drought tolerant maize transgenic due to constitutive expression of tobacco *MAPKKK* (NPK1) and it was achieved due to protection of photosynthetic machinery from drought stress. Transcription Factors (TFs) area also targeted for protein engineering owing to their important role in providing response under abiotic stresses. AP2/ERF, bZIP, NAC, MYB, MYC, Cys2His2 zinc-finger and WRKY are examples of some of the large families of TFs. Phytohormone engineering is also an innovative engineering approach for developing stress tolerant cultivars due to important role played by phyto-hormones during abiotic stresses (Sreenivasulu *et al.*, 2012).

Potential of *omics* technologies for abiotic stress breeding

Omics approaches have emerged as leading tool in crop improvement as it has made possible to study the ups and downs of temporal and spatial protein metabolites/transcripts expression under different abiotic stresses. Transgenics showing good growth under heat and drought stress have been developed in maize by the enhanced expression of *OsMYB55* and it was achieved because of higher plant biomass and reduced leaf damage in transgenic lines. Further studies based on transcriptomic analysis using RNA sequencing attributed the tolerance development in transgenic maize to constitutive up-regulation of *OsMYB55* (Casaretto *et al.*, 2016). ABA dependent expression of drought response-related proteins or genes have been identified in independent studies such as *ZmTPA* (Huang *et al.*, 2012), *ZmRFP1* (Xia *et al.*, 2012) and *ZmCPK4* (Jiang *et al.*, 2013). Stable-isotope labeling based maize proteome analysis and phospho-proteome dynamics under moisture stress led to the identification of drought stress related proteins such as protective proteins, late embryogenesis-abundant proteins (Benesova *et al.*, 2012) and stress response-related proteins (Hu *et al.*, 2012). Frey *et al.* (2015) studied a set of European maize inbred lines for heat stress tolerance in seedling stage and identified 607 heat responsive genes and 39 heat tolerant genes. MicroRNA (miRNA) studies in maize revealed that there is differential expression of miRNAs in the roots of maize and miR159, miR164, miR167, miR393, miR408, and miR528 were found to be key regulators in water-logging stress (Liu *et al.*, 2012). Transcriptomic analysis of maize roots under water-logging stress revealed the up regulation of genes responsible for providing the adaptation to excess moisture. Similarly, Thirunavukkarasu *et al.* (2013) attributed the water-logging stress tolerance in HKI 1105 to up regulation of genes involved in ethylene and auxin synthesis, cell wall metabolism and aerenchyma and adventitious root formation. Metabolomics can further help in identification of metabolite markers and thereby helping to enhance the pace of abiotic stress breeding. Metabolic profiling of maize leaves under drought and heat stress revealed that photorespiration and raffinose family oligosaccharide metabolism plays an important role for attaining better grain yields (Obata *et al.*, 2015). Therefore, a single metabolite or a combination of several metabolites can be utilized for developing abiotic stress tolerant maize cultivars.

Conclusion

In this era of climate change, breeding for abiotic stress tolerance is the priority of any crop improvement programme. Selection of target environments for genotypes evaluation, providing effective stress at critical growth stages, and scientific ways of identification of tolerant genotypes are the critical points which play major role in the success of abiotic stress breeding programme. Managed stress screening approaches provide an opportunity to keep heritability high and adequately representing abiotic stress factors that are relevant in the target environment. The integrated approach of conventional and molecular breeding can certainly enhance the efficiency and effectiveness of abiotic stress breeding programme. Further, the most important point is the validation of identified tolerant genotypes. Considering all these points while breeding can results in significant progress of abiotic stresses breeding.

References

Ahmed, H. M., Malik, T. A., and Choudhary, M. A. (2000). Genetic analysis of some physiomorphic traits in wheat under drought. *J. Agri. Plant Sci.* 10, 5–7.

Almeida, A.M., Cardoso, L.A., Santos, D.M., Torné, J.M. and Fevereiro, P.S. (2007). Trehalose and its applications in plant biotechnology. *In Vitro Cellular Dev Bio-Plant.* 43(3): 167-177.

Almeida, G.D. *et al.* (2013). QTL mapping in three tropical maize populations reveals a set of constitutive and adaptive genomic regions for drought tolerance. *Theor Appl Genet.* 126: 583-600.

Almeida, G.D., Nair, S. *et al.* (2014). Molecular mapping across three populations reveals a QTL hotspot region on chromosome 3 for secondary traits associated with drought tolerance in tropical maize. *Mol Breed.* 34(2): 701-715.

Bailey-Serres, J., Lee, S.C. and Brinton, E. (2012) Waterproofing crops: effective flooding survival strategies. *Plant Physiol.* 160: 1698-1709.

Benesova, M., Hola, D., Fischer, L. *et al.* (2012). The physiology and proteomics of drought tolerance in maize: Early stomatal closure as a cause of lower tolerance to short-term dehydration? *PLoS One.* 7: e38017.

Bernardo, R. (2008). Molecular markers and selection for complex traits in plants: learning from the last 20 years. *Crop Sci.* 48(5): 1649-1664.

Beyene, Y., Semagn, K., Crossa, J., Mugo, S., Atlin, G. N., Tarekegne, A. and Alvarado, G. (2016). Improving maize grain yield under drought stress and non- stress environments in sub-saharanafrica using marker-assisted recurrent selection. *Crop Sci.* 56(1): 344-353.

Beyene, Y., Semagn, K., Mugo, S., Tarekegne, A., Babu, R., Meisel, B. and Gakunga, J. (2015). Genetic gains in grain yield through genomic selection in eight bi-parental maize populations under drought stress. *Crop Sci.* 55(1): 154-163.

Cairns, J.E., Hellin, J., Sonder, K., Araus, J.L., MacRobert, J.F., Thierfelder, C. and Prasanna, B.M. (2013). Adapting maize production to climate change in sub-Saharan Africa. *Food Security.* 5(3): 345-360.

Cairns, J.E., Sonder, K., Zaidi, P.H., Verhulst, N., Mahuku, G., Babu, R. and Rashid, Z. (2012). Maize Production in a Changing Climate: Impacts, Adaptation, and Mitigation Strategies. *Adv Agron.* 114(1): 1-65.

Casaretto, J.A. *et al.* (2016). Expression of OsMYB55 in maize activates stress-responsive genes and enhances heat and drought tolerance. *BMC Genomics.* 17(1): 312.

Chen, T.H. and Murata, N. (2008). Glycinebetaine: an effective protectant against abiotic stress in plants. *Trends Plant Sci.* 13(9): 499- 505.

Collard, B.C.Y. and Mackill, D.J. (2008). Marker-assisted selection: an approach for precision plant breeding in the twenty-first century. *Philos Trans R Soc Lond B Bio Sci.* 363: 557-572.

Cooper, M. *et al.* (2014). Breeding drought-tolerant maize hybrids for the US corn-belt: discovery to product. *J Experimental Botany.* eru064.

Coraggio, I. and Tuberosa, R. (2004). Improving crops tolerance to abiotic stresses. pp. 413- 468. In: P. Christou, H. Klee (Eds.), Handbook of plant biotechnology. John Wiley and Sons, Ltd., Chichester, UK.

Cramer, G.R., Urano, K., Delrot, S., Pezzotti, M. and Shinozaki, K. (2011). Effects of abiotic stress on plants: a systems biology perspective. *BMC Plant Biology.* 11(1): 163.

Cushman, J.C. and Bohnert, H.J. (2000). Genomic approaches to plant stress tolerance. *Current Opin Plant Biol.* 3: 117-124.

Delannay, X., McLaren, G. and Ribaut, J.M. (2012). Fostering molecular breeding in developing countries. *Mol Breed.* 29: 857- 873.

Eathington, S.R., Crosbie, T.M., Edwards, M.D., Reiter, R.S. and Bull, J.K. (2007). Molecular markers in commercial breeding. *Crop Sci.* 47: 154-163.

Edreira, J. R. and Otegui, M. E. (2013). Heat stress in temperate and tropical maize hybrids: A novel approach for assessing sources of kernel loss in field conditions. *Field Crops Res.* 142: 58-67.

Frey, F.P., Urbany, C., Hüttel, B., Reinhardt, R. and Stich, B. (2015). Genome-wide expression profiling and phenotypic evaluation of European maize inbreds at seedling stage in response to heat stress. *BMC Genomics*. 16(1): 123.

Giuliani, S., Sanguineti, M.C., Tuberosa, R., Bellotti, M., Salvi, S. and Landi, P. (2005). Root-ABA1, a major constitutive QTL, affects maize root architecture and leaf ABA concentration at different water regimes. *J Experimental Botany*. 56(422): 3061-3070.

Goffinet, B. and Gerber, S. (2000). Quantitative trait loci: a meta-analysis. *Genetics*. 155(1): 463-473.

Hasanuzzaman, M. *et al.* (2012). Plant responses and tolerance to abiotic oxidative stress: antioxidant defenses are a key factor. In crop stress and its management: Perspectives and Strategies Bandi, V., Shanker, A.K., Shanker, C., Mandapaka, M., Eds.; Springer: Berlin, Germany. pp. 261-316.

Hasanuzzaman, M., Nahar, K. and Fujita., M. (2013). Extreme Temperatures, oxidative stress and antioxidant defense in plants. In *Abiotic Stress—Plant Responses and Applications in Agriculture*; Vahdati, K., Leslie, C., Eds.; *InTech:* Rijeka, Croatia. pp. 169-205.

Hu, X., Wu, X., Li, C., Lu, M., Liu, T., Wang, Y. and Wang, W. (2012). Abscisic acid refines the synthesis of chloroplast proteins in maize (*Zea mays*) in response to drought and light. *PloS One*. 7(11): e 49500.

Huang, H., Molle, I.M. and Song, S.Q. (2012). Proteomics of desiccation tolerance during development and germination of maize embryos. *J Proteomics*. 75: 1247-1262.

Hussain, I., Ahsan, M., Saleem, M. and Ahmed, A. (2009). Gene action studies for agronomic traits in maize under normal and water stress conditions. *Pak J Agri Sci.* 46: 108-112.

IPCC (2007). Sunthesis report-intergovermental panel on climate change. Cambridge: Cambridge university press.

Iqbal, A.M. *et al.* (2007). Combining ability analysis for yield and yield related traits in maize (*Zea mays* L.). *Int J Plant Breed Genet.* 1: 101–105. 10.3923/ijpbg.2007.101.105.

Jiang G.L. (2013). Molecular markers and marker assisted breeding in plants. In: S.B. Anderson eds., Plant Breeding from Laboratories to Fields. *InTech*, Croatia. pp. 45-83.

Jiang, S. *et al.* (2013). A maize calcium dependent protein kinase gene, ZmCPK4, positively regulated abscisic acid signaling and enhanced drought stress tolerance in transgenic A*Rabi*dopsis. *Plant Physiol Bioch.* 71: 112-120.

Kaur, R., Saxena, V.K. and Malhi, N.S. (2010). Combining ability for heat tolerance traits in spring maize (*Zea mays* L.). *Maydica*. 55: 195–199.

Khodarahmpour, Z. (2011). Effect of drought stress induced by polyethylene glycol (PEG) on germination indices in corn (*Zea mays* L.) hybrids. *African J Biotech.* 10(79): 18222- 18227.

Kole, C., Muthamilarasan, M., Henry,R., Edwards, D., Sharma, R. *et al.* (2015). Application of genomics-assisted breeding for generation of climate resilient crops: progress and prospects. *Front. Plant Sci.* 6: 563.

Ku, L.X. *et al*. (2012). QTL mapping and epistasis analysis of brace root traits in maize. *Mol Breed.* 30(2): 697-708.

Kugblenu, Y.O. *et al.* (2013). Screening tomato genotypes in Ghana for adaptation to high temperature. *Acta Agric Scand Sect B Soil Plant Sci.* 63: 516–522.

Kumar, B., Guleria, S.K., Khanorkar, S.M., Dubey, R.B., Patel, J., Kumar, V., Parihar, C.M., Jat, S.L., Singh, V., Yatish, K.R., Das, A., Sekhar, J.C., Bhati P., Kaur, H., Kumar, M., Singh, A.K., Varghese, E. and Yadav, O.P. (2016a). Selection indices to identify maize (*Zea mays* L.) hybrids adapted under drought-stress and drought-free conditions in a tropical climate. *Crop and Pasture Sci.* http://dx.doi.org/10.1071/CP16141.

Kumar, B., Srivastava, S., Kaur, H., Yadava, P., Jat, S.L., Parihar, C.M., Singh, A.K., Tiwari, T., Kumar, V., Sharma, S. and Kaul, J. (2016b). Hydro and aeroponic technique for rapid drought tolerance screening in maize (*Zea mays*). *Indian J Agron*. 61(4): 509-511.

Landi, P., Sanguineti, M. C., Salvi, S., Giuliani, S., Bellotti, M. *et al.* (2005). Validation and characterization of a major QTL affecting leaf ABA concentration in maize. *Mol Breed.* 15(3): 291-303.

Li, C., Sun, B., Li, Y. *et al.* (2016). Numerous genetic loci identified for drought tolerance in the maize nested association mapping populations. *BMC Genomics*. 17(1): 894.

Liu, Z., Kumari, S., Zhang, L., Zheng, Y. and Ware, D. (2012). Characterization of mi RNAs in response to short-term water logging in three inbred lines of Zea mays. *PLoS One*. 7(6): e39786.

Mano, Y., Muraki, M., Fujimori, M., Takamizo, T. and Kindiger, B. (2005). Identification of QTL controlling adventitious root formation during flooding conditions in teosinte (*Zea mays* ssp. huehuetenangensis) seedlings. *Euphytica*. 142(1): 33- 42.

Mano, Y., Omori, F., Kindiger, B. and Takahashi, H. (2008). A linkage map of maize x teosinte *Zea luxurians* and identification of QTLs controlling root aerenchyma formation. *Mol Breed.* 21(3): 327- 337.

Messmer, R. *et al.* (2011). Drought stress and tropical maize: QTLs for leaf greenness, plant senescence, and root capacitance. *Field Crops Res*. 124(1): 93-103.

Munns, R. and Tester, M. (2008). Mechanisms of salinity tolerance. *Ann Rev Plant Biol.* 59: 651-681.

Muraya, M.M., Ndirangu, C.M. and Omolo, E.O. (2006). Heterosis and combining ability in diallel crosses involving maize (*Zea mays* L.) S1 lines. *Aus J Exp Agri*. 46: 387–394. 10.1071/EA03278.

Obata, T., Witt, S., Lisec J. *et al.* (2015). Metabolite profiles of maize leaves in drought, heat, and combined stress field trials reveal the relationship between metabolism and grain yield. *Plant Physiol.* 169(4): 2665-2683.

Osman, K.A., Tang, B. *et al.* (2013). Dynamic QTL analysis and candidate gene mapping for waterlogging tolerance at maize seedling stage. *PLoS One*. 8(11): e79305.

Ottaviano, E., Gorla, M.S., Pe, E. and Frova, C. (1991). Molecular markers (RFLPs and HSPs) for the genetic dissection of thermo tolerance in maize. *Theor Appl Genet.* 81(6): 713-719.

Qin, F., Shinozaki, K., and Yamaguchi-Shinozaki, K. (2011). Achievements and challenges in understanding plant abiotic stress responses and tolerance. *Plant Cell Physio.* 52(9): 1569-1582.

Rahman, H. *et al.* (2011). Molecular mapping of quantitative trait loci for drought tolerance in maize plants. *Genet Mol Res.* 10(2): 889- 901.

Rattalino Edreira, J.I. and Otegui, M.E. (2013). Heat stress in temperate and tropical maize hybrids: A novel approach for assessing sources of kernel loss in field conditions. *Field Crops Res.* 142: 58–67.

Ribaut, J.M. and Ragot, M. (2007). Marker-assisted selection to improve drought adaptation in maize: the backcross approach, perspectives, limitations and alternatives. *J Exp Bot*. 58: 351-360.

Rutkoski, J.E., Heffner, E.L. and Sorrells, M.E. (2011). Genomic selection for durable stem rust resistance in wheat. *Euphytica*. 179(1): 161-173.

Saleem, M. A., Malik, T. A., and Shakeel, A. (2015). Genetics of physiological and agronomic traits in upland cotton under drought stress. *Pak. J. Agri. Sci.* 52, 317-324.

Salvucci, M.E. and Crafts Brandner, S.J. (2004). Inhibition of photosynthesis by heat stress: the activation state of Rubisco as a limiting factor in photosynthesis. *Physiologia Plantarum*. 120(2): 179-186.

Sanguineti, M.C. *et al.* (1999). QTL analysis of drought-related traits and grain yield in relation to genetic variation for leaf abscisic acid concentration in field-grown maize. *J Expt Bot*. 50(337): 1289-1297.

Setter, T.L. *et al.* (2010). Genetic association mapping identifies single nucleotide polymorphisms in genes that affect abscisic acid levels in maize floral tissues during drought. *J Expt Bot*. erq308.

Shou, H., Bordallo, P. and Wang, K. (2004). Expression of the Nicotiana protein kinase (NPK1) enhanced drought tolerance in transgenic maize. *J Expt Bot*. 55(399): 1013-1019.

Sreenivasulu, N., Harshavardhan, V.T., Govind, G., Seiler, C. and Kohli, A. (2012). Contrapuntal role of ABA: does it mediate stress tolerance or plant growth retardation under long-term drought stress? *Gene.* 506: 265-273.

Steven, J.C., Brandner, M. and Salvucci, M. (2002). Sensitivity of photosynthesis in C4 maize plant to heat stress. *Plant Physiol*. 129: 1773-1780.

Swamy, B.M., Vikram, P., Dixit, S., Ahmed, H.U. and Kumar, A. (2011). Meta-analysis of grain yield QTL identified during agricultural drought in grasses showed consensus. *BMC Genomics*. 12:319. doi: 10.1186/1471-2164-12-319.

Thirunavukkarasu, N., Hossain, F., Mohan, S. *et al.* (2013). Genome-wide expression of transcriptomes and their coexpression pattern in subtropical maize (*Zea mays* L.) under waterlogging stress. *PLoS One.* 8: e70433.

Tilman, D., Balzer, C., Hill, J. and Belfort, B.L. (2011). Global food demand and the sustainable intensification of agriculture. *Proc Natl Acad Sci U.S.A.* 108: 20260–20264.

Tuberosa, R. and Salvi, S. (2004). QTLs and genes for tolerance to abiotic stress in cereals. pp. 253- 315. In: P.K. Gupta, R. Varshney (Eds.), *Cereal Genomics*. Kluwer, The Netherlands.

Tuberosa, R. *et al.* (1998). RFLP mapping of quantitative trait loci controlling abscisic acid concentration in leaves of drought stressed maize (*Zea mays* L.). *Theor Appl Genet.* 97: 744–755.

Valliyodan, B. and Nguyen, H.T. (2006). Understanding regulatory networks and engineering for enhanced drought tolerance in plants. *Curr Opin Plant Bio.* 9: 189–195

Wang, W., Vinocur, B. and Altman, A. (2003). Plant response to drought, salinity and extreme temperature: towards genetic engineering for stress tolerance. *Planta.* 218: 1-14.

Wani, S.H. and Sah, S.K. (2014). Biotechnology and abiotic stress tolerance in rice. *J.Rice Res.*2: 2e105.

Whitford, R., Gilbert, M. and Langridge, P. (2010). Biotechnology in agriculture. In: Reynolds MP. Climate Change and Crop Production. CABI Series in Climate Change Vol. 1, *Global Plant Clinic* (CABI), Oxfordshire, UK. pp. 219-244.

Wu, G.H. (1987). Analysis of genetic effects for quantitative characters at different developmental states in maize. *Genet.* 18: 69.

Xia, Z., Liu, Q., Wu, J. and Ding, J. (2012). ZmRFP1, the putative ortholog of SDIR1, encodes a RING-H2 E3 ubiquitin ligase and responds to drought stress in an ABA-dependent manner in maize. *Gene.* 495: 146-153.

Yamamoto, K., Sakamoto, H. and Momonoki, Y.S. (2011). Maize acetyl-cholinesterase is a positive regulator of heat tolerance in plants. *J Plant Physiol.* 168: 1987-1992.

Zaidi, P.H. *et al.* (2015). QTL mapping of agronomic waterlog ging tolerance using recombinant inbred lines derived from tropical maize (*Zea mays* L) germplasm. *PloS One. 10*(4): e0124350.

Zhang, X., Tang, B. *et al.* (2013). Identification of major QTL for waterlogging tolerance using genome-wide association and linkage mapping of maize seedlings. *Plant Mol Bio Reporter*. 31(3): 594-606.

25

Improvement of Industrially Important Traits in Maize

Vishal Singh[*], Mukesh Choudhary, Alla Singh, Pravin Kumar Bagaria Mamta Gupta, Bhupender Kumar, Chikkappa G. Karjagi Shankar Lal Jat, Vinay Mahajan and Pardeep Kumar

ICAR-Indian Institute of Maize Research, Ludhiana, Punjab
**Corresponding Author's Email: vishaliari.singh@gmail.com*

Maize is a miracle crop and is popularly known as "Queen of Cereals" due to its very high yield potential. It's a leading crop worldwide in terms of area, production and productivity and is suitable to diverse agro-climates. Its extent of adaptability is unmatched by any other crop. Due to this adaptability to diverse climates, maize had been grown by many generations in different parts of the world. Earlier, the primary utilisation of maize was focused towards human consumption and even today it is a staple food of many underprivileged societies especially in African countries and certain tribal pockets in India. It is playing an important role as a cheap source of carbohydrate in daily meals due to high productivity and wider adaptability. With the passage of time the productivity increased due to conscious selection towards high yield by farmers and researchers. This enhanced production and productivity opened new doors of utilisation of this wonderful crop. Maize found its place as a feed crop for economically important animals like chicken and swine. This shift towards feed industry got encouragement by change in the food habits of people. Looking at the recent consumption pattern of maize as a food crop displays declining trend. As per a NSSO survey (2014) maize consumption in rural areas in India has seen a decline of around 35% from 3.7 kg/year in 2004-05 to 2.4 kg per year in 2009-10, and further declined to 1.56 kg during the year 2011-12. Continued efforts towards yield enhancement and decline in human consumption made its way to utilise surplus maize in economically viable industrial products. Maize is a crop where every part of its plant can be utilised to produce food and non-food products. Oil, starch, glucose, dextrose and sorbitol are few important products produced by processing of maize grains. Apart from these, maize has been very widely used for production of ethanol which is used as a fuel in many developed countries. Direct utilisation of maize grains in poultry and other feed industries is on the rise with increasing demand of such foods. Maize oil is of high quality and is being widely used in developed countries. Maize is a major source of industrial starch which finds its way to diverse utilization such as food and beverage, pharmaceutical, textile, paper and other industries. Processing of maize is done by two ways- Dry Milling and Wet Milling. Wet milling produces starch, gluten and husk and Dry Milling produces corn meal, germ, grits and animal feed. Apart from these processed products, maize is also being developed to be utilised as a biopharmaceutical crop. Maize crop is being bioengineered to extract products of therapeutic and medicinal importance. Such a diverse utilisation of maize has generated a necessity to improve morpho-physiological traits which are important with respect to its industrial utilisation. In the present chapter, we will discuss various aspects of maize improvement for industrially important traits which will include oil, starch, methionine, biofuel and biopharmaceuticals.

High oil maize

Oil in cereals ranges from 1.5 to 5.9% with oat having the highest grain oil. Normal corn has usually 3-4 % oil, majority of which lies inside the germ. Cultivars which possess more than 6% oil are considered as high oil corn. High oil corn carries relatively higher kernel protein content than the normal corn. Oil has higher calorific value than carbohydrates thus important to fulfil the energy needs of both human as well as animal diet. Increasing oil upto 8 percent without substantially affecting yield levels can increase the value of maize for both feed and processing industries. Potential of corn oil was not realised earlier and non-starch portion was considered as a waste during processing. After realising the potential of corn oil extraction from germ its first commercial production took place in 1889. Corn oil is extracted from germ portion of the kernel. There are six major divisions in maize kernel namely pericarp, aleurone, horny endosperm, soft starch and germ. Normal corn has 82% endosperm and 12% germ. Majority of kernel oil (84%) is located in the germ and rest of it is found in endosperm (5%) and aleurone (12%). Apart from oil, germ is also richer in protein content than endosperm. Extraction of con oil is not an isolated process and it is done along with extraction of starch and other products.

Improvement of oil content in corn is a success story. A number of researchers have worked on this aspect with significant improvement in kernel oil content. Since the oil content is a quantitative trait, an optimum strategy to improve it lies in increasing the frequency of favourable alleles trough population improvement methods. Among the long term experimentations with oil content in maize, long term selection experiments at Illinois, USA are the most extensive and highly informative. In 1896, Hopkins initiated these experiments in a local variety

(Hopkins, 1899) and developed four different strains {Illinois High Oil (IHO), Illinois Low Oil (ILO), Illinois High Protein (IHP), and Illinois Low Protein (ILP) strains)} based on their oil and protein content to carry out further selections. Upto 100 generations were subjected to selection under different segments. To determine the availability of extent of residual variation, reverse selection was also performed. Evidences suggested that significant genetic variance for oil existed even after 98 generations for IHO strains and the *per se* evaluation trials showed significant increases in oil during the last five generations measured. These experiments showed that genetic variability did not exhaust even after 100 generations. Gain during selection after 100 generation of selection was 17.1% in IHO population. Such long term experiments provide important information for the improvement of the trait. Mass selection was the major strategy in such studies. Recurrent selection had reported significant gain during selection process for oil content. Serna-Saldivar *et al.* (2013) conducted eight cycles of recurrent selection and recorded 33 to 60 % increase in oil content over the initial values after last generation of recurrent selection. Absence of correlation between oil content and yield reported by them is a hope to find out segregants combining high oil and high yield traits together in same background. A negative linkage between yield and germ oil content has been reported (Mosse *et al.*, 2004)

Apart from several conventional breeding efforts to enhance oil content in maize, studies have been undertaken to understand the molecular and biochemical architecture of the trait. Molecular mapping of oil trait has shown a large number of QTLs governing the traits. Zheng *et al.* (2008) identified a QTL encoding an acyl-CoA:diacylglycerol acyltransferase (*DGAT1-2*), affecting the oil content and oleic acid content. More than 50 QTLs were mapped in a study (Laurie *et al.*, 2004) for lipid accumulation making it a difficult trait to improve. Expression of wheat *Purindoline* a and b (*PINA and PINB*) genes in maize through recombinant DNA technology has increased oil content by 25% (Zhang *et al.*, 2010). Hao *et al.*, (2004) did efforts to transfer major QTL (*qHO6)* which is located on chromosome 6, from a high oil donor inbred By804 to two inbred lines (Zheng58, and Chang7-2). Six generation of marker assisted backcrossing yielded Zheng58-qHO6 and Chang7-2-qHO6 which showed 1% increase in oil content without any change in grain weight. Hybrids generated out of crossing of these two parents showed 0.17% of absolute increase in oil content. Interestingly, the yield level of the new hybrid was similar to the one developed from original inbred parents. This way, we can see that multiple approaches are available to enhance oil content in maize kernels. Population improvement methods are slow to respond but are promising when no much information regarding molecular architecture is available or the breeding program is low in resources to be spent on sophisticated molecular techniques. In the era of genomics, application of advanced selection methods like genomic selection approach has immense potential to enhance the amount of gain per selection cycle. Genetic enhancement of corn oil may be helpful for its enhanced industrial output and its wider consumption in an increasingly health conscious society.

Maize for high starch

Starch is a polysaccharide comprising amylose and amylopectin as its components. Starch is an energy source in all green plants. In human diet starch is the most common source of carbohydrates. Major sources of food starch are potato, wheat, maize, rice and cassava which are staple foods at different places across the globe. Apart from a source of energy in its native form, extracted starch is a well established industrial product. In industry starch is converted into several different products. Modified starch acquires several properties which are not available in native starch but are essential to meet the requirements of food processing. Extracted starch is utilised in a wide variety of products such as desserts and dairy products, bakery products, confectionery and chocolates, beverages, processed foods, paper and board, pharmaceuticals and cosmetics, pet food and aqua-feed. Maize endosperm, the main energy reserve, makes around 80% of the whole kernel. Around 90% of the endosperm is starch and 7% is protein. Starch in maize like other grains is composed of amylose and amylopectin. The ratio of amylopectin to amylose is roughly 75:25 in case of normal maize. Genetic variation for this ratio is found in nature where amylopectin content grows up to 100% in case of waxy maize genotypes. Both amylose and amylopectin components of starch vary in their branching pattern. This variation in branching pattern expresses different physico-chemical properties of starch. Starch with high amylose content are hard to gelatinize as extra energy is needed to hydrate and disintegrate the firmly- bonded aggregates of amylose. Such starches give firm gels and tough films. In case of starches containing 100% amylopectin (waxy maize), gelatinization is easier and gives transparent pastes. With conscious breeding efforts, ratio between amylose and amylopectin can be changed upto certain extent to get a desirable starch profile according to the need. A brief discussion about these variants is given further.

There have been many efforts to enhance total starch content and amylose content in maize. During the late 1930s intensive efforts were done to search for genetic variation which can provide starch similar to that obtained from tapioca. In the year 1908, such variant was found in china and popularly known as waxy corn due to its peculiar appearance. Waxy maize is a result of a specific mutation in normal maize and controlled by a recessive gene *'wx'* on chromosome 9 (Coe et al., 1988). Introgression of waxy trait in diverse genetic background has been attempted by many workers. Researcher tried to develop markers which are linked to waxy gene (Liu *et al.,* 2007) but the more popular and reliable method to screen waxy phenotype is staining by iodine (I_2/KI) of starch contained in pollen, embryo sac or in the endosperm. In contrast to the black colour shown by the amylose containing grains, waxy grains show brown colour. If we look at the biochemical pathway of amylose synthesis, Granule Bound Starch Synthase I (Hylton et al., 1996; Nelson and Rines, 1962) is known to be linked with amylose synthesis in the endosperm and is reported to be highly active in non-waxy maize, but almost absent in waxy maize. Major genetic variation of waxy corn has been reported (Brewbaker *et al.,* 2007) to occur in south-east Asia and temperate regions of China, Vietnam and Korea. Numerous efforts have been made to increase quality and yield level of waxy corn in china as it is an important vegetable in that region. The potential of waxy maize for industrial utilisation may come through the route of contract farming of specialised varieties.

In contrast to the waxy corn where the complete starch is made up of amylopectin, amylomaize is a type of corn which possesses higher percentage of amylose in endosperm starch than that is found in normal maize kernels. Amylose is comprised of straight chain monomers and exhibits a phenomenon called retrogradation where amylose molecules stick to each other when cooled resulting in gelled paste. After gelatinisation such starches form tough films. High amylose starch is used in textiles industries and to produce quick setting confectionaries gum. Physicochemical properties of amylose starch allow it to be converted into threaded structures which may find its place in biodegradable plastic formation. Food uses of high amylose starch include coating canned foods, cooked food thickener, tomato paste texturing, deep fried food coating, low calorie bread, edible foams, pudding thickener etc. Non food utilization of high amylose starch include amylose yarn, paint thickener, water soluble film, corrugating adhesive, encapsulating agent, biodegradable film, Hydrogels for slow release drugs, glass fibre sizing and reinforcement and so many others. Improvement in proportion of amylose content of normal maize starch has been tried through systematic breeding efforts with a significant success. Significant plant breeding and genetics research effort had been undertaken to develop high amylose maize in USA between 1946 and 1970. Mutation at *amylose extender* or *ae* locus is reported to enhance amylose content in maize starch. This increment in amylose content may range from 25 to 70% (Shannon and Garwood, 1984). Many workers reported about useful recessive mutations of *ae* locus which can enhance amylose percentage when homozygocity is achieved for the recessive mutant (Vineyard *et al.*, 1958; Helm *et al.*, 1967; Garwood *et al.*, 1976). In 1991, Robertson and Stinad patented (US Patent no: 5004864) a dominant mutant of *ae* locus which significantly enhances amylose fraction in endosperm starch. The dominant nature of the mutant made its transfer easier through backcrossing by reducing number of generations required to recover recipient parent genome. Extensive efforts have been made to improve amylose content through population improvement methods in Australia after keeping in mind the industry requirement for amylomaize (McWhirter and Dunn, 1986). Current goal regarding amylomaize is to enhance amylose content beyond 80% and to achieve 100% amylose corn in future. Thus, alternation of the proportion of the two different types of starches may help industry enhance its extraction efficiency and simultaneously may help farmers to get benefitted from specialised corn cultivation through contractual farming approach.

High methionine corn

In India, maize is primarily used as a feed crop for poultry. From nutritional point of view maize protein is of low quality and is deficient in three essential amino acids methionine, lysine and tryptophan (Vasal, 1998). Supplementation of such amino acids in feed with synthetic amino acids is one way to improve the quality of feed. Unlike lysine which is available at low cost in synthetic form, synthetic methionine is a costly additive in poultry feed. Improvement of natural methionine in maize kernels can reduce the cost of feed. Increase in the level of seed methionine can be achieved through three approaches. The first approach is to enhance 10-kDa deltazein (*dzs10*) fraction of seed protein which is rich in methionine in comparison to other fractions. Post-transcriptional regulation of *dzs10* transcript by *dzr1* (*delta zein regulator 1*) affects the accumulation of methionine in kernels (Cruz-Alvarez *et al.*, 1991). Lai and Messing (2002) enhanced the stability of *dzs10* transcript by modifying genomic region around *dzs10* coding region. Another approach to enhance methionine concentration is to introgress a naturally occurring mutant *floury-2*(*fl-2*) which enhances methionine accumulation in grains (Nelson *et al.*, 1965). Precise role of this mutation in increasing methionine concentration is poorly known. Another important approach for enhancing methionine level is population improvement. Population improvement methods have been successfully applied in maize for improving seed related traits (reviewed by Moose *et al.*, 2004). Scott *et al.* (2008) altered grain methionine level through recurrent selection approach. Integrated approach to enhance methionine level has been suggested by Huffman *et al.* (2016) instead of going with one approach alone. Thus, with systematic breeding efforts, methionine concentration in maize grain can elevated which is a trait of significance for poultry industry.

Maize for bio-ethanol

In the present era of climate change and its impact on environment, there is an immense need at global level to make a shift from finite fossil fuels (mineral oil based economy) to ecofriendly energy alternatives like cellulosic ethanol (Schubert, 2006; Vermerris *et al.*, 2007; Huber and Dale, 2009). Maize grain is a good source of starch and is being widely used for starch based ethanol production. Maize can be stored over a longer period compared to sugarcane and thus large scale processing plants can lead to cost effective production of ethanol generating higher scale economies. Tropical maize can be considered unique among the C4 grasses due to availability of diverse of germplasm for crop improvement as well as its ability to adapt in diverse environments. The need of the hour is to assess the benefits of tropical maize as an alternative to grain corn for efficient ethanol production. In comparison to the starch as the only harvestable carbon form in corn grain, tropical maize can produce three distinct types of biomass feedstocks-sugar, starch and lignocellulosic biomass. Tropical maize exhibits around 2.5 times more total annual biomass than grain corn. The fuel conversion efficiency of the tropical maize based ethanol is far better than the corn grain based ethanol. Hence, the growing bioenergy industry demands the use of tropical maize as a new biofuel feedstock. Considering the potentiality of maize as the future bioenergy crop, we need to look at the genes involved in the synthesis of the main components of maize cell walls (cellulose, hemicellulose and lignin) along with their regulation so that strategies can be designed to improve biomass composition for bioethanol production. The most important strategy for improving the lignocellulosic feedstock may be improving the relative content and industrial quality of cellulose. It will help in attaining the increased amount of harvestable energy per unit of land. A mutant named as *brittle stalk-2* (*bk-2*) is characterized by stalks which break easily under mechanical pressure can be utilized for ethanol production. Besides this, characterization of allelic variants of genes crucial for cellulose biosynthesis can be done for their direct utilization in classical breeding. The successful example of this strategy can be demonstrated from A*Rabi*dopsis mutant (*irx 1-2*) displaying lower cellulose crystallinity (<~30%) and improved cell wall digestibility relative to wild type, without significantly affecting growth and fitness. The mutant phenotype arose from a point-mutation at the C-terminal transmembrane region of the *CesA3* (Harris *et al.*, 2009).

Modification of hemicellulosic component may also help in improving cell wall digestibility resulting in improvement in ethanol yield. It has been successfully proven that the enzymatic conversion efficiency of mildly pre-treated maize stem materials can be significantly improved by the compounded effect of reduced cell wall lignin and high GAX a*Rabi*nose-to-xylose ratio. Thus, the existing untapped genetic variability in maize in the form of the degree and distribution of GAX substitution patterns can be explored to modify maize cell wall hemicelluloses.

Lignin has been found to protect the cellulose microfibrils (cell wall polysaccharides) from enzymatic attack by binding to the hydrolytic enzymes and therefore reducing the effectiveness of enzymatic saccharification process. Therefore, reduction of lignin content in cell wall has been found to be significantly improving the cell wall digestibility. Classical breeding have proven its worth to alter the lignin composition for easy cell wall digestibility. Extensive variation for lignin content and enzymatic digestibility properties have been found in maize experimental populations and mutant panels (Argillier *et al.*, 1996; Fontaine *et al.*, 2003; Marita *et al.*, 2003; Mechin *et al.*, 2005; Barriere *et al.*, 2010, 2013) and these resources have been utilized for quantitative trait loci (QTL) mapping of underlying lignification characteristics relevant to cellulosic ethanol production (Mechin *et al.*, 2001; Cardinal *et al.*, 2003; Barriere *et al.*, 2008, 2009; Thomas *et al.*, 2010; Courtial *et al.*, 2013). Studies reported that lignin content variation for improved bioconversion efficiency is highly heritable, which can be further utilized to select and advance the promising maize feedstocks for efficient and cost-effective production of ethanol (Torres *et al.* 2013). A combined approach of breeding and biotechnology can help in identification and modification of genes for high biomass production and improving the fermentation efficiency in maize. Maize possesses an advantage of non-invasiveness as compared to other perennial sources of bioenergy. The untapped vast genetic diversity of tropical maize has opened an avenue for rapid advances in biofuel industries and through genetic modification of lignocellulosic and stalk sugar traits, it is possible to meet the current cellulosic ethanol demand.

Maize for bio-pharmaceuticals

Majority of world production depends on plant-based medicines for healthcare (Khan *et al.*, 2009). The naturalness of plant products is responsible for the increasing popularity and demand of medicines obtained from plants. As such, plant-based medicines represent an important industrial product from plants. Bioactive compounds, which are primarily plant secondary metabolites, have profound health benefits. Efforts have been directed towards increasing the proportion of these health benefit-conferring compounds in plants. Various tools and approaches in plant biology aid in the development of plants specifically for medicinal purposes. These include, but are not limited to plant breeding, plant tissue culture, recombinant DNA technology, genome editing, etc. Plant breeding, efforts have largely been responsible for increased secondary metabolites in plants, for increased resistance to pests and pathogens. The advantages of medicinal protein production in plants include low contamination risk, capability of scale-up, appropriate glycosylation pattern, low cost of production as well as storage (Kamenarova *et al.*, 2005). Kaushik *et al.* (2015) have reviewed the use of vegetable breeding for increasing the amount of bioactive phenolic acids.

The authors note that a large diversity of phenolic acids is observed among vegetables. Increase in phenolic content has been brought about by various means that include development of improved varieties and growth under well-defined specific conditions. Apart from this, postharvest treatments also results in enhanced phenolic content (Balyan *et al.* 2013). Conventional breeding has shown promise in achieving cultivars with traits of medicinal importance. Plant cell culture is emerging as an alternative bio-production system for recombinant pharmaceuticals. Recently, food and Drug Administration of United States approved plant-produced β-glucocerebrosidase. Xu *et al.* (2014) have presented review of the technology platforms available in cell and tissue cultures for biopharmaceutical. Plant tissue culture offers special advantages in that it is amenable to large scale and offers better control of quality parameters than those in field plantations with the possibility of producing biopharmaceuticals in industrial-scale plant cell cultures has opened new avenues in the field.

Among cereals, maize is the major crop for production of biopharmaceuticals. The high biomass yield and capacity of scale-up are major benefits in using maize. A number of proteins have been recombinantly expressed in maize. These include approtinin, avidin, trypsin, collagen, vaccines, etc. The recombinant protein expressed in maize are functionally active. Hood *et al.* (1999) analyzed production of avidin and â-glucuronidase in maize and reported their purification with any major obstacles in transgenic maize. One big advantage is that the biopharmaceuticals are expressed in high amounts and can be recovered from the seeds. Usually, maize-derived products are free of any bacterial pathogen or contaminants and are thus, easier to purify. According to data derived from Daniell *et al.* (2001), maize has the highest biomass yield amongst cereals, a feature very important for the industrial-scale production of biopharmaceuticals. Apart from this, protocols have been standardized for extraction of biopharmaceuticals with efficient recoveries in maize (Azzoni *et al.*, 2002 and Zhong *et al.*, 2007). Vaccines produced in maize have been used in clinical studies (Streatfield *et al.*, 2001 and Tacket *et al.*, 2004). The recombinant antigen has been found to be stable and homogenous and can be consumed in various forms without destruction of the antigen.

Hence, maize is a very attractive option for its use in biopharmaceutical industry. This trait can be exploited through various approaches. The possibility of wide-scale dissemination of the produced biopharmaceutical, low cost of production, stability of the product are some of the key features that make maize a potent crop for medicine industry.

Improvement in industrially important traits in maize has the potential to enhance the income level of farmers by fetching special prices for special corn types. It may also help industries in attaining higher efficiency in production of related products by using more suitable raw material as per their requirements. In Indian context, a more focused approach is needed to undertake research in these new avenues of crop improvement. A broader profile of maize cultivars suiting different processing requirement will certainly benefit farmers, industry and economy as a whole.

References

Argillier, O., Barriere, Y., Lila, M., Jeanneteau, F., Gelinet, K. and Menanteau, V. (1996). Genotypic variation in phenolic components of cell-walls in relation to the digestibility of maize stalks. *Agronomy.* 16: 123-130.

Azzoni, A.R., Kusnadi, A.B., Miranda, E.A. and Nikolov, Z.L. (2002). Recombinant aprotinin produced in transgenic corn seed: extraction and purification studies. *Biotech Bioeng.* 80(3): 268–276.

Balyan, H.S., Gupta, P.K., Kumar, S., Dhariwal, R., Jaiswal, V., Tyagi, S., Agarwal, P., Gahlaut, V. and Kumari, S. (2013). Genetic improvement of grain protein content and other health-related constituents of wheat grain. *Plant Breed.* 132: 446-457.

Barriere, Y., Charcosset, A., Denoue, D., Madur, D., Bauland, C. and Laborde, J. (2010). Genetic variation for lignin content and cell wall digestibility in early maize lines derived from ancient landraces. *Maydica.* 55: 65.

Barriere, Y., Chavigneau, H., Delaunay, S. *et al.* (2013). Different mutations in the *ZmCAD2* gene underlie the maize brown-midrib1 (bm1) phenotype with similar effects on lignin characteristics and have potential interest for bio-energy production. *Maydica.* 58: 6-20.

Barriere, Y., Mechin, V., Lafarguette, F. *et al.* (2009). Toward the discovery of maize cell wall genes involved in silage quality and capacity to biofuel production. *Maydica.* 54: 161.

Barriere, Y., Thomas, J. and Denoue, D. (2008). QTL mapping for lignin content, lignin monomeric composition, p- hydroxycinnamate content and cell wall digestibility in the maize recombinant inbred line progeny F838 x F286. *Plant Sci.* 175: 585- 595.

Brewbaker, J.L., Martin, I.F. and Pulam, T. (2007). Genetics and breeding of sweet corn adapted to the tropics. Conference on Genetics, Breeding, Planting and Industrialization of Sweet and Waxy Corn, Guangzhou, Guangdong, China, Nov. 26-28, 2007. Pp. 1- 19.

Cardinal, A., Lee, M. and Moore, K. (2003). Genetic mapping and analysis of quantitative trait loci affecting fiber and lignin content in maize. *Theor Appl Genet.* 106: 866- 874.

Coe, E. H., Neuffer, M. G. and Hoisington, D. A. (1988). The genetics of corn and corn improvement. Sprague, G. F. and Dudley, J. W., *American Society of Agronomy.*

Courtial, A., Mechin, V., Reymond, M., Grima-Pettenati, J. and Barriere, Y. (2013). Colocalizations between several QTLs for cell wall degradability and composition in the F288x F271 early maize RIL progeny raise the question of the nature of the possible underlying determinants and breeding targets for bio-fuel capacity. *Bioeng Res.* DOI: 10.1007/s12155-013-9358- 8.

Cruz-Alvarez, M., Kirihara, J.A. and Messing, J. (1991). Post-transcriptional regulation of methionine content in maize kernels. *Mol Genet.* 225: 331–339.

Daniell, H., Streatfield, S.J. and Wycoff, K. (2001). Medical molecular farming: production of antibodies, biopharmaceuticals and edible vaccines in plants. *Trends Plant Sci.* 6(5): 219–226.

Dudley, J.W. and Lambert, R.J. (2004). 100 Generations of Selection for Oil and Protein in Corn. *Plant Breed Rev.* Vol- 24, Part- 1.

Fontaine, A.S., Bout, S., Barriere, Y. and Vermerris, W. (2003). Variation in cell wall composition among forage maize (*Zea mays* L.) inbred lines and its impact on digestibility: analysis of neutral detergent fiber composition by pyrolysis-gas chromatography mass spectrometry. *J Agric Food Chem.* 51: 8080-8087.

Garwood, D.L., Shannon, J.C. and Creech, R.G. (1976). Starclies of endosperms processing different alleles at the amylose-extender locus of *Zea maize* L *Cereal Chem.* 53: 355-364.

Hao, X., Li, X., Yang, X. *et al.* (2004). Transferring a major QTL for oil content using marker-assisted back crossing into an elite hybrid to increase the oil content in maize. *Mol Breed.* 34: 739.

Harris, D., Stork, J. and Debolt, S. (2009). Genetic modification in cellulose-synthase reduces crystallinity and improves biochemical conversion to fermentable sugar. *GCB Bioeng.* 1: 51–61.

Helm, J.L., Fergason, V.L. and Zuber, M.S., (1967). Development of high-amylose corn (*Zea maize* L.) by the backcross method. *Crop Sci.* 7: 659-662.

Hood, E.E., Kusnadi, A., Nikolov, Z. and Howard, J.A. (1999). Molecular farming of industrial proteins from transgenic maize. *Adv Exp Med Bio.* 464: 127-147.

Hopkins, C.G. (1899). Improvement in the chemical composition of the corn kernel. p. 1–31. In J W. Dudley (ed.), Seventy generations of selection for oil and protein in maize. ASA, CSSA, and SSSA, Madison, WI.

Huber, G.W. and Dale, B.E. (2009).Grassoline at the pump. *Scientific American.* 301: 52–59.

Huffman, R.D., Edwards, J.W., Pollak, L.M. and Scott, M.P. (2016). Interaction of genetic mechanisms regulating methionine concentration in maize grain. *Crop Sci.* 56: 1–11.

Hylton, C.M., Denyer, K., Keeling, P.L., Chang, M.T. and Smith, A.M. (1996). The effect of waxy mutations on the granule-bound starch synthases of barley and maize endosperms. *Planta* 198: 230-237.

Kamenrova, K., Abumhadi, N., Gecheff, K. and Atanassov, A. (2005). Molecular farming in plants: An approach of agricultural biotechnology. *J Cell Mol Biol.* 4: 77- 86.

Kaushik, P., Anujar, I., Vilanova, S., Plazas, M, Gramazio, P., Herrair, F.J. *et al.* (2015). Breeding vegetables with Increase in Phenolic Acids. *Molecules.* 20: 18464-18481.

Khan, M.Y., Aliabbas, S., kumar, V. and Rajkumar, S. (2009). Recent advances in medicinal plant biotechnology. *Ind J Biotech.* 8: 9-22.

Lai, J. and Messing, J. (2002). Increasing maize seed methionine by mRNA stability. *The Plant J.* 4: 395-402.

Laurie, C.C., Chasalow, S.D., LeDeaux, J.R., Mc Carroll, R., Bush, D. *et al.* (2004). The genetic architecture of response to long-term artificial selection for oil concentration in the maize kernel. *Genet.* 168: 2141–2155.

Liu, J., Rong, T. and Li, W. (2007). Mutation loci and intra-genic selection marker of the granule-bound starch synthase gene in waxy maize. *Mol Breed.* 20: 93-102.

Marita, J.M., Vermerris, W., Ralph, J. and Hatfield, R.D. (2003). Variations in the cell wall composition of maize brown midrib mutants. *J Agric Food Chem.* 51: 1313-1321.

Mc Whirter, K.S. and Dunn, C.F., (1986). Development of high amylose maize production in Australia. *DSIR Plant Breed Symposium*. Pp. 86-91.

Mechin, V., Argillier, O., Hebert, Y., Guingo, E., Moreau, L. Charcosset, A. and Barriere, Y. (2001). Genetic analysis and QTL mapping of cell wall digestibility and lignification in silage maize. *Crop Sci.* 41: 690-697.

Mechin, V., Argillier, O., Rocher, F. *et al.* (2005). In search of a maize ideotype for cell wall enzymatic degradability using histological and biochemical lignin characterization. *J Agric Food Chem.* 53: 5872-5881.

Moose, S., Dudley, J. and Rocheford, T. (2004). Maize selection passes the century mark: A unique resource for 21st century genomics. *Trends Plant Sci.* 9: 358– 364.

Nelson, O.E. and Rines, H.W. (1962). The enzymatic deficiency in waxy mutant of maize. *Biochem Biophys Res Commun.* 9: 297- 300.

Nelson, O.E., Mertz, E.T. and Bates. L.S. (1965). Second mutant gene affecting the amino acid pattern of maize endosperm proteins. *Sci.* 150: 1469– 1470.

N.S.S.O. (NSSO), Employment and Unemployment Situation in India, (July 2011 - June 2012), vol. 554 of 68th round. National Sample Survey Office, Ministry of Statistics & Programme Implementation, Government of India, January 2014.

Robertson, D.S., and Stinard, P.S. (1991). Dominant amylose-extender mutant of maize. *Iowa State Univ Patents* paper 182.

Schubert, C. (2006). Can biofuels finally take center stage? *Nat Biotech.* 24: 777– 784.

Scott, M.P., Darrigues, A.T., Stahly, S., and Lamkey, K.R. (2008). Recurrent selection to control grain methionine content and improve nutritional value of maize. *Crop Sci.* 48: 1705– 1713.

Serna-Saldivar, S.O., Ortega-Corona, A., Ortiz-Islas, S., Garcia-Lara, S., and Preciado-Ortiz, R. (2013). Response of recurrent selection on yield, kernel oil content and fatty acid composition of subtropical maize populations. *Field Crops Res.* 142: 27-35.

Shannon, J.C. and Garwood, D.L. (1984). Genetics and physiology of starch development in starch: Chemistry and Technology, 2nd edition, ed. E. F. Faschall, Academic Press, Inc. Orlando. Pp. 25-86.

Streatfield, S.J. (2005). Mucosal immunization using recombinant plant-based oral vaccines. *Methods.* 38(2): 150–157.

Tacket, C.O., Passeti, M.F., Edelman, R.E., Howad, J.A. and Streeatfield, S.J. (2004). Immunogenicity of recombinant LT-B delivered orally to humans in transgenic corn. *Vaccine. 22*. 4385–4389.

Thomas, J., Guillaumie, S., Verdu, C., Denoue, D., Pichon, M. and Barriere, Y. (2010). Cell wall phenyl-propanoid related gene expression in early maize recombinant inbred lines differing in parental alleles at a major lignin QTL position. *Mol Breed.* 25: 105-124.

Torres, A.F., Van Der Weijde, T., Dolstra, O., Visser, R.G. and Trindade, L.M. (2013). Effect of maize biomass composition on the optimization of dilute-acid pre-treatments and enzymatic saccharification. *Bioeng Res.* 6: 1038-1051.

Vasal, S. (1998). Quality protein maize story. Special Issue on Improving Human Nutrition Through Agriculture. 21: 445.

Vermerris, W., Saballos, A., Ejeta, G., Mosier, N.S., Ladisch, M.R. and Carpita, N.C. (2007). Molecular breeding to enhance ethanol production from corn and sorghum stover. *Crop Sci.* 47: 142–153.

Vineyard, M.L., Bear, R.P., Mac Masters, M.M. and Deatherage, W.L. (1958). Development of "Amylomaize"-corn hybrids with high amylose starch: I. Genetic considerations. *Agron J.* 50: 595-598.

Xu, J. and Zhang, N. (2014). On the way to commercializing plant cell culture platform for biopharmaceuticals: present status and prospect. *Pharm Bioprocess.* 2(6): 499- 518.

Zhang, J., Martin, J.M., Beecher, B., Lu, C., Hannah, L.C., Wall, M.L., Altosaar, I. and Giroux, M.J. (2010). The ectopic expression of the wheat Puroindoline genes increase germ size and seed oil content in transgenic corn. *Plant Mol Biol.* 74: 353–365.

Zheng, P., Allen, W.B., Roesler, K., Williams, M.E., Zhang, S. *et al.* (2008). A phenylalanine in DGAT is a key determinant of oil content and composition in maize. *Nat Genet.* 40: 367–372.

Zhong, Q., Xu, L., Zhang, C., and Glatz, C.E. (2007). Purification of recombinant aprotinin from transgenic corn germ fraction using ion exchange and hydrophobic interaction chromatography. *Appl Microbio Biotech.* 76(3): 607–613.

26

Quality Seed Availability of Maize in India: Status and Strategy

Bijendra Pal

Bioseed Research India a Division of DCM Shriram Ltd., Hyderabad
**Corresponding Author's Email: bijendra.pal@bioseed.com*

Seed in agriculture recognized as one of the most vital and crucial input in agriculture ever since the human kind has started tilling the land to produce the food for his existences. Seed technologist defined seed as matured fertilized ovule consisting of an intact embryo, endosperm and or cotyledon with protective covering (seed coat). A quality hybrid maize seed production system ensures design of the production technology which enables to maintain the healthy embryo, endosperm and /or cotyledon with protective covering. In fact the search of quality seed had acted as a major force to evolve the science of plant breeding. Ever since the farmers understood the importance of seed, they started trading the quality seed within community and now between geographies. Self-pollinated crops possess the natural mechanisms to maintain the intactness of genetic constitution of a seed during multiplication as there is very minimal chance of natural crossings. On the contrary, cross pollinated crops like maize requires special techniques in seed multiplication to maintain the genetic constitution of a F_1 hybrid. Hybrid Maize seed production techniques are well defined and are used by public and private sectors to produce the required hybrid seed in India. In this chapter, I will be highlighting the process, potential agro-climatic zones, hybrid maize seed production capacity and its future perspective in India. In addition, it also reviews the current system and possible strategic move in order to support the maize seed production system and steps required to meet future demand.

Maize Seed production system in India

Almost 80% of the total hybrids maize seed requirement of India is produced and distributed by private sector. Majority of these hybrids are propriety genetics of different seed companies. However, a small portion of hybrids bred by public sector are also produced and distributed by smaller private seed companies which have licensed to produce and market. Production system has been evolved and refined in last 15- 20 years and used by different seed companies in different seed production area (see picture below).

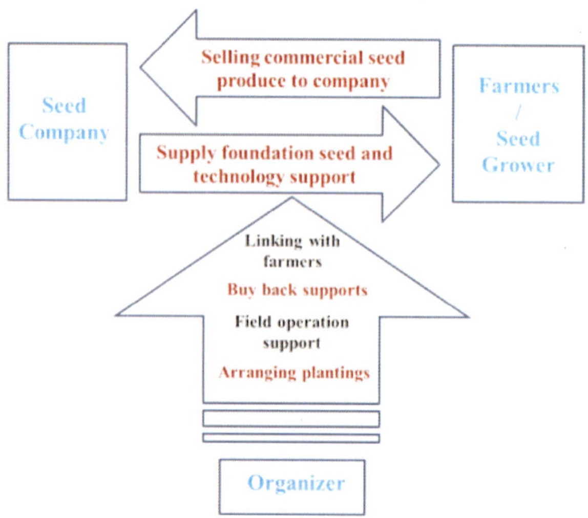

Indian maize seed industry is using contract farming system to produce the seed through farmers. There are two most commonly used processes in the contract seed production: 1) company procure the fresh ear which are harvested at 28-30% moisture after maturity and arrange the mechanical drying and shelling etc. before final processing and seed packing; this process is known as "**cob buying system**". 2) Farmers arrange their own harvesting, sun drying and shelling of the seed produce at their farm in supervision of technical personnel of the company and company procure the **dry seed** and bring it to their processing plants for final processing and packing; this system is known as **"grain buying"**. In both the system, payment for the seed produced by framers made on mutually agreed price which is 20-30% higher than whatever farmers can get from growing the commercial maize or other crop in the same field. Organizers also charge his service fee for providing the support to seed company for managing their production in that area.

Maize seed production Highlights

In India, *Kharif* is the main maize growing season, which covers almost 80% of the total maize area in the country. On the other hand, more than 80% of the maize seed is being produced during *Rabi* season. Seed production during *Rabi* season forms a very intelligent, effective and economic maize seed production and seed supply system in country. Remaining 20% maize seed produced in *Kharif* or late *Kharif* season, which serves as a contingency production season.

There are number of advantages listed as below for producing the seed during *Rabi* season over *Kharif* season.

1. *Rabi* season is a dry season and seed produced during dry season have no or very low infestation from fungal disease which is very common in the *Kharif* season.
2. *Rabi* season is not a main maize growing season, finding the isolation is easier than in *Kharif* season, which enables the condition to produce the genetically pure seed.
3. Climatic conditions are more clearer (no cloud, no rains) during pollination time in *Rabi* season, whereas this is the main problem in *Kharif* season during pollination time and this affects the % seed setting in female rows.
4. The seed supply chain will be more effective as newly harvested seeds are just ready to use before the start of planting season, which enables the farming system to use the genetically pure and physically very healthy seed.

Seed production plot in Eluru-*Rabi* 2016-17

Indian seed sector is using the cluster approach in maize seed production, which helps the industry to get the trained manpower and farmer who are producing and/ or involved in maize seed production on a regular basis. Approximately 70% of the maize seed produced in India comes from five main clusters (Eluru, Nandyal, Rajahmundary, Sattupally and Karimnagar) in Andhra Pradesh and Telangana states (Table 1).

Table 1. Seed production status in 5 main clusters in India – *Rabi* 2012-13 to *Rabi* 2016-17.

Districts	Details	2012-13	2013-14	2014-15	2015-16*	2016-17
Eluru	Area (Acres)	45500	76280	37,878	24700	58550
	Wet cob (MT)	89635	167816	89,842	55680	128810**
	Process seed (MT)	44818	83908	44,921	27840	64405**
Nandyal	Area (Acres)	37600	25991	15000	6400	21100
	Wet cob (MT)	74072	57180	31000	14080	46420**
	Process seed (MT)	37036	28590	15500	7040	23210**

Rajahmundry	Area (Acres)	4500	12100	4700	2100	6500
	Wet cob (MT)	8865	26620	10850	5010	14300**
	Process seed (MT)	4433	13310	5425	2505	7150**
Sattupally	Area (Acres)	18250	17450	13100	7550	24000
	Wet cob (MT)	35953	38390	27950	15275	52800**
	Process seed (MT)	17976	19195	14975	7638	26400**
Karimnagar and	Area (Acres)	17600	10500	10800	4800	6400
Nizambad	Wet cob (MT)	34672	23100	26250	11520	14080**
	Process seed (MT)	17336	11550	13150	5760	7040**
Total	Area (Acres)	112350	142321	81478	45550	11650
	Wet cob (MT)	221329	313106	185533	101565	256300**
	Process seed (MT)	110665	156553	92767	50783	128150**

* In year 2015-16, there was shortage of irrigation water in main production cluster and lot of production area was moved to non-traditional production area which is not listed in this table.
** *Rabi* 2016-17: A maize seed production acre is actual and other figures are estimated.
Source: These information generated from internal discussion and estimation based on several source of information such as production organizers, different company representatives in the production area and cob drying facility provider.

National Seed Corporation (NSC) and other State Seed Corporations (SSCs) also take maize seed production in their respective states, most noticeable areas are in the states of Rajasthan, Karnataka, Madhya Pradesh, Maharashtra, Tamil Nadu and West Bengal.

Maize seed drying and processing

Seed drying and processing before packing play a significant role in the quality of seed. Indian seed industry has made good investments to build up required infrastructure for drying and seed processing, which also supplement public sector facility. Both public and private seed drying and processing facility are sufficient to produce the seed not only for domestic consumption but also to support the maize seed export. Most of these facilities (more than 15 drying and processing plants) are located in the main production cluster areas.

Most of the single cross hybrid seed production goes under the cob buying system and seed companies arrange the mechanical drying and gentle shelling before processing and packing the seed for sale. Some processing plants completes drying until processing and packing the seed, whereas there are some private parities who provide the drying facility. The seed companies who avail the private parties drying facility will bring the wet ear and complete the drying and shelling and then take the shelled seed for final processing and packing in their respective processing plant. More than 15 private parties own the dryers and provide the drying facility to seed companies in Andhra Pradesh and Telangana states. The drying plants run 3 months and completes up to 15 drying cycles. In general, these drying and processing units dry approximately 44700 MT wet ear per cycle.

Maize seed supply and consumption

According to seed industry estimate, India consumes around one lakh MT seed of maize per year in *Kharif* and *Rabi* seasons together. Approximately 80 % of the total seed produced planted in *Kharif* and around 20% of it planted in *Rabi* season where majority (>95%) of the seed is supplied by private seed companies. In India, around 41 seed companies are engaged in hybrids seed research, production and seed supply. India is producing more maize seed than what is consumed in India (Table 1). There are number of multi-national and Indian seed companies who produce the maize seed for export and a sizable quantity of maize seed get exported every year by different seed companies to south east Asian and African countries.

Future outlook and challenges

There is need to improve the present strategy for maize seed production in order to have sustainable maize seed production system. The present cluster approach has its own limitations and risk factors that need to be understood (listed below).

1. **Crop shift** – Indian farming system is moving from conventional agriculture to commercial farming system and famers have started changing the crops very often based on their commercial return from a particular crop. Sometime getting enough land for seed production is difficult due to crop shift.
2. **Availability of trained manpower for detasseling** –the planting window is very narrow in these production clusters and the maximum crop reach at flowering in a very short window thus requires huge manpower for detasseling which is an essential activity.
3. **Isolation** – maximum companies take seed production of several hybrids in the same area, finding the proper isolation distance for different hybrids in the same cluster becoming very challenging.
4. **Disease epidemic** - There is a very high possibility to risk the required seed quantity due to disease, if any particular disease appear in the area and reach to an epidemic level.

The above challenges need to be addressed and as a future strategy, it's required to explore the new seed production zone. Public sector institute can really play a significant role in identifying and stabilizing new seed production zones in different states, which have similar climatic conditions as exists in Andhra Pradesh and Telangana. Maharashtra, Madhya Pradesh, Karnataka, Gujarat, Rajasthan and Chhattisgarh have good scope to serve as new maize seed production cluster in *Rabi* season.

27

Frontier Technologies in Maize Improvement

Krishan Kumar[1], Mamta Gupta[1], Alla Singh[1],Chetana Aggarwal[1] Mukesh Choudhary[1], C.M. Parihar[2], Ishwar Singh[1] and Pranjal Yadava[1,2,3]*

[1]*ICAR- Indian Institute of Maize Research, Pusa Campus, New Delhi-110012, India*
[2]*ICAR- Indian Agricultural Research Institute (IARI), New Delhi-110012, India*
[3]*Department of Biology, Stanford University, 385 Serra Mall, Stanford, CA 94305-5020, USA*
**Corresponding Author's Email*: pranjal.yadava@icar.gov.in, pranjaly@stanford.edu

Maize or corn (*Zea mays* L.), which was first domesticated in Mexico, is the most widely grown cereal crop in the world. Maize is consumed as feed and food, and also has myriads industrial applications, including the production of bioethanol and starch. Due to its highest genetic potential, production and productivity among the cereal crops, maize is called as "Queen of the Cereals". Noble Laureate Dr. Norman E. Borlaug, the Father of Green revolution stated 15 years ago that *"The past two decades saw the revolution in rice and wheat, the next few decades will be known as maize era"*. In India, maize is the third largest grain crop and its production and demand is increasing rapidly. In India, it is primarily a non-food cereal crop, with only 10% being consumed as direct human food, while the rest being used for feed, industrial and export purposes. The effective deployment of various frontier technologies *viz.* genetic engineering, doubled haploids, molecular markers, phenomics etc.has greatly helped in raising productivity of maize crop worldwide in the last decade.The present chapter provides an overview on role of frontier technologies in maize crop improvement.

1. Genetic engineering

1.1. Transgenic

Classical plant breeding has contributed remarkably in terms of increase in maize yield and quality improvement However; it has limitation to introduce required characters into plants by genetic crossing because of requirements of species compatibility and non-availability of diverse germplasm within the species. The advent of genetic engineering technology has made possible the insertion of desired foreign gene(s) (also known as transgene) there by overcoming problems of sexual incompatibility and species barriers between organisms. Therefore, genetic modification of crop plants by using genetic engineering technology facilitate introduction of a novel traits which may not occur naturally in the species.

The first transgenic maize cultivars were released for commercial cultivation in 1996. Since then, maize has become the main target crop for plant genetic engineering. Among all crops, maize has highest number of transgenic events that have been commercialized till today. Transgenic maize is already cultivated on a very large scale outside India. Thus, maize transformation to develop genetically modified (GM) or transgenic maize has been a forefront technology for genetic improvement of maize. Table:1 highlights the list of genes used for developing commercially cultivated transgenic maize till date.So far, 188 different transgenic events in maize have been approved for food/feed/commercial cultivation across 30 different countries (with European Union counted as one country) (http://www.isaaa.org/). The released transgenic events belong to six major trait groups- herbicide tolerance (162 events), insect resistance (158 events), modified product quality such as high lysine and enhanced ethanol production capacity (12 events), abiotic stress tolerance (9 events) and pollination control system (6 events) with stacking of events being a common phenomenon (James, 2015). Several other novel traits are under research and development. Transgenic maize is planted in 17 countries while several other countries have authorized import of transgenic maize to meet domestic demand. In 2014, out of 184 million ha of global maize area, 55.20 million ha was planted with cultivars with transgenic traits (James, 2014). The transgenic maize cultivars aid in better weed and insect-pest management, leading to significant increase in yield.

Although, transgenic maize is not approved for commercial cultivation in India, but lot of transgenic maize germplasm have been imported every year from other countries for introgression breeding purpose. Beside this, multi-location field trials of transgenic maize are going on for herbicide tolerance and insect resistance traits (http://igmoris.nic.in/field_trials.asp).

1.2. Genome editing

Targeted genome modifications or gene editing is the new concept which enabled the scientists to modify the crop genomes with an unprecedented ease and accuracy. A set of new editing techniques, using various site specific nucleases (SSNs) *viz.* Zinc finger nucleases (ZFN), Transcription activator-like effector nucleases (TALENs) and the newest one Clustered regularly interspaced short palindromic repeats (CRISPR) /Cas system, have by and large succeeded in overcoming the unpredictability and inefficiency associated with conventional random mutagenesis and transgenesis. ZFNs and TALENs work almost similarly and are artificial proteins composed of a non specific DNA cleavage domain fused with the specific DNA-binding domain. In both of these approaches, a nuclease has to be designed that can specifically target and create double strand breaks (DSBs) at targeted sites in the DNA. The DSBs are immediately repaired by two endogenous repair mechanisms, *viz.* non-homologous end joining (NHEJ), which lead to imperfect repair and hence, results in mutation by Homology direct repair (HDR).These molecular repair processes, modify the targeted locus by disrupting specific genes either byinserting exogenous DNA elements into desired genomic sites or by introducing single-nucleotide substitutions (Weeks *et al.*, 2016).

Table 1. Details of various genes used for developing commercially cultivated transgenic/ GM maize.

Gene name	Gene source	Gene product	Gene Function
Trait - Herbicide tolerance			
dmo	*Stenotrophomonas maltophilia* strain DI-6	dicamba mono-oxygenase enzyme	confers tolerance to the herbicide dicamba (2-methoxy-3,6-dichlorobenzoic acid) by using dicamba as substrate in an enzymatic reaction
bar	*Streptomyces hygroscopicus*	phosphinothricin N-acetyltransferase (PAT) enzyme	eliminates herbicidal activity of glufosinate (phosphinothricin) herbicides by acetylation
pat	*Streptomyces viridochromo genes*	PAT enzyme	eliminates herbicidal activity of glufosinate (phosphinothricin) herbicides by acetylation
cp4 epsps (aroA:CP4)	*Agrobacterium tumefaciens* strain CP4	herbicide tolerant form of 5-enolpyruvulshikimate-3-phosphate synthase (EPSPS) enzyme	decreases binding affinity for glyphosate, thereby conferring increased tolerance to glyphosate herbicide
aad-1	synthetic form of the aad-1 gene from *Sphingobiumh erbicidovorans*	aryloxyalkanoate dioxygenase 1 (AAD-1) protein	detoxifies 2,4-D herbicide by side-chain degradation and degrades the R-enantiomers of aryloxyphenoxypropionate herbicides
gat 4621	*Bacillus licheniformis*	glyphosate N-acetyltransferase enzyme	catalyzes the inactivation of glyphosate,conferring tolerance to glyphosate herbicides
goxv 247	*Ochrobactrum anthropi* strain LBAA	glyphosate oxidase	confers tolerance to glyphosate herbicides by degrading glyphosate into amino methyl phosphonic acid (AMPA) and glyoxylate
2mepsps	*Zea mays*	5-enolpyruvyl shikimate-3-	decreases binding affinity for glyphosate, thereby increasing

		phosphate synthase enzyme (double mutant version)	tolerance to glyphosate herbicide
mepsps	*Zea mays*	modified 5-enolpyruvylshikimate-3-phosphate synthase (EPSPS) enzyme	confers tolerance to glyphosate herbicides
zm-hra	*Zea mays*	herbicide tolerant acetolactase synthase (als) enzyme	confers tolerance to acetolactate synthase-inhibiting herbicides such as sulfonylurea and imidazolinone
Trait - Insect resistance			
cry1A	*Bacillus thuringiensis*	delta-endotoxin of the Cry1A group	confers resistance to lepidopteran insects by selectively damaging their midgut lining
cry1A.105	*Bacillus thuringiensis* subsp. kumamotoensis	Cry1A.105 protein which comprises the Cry1Ab, Cry1F and Cry1Ac proteins	confers resistance to lepidopteran insects by selectively damaging their midgut lining
cry1Ab	*Bacillus thuringiensis* subsp. Kurstaki	Cry1Ab delta-endotoxin	confers resistance to lepidopteran insects by selectively damaging their midgut lining
ecry3.1Ab	synthetic form of Cry3A gene and Cry1Ab gene from *Bacillus thuringiensis*	chimeric (Cry3A-Cry1Ab) delta endotoxin protein	confers resistance to coleopteran and lepidopteran insects by selectively damaging their midgut lining
cry1Ac	*Bacillus thuringiensis S*ub sp. Kurstaki strain HD73	Cry1Ac delta-endotoxin	confers resistance to lepidopteran insects by selectively damaging their midgut lining
cry1F	*Bacillus thuringiensis* var. Aizawai	Cry1F delta-endotoxin	confers resistance to lepidopteran insects by selectively damaging their midgut lining
cry2Ab2	*Bacillus thuringiensis* subsp. kumamotoensis	Cry2Ab delta-endotoxin	confers resistance to lepidopteran insects by selectively damaging their midgut lining
cry2Ae	*Bacillus thuringiensis* subsp. Dakota	Cry2Ae delta-endotoxin	confers resistance to lepidopteran insects by selectively damaging their midgut lining
cry9C	*Bacillus thuringiensis* subsp. tolworthi strain BTS02618A	Cry9C delta endotoxin	confers resistance to lepidopteran insects by selectively damaging their midgut lining
pinII	*Solanum tuberosum*	protease inhibitor protein	enhances defense against insect predators by reducing the digestibility and nutritional quality of the leaves
dvsnf7	Western Corn Rootworm (*Diabrotica virgifera*)	double-stranded RNA transcript containing a 240 bp fragment of the WCR Snf7 gene	RNAi interference resulting to down-regulation of the function of the targeted Snf7 gene leading to Western Corn Rootworm mortality.

cry9C	*Bacillus thuringiensis* subsp. tolworthi strain BTS02618A	Cry9C delta endotoxin	confers resistance to lepidopteran insects by selectively damaging their midgut lining
pinII	*Solanum tuberosum*	protease inhibitor protein	enhances defense against insect predators by reducing the digestibility and nutritional quality of the leaves
dvsnf7	Western Corn Rootworm (*Diabrotica virgifera*)	double-stranded RNA transcript containing a 240 bp fragment of the WCR Snf7 gene	RNAi interference resulting to down-regulation of the function of the targeted Snf7 gene leading to Western Corn Rootworm mortality.
cry9C	*Bacillus thuringiensis* subsp. tolworthi strain BTS02618A	Cry9C delta endotoxin	confers resistance to lepidopteran insects by selectively damaging their midgut lining
vip3Aa20	*Bacillus thuringiensis* strain AB88	vegetative insecticidal protein (vip3Aa variant)	confers resistance to feeding damage caused by lepidopteran insects by selectively damaging their midgut
cry34Ab1	*Bacillus thuringiensis* strain PS149B1	Cry34Ab1 delta-endotoxin	confers resistance to coleopteran insects particularly corn rootworm by selectively damaging their midgut lining
Trait - Modified product quality			
phyA2	*Aspergillus niger* strain 963	phytase enzyme	Degrade phytase phosphorus in seeds into inorganic phosphate to be available to animals when used as feed
cordapA	*Corynebacterium glutamicum*	dihydrodipicolinate synthase enzyme	Increase the production of amino acid lysine
amy797E	synthetic gene derived from *Thermococcales* spp.	thermostable alpha-amylase enzyme	enhances bioethanol production by increasing the thermostability of amylase used in degrading starch
Trait- Pollination control			
ms45	*Zea mays*	ms45 protein	restores fertility by restoring the development of the microspore cell wall that gives rise to pollen
zm-aa1	*Zea mays*	alpha amylase enzyme	hydrolyses starch and makes pollen sterile when expressed in immature pollen
barnase	*Bacillus amyloliquefaciens*	barnase ribonuclease (RNAse) enzyme	Causes male sterility by interfering with RNA production in the tapetum cells of the anther
Trait - Abiotic stress tolerance			
cspB	*Bacillus subtilis*	cold shock protein B	maintains normal cellular functions under water stress conditions by preserving RNA stability and translation
Trait - Increased ear biomass			
athb17	*Arabidopsis thaliana*	a protein of the class II family of the homeodomain-leucine zipper (HD-Zip) transcription factors.	Modulates plant growth and development and regulates gene expression

Source: ISAAA, GM Approval Database (http://www.isaaa.org/)

First successful use of ZFNs in maize was reported by Shukla *et al.*, 2009 and in this study authors disrupted the endogenous *IPK*1 gene, which is involved in phytate accumulation, by the insertion of the *PAT* gene. This results in both herbicide tolerance and alteration of the inositol phosphate profile in developing seeds. ZFN modified maize plants faithfully transmit these genetic changes to the next generation (Table 2).

Char *et al.*, 2015 reported stable and heritable TALEN-induced mutations at the maize glossy 2 (*gl2*) locus in transgenic lines containing mono- or di-allelic mutations at a frequency of about 10% (nine mutated events in 91 transgenic events). Liang *et al.*, 2014 designed TALENs to target the four genes *ZmPDS, ZmIPK1A*, *ZmIPK* and *ZmMRP4* encoding enzymes for phytic acid biosynthesis pathway (Table 2).

Recently, a newest targeted genome modification system called CRISPR/Cas9 has been widely adopted to make precise, targeted changes in DNA. This technology evolved from bacterial adaptive immune system that includes clustered palindromic repeats with cas endonuclease which protect the bacteria from invading plasmid and viral infection (Barrangou *et al.*, 2007). Targeted genome modified products are essentially mutation based products.

Table 2. Different agronomic traits modified through the application of various genome editing approaches in maize crop.

Trait	Gene	Technology	Reference
Phytase biosynthesis	*IPK1/PAT*	ZFN	Shukla *et al.* (2009)
Leaf epicuticular wax composition	*glossy2 (gl2)*	TALEN	Char *et al.* (2015)
Phytase biosynthesis	*ZmIPK1,ZmIPK, ZmMRP4*	TALEN	Liang *et al.* (2014)
Leaf development	*L1G1*	CRISPR/Cas9	Svitashev *et al.* (2015)
Male fertility	*Ms26, Ms45*	CRISPR/Cas9	Svitashev *et al.* (2015)
Herbicide resistance	*ALS 1, ALS 2*	CRISPR/Cas9	Svitashev *et al.* (2015)
Phytase biosynthesis	*ZmIPK*	CRISPR/Cas9	Liang *et al.* (2014)
Carotenoid biosynthesis	*PSY1*	CRISPR/Cas9	Zhu *et al.* (2016)
MEP pathway	*Zmzb7*	CRISPR/Cas9	Feng *et al.* (2016)
Grain yield	*ARGOS8*	CRISPR/Cas9	Shi *et al.*(2017)

Therefore, plants developed using these technologies may be treated at par with mutation breeding products and could be regulated as such.In last three years, the CRISPR/Cas technique has been much explored in plant system with numerous successful attempts of targeted mutagenesis which clearly demonstrates the revolutionary aspects of this novel system (Khatodia and Khurana, 2014). Svitashev *et al.*, 2015 reported targeted mutagenesis by using RNA guided Cas9 endonucleaes in maize. ARGOS8 variants generated by CRISPR-Cas9 improve maize grain yield under field drought stress conditions (Shi *et al.*, 2017). Table 2 summarizes the different genes manipulated using genome editing techniques in maize till date. Qi *et al.*, 2016 optimized and introduced the multiplex gene editing strategy based on the tRNA-processing system into maize. Maize glycine-tRNA was selected to design three multiple tRNA-gRNA units for multiplex editing. They found the tRNA-processing system-based method enhanced mutagenesis efficiency of CRISPR-Cas9 editing in maize and proposed an advanced gRNA selection strategy for chromosomal fragment deletion purpose (Qi *et al.*, 2016). Zhu *et al.*, 2015 studied impact of CRISPR-Cas9 for maize plants through transcriptome analysis and demonstrated that CRISPR-Cas9 is an effective tool for the targeted gene mutagenesis in maize.

2. Doubled Haploid (DH) Technology

The term haploid is generally used to designate sporophytes having the gametic chromosome number, and the first haploids in flowering plants were reported by Belling and Blakeslee, 1922 more than 80 years ago. Later, Guha and Maheshwari (1964) standardized the technique for *in vitro* culture of anther-derived haploids from *Datura stramonium.* Further, their potential in crop improvement was sincerely recognized due to rapid attainmentof homozygous lines.The generated haploid plants remain sterile due to single copy of genome hence there is a need to induce chromosome doubling for the restoration of their fertility. The genome duplication of the haploid individuals has been thought to occur through three mechanisms: endo-reduplication, nuclear fusion and c-mitosis (Segui-Simarro and Nuez, 2008). DH technique has many applications in plant research for example production of large numbers of homozygous plants at any stage of breeding program, which significantly reduce the time to release new cultivars. The DH lines are nearly homozygous at all the loci representing the parental inbred line in the hybrid breeding program of cross-pollinated crops. Now, this technology facilitates the construction of genetic maps to identify marker-trait associations, loci/gene responsible for economically important agronomic traits (Forster and Thomas, 2005). In the last few years, DH lines have increasingly been used in maize (*Zea mays* L.) research and hybrid breeding (Niroula and Bimb, 2009). This became possible by substantial progress in *in vivo* haploid induction technology (Rober *et al.*, 2005). Doubled haploid combined with the marker- assisted selection provides a shortcut for backcross method to correct the defect of elite variety (Forster *et al.*, 2007). Thus, DH technology provides an exciting path to enhance the efficiency and progress rate in maize improvement programs.

2.1 Doubled Haploid production in maize breeding program

DH lines development in maize involves the *in vivo* haploid induction; identification and selection of haploids; chromosomal doubling and seed multiplication of DH plants (Prigge and Melchinger, 2012).In maize, haploids can be obtained by *in vitro* and *in vivo* approaches. *In vitro* approach can be referred as androgenesis. However in the case of maize, androgenesis has limited use due to non-responsiveness and often dependency on donor genotype, anther stage and pretreatment conditions (Spitko *et al.*, 2006). Therefore, *in vivo* haploid induction is a successful method for maize and can be used in maize breeding program due to its operational feasibility (Seitz, 2005). Success of *in vivo* haploid induction in maize depends upon the availability of inducer lines. Firstly, a spontaneous haploid induction rate (HIR) of 0.1% was reported in maize (Chase, 1951) which was too low to be utilized for practical breeding in maize. But later a higher range of induction rate (up to 2.3%) was detected by Coe (1959) in crosses with inbred line Stock 6. This line became the ancestor of all subsequently developed inducer lines (Geiger, 2009). Induction frequency of 8-23% was achieved by Rober *et al.*, 2005 in RWS (cross of KEMS X WS14) inducer line. About 6% induction rate was observed in PK6 line and >8% in UH400 (Chang and Coe, 2009). The lines RWS, UH400 and their cross RWS X UH400 have been widely adopted in temperate maize breeding programmes. Prigge *et al.*, 2011 has developed inducer lines (TAILs) with 8-10% haploid induction rate for tropical germplasm. Inducers for both tropical and temperate maize with 6% to 15% HIR are now available for integration in maize breeding (Prigge *et al.*, 2012a; Prigge *et al.*, 2012b).

After haploid induction, identification of haploids is the most critical step in DH technology. The phenotypic seed coloration marker system is a promising method for direct identification of harvested haploid seeds. The basis of induction of identifiable haploids has been the presence of dominat anthocyanin marker *R1-Navajo (R1-nj)* known as 'red crown' or 'navajo' kernel trait encoded by the dominant mutant allele *R1-nj* of the 'red color' gene *R1*. In the presence of the dominant pigmentation genes *A1* or *A2* and *C2*, *R1-nj* conditions the deep pigmentation of the aleurone layer (endosperm tissue) in the crown (top) region of the kernel. In addition, it causes pigmentation of the scutellum (embryo tissue) (Sarkar and Coe, 1966). Pigmentation varies in extent and intensity depending on the genetic background of the particular donor genotype. To be effective, the donor has to have colorless seeds and the inducer needs to be homozygous for *R1-nj* and the aforementioned dominant pigmentation genes. A kernel resulting from haploid induction then has a red crown (regular triploid endosperm) and a non-pigmented scutellum, whereas a regular F_1 kernel displays pigmentation of both the aleurone (outermost layer of endosperm) and scutellum (Geiger, 2009). The red crown marker does not work if the donor genome is homozygous for *R1* or for dominant anthocyanin inhibitor genes such as *C1-I*, *C2-Idf* and *In1-D*, inhibit anthocyanin synthesis of the marker for the selection of haploids

(Coe, 1994). In a situation, where at least one of the genes is present in a maternal genotype, the isolation of haploids is very difficult due to a lack of pigmentation of endosperm and embryo. In general, genes, that inhibit anthocyanin synthesis, are rather rare in dent maize and, therefore, anthocyanin marker genes in dent maize mostly are well expressed as compared to flint maize which possess higher frequency of anthocyanin inhibitors (Rober, 1999). This drawback of anthocyanin inhibitors in donor lines has been negated in the RWS inducer line, which carries a dominant sun-independent purple-stem marker. In this case, the inducer is homozygous for the anthocyanin genes *B1 (Booster 1)* and *Pl1 (Purple 1)* which gives light-independent pigmentation in the coleoptile and root of the F_1 seedlings. Thus, the colourless coleoptiles or root can be identified as haploid type, while the purple coloured ones are diploids at the early development stage (Geiger, 2009). In the late seedling stage also, haploid can be differentiated from the F_1 plants by less vigour and growth, which also increases the efficiency of haploid selection. A Stock 6-derived inducer CAUHOI (with 2% HIR and higher kernel oil content of 78 g kg^{-1}) have been developed to identify haploids based on basis of absence of scutellum coloration of *R1-nj* scheme and low embryo oil content (Li *et al.*, 2009). Hence, the utility of Near-Infrared Spectroscopy has also been explored for differentiation between haploids and diploid types (Jones *et al.*, 2012).

The chromosome doubling helps in attaining the fertility in haploid plants and therefore attaining the multiplication of haploid plants.Therefore, a protocol with mitotic inhibitor colchicine was developed (Gayen *et al.*, 1994) for efficient artificial chromosome doubling. Mitotic inhibitors bind to the microtubular protein tubulin and it prevents the microtubules from pulling of chromatids towards the poles and thereby resulting into duplicated haploid genome without cell division. Due to carcinogenic property of colchicine breeding institutes are now using less toxic substrates like herbicides *viz.* pronamid, trifluralin, oryzalin for artificial chromosome doubling treatments (Hantzschel and Weber, 2010). Colchicine treated D_0 seedlings should be under the proper supervision and agronomic practices for the successful DH line development.

The colchicine treated D_0 seedlings produces D_1 seeds under field condition, which represents the completely homozygous line. Five percent of the haploid seeds from source germplasm will results into DH lines. This rate depends upon germplasm genotype, an accuracy of haploid identification system and chromosome doubling procedure, agronomic practices in the green house and field conditions. Hence, DH system has increased the speed and precision of maize breeding. The merits and demerits of the DH technology have been summarized in Table 3.

Table 3. Merit and demerit of doubled haploid technique.

Merit	Demerit
Develops near about 99% homozygosity for all loci in one step	More expensive, expertise, facilities required
Shortens the time to cultivar release and increases the efficiency of maize breeding program	Mutagenic and carcinogenic chemical treatment involved
Improves the precision of genetic and mapping studies	Genotype dependency during anther or microspore culture
Perfect compliance with DUS criteria for variety protection	Less haploid regeneration
It reduces expenses for selfing and maintenance in breeding program	Chances of gametoclonal variation lead to contamination
Facilitates marker assisted selection and reverse breeding	Over-usage of DH may reduce genetic variation in breeding germplasm
Genome-wide selection can easily be integrated with DH lines based research programs.	-

The potential applications of DH technology are as follows:

1. DH lines can be used as ideal mapping population and can be effectively used for development of inbred lines
2. It can be used in reverse breeding for reconstitution or derivation of complementary homozygous parental lines
3. It can help in gene stacking/pyramiding
4. It can aid in hybrid breeding through utilization of untapped genetic diversity of maize genetic resources
5. It has potential application in plant varietal protection
6. DH lines can be used as raw material for development of immortalized F_2 (IF_2) population.

3. Molecular Markers

Molecular marker is the first technology among molecular breeding strategies to improve the maize populations. DNA of various generations can be detected using markers such as Restriction Fragment Length Polymorphism (RFLP), Random Amplified Polymorphic DNAs (RAPDs), Amplified Fragment Length Polymorphism (AFLP), Simple Sequence Repeats (SSRs) and Single Nucleotide Polymorphisms (SNP'S).

NGS (Next generation sequencing) technologies increased the pace for development of SSR and SNP markers. These sequencing platforms in combination with restriction enzymes as Restriction site associated DNA (RAD)-tag sequencing, Genotyping by sequencing (GBS) are now assisting the modern breeding strategies like GWS (Genome wide selection) and GWAS (Genome wide association studies) by providing highly dense markers in the whole genome of a crop (Mir and Varshney, 2012).

3.1 Amplified fragment length polymorphisms (AFLPs)

Vos *et al.* (1995) developed AFLP technology for DNA fingerprinting. This technology is based on selective PCR amplification of restriction fragments from genomic DNA without prior knowledge about sequence (Gonzalez-Chavire *et al.*, 2006). AFLP markers have vast applications in fingerprinting, gene identity, phylogenetic studies, identification of clones and cultivars. This technique can differentiate the individuals at subspecies level (Agarwal *et al.*, 2008). It is also used to analyze similarity among different accessions of maize inbreds on genetic basis (Heckenberger *et al.*, 2003). This marker is locus specific at species level, but dominant in nature, required high cost, technical expertise and not suitable for quick screening (Yuste-Lisbona *et al.*, 2008).

3.2 Simple sequence repeats (SSRs)

Among the DNA markers, SSRs are most widely used DNA markers for characterization of maize germplasm (Kostova *et al.*, 2006). These are also known as microsatellites which consists tandem arrangement of 10-100 repeated sequences of di, tri and tetra nucleotides (Santana *et al.*, 2009).Different repeat numbers in SSRs can be treated as separate "alleles" and the site can be treated as a highly polymorphic site with multiple alleles for the detection of variations in populations (Akkaya *et al.*, 1992).

3.3 Single nucleotide polymorphisms (SNPs)

Agarwal *et al.* (2008) stated that single nucleotide variations constitute the most abundant molecular markers. These are widely distributed throughout genomes although their occurrence and distribution varies among species (Yan *et al.*, 2009). A considerable amount of screening effort is required for the development of SNPs. A large number of loci need to be assessed due to the low amount of information per marker, thus it requires high developmental costs (Ibitoye and Akin-Idowu, 2010). With the recent development of several high throughput genotyping technologies that take advantage of the wealth of SNPs in eukaryotic genomes and the studies of nucleotide diversities using SNPs in wheat and maize (Van Inghelandt *et al.*, 2010) SNPs are becoming popular in today's era of biotechnology (Deulvot *et al.*, 2010).

3.4 Application of molecular markers in maize improvement program

3.4.1 DNA profiling

Study of genetic variation on the morphological basis is very difficult and markers can detect these differences at DNA level plants (Meng *et al.*, 1998). SSR markers have been used for diversity analysis and DNA profiling of maize germplasm in different countries (Prassana and Pixley, 2010). Microsatellite markers were used for assessing diversity and evolutionary studies in maize germplasm (Matsuoka *et al.*, 2002) as they showed broad variations in different accessions due to their frequent occurrence in the plant genome (Akkaya *et al.*, 1992).

3.4.2 Marker assisted selection

The selection or identification of particular gene/loci responsible for specific trait (for which direct selection is ineffective) with the help of linked marker is known as marker assisted selection (MAS). Therefore, selection of plants with desired attributes has been accomplished by the use of genetic markers in the MAS, which can accelerate the breeding process and helpful in early release of new improved cultivars for the benefit of farmer community. In early 1970s, this technique has been used for association studies between trait and markers which led to the development of genetic maps.

MAS have great advantage in early generations because plants with undesirable gene combinations can be eliminated and thereby allowing breeders to focus on a lesser number of high-priority lines in subsequent generations. The major disadvantage of applying MAS at early generations is the cost of genotyping a larger number of plants. This can be rectified through use of co-dominant DNA markers for selection of fixed homozygous alleles in as early as the F_2 generation.

This technique also requires accurate phenotypic data and statistical tool for analysis of gene/marker-trait association, QTL mapping, etc. (Edmeades, 2013). Various genes/loci of important traits have been tagged with molecular markers in maize at global level (Prasanna and Pixley, 2010). Stuber *et al.*, 1987 firstly reported QTLs for yield related attributes in maize. Further QTLs have been identified for many important traits in maize *viz.*, plant height, high oil content, tolerance to drought stress and water-logging, nutrient composition under low level of nitrogen, resistance to downy mildew, head smut and Maize dwarf mosaic virus. In maize, this approach has been used to develop QPM lines from normal maize lines by using opaque 2-specific SSR markers, which contains more lysine and tryptophan than the normal lines (Babu *et al.*, 2004). The "Vivek QPM hybrid 9" is the first successful example of MAS which developed by transferring *O2* gene (Buba *et al.*, 2005). Using composite interval mapping, sorghum downy mildew (SDM) resistance associated 3 putative QTLs in maize have been reported on chromosomes 2, 3 and 6 (Lohithaswa *et al.*, 2015).

MAS technique can be further divided as Marker assisted backcross breeding (MABB), Marker assisted recurrent selection (MARS) and Genomic Selection (GS). Marker assisted gene pyramiding can be attained by using this approach by MARS and GS (Bernardo, 2010). Assembling the desired genes into single parent from multiple parents is known as Marker assisted gene pyramiding (Collard and Mackill, 2008).

3.4.2.1 Marker assisted backcross breeding (MABB)

The incorporation of one or more genes/loci to adapted or elite cultivar having desirable attributes but deficient in few traits/characters is termed as marker assisted backcrossing (MAB). This approach involves foreground selection, recombinant selection and background selection (Holland, 2004). Initially, "foreground selection" involves the markers to screen the target/desired gene in the individual plants, which has been useful for traits require difficult and time consuming phenotyping, selection at early seedling stage (Collard and Mackill, 2008). Secondly, selection of backcross (BC) progeny having target gene and recombinant events between target loci and linked flanking markers has been performed during "recombinant selection" to reduce the linkage drag (Hospital, 2005). This linkage drag can be minimized by using flanked markers (less than 5cM on either side) of target gene. Since, there are rare chances of occurrence of double recombination on both sides of target loci, at least two BC generation should be used for this selection (Frisch *et al.*, 1999). Lastly, "background selection" is used for recovery of recurrent parent (RP) genome. This selection approach has potential to accelerate the RP genome recovery as compare to conventional backcrossing which require about 6 generations for same task (Visscher *et al.*, 1996).

A donor line CML176 having *opaque 2* gene was used to introgress it in the elite normal line of maize V25, which is high yielding and extra early maturing with low nutrient content through several backcrosses to develop Quality Protein Maize (QPM) (Babu *et al.*, 2005). A drought susceptible maize tropical inbred line CML 247 was converted into drought tolerant line by backcrossing with tolerant line Ac7643 (Ribaut and Ragot, 2007). Scientists at IARI (Indian Agricultural Research Institute) have performed the gene pyramiding in five elite cultivars of India such as CM 137, CM138, CM139, CM150 and CM151 for development of resistance against Turcicum leaf blight and Polysora rust. Seven backcross populations were generated to pyramid *Htn1*and *Ht2* genes of Turcicum leaf blight resistance and RppQ QTL for Polysora resistance from four resistant donors *viz.*, NAI 147, SKV 21, NAI 112 and SKV18 (Prassana *et al*, 2009b). To develop resilience in maize susceptible cultivars, CM139 a sorghum downy mildew resistant line and QTL-NILS were generated through this approach (Prasanna 2009a).

3.4.2.2 Marker assisted recurrent selection (MARS)

Phenotypic recurrent selection focus on increasing the frequency of favorable alleles within the population, through repeated cycles of selection, evaluation and recombination. Now, markers can be used for this process to enhance the efficiency of phenotypic recurrent selection which is known as MARS. This approach can be used to combine several gene/QTL into single individual/genotype for traits like drought tolerance, disease tolerant, yield, etc in maize resilience breeding programs.

In this approach, trait linked markers are first identified and numerous genes/ loci responsible for trait expression are selected to integrate them to superior genotype within the population (Ribaut *et al.*, 2010). It involves the genotyping of F_2 or F_3 population and phenotyping of F_2 derived F_4 or F_5 individuals to evaluate the marker effects, followed by recombination cycles based on marker linked to minor QTLs to evaluate the marker effects (Eathington *et al.*, 2007). Initially, identification of QTLs in the population is performed. After that, pyramiding of superior alleles in single genetic background is carried out by crossing with lines having superior alleles for highest number of QTLs. Further, recombinant lines are selected on the basis of accurate phenotypic screening for multilocation field trials to release them as varieties (Bernardo and Charcosset, 2006).

Major use of this strategy in maize is to build up stable disease tolerance, as new pathogen races emerges to conquer the disease resistance developed by incorporation of single gene (Collard and Mackill, 2008). MARS has been used to make an early flowering line, NSE331 which was having advance agronomical attributes but was late in flowering and it has been developed by crossing with donor NSE626 which flowers 8 to 10 days earlier (Ragot *et al.*, 2000). CIMMYT also used this approach to develop commercial drought tolerant genotypes and developed 10 populations based on SNP-based recurrent selection of crosses between commercial lines CML and drought tolerant donor lines CZL or VL (Beyene *et al.*, 2016).

3.4.2.3 Genomic selection (GS)

Simultaneous selection of many markers (tens or hundreds or thousands) densely covering the whole genome which anticipated that all the genes are in linkage disequilibrium with at least some of markers is defined as Genomic Selection (Meuwissen, 2007). This strategy is still at its early stage for its use in plants as compared to other marker assisted approaches (Eathington *et al.*, 2007). This approach involves the prediction of complex traits by using genetic markers covering the entire genome and desirable individuals can be selected through genomic estimated breeding value (GEBV), which can be calculated by the information based upon dense markers available on the genome. It does not phenotyping of the population, but a GS model has to be developed to estimate the GEBV based on genomic information (Nakaya and Isobe, 2012).

GS was first demonstrated in maize, barley and A*Rabi*dopsis (Lorenzana and Bernardo, 2009) and accuracy was greater in GS as compared to pedigree information alone (Jannink *et al.*, 2010). Maize, which involves doubled haploids and off season nurseries, at least two years are required for selection of test cross performance (Bernardo and Yu, 2007), GS provides a chance to trim down the time per selection cycle due to less need of progeny test data per cycle. Evidently, molecular markers are always advantageous than morphological and biochemical markers due to their immediate application in enhancing the development pace of improved crop varieties. Different strategies involving molecular markers *viz.*, MAS, MABB, MARS and GS provides enormous opportunities to breeder to develop resilient cultivars of maize to meet the demand of crop from ever growing human population.

4. Phenomics

Phenotype refers to the observable features that result from gene-environment interaction. Phenome is the sum total of all phenotypic traits and the study related with measurement of phenomes is referred to as Phenomics. Prediction of plant performance under varying environmental conditions necessitates the development of robust phenotyping platforms, which in conjunction with plant genetic resources can provide a deeper, mechanism-based understanding of the gene action and its influence on phenotype.

4.1 Need for robust phenotyping platforms

To meet the rapidly rising demands for increased food production, as much as 70% increase in cereals grain yield is required by 2050 (Tester *et al.*, 2010). Similarly, the demand for biofuel feedstocks is expected to increase (Sticklen, 2007). Conventional breeding strategies have not been able to fulfill the above requirements as of now. Due to this reason, scientists have focused their attention on the use of 'omics' technologies to dissect the underlying molecular relationships responsible for various traits or phenotype. Genomics, transcriptomics and proteomics deal with the study of functions and interactions between genes, expressed mRNAs and proteins, respectively. The research in 'omics' resources and the amount of information there has expanded enormously over the years. In order to utilize this information, it must be supplemented and coupled with precise description of the 'phenotype' of the plant (Furbank *et al.*, 2011). Phenotypic description of genomics resources is the limiting factor in linking genetic action to a physiological mechanism. Prediction of plant performance under varying environmental conditions necessitates the development of robust phenotyping platforms, which in conjunction with plant genetic resources can provide a deeper, mechanism-based understanding of the gene action and its influence on phenotype.

4.2 Instrumentation and steps towards field phenotyping

A variety of plant traits from plant pigments to crop biomass can be analyzed by different instruments. Garrity *et al.* (2010) used photodiodes for measurement if different indices in vegetative canopies. Yeh and Chung (2009) exploited high intensity light emitting diodes (LED) for active sensing. Similarly, infrared imagery has been used at large in proximal sensors (French *et al.*, 2007). Other technologies include near infrared spectrosopy, acoustic-based sensing, stereo image analysis, etc. (Andrale-Sanchez *et al.*, 2012). Different options to mount and carry phenomics sensors are available. These include high-clearance tractors, cranes (Haberland *et al.*, 2010), cable robots and aerial vehicles. The advances being made in field-based phenomics are expected to lead to a realistic characterization of promising genotypes in real-world situations and hold a huge potential to enable increase in crop value and productivity.

4.3 Phenotyping in maize

Many research efforts have been directed towards phenotying of maize. Campbell *et al.* (1997) used near-infrared transmittance spectroscopy to analyze amylose content in whole grains. Brenna *et al.* (2004) used near-infrared reflectance spectroscopy to evaluate carotenoids content in maize kernels. Romano *et al.* (2011) evaluated thermography as method to screen for drought adaptation in maize. Thermal images of grain filling 1 stage were acquired and it was shown that genotypes adapted to drought stress displayed lower canopy temperatures. Grift *et al.* (2011) developed high-throughput measurement technique for phenotyping maize roots. Using a semi-automated imaging box, root top angle (RTA) was measured and the complexity of maize roots was expressed in Fractal Dimension (FD). The probable gene action governing the root top angle and root complexity were deciphered. The methods employed were capable of imaging up to seven hundred roots per day, representing the high throughput achievable through the above analysis.

In another study, Trachsel *et al.* (2011) developed a method to visualize ten traits related to root architecture in a matter of few minutes. The authors concluded that visual scoring of brace roots and root crowns provides a realistic assessment of various root architectural traits. Pace *et al.* (2014) used a high-throughput trait capture software system to link relationships between developmental stages from seedling to adult traits of maize root system. Liebisch *et al.* (2015) used an aerial experimental sensor system to monitor maize traits. The plants were continuously analyzed using colour (RGB), imaging (Near Infrared) and thermography. Parameters were standardized to compare the remotely sensed data with the one found at ground. The methodology developed successfully identified the plots from data of different sensors at different points of time.

Apart from research on methodologies for phenotyping, efforts have been made to make the data available in public domain. Colmsee *et al*. 2012 have worked on OPTIMAS-Data Warehouse database, which hosts a comprehensive collection of omics resources that includes phenomics. Under the OPTIMAS project, maize plants have been characterized for phenotypes under a large set of laboratory-controlled stress conditions. Biogen Ben is another resource that provides information on maize phenomics. Other resources like MaizeGDB and Gramene provide information on maize phenotypes arising from different mutagenesis projects.

5. Conclusion

It is estimated that maize demand will continue to increase worldwide including in India because of its myriad uses and ever-increasing population. To meet the growing demand, enhancement of maize yield in coming years is a big challenge in the era of climate change. Despite the challenges, effective deployment of frontier technologies based interventions offer promising potential to improve maize quality and increase yield in eco-friendly way.

References

Agarwal, M., Shrivastava, N. and Padh, H. (2008). Advance in molecular marker techniques and their applications in plant sciences. *Plant Cell Rep.* 27: 617-631.

Akkaya, M.S., Bhagwat, AA. and Cregan, P B. (1992). Length polymorphisms of simple sequence repeat DNA in soybean. *Genet.* 132: 1131-1139.

Andrade-Sanchez, P., Heun, J.T., Gore, M.A. *et al.* (2012). Use of a moving platform for field deployment of plant sensors. In: Proceedings of the 2012 ASABE Annual International Meeting, Dallas, TX.

Babu, R., Nair, S.K., Kumar, A. *et al.* (2005). Two-generation marker aided backcrossing for rapid conversion of normal maize lines to Quality Protein Maize (QPM). *Theor Appl Genet.* 111: 888-897.

Babu, R., Nair, S.K., Prasanna, B.M. and Gupta, H. S. (2004). Integrating marker-assisted selection in crop breeding – Prospects and challenges. *Curr Sci.* 87: 607-619.

Barrangou, R., Fremaux, C., Deveau, H. *et al.* (2007). CRISPR provides acquired resistance against viruses in prokaryotes. *Sci.* 315(5819): 1709-1712.

Belling, J. and Blakeslee, A.F. (1922). The assortment of chromosomes in triploid *Daturas. Am Nat.* 56: 339-346.

Bernardo, R. (2010). Genomewide selection with minimal crossing in self-pollinated crops. *Crop Sci.* 50:624–627.

Bernardo, R. and Charcosset, A. (2006). Usefulness of Gene Information in Marker-Assisted Recurrent Selection: A Simulation Appraisal. *Crop Sci.* 46: 614-621.

Bernardo, R. and Yu, J. (2007). Prospects for genomewide selection for quantitative trait in maize. *Crop Sci.* 47:1082–1090.

Beyene, Y., Semagn, K., Crossa, J. *et al.* (2016). Improving maize grain yield under drought stress and non-stress environments in sub-saharan Africa using marker-assisted recurrent selection. *Crop Sci.* 56: 344-353.

Brenna, O.V. and Berardo, N. (2004). Application of near-infrared reflectance spectroscopy (NIRS) to the evaluation of carotenoids content in maize. *J Agric Food Chem.* 52: 5577-5582.

Campbell, M.R., Brunn, T.J. and Glover, D.V. (1997). Whole grain amylose analysis in maize using near-infrared transmittance spectroscopy. *Cereal Chem.* 74: 300-303.

Chang, M.T. and Coe, E.H. (2009). Doubled Haploids. In: AL Kriz, BA Larkins (eds). Biotechnology in Agriculture and Forestry. Vol 63. Molecular Genetic Approaches to Maize Improvement. Springer Verlag, Berlin, Heidelberg. Pp. 127-142.

Char, S.N., Unger Wallace, E. Frame, B. *et al.* (2015). Heritable site specific mutagenesis using TALENs in maize. *J Plant Biotech.* 13(7): 1002-10.

Chase, S.S. (1951). Production of homozygous diploids of maize from monoploids. *Agron J.* 44: 263-267.

Coe, E.H. (1959). A line of maize with high haploid frequency. *Am Nat.* 93: 381-382.

Coe, E.H. (1994). Anthocyanin genetics. In : M Freeling, V Walbot (eds) The maize handbook. Springer-Verlag, New York. Pp. 279-281.

Collard, B.C. and Mackill, D.J. (2008). Marker-assisted selection: an approach for precision plant breeding in the twenty-first century. *Philos Trans R Soc Lond B Biol Sci.* 363(1491): 557-72.

Colmsee, C., Mascher, M., Czauderna, T. *et al.* (2012). OPTIMAS-DW: a comprehensive transcriptomics, metabolomics, ionomics, proteomics and phenomics data resource for maize. *BMC Plant Biol.* 12(1): 245.

Deulvot, C., Charrel, H., Marty, A. *et al.* (2010). Highly-multiplexed SNP genotyping for genetic mapping and germplasm diversity studies in pea. *BMC Genomics.* 11: 468-478.

Eathington, S.R., Crosbie, T.M., Edwards, M.D., Reiter, R.S. and Bull, J.K. (2007). Molecular markers in commercial breeding. *Crop Sci.* 47: 154-163.

Edmeades, G.O. (2013). Progress in achieving and delivering drought tolerance in maize: an update. International Service for the Acquisition of Agri-Biotech Applications (ISAAA), Ithaca, NY.

Feng, C. Yuan, J. Wang, R. Liu, Y. Birchler, J.A. and Han, F. (2016) Efficient targeted genome modification in maize using CRISPR/Cas9 system. *J Genet Genomics.* 43: 37–43.

Forster, B.P. and Thomas, W.T.B. (2005). Doubled haploids in genetics and plant breeding. In: Janick J(ed) *Plant Breed Rev.* 25: 57-88.

Forster, B.P., Herberle-Bors, E., Kasha, K. N. and Touraev, A. (2007). The resurgence of haploids in higher plants. *Trends Plant Sci.* 12(8): 368-375.

French, A.N., Hunsaker, D.J., Clarke, T.R. *et al.* (2007). Energy balance estimation of evapotranspiration for wheat grown under variable management practices in central Arizona. *Trans ASABE.* 50: 2059-2071.

Frisch, M., Bohn, M., Melchinger, A.E. (1999). Minimum sample size and optimal positioning of flanking markers in marker-assisted backcrossing for transfer of a target gene. *Crop Sci.* 39: 967-975.

Furbank, R. T. and Tester, M. (2011). Phenomics–technologies to relieve the phenotyping bottleneck. *Trends in plant Sci.* 16 (12): 635- 644.

Garrity, S.R., Vierling, L.A. and Bickford, K. (2010). A simple filtered photodiode instrument for continuous measurement of narrowband NDVI and PRI over vegetated canopies. *Agric For Meteorol.* 150: 489- 496.

Gayen, P., Madan, J.K., Kumar, R., Sarkar, K.R. (1994). Chromosome doubling in haploids through colchicine. *Maize Genet Coop News Lett.* 68-65.

Geiger, H.H. (2009). Doubled haploids. *In:* J.L. Bennetzen, S. Hake (Eds.), Maize Handbook. Vol. II. Genetics and Genomics. Springer Verlag, Heidelberg, New York. Pp. 641-659.

Gonzalez-Chavire, M.M., Torres-Pacheco, I. and Villordo-Pineda, E. (2006). DNA markers. *Adv Agri Food Biotech.* 2: 99-134.

Grift, T. E., Novais, J. and Bohn, M. (2011). High-throughput phenotyping technology for maize roots. *Biosyst Eng.* 110(1): 40-48.

Guha, S. and Maheshwari, S.C. (1964). "*In vitro* production of embryos from anthers of Datura." *Nature.* 204: 497.

Haberland, J.A., Colaizzi, P.D., Kostrzewski, M.A. *et al.* (2010). AgIIS, Agricultural Irrigation Imaging System. *Appl Eng Agric.* 26: 247-253.

Häntzschel, K.R. and Wber, G. (2010). Blockage of mitosis in maize root tips using colchicines-alternatives. *Protoplasma.* 241: 99-104.

Heckenberger, M., Van der Voort, J.R., Melchinger, A.E. and Peleman, J. (2003). Variation of DNA fingerprints among accessions within maize inbred lines and implications for identification of essentially derived varieties: II. Genetic and technical sources of variation in AFLP data and comparison with SSR data. *Mol Breed.* 12: 97-106.

Holland, J.B. (2004). Implementation of molecular markers for quantitative traits in breeding programs, challenges and opportunities. In New Directions for a Diverse Planet: Proceedings for the 4th International Crop Science Congress. Regional Institute, Gosford, Australia, www. cropscience. org. au/ics.

Hospital, F. (2005). Selection in backcross programmes. *Philos Trans R Soc Lond B Bio Sci.* 360: 1503-1511.

http://igmoris.nic.in/field_trials.asp.

Ibitoye, D.O. and Akin-Idowu, P.E. (2010). Marker-assisted selection (MAS): A fast track to increase genetic gain in horticultural crop breeding. *Afr J Biotech.* 9: 8889-8895.

James C. (2015). 20th Anniversary (1996 to 2015) of the Global Commercialization of Biotech Crops and Biotech Crop Highlights in 2015. ISAAA Brief No. 51 Ithaca, NY: ISAAA.

James, C. (2014). Global Status of Commercialized Biotech/GM Crops: 2014, ISAAA.

Jannink, J.L., Lorenz, A.J. and Iwata, H. (2010). Genomic selection in plant breeding: from theory to practice. *Brief Funct Genomics*. 9: 166-177.

Jones, R.W., Reinot, T., Frei, U.K., Tseng, Y., Lubberstedt, T. and Mc Clelland, J.F. (2012). Selection of haploid maize kernels from hybrid kernels for plant breeding using near-infrared spectroscopy and SIMCA analysis. *Appl Spectrosc*. 66: 447-450.

Khatodia, S. and Khurana S. M. P. (2014). Trending: the Cas nuclease mediated genome editing technique. *Biotech Today.* 4:46–49. 10.5958/2322-0996.2014.00019.2

Kostova, A., Todorovska, E., Christov, N., Sevov, V. and Atanassov. A.I. (2006). Molecular characterisation of Bulgarian maize germplasm collection via SSR markers. *Biotechnol Equip*. 20: 29-36.

Li, L., Xu, X., Jin, W. and Chen, S. (2009). Morphological and molecular evidences for DNA introgression in haploid induction via a high oil inducer CAUHOI in maize. *Planta*. 230(2): 367-376.

Liang, Z., Zhang, K., Chen, K. and Gao, C. (2014). Targeted mutagenesis in Zea mays using TALENs and the CRISPR/Cas system. *J Genet Genome*. 41(2): 63-68.

Liebisch, F., Kirchgessner, N., Schneider, D., Walter, A. and Remote, A.H. (2015). Aerial phenotyping of maize traits with a mobile multi-sensor approach. *Plant Methods*. 11: 9.

Lohithaswa, H.C., Jyothi, K., Kumar, K.S. and Hittalmani S. (2015). Identification and introgression of QTLs implicated in resistance to sorghum downy mildew (*Peronosclerospora sorghi*) in maize through marker-assisted selection. *J. Genet*. 94(4): 741-8.

Lorenzana, R. E. and Bernardo, R. (2009). Accuracy of genotypic value predictions for marker-based selection in biparental plant populations. *Theor Appl Genet*. 120: 151-161.

Matsuoka, Y., Mitcheli, S.E. and Kresovich, S. (2002). Microsatellites in *Zea*-variability, patterns of mutations, and use for evolutionary studies. *Theor Appl Genet*. 104: 436-450.

Meng, E.C.H., Smale, M., Bellon, M. and Grimnaelli, D. (1998). Definition and measurement of crop diversity for economic analysis. In: M. Smale (Ed.), Farmers, gene banks and crop breeding: economic analyses and diversity in wheat, maize and rice. *IPGRI publication*. Pp. 19-29.

Meuwissen, T. (2007). Genomic selection: marker assisted selection on a genome wide scale. *J Anim Breed Genet.* 124: 321-322.

Mir, R.R. and Varshney, R.K. (2012). Future Prospects of Molecular Markers in Plants. Molecular Markers in Plants. John Wiley & Sons, Pp.169-190.

Nakaya, A. and Isobe, S.N. (2012). Will genomic selection be a practical method for plant breeding? *Ann Bot.* 110(6): 1303-1316.

Niroula, R. K. and Bimb, H. P. (2009). Overview of wheat x maize system of crosses for dihaploid induction in wheat. *World Appl Sci J.* **7**(8): 1037-1045.

Pace, J, Lee, N, Naik, H.S, Ganapathysubramanian, B. and Lübberstedt, T. (2014). Analysis of Maize (*Zea mays* L.) Seedling Roots with the High-Throughput Image Analysis Tool ARIA (Automatic Root Image Analysis). *PLos One.* 9(9): e108255.

Prasanna, B.M. and Pixley, K. (2010). Molecular marker-assisted breeding options for maize improvement in Asia. *Mol Breed.* 26: 339-356.

Prasanna, B.M., Beiki, A.H., Sekhar, J.C., Srinivas, A. and Ribaut, J.M. (2009a). Mapping QTLs for component traits influencing drought stress tolerance of maize in India. *J Plant Biochem Biotech.* 18: 151-160.

Prasanna, B.M., Hettiarachchi, K. and Mahatman, K. (2009b). Molecular marker-assisted pyramiding of genes conferring resistance to Turcium leaf blight and polysora rust in maize in bred lines in India. In: Proceedings of 10th Asian region Maize workshop (October 20-23, 2008, Makassar, Indonesia). CIMMYT, Mexico, DF.

Prigge, V. and Melchinger, A.E. (2012). Production of haploids and doubled haploids in maize. In: V.M. Loyola-Vargas, and N. Ochoa-Alejo, editors, Plant cell culture protocols. 3rd ed. Humana Press, Totowa, NJ. Pp. 161-172.

Prigge, V., Sánchez C., Dhillon B. S. *et al.* (2011). Doubled haploids in tropical maize: I. Effects of inducers and source germplasm on in vivo haploid induction rates. *Crop Sci.* 51(4): 1498-1506.

Prigge, V., Schipprack, W., Mahuku, G., Atlin, G.N. and Melchinger, A.E. (2012a). Development of in vivo haploid inducers for tropical maize breeding programs. *Euphytica.* 185(3): 481-490.

Prigge, V., Xu, X., Li, L., Babu, R., Chen, S., Atlin, G.N. and Melchinger, A. E. (2012b). New insights into the genetics of in vivo induction of maternal haploids, the backbone of dou-bled haploid technology in maize. *Genet.* 190(2): 781-793.

Qi, W., Zhu, T., Tian, Z., Li, C., Zhang, W. and Song, R. (2016). High-efficiency CRISPR/Cas9 multiplex gene editing using the glycine tRNA-processing system-based strategy in maize. *BMC Biotech.* 16(1): 58.

Ragot, M., Gay, G., Muller, J.P. *et al.* (2000). Efficient selection for the adaptation to the environment through QTL mapping and manipulation in maize, Molecular approaches for the genetic improvement of cereals for stable production in water-limited environments. México, FCIMMYT. Pp. 128-130.

Ribaut, J. M., De Vicente, M. C. and Delannay, X. (2010). Molecular breeding in developing countries: challenges and perspectives. *Curr Opin Plant Biol.* 13(2): 213-218.

Ribaut, J.M. and Ragot, M. (2007). Marker-assisted selection to improve drought adaptation in maize: The backcross approach, perspectives, limitations, and alternatives. *J Exp Bot.* 58: 351-360.

Röber, F.K. (1999). Fortpflanzungsbiologische und genetische Untersuchungen mit RFLP-Markern zur in vivo Haploideninduktion bei Mais. Ph.D. dissertation, University of Hohenheim, Stuttgart, Germany.

Röber, F.K., Gordillo, G.A. and Geiger, H.H. (2005). In vivo hap-loid induction in maize-performance of new inducers and significance of doubled haploid lines in hybrid breeding. *Maydica*. 50(3/4): 275-283.

Romano, G., Zia. S., Spreer, W. *et al.* (2011). Use of thermography for high throughput phenotyping of tropical maize adaptation in water stress. *Comput Electron Agric*. 79: 67-74.

Santana, Q.C., Coetzee, M.P.A., Steenkamp, E.T. *et al.* (2009). Microsatellite discovery by deep sequencing of enriched genomic libraries. *Biotech*. 46: 217-223.

Sarkar, K.R. and Coe, Jr, E.H. (1966). A genetic analysis of the origin of maternal haploids in maize. *Genet*. 54: 453-464.

Segui-Simarro, J.M. and Nuez, F. (2008). Pathways to doubled haploidy: chromosome doubling during androgenesis. *Cytogenet Genome Res*. 120: 358-369.

Seitz, G. (2005). The use of doubled haploids in corn breeding. In: Proc. 41st Annual Illinois Corn Breeders' School 2005. Urbana-Champaign, Illinois. pp. 1-7.

Shi, J., Gao, H., Wang, H. *et al.* (2017). ARGOS8 variants generated by CRISPR Cas9 improve maize grain yield under field drought stress conditions. *Plant Biotech J*. 15(2): 207-216.

Shukla, V.K., Doyon, Y., Miller, J.C. *et al.* (2009). Precise genome modification in the crop species Zea mays using zinc-finger nucleases. *Nature*. 459(7245): 437-441.

Spitkó, T., Sági, L., Pintér, J., Marton, L.C. and Barnabás, B. (2006). Haploid regeneration aptitude maize (*Zea mays* L.) lines of various origin and of their hybrids. *Maydica*. 51: 537-542.

Sticklen, M. B. (2007). Feedstock crop genetic engineering for alcohol fuels. *Crop Sci*. 47(6): 2238-2248.

Stuber C.W., Edwards, M.D. and Wendel J.F. (1987). Molecular markers facilitated investigations of QLT in maize and factors influencing yield and its component traits. *Crop Sci*. 27: 638-648.

Svitashev, S., Young, J.K., Schwartz, C., Gao, H., Falco, S.C. and Cigan, A.M. (2015). Targeted mutagenesis, precise gene editing, and site-specific gene insertion in maize using Cas9 and guide RNA. *Plant Physiol*. 169(2): 931-945.

Tester, M. and Langridge, P. (2010). Breeding technologies to increase crop production in a changing world. *Sci*. 327(5967): 818-822.

Trachsel, S., Kaeppler, S. M., Brown, K. M. and Lynch, J. P. (2011). Shovelomics: high throughput phenotyping of maize (*Zea mays* L.) root architecture in the field. *Plant Soil*. 341(1-2): 75-87.

Van Inghelandt, D., Melchinger, A.E., Lebreton, C. and Stich, B. (2010). Population structure and genetic diversity in a commercial maize breeding program assessed with SSR and SNP markers. *Theo App Genet.* 120: 1289-1299.

Visscher, P. M., Haley, C. S. and Thompson, R. (1996). Marker assisted introgression in backcross breeding programs. *Genet.* 144: 1923-1932.

Vos, P., Hogers, R., Bleeker, M. *et al.* (1995). AFLP: A new technique for DNA fingerprinting. *Nucleic Acids Res.* 23: 4407-4414.

Weeks, D. P., Spalding, M. H. and Yang, B. (2016). Use of designer nucleases for targeted gene and genome editing in plants. *Plant Biotech J.* 14(2): 483-495.

Yan, J., Yang, X., Shah, T. *et al.* (2009). High-throughput SNP genotyping with the Golden Gate assay in maize. *Mol Breed.* 25: 441-451.

Yeh, N. and Chung, J.P. (2009). High-brightness LEDs energy efficient lighting sources and their potential in indoor plant cultivation. *Renew Sust Energy Rev.* 13: 2175-2180.

Yuste-Lisbona, F.J., Capel, C., Capel, J. *et al.* (2008). Conversion of an AFLP fragment into one dCAPS marker linked to powdery mildew resistance in melon1. *In* M. Pitrat (ed.) *EUCARPIA Meeting on Genetics and Breeding of Cucurbitaceae.* INRA, Avignon.

Zhu, J., Song, N., Sun, S. *et al.* (2016). Efficiency and inheritance of targeted mutagenesis in maize using CRISPR-Cas9. *J Genet Genome Res.* 43(1): 25-36.

28

Value Addition in Maize for Nutritional Security

Shobha, D.[1] *and Usha Singh*[2]

[1]*AICRP (Maize), Zonal Agricultural Research Station, V.C. Farm, Mandya, UAS Bangalore*
[2]*AICRP (Maize), DRPCAU, Pusa, Samastipur, Bihar*
**Corresponding Author's Email: usha_pusa@yahoo.co.in*

Foodgrains are a fine example of diversity of nature with many varieties and varied features. They are storable food commodities and lend themselves to numerous processing manipulations. Nutritionally foodgrains are rich source of almost all known nutrients and several phyto-chemicals. The increasing incidence of metabolic diseases are attributed to unbalanced energy rich diets lacking in fiber and protective bioactive compounds such as micronutrients and phyto-chemicals (Prakash, 2013). Cereals along with pulse combinations constitute an important part of human diet in many parts of the world because of easy availability, low cost, long shelf life and nutritional balance. During recent past, India has made significant progress in improving the food security of its masses. The green revolution of 60's helped the country in achieving the food security by improving the availability and access components (Kaul *et al.*, 2016). However, surplus quantity of food produced is not uniformly distributed and the quality component is not addressed properly which has a direct impact on nutrition. In spite of surplus production of foodgrains still malnutrition persists. However, it can suitably be overcome by combining cereals with good quality pulses and other value addition techniques which will go long way in solving the problem of malnutrition. Even though country is producing large quantity of coarse grains particularly maize but its utilization for direct consumption is very low which is mainly due to poor quality of protein. Although maize contains higher proportion of protein (10-12 %), its quality is poor as compared to that of protein in rice due to presence of high concentration of alcohol soluble protein fraction 'prolamine' also known as 'Zein' in the endosperm. Zein is very low in some essential amino acids, mainly lysine and tryptophan which contribute to more than 50% of the total protein. On the other hand, very high amount of leucine and imbalanced proportion of isoleucine contribute to poor quality of protein in normal maize. To overcome such deficiencies, scientists developed Quality Protein Maize (QPM) where the quality of protein is not only higher than the common maize, but also significantly higher than that of other cereal grains. In the recent past QPM has got special distinction among the cereals due to presence of high amount of essential amino acids like lysine and tryptophan. QPM can be utilized for diversified purposes in food and nutritional security as infant food, health food, convenience food, specialty food and emergency ration. In India maize is consumed mainly in the form of boiled corn or roasted as pop corn. In order to increase the proportion of maize for direct consumption, several traditional food-processing and preparation methods can be used at household level to enhance the bioavailability of micronutrients in maize based diets. These methods include alkali cooking, thermal processing, mechanical processing, soaking, germination/ malting and fermentation. In countries like Central America and Africa, it is converted into food products by grinding, alkali processing, boiling, cooking and fermentation (Srilakshmi, 2008).

Maize is next to rice and wheat in terms of acreage and ranks second in terms of total production and productivity which accounts for approximately nine per cent of total foodgrain production in India (Anonymous, 2015) Being a potential crop in India, maize occupies an important place as a source of human food (23%), animal feed (12%), poultry feed (51%), industrial products mainly as starch (12%), and 1% each in brewery and seed (Parihar *et al.,* 2011). There are many products from maize that have been taken over by industry and manufactured and marketed at commercial scale. Several of these products already mentioned are now industrialized on a small or large scale. In USA, over 1000 different items can be found on the shelves of a typical supermarket and they are derived wholly or partially from maize. These product includes tortillas, maize flours (masa), chips and several types of snack, breakfast cereal, thickness, pastes, syrups, sweeteners, grits, maize oil, soft drinks, beer, whisky etc (FAO, 2009).

Nutritional value of maize

Maize also known as corn, an important foodgrain in India, has many industrial and food applications apart from its use as a feed ingredient. It is rich in starch and bland to taste and hence lends itself as an ingredient in several traditional foods. Like other cereals, maize contains large proportion of carbohydrates (60-65%) and thus provides bulk of energy in the diets based on it. Apart from carbohydrates it is also a good source of protein (10-12%), fat (3.5-5.5%) and crude fiber (2.0-2.5 %). By virtue of its composition it is quite comparable to rice and wheat in its nutritional value (Table 1 and 2). Maize is also rich in vitamins and minerals (Anonymous, 1998).

Table 1. Nutritional quality of maize compared to rice and wheat (*Source:* Anonymous, 1998).

Grain	Nutrient content (per 100g)							
	Protein (g)	Fat (g)	Carbo-hydrates (g)	Crude fiber (g)	Minerals (g)	Calcium (mg)	Phosphorus (mg)	Iron (mg)
Maize	11.4	4.6	66.2	2.7	1.5	10	348	2.3
Rice (milled)	6.8	0.5	77.6	0.2	0.7	10	100	0.7
Wheat	11.8	1.5	71.2	1.2	1.5	41	306	5.3

Table 2. Vitamin and mineral content of maize.

Nutrients	Per 100 g
Magnesium (mg)	139.00
Sodium (mg)	15.90
Potassium (mg)	286.00
Copper (mg)	0.41
Manganese (mg)	0.48
Zinc (mg)	2.80
Chromium (mg)	0.004
Carotene (µg)	90.00
Thiamine (mg)	0.42
Riboflavin (mg)	0.10
Niacin (mg)	1.80

Source: Gopalan *et al.*, 2012

Processing and consumption of maize

Maize is being processed in various ways in different parts of the world before consumption. However, the soaking and nixtamalization are the major pre treatments followed around the world.

Soaked grain

The grain is soaked and cooked in water or lime solution then the grain is ground to make dough which is used as the base for different preparations. Eventually, the grain soaked and cooked is dehulled and the germ removed partially or totally. This product can be pounded to obtain the grits and then cooked and eaten like boiled rice or can be used to prepare special type of breads such as the **arepas** in Venezuela or the **sopas** in Paraguay. Likewise, maize gruel can be transformed into sweet or sour drinks. Fermented drinks are popular in Africa and Latin America. An example of this product is the **chicha** and also the **pozol,** a fermented masa in Central America is used to prepare a typical beverage just by adding to the portions of fermented masa water or milk and sugar followed by shaking. Pozol is made in Nicaragua with maize purple coloured grains (called pujagua), although any maize variety can be used (Cuevas *et al.,* 1985; Camacho *et al.,* 1992).

Nixtamalized maize or lime treated maize or alkali treated maize

This process was developed by Native American Indians. The nixtamalization consists of mixing one third part of whole maize with two third part of lime (calcium hydroxide) solution between 1 to 2 percent of concentration. In general the cooking time may vary from 15 to 45 minutes and the temperature of cooking is held above 68^{0}C. The grinding of nixtamalized kernels is carried out by simple pounding with a hand operated or electric kitchen mixer-grinder for large scale masa production. This masa is the base for preparing several traditional products such as tortillas, tamales etc. (FAO, 2003). The dry masa flour is more stable against rancidity and the shelf life can be until one year in comparison with the whole kernel ground maize flour.

Maize

↓ After cleaning

Soak 500 g grains in 1% lime water for 5 minutes (10 g of lime in 1 litre water)

↓

Give heat treatment at simmering temperature for 30 minutes

↓

Remove the vessel and leave it for overnight

↓

Wash 3-4 times to remove lime

↓

Dry in sunlight (Moisture level should be 9-10%)

↓

Store in air tight container

Nixtamalization treatment advantages: It facilitates the pericarp removal, controls microbial activity, enhance water uptake, increases gelatinization of starch with improvement in nutritional value through an increased availability of niacin (FAO, 2003). The research conducted at AICRP (maize) Mandya centre, indicated that the lime treated maize flour can be kept up to three months in LDPE covers without affecting its flavor and roti making quality (Shobha *et al.*, 2012). The process of lime treatment and its advantages has been disseminated through training to maize growers, SHG's and mill owners. The nixtamalised corn or the lime treated corn is being used in India for the preparation of various products such as dumpling, dry pan cake, *upama, idli, dosa*, *dhokla,* snack foods such as *sev, muruku* and *laddu* by suitably combing with pulse and other adjuncts.

In India efforts are being made to intensify the production as well as the utilization of maize in order to alleviate pressing demand for rice and wheat. Grains can be processed into a number of products in dry milling process using mini grain mill such as grits, semolina (suji) and flour. These dry milled products are being used in numerous ways.

Uses of dry-milled maize products

1. **Large grits**: These are used for the manufacture of breakfast cereal corn flakes, for which grits from yellow maize are preferred.
2. **Coarse grits and medium grits**: These are used in the manufacture of cereal products and snack foods.
3. **Fine grits**: In industry it is used as a brewing adjunct. At domestic level, maize grits or hominy grits are used to prepare porridge by boiling with water, preparation of traditional breakfast items such as *kesaribath* (*sweet bath*), *upama* (*khara bath*), *idli* and *dhokla* preparations.
4. **Coarse or granulated maize**: This is used in pancake and muffin mixes, corn snacks, cereal products and other baking products.
5. **Fine meal or corn flour**: It is used to make maize bread, bakery products, mixed infant's foods and breakfast cereal, noodles, pasta and other snack foods.
6. **Corn (maize) flour**: Is used for making roti (dry pan cake), dumpling, bread, wet pancake, infant's foods, biscuits, wafers, as filler and carriers in meat products, and breakfast cereals. Research at AICRP (maize) centre has shown that judicious refining of maize grain will enhance the appearance and eating quality, texture and other organoleptic qualities without affecting the nutritional quality.

Maize semolina

Semolina (Suji) from maize can be used for making many preparations where wheat or rice are conventionally used for different products such as *upama, kesari bath*, *idli*, *dhokla* like soft cooked preparations. Study conducted at AICRP (Maize) centre revealed that *idli* and *dhokla* prepared from maize suji are accepted organoleptically by semi trained judges as well as consumers due to its creamish yellow colour and are highly porous in texture due to sparse distribution of pores in case of maize products compared to rice products. Maize suji of medium and large particle size (-20 B.S. mesh) require more cooking time (25-30 min) compared to rice and wheat suji. Hence, fine suji (-40 B.S. mesh) is preferred in order to reduce the cooking time to 10 minutes. The swelling capacity and water absorption capacity of maize suji and flour are quite high compared to rice and wheat products. Hence, it requires more water for cooking. The above properties of maize would help to produce and promote maize suji usage, particularly in south India, where suji finds ready use in their culinary practices for sweet and savoury dish preparations.

Maize flour

Many conventional dishes such as thick and thin *porridges*, *pan cakes, roti, thallipattu*, sweet and savoury preparations etc. are commonly prepared in India. Compared to *porridges*, *roti* in the form of dry pan cake is the most preferred method of use. However, the *roti* of maize cannot be rolled thin due to lack of gluten which contributes for adhesive and extensive properties. Hence, the flour of maize needs to be gelatinized by keeping a heap of maize flour in a boiling water and cooking under low flame for 8-10 minutes. Thus the dough acquires some adhesive property and can be rolled into quite thin products. A unique property of the maize dough is its ability to hold more water than other cereal dough (maize requires 115 ml water/ 100 g flour, sorghum; 90ml, pearl millet; 85 ml, and wheat; 65 ml/100 g flour). The maize *roti* will have higher moisture content than other cereal *rotis*. Even for getting better textural properties, maize can be mixed with wheat or sorghum flour between 30 and 40 per cent.

Nutritional enrichment or fortification of maize with other cereal/pulse combinations

Since the need for nutritive food at low cost is in demand and a majority of the population in different geographical area depend on cereals for their daily needs. Quality Protein Maize (QPM) may not be available in all the places, hence the better alternative for this is to combine maize with good quality pulses such as soybean, green gram, bengal gram, black gram, groundnut along with wheat, rice and millets in suitable combinations. The concept of multi grains or multigrain flour is more ideal not only for nutritional and cost benefits but also to improve the texture and shelf life of the flour. Since maize flour is gluten free, it can be recommended for people intolerant to gluten, where wheat chapati can be replaced with chapati or roti of maize.

Use of maize flour, either singly or in combination with the flours of other cereal or legume flours, for making many types of sweet and savory snack foods has great possibilities in the Indian context. Different types of deep fried products are possible and have been tried at the AICRP (Maize) centre such as *muruku, nippattu*, *sev* and *crispies* by adding maize flour to the tune of 50 to 60 per cent. In this respect, maize flour would be a good extender of gram flour (*Cicer arietinum*) for such preparations. A large variety of sweet dishes – *laddu, vada, mysorepak*, *burfi* and *chocolate* and such other products can also be prepared from maize flour by combining with gram flour and other food additives.

Table 3. Nutritional composition of maize foods (per 100g) on dry weight basis.

Nutrients	Regular foods (cooked)		Convenience foods (dry samples)		
	Maize Upama	Maize Roti	Maize Papad	Maize Vermicelli	Maize Noodles
Carbohydrates(g)	70.22	88.20	67.50	80.20	84.04
Proteins(g)	9.37	7.21	7.55	14.40	6.65
Fats(g)	13.38	3.09	1.40	1.56	4.28
Soluble fiber(g)	5.98	2.27	1.38	7.23	3.76
Insoluble fiber(g)	24.42	18.12	24.10	18.88	21.87
Niacin (mg)	1.47	0.14	0.01	1.06	1.72
Beta carotene (μg)	1380	2195	1498	2209	1298
Iron (mg)	2.35	1.98	5.50	4.36	18.60
Calcium (mg)	75.02	111.70	92.70	108.80	117.70
Zinc (mg)	1.77	2.41	0.68	3.12	27.70

Convenience foods

Many convenience foods such as *papad*, *vermicelli* and *noodles* were prepared by combining maize flour to the tune of 50 per cent along with rice, wheat and soya flour revealed that the products were nutritionally quite comparable to any other cereal products (Table 3).

The storage studies of the dehydrated products like *vermicelli*, *papad,* and *noodles* revealed that the products were safe in terms of biochemical, physical and microbiological attributes for a period of six months.

Specialty Corns

Recently, maize is being grown for diverse uses and specialty purposes. Such maize for specialty and value added purposes are collectively called specialty corn. Compared to field corns, specialty corns possess additional and characteristic features. Their global spread, increasing demand and premium price make them an attractive option for the farmers in many countries including India. Specialty corns are amenable to numerous options pertaining to harvest time and various economical products (Rajendran *et al.,* 2013). The major specialty corns which occupied significant portion in food shelves include baby corn, sweet corn, pop corn and quality protein maize.

Baby Corn

Baby corn (*Zea mays* L.) refers to the young cobs of maize harvested within 1-4 days of silk emergence. Baby corn is highly nutritive and its nutritional quality is at par or even superior to some of the seasonal vegetables. Besides protein, vitamins and iron, it is one of the richest source of phosphorus. It is a good source of fibrous protein and easy to digest. It is also free from the residual effects of pesticides as the young cobs are wrapped up within the husk and well protected from diseases, insects, fungicides and insecticides (Kawatra *et al.*, 2007).

Table 4. Nutritional value of baby corn compared with other vegetables per 100 g (Yodpet, 1979).

Components	Baby Corn	Cauliflower	Cabbage	Tomato	Eggplant	Cucumber
Moisture (%)	89.1	90.3	92.1	94.1	92.5	96.4
Fat (g)	0.2	0.04	0.2	0.2	0.2	0.2
Protein (g)	1.9	2.4	1.7	1	1	0.6
Carbohydrate (mg)	8.2	6.1	5.3	4.1	5.7	2.4
Ash (g)	0.06	0.8	0.7	1.6	0.6	0.4
Calcium (mg)	28	34	64	18	30	19
Phosphorus (mg)	86	50	26	18	27	12
Iron (mg)	0.1	1	0.7	0.8	0.6	0.1
Vitamin (fu)	64	95	75	735	130	0
Thiamine (mg)	0.05	0.06	0.05	0.06	0.1	0.02
Riboflavin (mg)	0.08	0.8	0.05	0.04	0.05	0.02
Ascorbic acid (mg)	11	10	62	29	5	10
Niacin (mg)	0.03	0.7	0.3	0.6	0.6	0.1

Baby corn may be consumed raw or used as an ingredient in various preparations. Different value added products such as *manchuirian*, *jam*, *pickle*, *pakoda*, *curry*, *salad*, *soups*, *halwa*, canned corns etc. are few examples under wide range of value added products. Recently a process has been standardized at AICRP (Maize) Mandya centre for preparing baby corn candy using 40, 50 and 60^0 brix sugar solution followed by dehydrating the same till the moisture level reaches between 10-12 per cent. Prepared cadies will have a shelf life of six months in MPP pouches.

Sweet corn

Sweet corn (*Zea mays* var. *saccharata*) is genotypes with specific endosperm mutation like su and sh. In India, sweet corn green ears are being consumed by direct toasting on fire or boiled in water. Sweet corn kernels often have a wrinkled appearance resulting from a sugary gene which retards the normal conversion of sugar to starch during endosperm development. The endosperm is composed of sweetish starch and characterized by translucent horny appearance during immature stage and after maturity the kernel becomes wrinkled. Sweet corn cob is harvested around 80-85 days after sowing (milky stage). It contains on an average 25-30% sugar; many sweet dishes such as *halva*, *kadabu*, *crunch*, *salad*, *jam*, *pakoda* can be prepared using sweet corn by combining with jaggery, vegetables and such other ingredients.

Popcorn

Among various types of corn, the most popular being the "Popcorn" (***Zea mays* var. *everata***), is a type of corn that expands from the kernel and puffs up when heated with light crunchy texture. Popcorns are usually consumed as a snack food with or without salt (regular), sweetened (caramel/chocolate corn) or butter like topping. Popcorn consumption has greatly increased in recent years because of the advent of microwavable popcorn and the proliferation of flavored ready to eat products.

Table 5. Nutritional composition of sweet corn (*Zea mays var. saccharata*).

Nutrients Value	Per 100g
Energy (Kcal)	86 Kcal
Carbohydrates (g)	18.70
Protein (g)	3.27
Total Fat (g)	1.35
Cholesterol (mg)	0
Dietary Fiber (g)	2.00
Folates (µg)	42.00
Niacin (mg)	1.77
Pantothenic acid (mg)	0.717
Pyridoxine (mg)	0.093
Riboflavin (mg)	0.055
Thiamin (mg)	0.155
Vitamin A (IU)	187 IU
Vitamin C (mg)	6.80
Vitamin E (mg)	0.07
Vitamin K (µg)	0.30
Calcium (mg)	2 .00
Copper (mg)	0.054
Iron (mg)	0.52
Magnesium (mg)	37 .00
Manganese (mg)	0.163
Selenium (µg)	0.60
Zinc (mg)	0.46

Source: USDA National Nutrient data base

Popping is simple and economical processing technique which is traditional and may be adopted easily with improvement in nutritional quality of grain. It is high temperature short time (HTST) treatment which sterilizes product, gelatinizes its starch and develops pleasant aroma to form a ready-to-eat food (RTE) at low processing cost. Popping process not only retains the actual nutritional profile of grains but also markedly enhances its protein digestibility, bio availability of iron and dietary fiber content due to the development of resistant starch. Popping also reduces some of the anti nutrients *viz.*, phytates, tannins, acid detergent fiber, lignin and cellulose (Reddy *et al.*, 1991).

Raw popcorn: Popcorn grains are normally smaller than the regular dent corns with a glossy coating. Moisture is one of the most important parameter which affects yields and stability during storage. Popcorn is less affected by moulds compared to regular dent corns. Popping quality especially expansion volume depends on moisture, hardness and average size of the grain. Generally lower sized kernels produce less expansion volume than larger ones. Stress cracks are created due to faulty harvesting with rapid drying and tend to produce higher incidence of unpopped kernels with less flake volume. For a good popping quality moisture content of the grains should be between 13-14 per cent. If the grains are too dry then moisture content is increased by adding known quantity of water followed by conditioning the grain until desired level of moisture is attained.

Popped grain: Crispiness is the major criteria for acceptability. The minimum moisture to preserve crispiness should be less than two %. Moisture also affects the texture of the pop corn and hence must be consumed at a moisture level of less than 2% (ideally 1.8-1.7%).

Mechanism of popping

Kernel of popcorn contains certain amount of moisture and oil. Unlike most other grains the outer hull of popcorn kernel is strong and impervious to moisture, while the starch is of hard and dense type. When the kernels are heated, moisture present in the kernel gets converted to pressurized steam; as a result the starch inside the kernel gelatinizes, softens and becomes pliable. At a pressure of about 135 psi (930 K. Pa) and a temperature of 180- 232^{0}C (356-450^{0}F) the breaking point of the hull is reached. The hull ruptures rapidly causing a sudden drop in pressure inside the kernel with a corresponding rapid expansion of the steam which expands the starch and proteins of the endosperm into crispy puffs popularly known as popping (Edmund and Lloyd, 2001). Only the strains of "*Zea mays* var. *everta*" which has special kind of flint corn with varied proportion of hard and soft starch in the endosperm will bear the characteristic popping quality (Ramachandrappa and Nanjappa, 2006).

Popping or cooking methods

Traditionally popcorn is popped on an iron tawa over an open flame. But during recent past a number of popcorn machines are available in the market in varied size and shapes. The most convenient one for small scale (house hold) consumers is "air poppers" which rapidly circulate heated air up through the interior part by keeping the un-popped kernels in motion to avoid burning and then blowing the popped kernels out through the chute. The majority of popcorn sold in the market is now packed in microwave bags which can be conveniently used in microwave oven. The common popping methods includes.

a. **Oil popping:** Traditionally salted popcorn is made by putting corn in a pot or iron tawa with oil and salt along with heating and shaking until popping is completed. The level of heat applied to the pot must be controlled so that the popping process takes about 2.5 minutes from a cold start with oil and corn (Fig 1.a).

b. **Dry popping:** Hot air poppers are popular in houses where consumers are concerned about fats and oils in their diet and also it requires little preparations. Small hot air-home poppers operate on the principle of forcing heated air up through a bed of popcorn until it pops (Fig 1.b). The popped corn may then be seasoned with salt, butter or any other flavors as desired.

c. **Microwave popping:** Microwaving is probably the easiest way to prepare popcorn which requires the least clean up but is the most expensive method. Raw popcorn and oil are packed in a specially designed package that is placed in a microwave oven and heated. Directions are printed on microwave popcorn packages and these packages contain disposable serving containers for the finished product (Fig 1.c).

(a) **(b)** **(c)**

Fig. 1. a- Oil popping, b - Air popping and c- Micro-wave popping.

Value addition to popcorn

Popped maize contains slightly higher proportion of fiber and hence is suitable for the preparation of high fiber specialty foods. Apart from normal snack eatable popcorn finds its use in the preparation of sweet popcorn, masala popcorn and popcorn laddus. Even the popcorns can be ground to make coarse powder, which can be used for the preparation of many traditional dishes such as burfi, popcorn gum laddu, popcorn holige, nutria-bars, chocolate coated bars etc.

Table 6. Nutritive value of popcorn.

Nutrients	Per 100 g
Energy	382 kcal
Carbohydrates	78.0 g
Dietary fiber	15.0 g
Fat	4.0 g
Protein	12.0 g
Thiamine (Vit. B_1)	0.2 mg (17%)
Riboflavin (Vit.B_2)	0.3 mg (25%)
Iron	2.7 mg (21%)

Source: USDA Nutrient Database

Table 7. Quality of Amino acids (Jat *et al.*, 2009).

Amino acids	Normal maize (%)	QPM (%)	FAO/WHO (%)
Isoleucine	3.77	3.10	4.00
Leucine	10.60	6.62	7.00
Sulphur amino acids	4.62	3.83	3.50
Tryptophan	0.48	1.20	1.00
Valine	5.02	5.46	5.00
Threonine	4.00	4.62	4.00
Lysine	2.73	4.92	5.50

Even the health mixes such as energy rich pop mix (prepared by combining with jaggery, copra gratings, poppy seeds and dry fruits with pop corn powder), protein rich pop mix (pop corn powder, puffed bengal gram powder and sugar powder), iron rich pop mixes (popcorn powder, besan powder, garden cress seeds, ragi malt powder and sugar powder) can be popularized by women SHG's and also marketing outlets need to be increased for such products.

Quality Protein Maize (QPM)

Quality Protein Maize (QPM) developed by lowering the concentration of zein by 30%. As a result the concentration of two essential amino acids *viz.*, lysine and tryptophan in grain was increased in QPM genotypes as compared to normal maize. The lower content of leucine in QPM further balances the ratio of leucine to isoleucine content, and the balanced proportion of all these essential amino acids in quality protein maize enhances the biological value of protein (Table 7). Even the protein quality of maize, particularly Quality Protein Maize (QPM) is quite comparable with that of other cereals (Fig. 2).

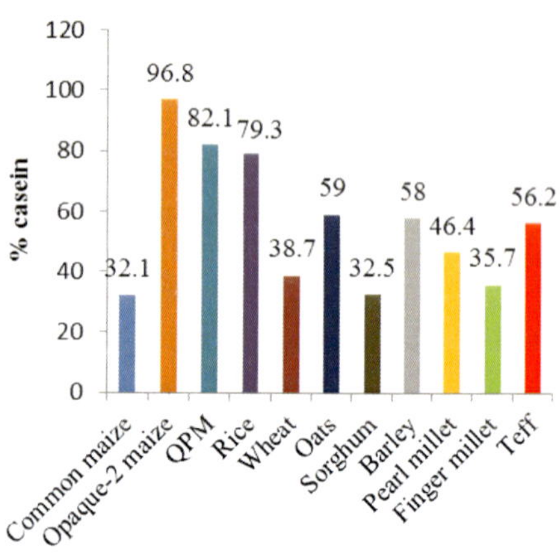

Fig 2. Protein quality of maize and other cereal grains.

QPM surpasses ordinary corn in biological value and true digestibility (Table 8). The studies of Eggum *et al.* (1979) and Bressani (1990) on children has showed that when QPM was the only protein source fed to children recovering from malnutrition, resulted in increased nitrogen balance and improvement in health condition. It was suggested that QPM can be a practical solution to homemade weaning foods. The QPM offers tremendous benefit to both human and animal nutrition, particularly in monogastric animals.

Table 8. Protein quality of normal and QPM grain.

Quality measures	Normal	QPM
True protein digestibility (%)	82-91	92
Biological value (%)	40-47	80
Amount needed for equilibrium	547	230

In human nutrition intervention, it may play a very important role in those countries where corn constitutes staple food in the diets of children and adults. In the words of Normal Borlaug (1992) "It is time, I believe, to make a serious effort to put it into commercial use to serve human needs." QPM, as an ingredient in animal feed, may have a far more important role to play. Increasing the use of corn as animal feed globally would provide indirect benefits to humans and would result in greater impact.

Value added products of QPM

Several value added products from Quality Protein Maize (QPM) were developed and popularized at AICRP (Maize) Pusa centre. The nutritious products from QPM includes recipes suitable for eastern region (*Chatpati, maize roll*, *Chutney*, *Hilsa corn chat*, *Pakodi*, *Jalebi* etc.) recipes for northern region (*Sweet dalia*, *Khichari*, *Kheer,* weaning mixture, *laddoo*, *sev*, *cheela*, *dhokla*, *pasta*, *fryums* and *poori*), recipes for western region (*dhokla*, *Bati*, *Rab*, *Papadi*, *Paratha, Gatta*, *Pakodi*, *Halwa*, *sev* and *muffins*) and the recipes for southern region (*Malt mix*, *Poustic mix*, *Chakkuli*, sweet and salt biscuits, *chutney* powder , health mix*, idli, dosa Sheera* and *besan laddo*) can be prepared by combining QPM flour or Suji to the tune of 50 to 60% or even 100% in some of the recipes.

Various QPM based products such as suji, flour, vermicelli, noodles, biscuits, cheese balls and crispies are some the products prepared by trained women SHG's and selling on large scale. Some products from QPM have already been developed and commercialized by some entrepreneurs. These commercialized products are Pusa Shakti (QPM chatpati), Dilkush (instant Kheer mix) and Proteino-H (Kadhi mix). There is a demand for QPM suji, QPM idli mix (ready to cook), QPM-vada mix (ready to cook) and Nutri mix from the people of Karnataka. Many other products from QPM are under process of screening on the basis of nutritive value, organoleptic quality as well as marketing potential (Kaul *et al*., 2016).

Utilisation of Quality Protein Maize (QPM): The major utilization of QPM is described as here under:

Food and Nutritional Security

The QPM can be utilized for diversified purposes in food and nutritional security as infant food, health mixes, convenience foods, specialty foods and emergency ration. It also finds its place in fulfilling the protein requirements of different sections of society (infants, lactating mothers, convalescing patients, kwashiorkor diseased, old persons etc.) to prevent malnutrition. Even for the food and nutritional security of the tribal population, which constitutes approximately 10 % of the total population, where in the majority of the tribal population depends on maize as their basic diet, in these areas the scope for QPM to ensuring food and nutritional security is paramount. Substituting maize with QPM is viable option for ensuring their nutritional requirement. Several studies conducted on human beings and animals revealed the positive health benefits of QPM in their daily diet. Study conducted by Gunaratna *et al.* (2010) reported that consumption of QPM instead of maize leads to increase in growth rate such as height and weight by 12 and 9% respectively in infants and young children who belong to mild to moderate nutritional background with maize as their major diet. The increase in height and weight is due to higher biological value of QPM as compared to maize (Table 8). Hence, it is high time that policy makers should think of introducing QPM in public distribution system (PDS), in mid day meal programmes in Schools and Aanganawadis to serve balanced food at low cost. Government of India has already started tribal- sub plan (TSP). Under the programme, Indian Institute of Maize Research has initiated the programmes like distribution of QPM seeds, inputs and training programmes on cultivation and preparation of maize value added products in order to create awareness regarding food utility of QPM.

Providing nutritious feed

Quality Protein Maize with its high carbohydrates, fats, better quality protein lends itself as nutritious feed for poultry, livestock, swine, fish etc, which leads to early development of broilers, save energy and feed and also extra cost incurred on lysine and tryptophan fortification.

Maize based entrepreneurship

The nutritious products developed from QPM can replace the highly priced non nutritious industrial products. These products can also be prepared in villages as a source of rural entrepreneurship (Jat *et al.,* 2009).

Value added products of maize with nutritive value

Maize papad

Nutritional composition (100g)

Carbohydrates (g)	67.5
Proteins (g)	7.55
Fat (g)	21.4
Soluble fibre (g)	1.38
Insoluble fibre (g)	24.1
Thiamine (mg)	0.01
Riboflavin (mg)	0.01
Niacin (mg)	0.01
Pyridoxine (mg)	0.01
Folic acid (mg)	0.01
ß carotene (µg)	1498
Calcium(mg)	92.7
Iron (mg)	5.5
Magnesium(mg)	53.4
Zinc (mg)	0.68

Maize laddu

Nutritional composition (100g)

Protein (%)	15.97
Fat (%)	28.41
Carbohydrates (%).	64.51
Energy (K cal)	537
ID fibre(%)	12.94
SDF (%)	1.21
Calcium (mg)	24.17
Magnesium (mg)	26.09
Copper (mg)	0.51
Iron (mg)	5.26
Zinc (mg)	1.92
Chromium (mg)	8.1
ß Carotene (µg)	9.61
Vit C (mg)	2.73
Thiamine (µg)	2.53
Riboflavin (µg)	0.1
Folic acid(/µg)	6.47
Niacin (µg)	5.97
Pyridoxin (µg)	1.21

Maize vermicelli

Nutritional composition (100g)

Carbohydrates (g)	80.2
Proteins (g)	14.4
Fat (g)	1.56
Soluble fibre (g)	7.23

Insoluble fibre (g)	18.22
Thiamine (mg)	0.03
Riboflavin (mg)	0.1
Niacin (mg)	1.06
Pyridoxine (mg)	0.74
Folic acid (mg)	0.1
ß carotene (µg)	2209
Calcium(mg)	108.8
Iron (mg)	4.36
Magnesium(mg)	89.49
Zinc (mg)	3.12

Maize Crispies

Nutritional composition (100g)

Protein (%)	18.18
Fat (%)	35.2
Carbohydrates (%)	51.82
Energy (K cal)	564
ID fibre (%)	12.51
SDF (%)	0.93
Calcium (mg)	61.16
Magnesium (mg)	45.48
Copper (mg)	0.52
Iron (mg)	3.63
Zinc (mg)	1.87
Chromium (mg)	4.2
ß Carotene (µg)	16.1
Vit C (mg)	5.4
Thiamine (µg)	10.38
Riboflavin (µg)	0.52
Folic acid (µg)	11.42
Niacin (µg)	1.87
Pyridoxin (µg)	9.66

Summary

Maize like other cereal grain is grown as rain-fed crop during *Kharif* and *Rabi* seasons as well. A number of high yielding varieties, hybrids, specialty corn (popcorn, baby corn, sweet corn and QPM) have been developed by Indian Institute of Maize Research (IIMR) with collaboration of AICRP centers across the country. Maize has great potential as a food extender for food security needs of the country. However, the consumption of maize has been limited only among traditional users. Hence, it must be popularized among the rice and wheat eaters. Proper refining and processing techniques need to be dispersed among larger mass. Diversification in food uses such as use of maize suji and flour for the preparation of traditional convenience foods would pave the way for their wider acceptance and utilization. Technology for maize flour suitable for roti making, maize *noodles*, *papad*, *pasta* and such extruded products from QPM are significant in this direction. Further efforts are needed to exploit the higher water holding ability of maize flour coupled with its bland taste and flavor can be made use in many traditional preparations for cost reduction and quality improvement.

Future Research perspective

Since whole grain products are treasure house of nutrients, inclusion of coarse grains like maize in daily diet not only bring variety to the diet, but contributes to energy, protein, fiber, and some of the vitamins and minerals also. However, maize like any other cereal grain is lacking in lysine and tryptophan and the same can be overcome by the use of Quality Protein Maize (QPM) or combining maize with good quality pulses. Popularization of QPM will not only aid in food security but to the nutritional security of the people also. However, the outreach activities for processing and usage of dry milled products in daily diet need to be strengthened. Huge scope exists for the value added products of maize may be in terms of baby corn, popcorn, sweet corn as well as normal maize. Transfer of technology from farm to plate should be carried out in maize growing areas. Training programmes and large scale demonstrations should be taken up in order to disseminate the method of processing and product preparation from maize.

References

Anonymous. (1998). Maize properties, processing and utilization: RESC Scientific series. 16: 10-11.

Anonymous. (2015). Area and production of maize in INDIA and Karnataka. Annual progress report of *Kharif* maize 2015. Indian Institute of Maize Research, IARI, New Delhi. Pp. 9.

Borlaug, N.E. (1992). Potential role of quality protein maize in Sub-Sahara Africa, in *Quality Protein Maize,* Mertz, E. T., ed., American Association of Cereal Chemists, St. Paul. MN 79.

Bressani, R. (1990). Nutritional value of high-lysine maize in humans, in *Quality Protein Maize,* Mertz, E.T., Ed., American Association of Cereal Chemists, St. Paul, MN, 205.

Cuevas, R., Figueroa, E. and Racca, E. (1985). The technology for industrial production of precooked corn flours in Venezuela. *Cereal Foods World*. 30: 707-712.

Camacho, L., Sierra, C., Campos, Guzman, E. and Marcus, D. (1992). Nutritional changes caused by germination of legumes commonly eaten in Chile. *Arch Latin Am Nutr*. 42:283-290.

Edmund, W.L. and Lloyd, W.R. (2001). *Snack Foods Processing*. CRC press. Pp. 388-389.

Eggum, B. O., Villegas, E. and Vasal, S. K. (1979). Progress in protein quality of maize. *J Sci Food Agric*. 30: 1148.

FAO. (2003). Maize in Human. Nutrition. Report series-25. Food and Agriculture Organization, Rome, Italy.

FAO. (2009). Food and Agricultural Organisation, Country statistical, Economic and Social Department: The Statistical Division, Rome, Italy.

Gopalan, C., Rama Sastri, B. V. and Balaubramanian, S.C. (2012). Nutritive value of Indian foods. National Institute of Nutrition, ICMR, Hyderabad. 47-95.

Gunarathna, N.S., De Groote, H., Nestel, P., Pixley, K.V. and McCabe, G.P. (2010). A meta analysis of Community level studies on quality protein maize. *Food Policy*. 35: 202-210.

Jat , M.L., Dass, S., Yadav, V.K., Sekhar , J.C. and Singh, D.K. (2009). Quality Protein Maize for food and nutritional security in India. *DMR Technical Bulletin*, 2009/4, Directorate of Maize Research, Pusa, New Delhi. Pp. 23.

Kaul, J., Kumar, R., Chaudhary, D.P., Neelam, S., Kumar, B. and Mahajan, V. (2016). Quality protein maize and Nutritional security In: Souvenir, 59th Annual Maize Workshop, All India Co-ordinated Maize Improvement Project, Organized by UAS, Bengaluru. Pp. 33-40.

Kawatra, A., Pal, Dharam, Yadav, V.K., Rakshit, S., Singh, R.P. *et al.* (2007). Baby Corn-Cultivation and Value addition , Directorate of Maize Research Publishing Co, Pusa campus, New Delhi. Pp. 1-2.

Parihar, C.M, Jat, S.L, Singh, A.K, Hooda, K.S, Chikkappa, G.K, Singh, D.K. and Saikumar, R. (2011). Maize production technologies in India. *In*: DMR Technical Bulletin 3. Directorate of Maize Research, Pusa campus, New Delhi. Pp. 36.

Prakash, J. (2013). Whole grain nutrition- rediscovering the hidden wealth, *Indian Food Ind. Mag.* 32(6): 47-48.

Rajendran, A.R., Singh, N. and Dhandapani, R. (2013). Scope of specialty corn for income generation. In: *AGROLOOK,* Oct-Dec. Pp. 13-18.

Ramachadrappa, B.K. and Nanjappa, H.V. (2006). *Specialty corns* published by Kalyani publisher's. Pp. 17-18.

Reddy, N.S., Kamble, R.M. and Khan, T.N.I. (1991). Evaluation of nutritional quality of maize and maize products. *Ind J Nutr Dietet.* 98: 90-94.

Shobha, D., Dileep Kumar, H.V., Sreeramasetty, T.A., Puttaramanaik, Pandurange Gowda, K.T. and Shivakumar, G.B. (2014). Storage influence on the functional, sensory and keeping quality of quality protein maize. *J Food Sci Technol.* 51(11): 3154-3162.

Srilakshmi, B. (2008). Food Science. Third Edition, New Age International (P) Ltd Publishers. Pp. 17-72.

Yodpet, C. (1979). Studies on sweet corn as potential young cob corn (*Zea mays* L.). PhD thesis, University of Philippines (unpublished).

29

Specialty Corn: A Rich Souce of Quality Fodder

D.P. Chaudhary[1] and Balwinder Kumar[2]

[1]ICAR-Indian Institute of Maize Research, PAU Campus, Ludhiana, Punjab
[2]KVK Taran Taran, Punjab
**Corresponding Author's Email: chaudharydp@gmail.com*

Shrinking land resources, vulnerable market of cash crops along with climatic aberrations have adversely affected the Indian agriculture and as a result, small farmers are increasingly alienating from this profession. To sustain their livelihood, farmers are increasingly adopting the alternative sources and dairy farming is one of the best-suited businesses, which can rightly fit at this stage. It is fast emerging as a potential business in rural India, particularly in the peri-urban belt. Among the major inputs required to set up a dairy unit, round the years availability of good quality fodder is the foremost requirement. Green forages are rich and cheapest source of carbohydrates, protein, vitamins and minerals for dairy animals. The importance of forages in our country is well recognized since feeding forages alone accounts for over 70% of the cost of milk production. Hence by providing sufficient quantities of green fodder instead of costly concentrates and feeds to the milch animals, the economics of dairy unit can easily be maintained. Milch animals need around 40 kg of green fodder daily for better milk production and for maintaining good health of the animal. There is a large number of fodder crops grown in our country, the majority being dominated by pearl millet, sorghum, maize and oats in the non-leguminous and berseem and lucerne in the leguminous category. Pearl millet and sorghum are extensively grown as cereals in the southern and dryer regions of the country and the stover remained is used as animal fodder. However, the nutritional quality of the stover is very poor due to higher content of indigestible lignin and low levels of cellulose, whereas, the cultivated green fodders are mostly confined to the North-Western plain zones of Punjab, Haryana and Western Uttar Pradesh. The availability of round the years quality green fodder, therefore, seems to be a tough task. At present, the country faces a net deficit of 61.1% green fodder, 21.9% dry crop residues and 64% feeds (IGFRI Vision, 2030).

Maize is considered to be an ideal forage crop as it produces high biomass, possesses high levels of nutrients, is palatable and contain negligible amounts of anti-quality components. The fodder quality of green maize is excellent among non-legume cultivated fodders e.g. pearl millet and sorghum. Both sorghum as well as pearl millet possesses anti-quality components which are deleterious to animal health. Crude protein (CP) and in-vitro dry matter digestibility (IVDMD) are two important parameters governing fodder quality. Crude protein as well IVDMD are significantly higher in maize compared to its competitive fodders. The biomass production of maize is also equivalent to sorghum and pearl millet. Pearl millet is a hardy crop cultivated in the dryer regions of the country, whereas, sorghum, though cultivated throughout the country, contains the most toxic anti-quality component called prussic acid (HCN). The toxicity of HCN is so severe that animal may die within minutes after consuming young sorghum fodder. Maize on the other hand is almost free from any anti-quality components. Maize is considered ideal forage because it grows quickly, produces high yields, is palatable, rich in nutrients, and helps to increase body weight and milk quality in cattle (Sattar *et al*., 1994). As fodder for livestock, maize is excellent, highly nutritive and sustainable (Hukkeri *et al*., 1977; Iqbal *et al*., 2006).

Specialty Maize

Specialty maize is referred to the maize grown for some special purposes. Baby corn and sweet corn are the two most important special types grown in our country. Compared to normal maize, specialty corns possess additional characteristic features. Their global spread, increasing demand and premium price make it an attractive options for farmers in many countries including India. With the increase of urbanization, changing food habits along with improved economic status, the specialty corn has gained significant importance in peri-urban regions of the country. Increasing cultivation of specialty corn played a vital role in enhancing profitability and livelihood security of the farmer thus reducing rural to urban migration. Suitable hybrids and production technology for specialty corns has already been developed. Over the years, the demand for specialty corn in Indian market is increasing. Although many types are included in the specialty corn category, but sweet corn and baby corn are majorly cultivated and also used as animal fodder in our country. The present chapter will be focused on discussing the effective use of these two types as a source of quality fodder.

Sweet corn

Sweet corn is consumed fresh when the kernels are in the milking stage. Immature sweet corn is considered desirable when kernels are succulent because of a mutant recessive sugary-1 gene (*su*-1) that retards the conversion of sugar into starch during endosperm development. In normal maize glucose produced in leaves during photosynthesis is passed on to developing kernels, where it is rapidly converted to dextrin for its onward conversion to starch. Normal maize contains approximately 3-4% sugars in the immature milky stage, whereas sweet corn at the same stage contains much higher content of sugars. In the past, sweet varieties have been developed by selections within the homozygous *su*-1 genotype. One such selection, 'Silver Queen', became the standard to which other varieties were compared. However, the sweetness breakthrough was accomplished with the discovery of the shrunken (*sh*-2) gene. Super sweet varieties with this gene exhibit sugar levels 2 to 3 times greater than standard sweet corn. Sweet corn must be grown Isolated, otherwise cross pollination with standard corn will make kernels tough and starchy. Germination in super sweet varieties is usually poor. For good stand establishment, planting depth should be shallower and soils should be warmer. In combination with the *su*-1 gene, the "sugar enhanced" (se) gene produces a variety with increased sweetness and creamy texture, requiring no isolation. Varieties with these gene have sugar levels that normally peak somewhere between *su*-1 and *sh*-2 varieties. For shipping purposes, *sh*-2 varieties are excellent because sugar conversion to starch is much slower.

Presently, one hybrid (HSC-1) of sweet corn, developed by public sector (CCS HAU, Hisar), is under cultivation to some extent. However, most of the area under sweet corn cultivation is dominated by private sectors hybrids. Earlier the varieties are cultivated, which lack uniformity, thus reducing the market value. The cultivation practices like roughing, harvesting, etc is also difficult in varieties as compared to hybrids. There is an urgent need to breed high yielding sweet corn hybrids. In spite of availability of promising hybrids and production technology, sweet corn cultivation is still limited to the peri-urban belt. Combined efforts including dissemination of production technology, availability of good quality hybrid seed and the establishment of processing units are required for large scale cultivation of sweet corn in India.

Baby corn

Baby corn is the young ear of female inflorescence of maize plant harvested before fertilization when the silk has just emerged. It can be eaten raw and included in diet in a number of ways as salads, *chutney*, *pakoda*, soup, *raita*, vegetables pickles, and *kheer*, etc. Baby corn cultivation is fast emerging in the peri-urban belt of India. The desirable size of baby corn is 6 to 11 cm in length and 1 to 1.5 cm in diameter with regular rows. The most preferred colour is generally creamish to very light yellow (Dass *et al*., 2009). It has great potential both for internal consumption as well as for export purpose as 3-4 crops of baby corn can be taken in one year. Baby corn is cultivated widely. Thialand is a major baby corn producing country which is earning huge sums from its export. Baby corn is nutritionally rich and its nutritional quality is at par or even superior to some of the seasonal vegetables. Besides proteins, vitamins and iron, it is one of the richest sources of phosphorus. It is a good source of fibrous protein and easy to digest. It is the most "safe" vegetable to eat as it is almost free from residual effects of pesticides as the young cob is wrapped up with husk and well protected from insects and diseases. Baby corn has played a significant role in ensuring livelihood security and augmenting income level of farmers in peri-urban regions. Hybrids of public as well as private sector are available for commercial cultivation in India. Maize hybrid HM 4 possesses all the desirable traits of ideal baby corn and produces good quality fodder as well as silage (Chaudhary *et al.*, 2016). Baby corn cultivation is fast emerging and many small as well as large scale processing plants are being set up where baby corn is canned and marketed. Fresh baby corn is also exported to the European countries. Considering the acceptability of baby corn by Western world a vast potential exists for its export.

Specialty corn as fodder

The increasing cultivation of specialty corn such as baby corn and sweet corn, results in the production of a large quantity of green biomass which could be preserved in the form of silage to be used for cattle feeding during lean periods when there is acute shortage or almost no availability of green fodder. The green biomass left out after the harvest of baby corn and sweet corn is nutritious enough to be used for cattle feeding. Yield potential and nutritional quality of some promising genotypes of baby corn (HM-4) and sweet corn (HSC-1) along with normal maize (DHM117) and fodder maize (J1006) were evaluated for nutritional quality of green fodder as well as silage. Green samples were taken immediately after harvesting baby corn and sweet corn and used for forage quality analysis and the data was compared with normal as well as fodder maize. The samples were analyzed for fodder quality parameters such as green fodder yield (GFY), dry matter (DM), crude protein (CP), fiber components *viz.* crude fiber (CF), neutral detergent fiber (NDF), acid detergent fiber (ADF), *in-vitro* dry matter digestibility (IVDMD) and total ash (TA) in forage as well as silage. The data shows that the nutritional quality of baby corn stalks is almost at par with the fodder maize (J1006) (Chaudhary *et al.*, 2016). Although, biomass from baby corn stalks was little less compared to fodder maize (J-1006 and African tall), there is little difference in terms of crude protein and in-vitro dry matter digestibility. The woodiness is also comparable (Chaudhary *et al.*, 2016).

Preservation of fodder

Forage preservation is the primary requisite in establishing an economically viable dairy farm. It permits a steady supply of good quality fodder when green forage production is low. Forage conservation also provides farmers with means of preserving forage when production is faster than its adequate utilization. This prevents loss of valuable nutrients in the lush green forage by natural maturity. Consequently, forage preservation provides a more uniform level of high quality forage for ruminant. Silage is the material produced by controlled fermentation of nutrients under an anaerobic condition. The fermentation process is governed by microorganism present in fresh herbage or by additives to maintain anaerobic conditions and discourage clostridial growth with minimum loss of nutrients.

Preservation of maize as silage

Maize is considered the best forage for ensiling. It possesses the appropriate concentrations of nutrients required for proper ensiling. For proper fermentation, the crop might possess sufficient quantities of moisture as well as soluble carbohydrates which are converted to lactic acid during the process of fermentation. Maize crop harvested at 50% flowering and the maize stalks remained after the harvest of baby corn and sweet corn possesses the moisture as well as soluble sugars in the required concentration needed for proper fermentation process. Maize silage is becoming more important in dairy rations particularly in Punjab where big dairy farms are utilizing silage as the only source of fodder for their cattle. Maize is valued because of its high yield, ability to make excellent silage, and it can be harvested in a single operation without significant leaf loss. Cows fed corn silage produced more milk and consumed more silage dry matter in both trials than those fed the sorghum silage (Lance *et al.*, 1964). Corn silage is used extensively for lactating dairy cows that require high energy feed for maximum milk production (Marsalis et al. 2010, Irlbeck *et al.* 1993).

Methodology of silage making

The stage at which the fodder is harvested is the major factor affecting the quality of silage. Maize is best suited to be ensiled when the grains are in the milking stage (60-70 days after sowing). However, nowadays little maturity stage (80-85 DAS) is preferred as the kernels at this stage are nutritious enough and no additional ration in the form of concentrates is required to feed cattle. In general, the dry matter content of whole plant should be around 25-30 percent. Harvesting at this recommended time will ensure optimum compaction properties, reduced tendency of mould formation and heating. Silage making is a simple process which can be carried out manually in the farm area by employing a few laborers.

The first activity in silage making is the digging of a pit. A rectangular pit is to be dug up near the cattle shed whose size depends upon the number of animals along with the availability of green fodder. If the fodder is sufficiently available then the time duration of feeding could be considered in deciding the size of the pit. Usually one cubic meter pit can accommodate roughly 5-6 quintals of green fodder. The size of some rectangular silo pits along with their capacity is given below:

Table 1. Dimensions of some most common silo pits

Length (m)	Breadth (m)	Depth (m)	Quantity of fodder to be filled (q)
3	3	2	90-95
7	3	2	225-235
10	3	1.5	245-250

According to thumb rule for determining the capacity of the pit, if an animal need 20 kg silage daily then to feed 5 animals for a period of 90 days, the size of the pit should be 3m X 3m X 2m. The shape of the silo pit is also important. It should have slanting walls with narrow base and broad opening. This type of shape is beneficial for filling the pit as it helps in maximum fodder compaction and removing silage from this type of pit is also easy. The walls of the pit may be plaster with cowdung or a cemented silo pit may also be prepared. Plastered pit may be covered with polythene sheet. The base of the pit should not be covered by plastic sheet, rather it should be covered by straw so that the excess moisture, if present, or juice could be absorbed efficiently. Fill the pit with chaffed fodder (5-7 cm length). For good silage the chop length should be kept shorter. Chaffed silage is more palatable to livestock and has little chance of secondary fermentation. The next step is to fill the pit with chaffed fodder. For this purpose fodder is spread up to a height of 1 foot in the pit followed by compression. This process is repeated till the pit is filled with fodder. The major precaution during this process is to exclude as much air as possible from the chaffed fodder by compressing it properly. This is executed by pressing the material through manual labor or mechanically by using a tractor. Care should be taken that material on the sides and edges are properly compressed. Raise the fodder heap above the ground level up to a height of around one meter. Finally add some more fodder in the central portion of the heap and then trample it. Packing is important to create anaerobic conditions. It should be thoroughly pressed so that no air pocket is left in the silo otherwise chances of mould formation will be there which will spoil the silage. After filling, silo should be covered with polythene sheet followed by a layer of soil, etc. Some cracks may develop in the covered soil over time. These are to be plugged immediately.

After 45 days of ensilage, the silage will be ready to use, therefore, can be removed for feeding to animals. Care should be taken in removing the silage from pit. It should not be opened from one side only. Cover should be kept firmly in place as long as possible and the minimum face should be exposed at one time. The sugars, proteins and lactic acid present in the silage are subject to attack by mould growth and oxidation as some air is allowed to fermentation and causes loss of feeding value and intake by the animals.

Filling a small silage pit

Recognition of well-fermented silage

Properly prepared silage is recognized from its color. The color of the well-fermented silage is bright light green or dull yellow, whereas that of the poorly fermented silage is olive, blue green or dark brown. The smell of the well-fermented silage is like that of vinegar whereas poorly fermented silage is foul smelling. In the properly fermented silage the soluble carbohydrates are converted to lactic acid, whereas in the poorly fermented silage, butyric acid is the end product of fermentation which is primarily responsible for the bad smell. The poorly fermented silage should not be fed to the animals and should be discarded.

Properly prepared silage can be preserved for a long period. If properly covered silage could be stored as long as 10-12 years or so. Once opened, it should be used regularly, and to be consumed within 3-4 months.

Feeding of silage

To start feeding pit should be opened from one side after removing the covered soil and straw. Each time, a uniform layer of silage is removed vertically (from top to bottom) depending upon the daily need. Do not open the whole pit at once. Cover the opened side immediately after removing the silage, to avoid any exposure by air/moisture. The top portion may contain moulds which should not be used for feeding. The animals may take some time (3-4 days) to adapt to the silage feeding, therefore feed 5-7 kg of silage along with some other fodder for the initial period. Once adapted, the cattle can be put on silage exclusively.

Silage quality

Silage quality is determined mainly by the physical state i.e. colour and odor. Good quality silage should be light brown in colour and have the smell of vinegar. Poorly fermented silage will be dark in colour and foul smelling due the production of butyric acid. Chemically it should be having the following characteristics: (i) pH 4.5-5.0, (ii) ammonical nitrogen of total N – less than 10% of total N, (iii) butyric acid- less than 0.2%, (iv) lactic acid 3 to 12%, etc.

Advantages of silage

In the present day dairy farming, silage plays a significant role. It acts as a fodder bank which ones made could be used throughout the year. There are numerous advantages of silage making. Some of these are listed below:

- The most important advantage of silage making is that we can use it during the scarcity of green forages called lean periods.
- It is a nutritious feed as it preserves the nutrients of green forages in their original form and hence is as good for animal feeding as green forages itself.
- The practice of silage making will immensely help in reducing the shortage of green fodder in the country.
- Silage making can immensely help towards fast expansion of dairy sector as the supply of nutritious fodder will be ensured round the year.
- The labor cost is significantly reduced as 4-5 persons can easily manage a flock of 40-50 cattle heads, since maximum labour is consumed in harvesting the green forages.
- The entire fodder crop is harvested in a single step for making silage as is the case with baby corn and sweet corn. Baby corn as well as sweet corn stalks are the best fit fodders for silage making as the entire field is harvested in one go. One time harvesting is beneficial in many ways since we can harvest the crop at the appropriate time and the field became available for the timely sowing of next crop.
- Hard stems when fermented into silage become soft and are better digested by the dairy animals.
- Most of the anti-quality components present in the green forages are either destroyed or lowered during silage fermentation, for example nitrates, if present, were reported to be lowered in silage as compared to the green forages.
- Lastly the seeds of the most common weeds are destroyed during silage fermentation thereby reducing the problem of dispersal of these seeds with cowdung as farm yard manure.

Properly made silage

Nutrient losses during ensilage and ways to Reduce Nutrient Losses

Although silage making is highly beneficial but some losses in the form of dry matter, carotenoids, carbohydrates and proteins occurs due to respiration, fermentation and aerobic deterioration. The other losses of nutrients arise from field, harvesting and affluents. The field losses may occur due to shattering of leaves and other nutritious portions because of poor harvesting managements. The extent of loss in dry matter depends on the time at which the forage is ensiled. Over the period of 48 hours, losses of dry matter as high as 6 % may occur. Loss of carbohydrates and protein also occur due to respiration and proteolysis by plant enzymes. It has been reported that loss of nutrients during ensilage was drastically minimized with increasing dry matter content of ensiling material (Honig, 1968). The fermentation losses chiefly depend upon the moisture content. The clostridial type fermentation is deleterious for most of the nutrients. The clostridia are responsible for the loss of protein. Fermentation losses, therefore, are dependent upon pH, moisture content of ensiling material and type of micro-organism growing during course of fermentation. Forages with low dry matter content (less than 22.9%) leads to effluent production with a considerable loss of nutrients (Castle and Watson, 1993). During aerobic degradation, the temperature and pH rises while lactic acid content reduces. Loss of dry matter (DM) and nitrogenous substances occur due to escape of volatile fatty acid, lactic acid and ammonia. Loss of nutrients arising out of secondary fermentation could be 0-15% and could be minimized by management practices such as use of cover, propionic acid etc (Wyss, 2000). Different nutrient losses occurred during preservation of herbages as silage are mentioned below.

Table 2. Nutritive losses during ensilage

Biological process	Judgment	Approx loss (%)
Respiration	Unavoidable	1-2
Fermentation	Unavoidable	1-4
Effluent	Mutual	5-7
Pre-wilting	Unavoidable	2-5
Secondary fermentation	Avoidable	0-5
Aerobic transformation	Avoidable	0-15
Total losses		**7-35**

Source: Mojumdar (2009)

Reduction in the nutritive value of silage fermentation with respiratory losses, silage heating and clostridial fermentation is minimized by limiting air and moisture contact with silage (Bolsen *et al.*, 1996). Minimizing oxygen exposure to silage is essential for obtaining good quality silage. Air allows the respiration process to continue using soluble carbohydrates essential for acid production, which generates heat and increases the temperature. Air exposure during preservation tends to progress towards mould formation and silage rott. The increase in the temperature of silage as a result of heating also reduces its palatability when fed to livestock (Pelz and Hoffman, 1997).

Dry matter concentration of the forages plays a vital role in minimizing the nutrient losses during ensilage. High moisture silage leads to effluent losses and clostridial fermentation, which cause excessive dry matter loss, high butyric acid concentration and lower nutrient intake (Henderson and McDonald, 1971). Proper stage of harvesting and dry mater content maximizes the nutritive value of silage (Mojumdar and Rekib, 1980). Wilting of high moisture forage to 30% dry matter is a safe way, which inhibits the clostridial fermentation. Clostridia bacteria degrade sugars and also convert lactic acid to butyric acid and elevate ammonia concentration and thus causing pH to rise (Nikolic and Jovanovic, 1986). It also breaks down protein to amines. Thus, clostridial fermentation has an undesirable effect on the nutrients leading to their decomposition to undesirable end products, dry matter loss and reduced palatability (Nikolic and Jovanovic, 1986).

The heat caused during fermentation plays important role in destruction of nutrients. Higher temperature silage (100°F) has been found to be poor in quality. The over-heated silage produced at a temperature above 120°F have been found to possess heat damaged protein having brown to dark brown colour with a tobacco type foul smell. Protein of heat-damaged silage forms a complex with carbohydrates and is not digestible. The part of protein and energy is not available to livestock and results in lower DCP and TDN values (Rodriguez et al, 1985). Higher temperature also increases aerobic spoilage and reduces stability of silage.

Water soluble carbohydrate of forages constitutes the primary nutrients that are fermented to lactic acid and acetic acid by *Lactobacillus* bacteria to produce a low pH (4.5) and stable silage. Maize, sorghum, oat and other cereal fodders usually have higher soluble sugar concentration and a good and stable silage is obtained while legume forages having low soluble sugar content do not produce stable and good quality silage chiefly because of low lactic acid production mostly below 3% of dry matter (Singh and Rekib, 1986a). Carbohydrates in the forages may be naturally occurring or may be added as a separate ingredient such as molasses obtained as sugar industry by-products (Evers and Carrell, 1998), which act as a fermentable substrate. Relatively more lactic acid is produced from glucose present in the ensiling forage than fructose. Hemi-cellulose after acid hydrolysis produces pentoses, which is then fermented to lactic acid and acetic acid. Besides carbohydrates, the protein content of the ensiling forage plays an important role in determining the quality and feeding value of silage. High CP content in the leguminous forages leads to ammonia production during fermentation resulting in rise of pH (5 and above), buffering action and temperature. The high moisture content (more than 75%) causes more protein loss due to proteolysis by clostridia. Nitrates present in the plant are reduced to nitrites which in turn release ammonia (Singh et al, 1983).

Summary

From the above it could be concluded that specialty maize has vast potential as a cash crop as well as a good quality livestock fodder. The fodder of sweet corn and baby corn stalk is of excellent quality and can easily be preserved in the form of silage to be used as animal fodder during lean periods. The silage made out of specialty corn could act as a fodder bank which can play a significant role in reducing the green fodder shortage of the country. Silage making may be promoted to a large extent in the maize growing areas by promoting specialty corn cultivation. Silage, packed in the form of bales, may be exported to drier regions of the country to be used as animal fodder. This will immensely help towards increasing the profitability of rural masses.

References

Bolsen, K.K., Ashbell, G. and Weinberg, Z.G. (1996). Silage fermentation and silage additives review. *Asian Aust J Anim Sci.* 9**:** 483-493.

Castle, M.E. and Watson, J.N. (1993). The relationship between the dry matter content of herbage for silage making and effluent production. *J British Grass Soc.* 28: 135-138.

Chaudhary, D.P., Kumar, A., Kumar, R., Singode, A., Mukri, G., Sah, R.P., Tiwana, U.S. and Kumar, B. (2016). Evaluation of normal and specialty corn for fodder yield and quality traits. *Range Mgmt Agroforestry.* 37: 79-83.

Dass, S., Yadav, V.K., Shekhar, J.C. and Yadav, Y. (2009). Baby corn production technology and value addition. Directorate of Maize Research, Pusa Campus, New Delhi-110012, India. *Tech Bulletin* 2009/11. Pp. 46.

Evers, DJ. and Carrell, D.J. (1998). Ensiling salt preserved waste with grass straw and molasses. *Anim Feed Sci Tech.* 61: 241-249.

Hendersan, A.R. and Mc Donald, P. (1971). Effect of formic acid and the fermentataion of grass of low dry matter content. *J Sci Food and Agri.* 22: 152-163.

Honig and H. (1968). *Die volkenroder bilangaulage. Seigene Futter.* 14: 304-315.

Hukkeri, S.B., Shukla, N.P. and Rajput, R.K. (1977). Effect of levels of soil moisture and nitrogen on the fodder yield of oat on two types of soils. *Ind J Agron.* 47: 204-209.

Iqbal, A., Ayub, M., Zaman, H., Ahmed, R. (2006). Impact of nutrient management and legume association on agro-qualitative traits of maize forage. *Pak J Bot.* 38: 1079-1084.

Irlbeck, N.A., ussell, J.R., Hallauer, A.R., and Buxton, D.R. (1993). Nutritive value and ensiling characteristics of maize stover as influenced by hybrid maturity and generation, plant density and harvest date. *Anim Feed Sci Tech.* 41: 51-64.

Lance, R.D., Foss, D.C., Krueger, C.R., Baumgardt, B.R., and Niedermeier, R.P. (1964). Evaluation of Corn and Sorghum Silages on the Basis of Milk Production and Digestibility. *J Dairy Sci.* 47: 254–257.

Marsalis, M.A., Angadi, S.V., and Contreras-Govea, F.E. (2010). Dry matter yield and nutritive value of corn, forage sorghum, and BMR forage sorghum at different plant populations and nitrogen rates. *Field Crops Res.* 116: 52–57.

Mojumdar, A.B. (2009) Technologies of quality silage and hay making from forages. In Forage for sustainable livestock production (Das N, Misra A K, Maurt S B, Singh K K and Das M.M eds.). SSPH, New Delhi. pp. 361-375.

Mojumdar, A.B. and Rekib, A. (1980). Physical and chemical changes in laboratory silage of *Kharif* cereal forages. *Forage Res.* 6: 75-82.

Nikolic, J.A. and Juvanoic, M. (1986). Some properties of apple pomace ensiled with and without additives. *Anim Feed Sci Tech.* 15: 57-67.

Pelz, D. and Hoffman and S. (1997). Dewatering, compacting and ensilaging of spent grains. *Brauwelt* 15: 436-439.

Rodriguez, A., Riley, JA. and Thorpe, W (1985). Animal performance and physiological disturbances in sheep fed diets based on ensiled sisal pulp. *Trop Anim Prod.* 10: 23-31.

Sattar, M.A., Haque, M.F., and Rahman, M.M. (1994). Intercropping maize with broadcast rice at different row spacing. *Bang J Agric Res.* 19: 159-164.

Singh, A.P. and Rekib, A. (1986). Effect of wilting and mixing berseem and oat fodder on fermentation pattern of silage. *Ind J Anim Sci.* 56: 85-88.

Singh, A.P., Prasad, J. and Rekib, A. (1983). Note on high nitrate paragrass silage given to rabbits. *Anim Feed Sci Techn.* 9: 325-331.

Wyss, Y. (2000). Silage additives and aerobic stability. *Revue Swiss di Agric.* 32: 29-32.

30

Maize Insect Pest Management Research in India: Progress and Future Thrust

*Suby S.B.[1], Pradyumn Kumar[1], Soujanya L.P.[1], Jaswinder Kaur[1] Chikkappa G.K.[1], Jawala Jindal[2], M.L.K. Reddy[3], Maha Singh[4] S.S. Mahadik[5], M.K. Mahala[6], T. Alam[7] and J.C. Sekhar[1]**

[1]*ICAR-Indian Institute of Maize Research (IIMR), Ludhiana*
[2]*AICRP-Entomology, Punjab Agricultural University (PAU), Ludhiana*
[3]*AICRP-Entomology, PJTSAU, Hyderabad*
[4]*AICRP-Entomology, CCSHAU, Karnal*
[5]*AICRP-Entomology, Maharashtra Shahu Agricultural School Campus, Kolhapur*
[6]*AICRP-Entomology, MPUAT, Udaipur*
[7]*AICRP-Entomology, Tirhut College of Agriculture, Dholi*
**Corresponding Author's Email: jcswnc@rediffmail.com*

Maize (Zea mays L.) has tropical origin and is traditionally grown in monsoon. The crop is grown in diverse geographical and climate conditions around the world and in India maize is grown in *kharif* (rainy season), *rabi* (winter season) and spring seasons where more than 70 per cent of the production happen in *kharif* alone. Of late, the area under *rabi* maize has increased because of higher productivity and ever increasing demand.

Insect pests of maize

Compared to most cereals, maize faces fewer biotic and abiotic constraints in production. Among biotic stresses, insect pests are major constraints in the production and productivity of maize crop. The earlier literature cites over 160 insect and mite species which attack maize crop (Fletcher, 1914, 1917; Ayyar, 1940; Bhutani, 1961; Pant and Kalode, 1964) but later Mathur (1987) observed over 250 species of pests associated with maize in field and storage conditions. Dick and Guthrie (1988) identified 87 species that directly or indirectly exert severe stress on corn culture in tropical and temperate regions throughout the world. Excluding stored grain insects, Luckman (1978) lists 34 pests or pest groups for which chemical controls are recommended on corn in the United

States. There have been more than 130 insect pests reported to cause damage to maize in India but only about a dozen cause economic losses (Sarup *et al.*, 1987).

The pyralid *Chilo partellus*, the noctuid *Sesamia inferens* and muscids *Atherigona soccata, A. orientalis* and *A. naqvi*, and rice weevil *Sitophilus oryzae* L., which attacks during storage are of major importance. Maize borers and shoot fly are the regular pests, occurring in three different growing seasons of the crop. *Chilo partellus* is a regular pest of *kharif* maize. *Sesamia inferens* is more prevalent in *rabi* crop but it also occurs in spring maize. *Atherigona* spp. is a regular pest of spring maize in northern part of India. Besides, there are nearly one dozen more pests, which occur sporadically and cause considerable crop losses at times. The loss caused by insect pests in maize crop ranges from 5-15%, the stemborers contributes a major share in it. Dead hearts, foliar damage and stem tunnelling are the major damage symptoms by stemborers that cause severe yield loss (Mathur and Rawat, 1981). Infestation of these pests at 4-8 leaf stage kill the growing point resulting in complete death of the plant. Moreover by the entry points created by stemborers, bacterial and fungal pathogens may enter within plant tissues and cause secondary losses (Ndiritu, 1999). Apart from stemborers, cob borer *Helicoverpa armigera* (Hubner), tobacco caterpillar *Spodoptera litura* (Fabricius) and chafer beetle *Chiloloba acuta* (Weidemann) are emerging pests of maize. Cob borer and tobacco caterpillar feed on silk, tassel and kernels at the tip of the ear are eaten down to the cob. Chafer beetles feed on the pollen and results in poor seed set. Storage pests cause severe economic damage either by direct consumption of grain or indirectly by creating favourable environment for the establishment of other pests/ fungi during storage (Tefera *et al.*, 2010).

Table 1. Insect pests of maize and distribution.

Common Name	Scientific Name	Family and order	Geographical Distribution
Major Pests			
Spotted stemborer	*Chilo partellus* (Swinhoe)	Pyralidae, Lepidoptera	Distributed throughout India
Pink Stemborer	*Sesamia inferens* Walker	Noctuidae, Lepidoptera	Andhra Pradesh, Assam, Bihar, Delhi, Karnataka, Madhya Pradesh, Maharashtra, Orissa, Punjab, Tamil Nadu, Telangana, Uttar Pradesh and West Bengal.
Shoot fly	*Atherigona* spp.	Muscidae, Diptera	Delhi, Haryana, Punjab, Rajasthan, Uttarakhand and Uttar Pradesh during spring season

Rice weevil	*Sitophilus oryzae* L.	Curculionidae, Coleoptera	Distributed in tropical and sub tropical regions
Emerging Pests			
Cob borer	*Helicoverpa armigera* (Hubner)	Noctuidae, Lepidoptera	Polyphagous and is widely distributed in the tropics and sub tropics
Tobacco caterpillar	*Spodoptera litura* (Fabricius)	Noctuidae, Lepidoptera	Distributed throughout India
Chafer beetle	*Chiloloba acuta* (Weidemann)	Cetoniidae, Coleoptera	Delhi, Punjab, Haryana, Uttar Pradesh, Telangana and Andhra Pradesh
Hairy Caterpillar	*Euproctis* spp	Lymantriidae, Lepidoptera	Punjab, Haryana, Uttar Pradesh, Bihar, Maharahtra, Tamil Nadu, Telangana and Andhra Pradesh

Rice weevil is a field to storage pest, lays eggs on mature grains, which serves as initial inoculum leading to higher infestation in storage and completes. The pest completes the entire life cycle in the seed itself as an internal feeder. It was reported that 80% loss occur to untreated maize grain stored in traditional structures depending on the period of storage (Boxall, 2002). The major insect pests of maize and their distribution are given in Table 1.

Among all the insect pests of maize, stemborer *C. partellus* is the most serious. It is an important pest in Asian and African countries like Afghanistan, Bangladesh, East Africa, Iraq, Japan, Indonesia, Nepal, Malawi, Pakistan, India, Sudan, Taiwan, Thailand and Uganda (Siddiqui and Marwaha, 1993; Arabjafari and Jalali, 2007). In India this pest is found in all the states. *C. partellus* has a wide host range of both cultivated and wild grasses (Khan *et al.*, 1997 a; Van den Berg *et al.*, 2001).

Yield losses due to major insect pests of maize

James (2003) reported that 9% of the world maize crop is lost annually due to insect pests. According to Khan *et al.* (1997 b) insect pests accounts for 25-40% yield losses in maize. The yield losses in maize (Rehman, 1940; Trehan and Butani, 1949; Reddy, 1968) are mostly estimated empirically rather than by experimentation in India. This crude estimate of loss due to *C. partellus* was 70-80 per cent. Similarly, Srivastava (1959) is of the opinion that at the very conservative estimate 10-15 per cent of the maize produce is lost annually in Rajasthan on account of insect pests alone. Reddy (1968) also put forth estimated gross loss caused by insect pests and diseases in India to be at ten per cent. These guess work estimates are generally covered under the accepted loss of 10-12 per cent. Chatterji *et al.* (1969) showed that the percentage of avoidable loss primarily due to *C. partellus* varied from 24.30 to 36.30 percent in different agro climatic regions of India. *S. inferens* causes loss in winter season in peninsular India, which varies from 25-80 per cent (Rao, 1983). The insect pests have increased due to the large scale cultivation of maize as sole crop and widespread use of pesticides for pest control (Mathur, 1992). Recently, *C. partellus* infestation was observed to range 15-97.5% in five different locations (AICRP Kharif, 2017). Grain yield loss reported due to *A. soccata* is 21.28% and *A. orientalis* is 20% (Panwar, 2005; Pathak *et al.*, 1971).

Management strategies

Several strategies have been adopted to control the losses caused by insect pests. Among various strategies, the use of chemical insecticides is the major one across the globe, but, it results in ecological damage, environmental pollution, human health hazards and development of resistance in the insect pests. The presence of a multitude of pest problems and the boring behaviour of maize pests where insecticides are unreachable, emphasises the need for Integrated Pest Management (IPM) in Maize based cropping system. Maize IPM includes both preventive and responsive pest management tactics. Preventive tactics mainly involve host plant resistance, augmented biological control and those cultural practices which reduce pest incidence. Responsive management involves certain cultural, physical, biological and chemical control measures individually or in combination.

1. Host Plant Resistance

The use of Host Plant Resistance is an ideal method of controlling insect pests of maize as resistant genotypes provide an inherent control that involves no environmental pollution and are generally compatible with other control strategies. Durbey and Sarup (1984) showed the antibiosis effect of rearing *C. partellus* on different maize germplasm, as expressed clearly on mortality of larvae. Maximum antibiosis resulted from rearing on resistant Antiqua Gr. I, Mex-17 and tolerant Ganga 5. The expression of antibiosis due to tolerant Ganga-5 revealed its intermediate behavior towards *C. partellus*. The average larval and pupal weights were significantly lower on resistant varieties (Antiqua Gr. I and Mex-17) compared to the susceptible ones (Basi local and Vijay composite). Panwar, *et al.* (2001) evaluated different maize lines for their reaction to *C. partellus* under artificial infestation. They made distinct categories namely; tolerant, moderately tolerant and highly susceptible. Several biological parameters of *Chilo partellus* viz., larval and pupal survival, larval and pupal recovery, larval and pupal weights and per cent pupation and plant parameter i.e. leaf injury rating (1-9) were used to study the germplasm susceptibility level (antibiosis) through artificial infestation using neonate larvae.

Sekhar *et al.*, 2016 reported several maize land races to have high levels of resistance to pink stemborer. *S. inferens* resistant/tolerant lines registered with ICAR-NBPGR are DMR-7(INGR 10077), DMR E9 (INGR 11028), DMRE 57(INGR 11029) and DMR E 63(INGR14014). Sekhar *et al.* (2015) found that resistance against *S. inferens* is governed by additive × additive (i) followed by dominance (d) and additive (a) gene effects/ variance. So far none of the lines screened were found to be resistant to rice weevil. However, very few maize genotypes namely Ent 2-3, LM14 have been found moderately resistant.

Screening of pre-released hybrids and inbreds against *C. partellus* and *S. inferens* under AICRP Entomology helped in the release of many tolerant maize hybrids. Entries pertaining to extra early, early, medium and late maturity of advanced varietal trials and speciality corn are evaluated against stemborers under artificial infestation at Delhi, Hyderabad, Karnal, Kolhapur, Ludhiana, Dholi and Udaipur centres. To identify the sources of resistance, 3221 maize accessions were evaluated against *C. partellus* (856), *S. inferens* (2018), and shoot fly *Atherigona* spp. (347) during 2005-06 to 2015-16 under artificial inoculated conditions for stemborers and at hot-spot locations for shoot fly. Based on the results of multi-year and multi-location screening, 149 accessions were identified to have some resistance to *C. partellus* (66), *S. inferens* (67) and shoot fly (16).

1.1. Biochemical basis of Host Plant Resistance

Host plants resort to direct and indirect defence strategies when attacked by the herbivores. Direct defences mechanisms include physical barriers and antibiotic metabolites while the indirect defence work by recruiting natural enemies of the attacking herbivores. Both the strategies can either be constitutive where the plants are constantly in defence mode or inducible where the system is activated upon attack (Wittstock and Gershenzon, 2002, Arimura *et al.*, 2005).

1.1.1. DIMBOA

The major biochemical factor of resistance in maize to several maize pests is the hydroxamic acid, 2,4-dihydroxy-7-methoxy- (2H)-1,4-benzoxazin-3- (4H)-one (DIMBOA), found at high levels during the early stages of plant development. It is released from vacuoles enzymatically following wounding, thus defend the attacking pathogens and herbivores. It is the most extensively studied secondary plant metabolite against the European corn borer *Ostrinia nubilalis* (Klun *et al.*, 1967), Asian corn borer *Ostrinia furnacalis* (Yan *et al.*, 1995) and Mediterranean corn borer *Sesamia nonagrioides* (Gutiérrez and Castañera 1986, Ortego *et al.* 1998). The role of DIMBOA was evaluated against *C. partellus* at ICAR-IIMR. In the course of this study a modified method of extraction of DIMBOA has been developed which was found economical and efficient (Chandra *et al.*, 2013). DIMBOA content was quantified in the hybrid HQPM 1 and the inbred HKI 335 at seven plant ages and found two day old maize plants has maximum amount of DIMBOA and its concentration drastically decreases as the plant ages to twenty days. The DIMBOA concentration from the leaves of 20 maize genotypes at 6 days after germination was quantified and it was found highly negatively correlated with insect damage (-0.71), proving an important factor in imparting resistance, however it is not true in each and every germplasm suggesting the role of other resistance factors (Chandra *et al.*, 2012).

1.1.2. Phenolic acids

The hydroxycinnamates in plant cell walls mediate cell wall cross-linking towards strengthening of cell walls which, thus contributes to low digestibility in herbivores. Ferulic acid is the major hydroxycinnamate derivative in young grass cell walls while p-Coumaric acid is mainly found in lignified cell walls. Maize cell walls can contain up to 5% ferulate monomers plus dimmers and up to 3% p-coumarate. The role these phenolics in insect resistance maize were found earlier against *O. nubilalis*, *Diatraea saccharalis and Sesamia nonagrioides* (Santiago *et al.*, 2013)

Role of cell wall bound ferulic and p-coumaric acids in imparting *C. partellus* resistance was studied at ICAR-IIMR in maize inbred lines. Quantity of these phenolics at different plant age was negatively correlated with damage parameters (LIR and tunnel length) caused by *C. partellus*. Maximum correlation was observed at 20 day old plants, 0.71 and 0.61 respectively for LIR and tunnel length with ferulic acid 0.78 and 0.62 for the same with p-coumaric acid (Gundappa, 2012). Diet incorporation bioassay in which p-coumaric acid revealed its antibiotic effect in terms of per cent increase in mortality (41.5% at 2.5mg/ml), reduction in the larval and pupal weight, whereas ferulic acid did not affect larval and pupal weight (Gundappa *et al.*, 2013). As larval feeding progress from one week to three weeks in infested plants, total soluble phenolic content and total flavanoid content were found to decrease while the total bound phenolics recorded a rise in its contents in the leaf tissue. The concentration of ferulic and p-coumaric acid was found to increase in the total cell wall bound phenolics fraction which can be attributed to the release of these molecules from the degrading cell walls due to insect feeding.

The secondary metabolites discussed above are directly involved in resisting pest attack. The indirect defense signals are the plant volatiles which are released upon injury by the pest and the natural enemies exploit these plant-provided cues to locate their victims, while several herbivores are repelled (Hoballah *et al.*, 2004). We have evaluated the indirect defence responses of twenty maize genotypes to *C. partellus*, where two weeks old potted maize plants were artificially infested with *C. partellus* neonate larvae. Twelve days after infestation, the pots were individually enclosed in transparent PE film to form open top chambers and approximately 800 female parasitoids were released along with males, everything confined in a large mosquito net. Six days after release, *C. partellus* was recovered from the plants and reared in baby corn. Cocoon masses of *C. flavipes* were harvested after rearing for 10 more days and the cocoons were counted once adult emergence was completed. Correlation analysis suggests that the parasitoids tend to lay more eggs on the host which is of higher weight. The leaf injury rating which is currently used as a measure to determine resistance wasn't significantly correlated with percent parasitization, even though with number of cocoons per cocoon mass does. The volatiles induced by *C. partellus* feeding, whose content and composition may vary among genotypes can also influence the attractiveness to *C. flavipes*. To investigate this, a multi-sampling dynamic volatile collection system was devised (Fig. 1) (Patent appl. No. 1235/DEL/2015) and the volatiles were collected in Tenax TA traps using the same genotypes. The preliminary analysis shows, the volatiles include common green leaf volatiles, long chain hydrocarbons and terpenoids which may have tremendous importance in host plant resistance assisted biological control leading to sustainable agriculture (Suby *et al.*, 2015).

Fig. 1. Volatile collection system

1.2. Screening techniques to evaluate host plant resistance in maize germplasm

The ability to develop resistant cultivars depends on the precision of resistance screening techniques. The effective and reliable screening technique helps in identifying the desired level of insect resistance in large number of genotypes. Hot spot locations can be screened under natural pest infestations. In these locations, planting date of the crop should be adjusted in such a way that the susceptible stage of the crop coincides with the peak activity period of the pest. This can be determined by conducting population dynamic studies either by using attractant traps or by monitoring pest infestation at regular intervals.

Screening under natural infestation is unreliable and takes long time to identify lines with stable resistance. Therefore, artificial infestation techniques have been standardized for evaluating maize germplasm against stemborers. Mass rearing of stemborers for artificial infestation was made efficient by two of the patented technologies of IIMR viz., Insect handling device (Fig. 2) (Patent No. 252363) and collapsible insect rearing cage (Fig. 3) (Patent Application No. 923/DEL/2011).

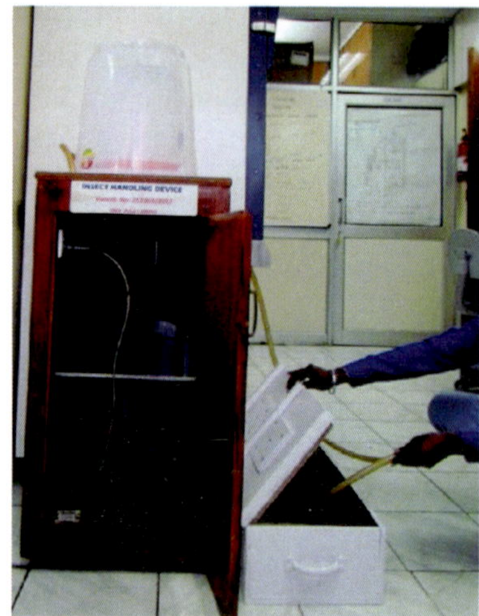

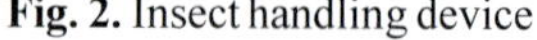

Fig. 2. Insect handling device

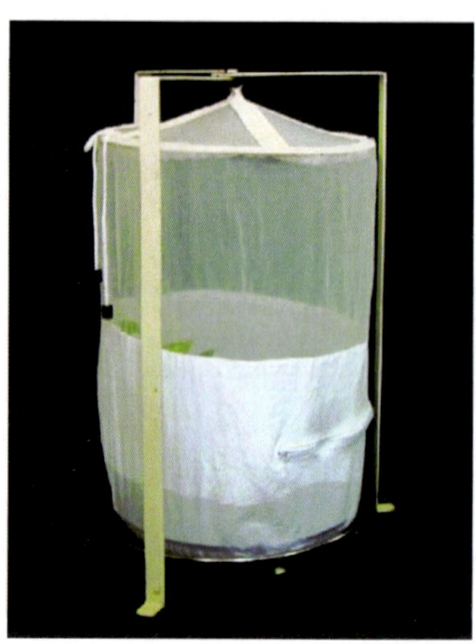

Fig. 3. Collapsible insect rearing cage

1.2.1. Screening for stemborer resistance

Under artificial infestation, 10 neonate larvae or 20-25 black head stage eggs should be released slowly into the whorl of the plant with a soft brush on 12-15 day old maize crop. Usually, *C. partellus* eggs laid on butter paper, cut into bits are pinned into the whirl, while a larval dispenser called bazooka is used to release *S. inferens* eggs. Field infestation is generally carried out after 5 pm to avoid larval mortality due to high temperatures and if necessary, a repeat infestation is done after heavy rains. The availability of eggs for mass infestation is ensured by advancing or slowing the eggs development by incubating at temperatures ranging from 15- 30°C in BODs. The resistance level of the genotypes is assessed by Leaf Injury Rating (LIR) 21-25 days after infestation (Table 2 & 3).

Table 2. Leaf Injury Rating scale for *Chilo partellus* (Sarup *et al.* 1978).

Rating	Description
1.	Apparently healthy plant
2.	Plant with parallel, oval or oblong holes, slightly bigger than pin sized (2-3 mm) on 1-2 leaves
3.	Plant with more elongated holes (4-5 mm or match stick head sized) or shot holes on 3-4 leaves
4.	Plant with injury (oval holes, shot holes and slits of 1-4 cm) in about 1/3 of total number of leaves
5.	Plants with about 50% leaf damage
6.	Plants with a variety of leaf injuries in about two thirds of the total number of leaves (ragged appearance) or one or two holes or slits at the base of the stem (> 10 cms streaks are observed)
7.	Plants with every type of leaf injury and almost all the leaves damaged (ragged or crimpled appearance), with tassel stalk boring or circular dark ring at the base of stem
8.	Plants with stunted growth in which all the leaves are damaged
9.	Plants with dead heart

Table 3. Leaf Injury rating scale for *Sesamia inferens* (Reddy *et al.* 2003).

Rating	Description
1.	Apparently healthy plant
2.	Plant with parallel, oval or oblong holes, slightly bigger than pin sized (2-3 mm) on 1-2 leaves
3.	Plant with more elongated holes (4-5 mm or match stick head sized) or shot holes on 1-2 leaves
4.	Plant with injury (oval holes, shot holes and slits of 1-4 cm) in about 1/3 of total number of leaves and midrib damage on 1-2 leaves
5.	Plants with about 50% leaf damage, oblong holes, shot holes, slits and streaks of 5-10 cms and midrib damage on leaves
6.	Plants with a variety of leaf injuries to about two thirds of the total number of leaves (ragged appearance) or one or two holes or slits at the base of the stem (> 10 cms streaks are observed)
7.	Plants with every type of leaf injury and almost all the leaves damaged (ragged or crimpled appearance), with tassel stalk boring or circular dark ring at the base of stem
8.	Plants with stunted growth in which all the leaves are damaged
9.	Plants with dead heart

1.2.2. Improved screening techniques for stemborers

The larvae of *C. partellus* and *S. inferens* tend to disperse in the field mostly as a response to deterioration of the plants where initial infestation occurs. Berger *et al* (1992) found *C. partellus* larvae leave the sorghum plants in mass after 10 days of initial infestation and this dispersal is much later in maize. This dispersal is a major factor for the spread of infestation in the field. Studies at IIMR revealed that the most vulnerable part of the plant in this secondary infestation is the second above ground internode of 6 to 10 leaf stage maize plants. Penetration resistance of this internode was found highly positively correlated with damage and negatively correlated with the performance of *C. partellus* larvae on various genotypes. This indicates tough stalk is an important trait of stemborer resistant maize, thus suggests the utility of the technique for screening stemborer resistant germplasm. Tough stalked genotypes can resist stemborers and stalk lodging, thus can tremendously reduce crop losses (Suby *et al.*, under publication).

1.2.3. Screening for shoot fly resistance

Screening of inbred lines against Shoot fly spp. cannot be done under artificial conditions, as it is difficult to mass multiply it in the laboratory. Adequate shoot fly density for resistance screening can be achieved by manipulating the sowing date and spreading fishmeal (which attracts the shoot flies) in the field. Planting four rows of a susceptible genotype 20 days prior to the sowing of test material will also ensure infestation. Data is recorded on number of plants with eggs, plants with dead hearts, total number of eggs and the total number of plants at 14 and 21 days after seedling emergence. Maize genotypes are screened in the hot spot locations of Delhi and Ludhiana since 2013 by Fish Meal Technique during spring season. The programme was later extended to Karnal and Ludhiana centre also. A susceptibility index (SI) to classify maize genotypes for shoot fly resistance was developed (Kumar *et al.*, 2014).

$$SI = A_{X1} + A_b/A_{X2} + A_d/A_n$$

Where, A_{X1} = number of plants oviposited/total number of plants, A_b = number of dead hearts/ number of plants oviposited, A_{X2} = total number of eggs/ number of plants oviposited, A_d = per cent dead hearts and A_n = per cent plants infested

The germplasm within mean susceptibility index + S.D. range are moderately resistant, while germplasm on left and right side represents resistant and susceptible germplasm, respectively.

1.2.4. Screening technique of rice weevil *Sitophilus oryzae* L.

Maize genotypes for *S. oryzae* resistance are screened by placing twenty five seeds of a genotype in a 250 ml plastic jar with screw lid at the top allowing ventilation and preventing the escape of weevils. Eight male and eight female adult weevils of four to five days old will be introduced into each jar containing maize kernels and kept for seven days for oviposition (Derera *et al.*, 2001). In adult mortality data assessment, all the dead and adult live insects are removed from each jar and the kernels of each genotype kept under the same experimental conditions to further assess F_1 progeny emergence for the subsequent two months period. Inspection of progeny is made every day and emerging progeny will be removed and counted per jar on each assessment day. The number of grains damaged by weevil feeding is assessed 63 days after introduction of the weevils and expressed as proportion of damaged grains on the total number of grains sampled. Grain weight loss is determined using the count and weight method of Gwinner *et al.* (1996).

Weight loss (%) = (Wu x Nd) - (Wd x Nu) X 100 / Wu x (Nd + Nu)

Where Wu = Weight of undamaged grain, Nu = Number of undamaged grain, Wd = Weight of damaged grain, and Nd = Number of damaged grain.

The median development period is calculated as the time (days) from the middle of the oviposition period to the emergence of 50% of the F1 progeny (Dobie, 1977). The index of susceptibility can be calculated using the method of Dobie (1974).

Index of susceptibility = 100 x [$\log_e$ (total number of F1 progeny emerged) / (median development time)]

The susceptibility index, ranging from 0 to 11, is used to classify the maize genotypes; where; 0 - 4 = resistant, 4.1 - 7 = moderately resistant, 7.1 - 10 = susceptible and e"10 = highly susceptible. Twenty-six maize inbred lines were screened against *S. oryzae* and significant differences were observed in their level of resistance (Soujanya *et al.*, 2015).

1.3. Host plant resistance through transgenic approach

Genetically engineered corn is developed to introduce resistance to herbicides and insect pests and is grown in many countries. As a pest control strategy, transgenic *Bacillus thuringiensis* (Bt) maize hybrids are undergoing regulatory trials in India. As part of a multi-institution consultancy project, we have determined the sensitivity of the target lepidopterans to the insecticidal Bt proteins Cry1A.105 and Cry2Ab2 expressed in Bt maize of event MON89034 as this determines product efficacy and the resistance management strategy to be adopted. Sensitivity profiles of 53 populations of *C. partellus*, 21 populations of *S. inferens* and 21 populations of *H. armigera*, collected between 2008 and 2013 from maize-growing areas in India were generated through dose–response assays. Cry1A.105 protein was the most effective to neonates of *C. partellus* (mean MIC_{90} range 0.30–1.0 µg mL-1) and *H. armigera* (mean MIC_{90} range 0.71–8.22 Mg mL-1), whereas Cry2Ab2 (mean MIC_{90} range 0.65–1.70 µg mL-1) was the most effective to *S. inferens* (Jalali *et al.*, 2015). Bioassay studies were also carried out for these pests against Cry 1Ab and Cry 1F later. The data generated will serve as pre-commercialization benchmarks for resistance monitoring purposes.

2. Introduction of agro-ecological techniques as an integral part of IPM

Crop cultural practices have profound effect on stemborer population and have great potential to minimize infestation. To destroy the over wintering stemborer larvae in the stubble, burning or removing the stubble is the most effective strategy. Deep summer ploughing is essential to kill the hibernating or aestivating larvae and pupae. The infestation of shoot fly can be escaped by sowing maize before the first week of February. In the irrigated area, sowing during mid June is most appropriate time for harnessing the optimum potential in Haryana, Punjab and Western Uttar Pradesh, were the yield was observed 29.1 per cent higher in unprotected crop as compared to unprotected normal sown crop (Sarup *et al*., 1978). For storage pest management, cleanliness and sanitation are the most important steps towards prevention of insect infestation. The crop should be harvested at the proper time to prevent egg laying by storage pests. The moisture content of grain should be less than 10%.

Conservation biological control through habitat manipulation can be very effective ecological approach in bringing the pest population at lower level of equilibrium, thereby obviating the reliance on chemical pesticides and contributing substantially towards sustainable agriculture. Intercrops, strip crops, trap crops as well as modification of agronomical practices are promising techniques to modify corn production systems to create of natural enemy resource habitats to increase natural control of maize insect pests without compromising the yield. Habitat management diversifies maize ecosystem and enhances the availability and diversity of prey, sheltering sites and mates for natural enemies.

Inter-cropping of maize with suitable varieties of cowpea in 2:1 proportion (Fig. 4) was found to reduce *C. partellus* infestation as the adult prefers cowpea over maize for oviposition while being a non host crop the young larvae does not feed on it. Additionally, there is gain of cowpea produce and the soil nitrogen enrichment for the next crop. Experiment on habitat manipulation was conducted during 2010 and 2011 at six locations viz. Delhi, Kolhapur, Ludhiana, Hyderabad, Srinagar and Udaipur. Based on percent plant infestation, leaf injury rating and yield, maize intercropped with cowpea in the ratio of 2:1 row was on par with maize pest control by the state recommended pesticide of the region. Later, maize-cowpea intercropping was compared with imidacloprid seed treatment; seed treatment and intercropping; treatment with Carbaryl at 10 and 20 days after germination; treatment with Neem Seed Kernel Extract (NSKE) at 7 and 14 days after germination and intercropping with NSKE treatment. This multi-location trial concluded that the combination of seed treatment and cowpea intercropping as well as cowpea intercropping with NSKE gives better yield than other interventions (AICRP Kharif 2010, 2011). Marigold for *H. armigera* (Fig. 5), cauliflower for *S. litura* and Napier millet for *C. partellus* and *S.*

inferens were observed to be good trap crops in various experiments conducted at our centres (AICRP Kharif 2009).

Fig. 4. *Cowpea intercropped in maize*

Fig. 5. *Marigold intercropped in maize*

3. Minimizing the use of chemical insecticides, promoting biopesticides

The reliance on pesticides brings significant environmental liabilities of off target drift, chemical residues and resistance. Side effects cannot be avoided totally, but they can be minimized by their proper use such as application of 2-3 granules of carbofuran per infested plant are sufficient for stemborer management than blanket application. The spray can be avoided if the infestation is low. Following the recommended dose and time of spray can mitigate harmful effect of pesticide drifting and leaching which can have prolonged impact on the ecosystem. Aphids and mites can be controlled by spraying jets of water or by sprinkler irrigation at the beginning of infestation. NSKE at 1.25% was found to have complete oviposition deterrent effect against gravid *C. pertellus* females at Delhi centre (AICRP Kharif 2010). The entomopathogenic nematode *Steinernema thermophilum* at 500IJs/ml level was found to have promising to control the chaffer beetle grubs causing up to 60 percent mortality (Dhanya, 2009). Seed treatment with Imidacloprid @ 5g/kg seed gave protection from shoot fly at all three places tested i.e. Ludhaina, Hoshiarpur and Gurdaspur. Imidacloprid residue when tested showed below detectable level at 60 days after sowing at 10g/kg seed treatment level. By the use of plant products such as *Adathoda vasica*, *Azadirachta indica*, *Vitex negundo* and *Catharanthus roseus* @ 2% w/w (20g /kg seed) rice weevil can be effectively controlled. Hermetic storage in combination with botanical i.e maize treated with *Ageratum conyzoides* leaf powder at the rate of 2% w/w and stored in completely air tight high density and double layered polythene bags protects stored maize effectively from rice weevil infestation. In our recent study, dichloromethane extracts of *Ixora coccinea* and its pure isolated compound tanacetene (2,6,11-trimethyl-dodeca-2,6,10-triene) was proved superior in killing *S. oryzae* (Lakshmi Soujanya *et al.*, 2016).

4. Conservation and augmentation of natural enemies as a major component in IPM

Conservation and augmentation of natural enemies is the most promising integrated component in maize ecosystem. The egg parasitoid *Trichogramma chilonis* was found to parasitize up to 34 percent of *C. partellus* eggs in Delhi and up to 51 percent in Udaipur. Larval parasitization by *Cotesia flavipes* ranged from 15-40 percent at various centres and it was recorded in stemborer infested plants of 40-60 day old crop (AICRP Kharif 2014, 2015). Release of *T. chilonis* in the form of 8 Trichocards/ha (1,50,000 parasitized eggs/ha), twice at 12 and 22 days after germination is the general recommended practice for the management of *C. partellus* in maize. Higher grain yield with lowest percent dead heart was recorded where egg parasite *Trichogramma chilonis* released at 7 DAG followed by 5.0% neem seed kernel spray 10-12 days after germination and a second release of the parasitoids at 15 DAG was recorded at Godhra centre. Tobacco caterpillar is managed by the release of egg parasitoid *Telenomus remus* @ 1,25,000/ha for 4 times at 7-10 days interval.

Conclusion

The demand for maize is increasing across the country and the globe, yet the yield potential of available cultivars is untapped largely due to the losses in biotic and abiotic stresses where insect pests have a considerable stake. The release of cultivars with insect resistance, identification new sources of resistance, greener technologies like habitat management and generation of knowledge on pesticidal plant secondary metabolites shows the current research in maize entomology is equipped to bridge this yield gap. However, the threatening new challenge of emerging pests like cob borers under the climate change scenario urges us to revamp the current research platforms.

Future research perspective

There have been sustained efforts in developing inbred lines for resistance against stemborers and their contribution to the development of maize cultivars. Not much of the cross-resistance has been observed in maize lines when a set of lines were challenged by *C. partellus*, *S. inferens* and *Atherigona* spp, revealing the complex genetic control of insect resistance. This makes the job of developing cultivars resistant to multiple pests difficult. While screening maize germplasm against stemborers, the plants are challenged by neonate larvae, thus only the antibiosis of the plants is tested. In the field, the plants are selectively chosen for oviposition by the female moth. The antixenosis ability of plants is bypassed in the current practice of screening. There is a need to develop screening technology which subsumes both antixenosis and antibiosis. The screening of maize germplasm against *Atherigona* spp. could not be taken up under artificial infestation condition since the pest is not amenable for laboratory rearing. The mass rearing technique of *Atherigona* spp. will improve the current screening methodology of germplasm against this pest.

The host plant resistance component of maize entomology in India is focussed on the identification of resistant sources by the traditional visual phenotyping. Development of high throughput phenotyping methods involving digital data capture will accelerate the resistance breeding programme. Studies on the biochemical bases of resistance are needed to be extended to molecular level to hasten the insect resistance breeding programme.

Though many achievements have been made in understanding the role of habitat management in enhancing natural enemy effectiveness, yet more options need to be explored to make it the robust component of IPM in maize ecosystem. The ecosystem services beyond the benefit of pest suppression should be incorporated into habitat management research.

The post-harvest losses caused by *Sitophillus oryzae* and *Sitotroga cerealella* are substantial. Research is required to reduce the post-harvest losses. Hermetic grain storage for small farmers needs to be worked upon. New research initiatives are need to be undertaken on emerging pests such as cob borers, chafer beetles and hairy caterpillars in maize.

References

AICRP Kharif (2009). AICRP reports, https://iimr.icar.gov.in/

AICRP Kharif (2010). AICRP reports, https://iimr.icar.gov.in/

AICRP Kharif (2011). AICRP reports, https://iimr.icar.gov.in/

AICRP Kharif (2017). AICRP reports, https://iimr.icar.gov.in/

Arabjafari, K.H. and Jalali, S.K. (2007). Identification and analysis of host plant resistance in leading maize genotypes against spotted stemborer, Chilo partellus (Swinhoe) (Lepidoptera: Pyralidae). *Pak J Biol Sci.* 10(11): 1885-1895.

Arimura, G.I., Kost, C., and Boland, W. (2005). Herbivore-induced, indirect plant defences. *BBA-Molecular and Cell Biology of Lipids. 1734*(2): 91-111.

Ayyar, T.V.R. (1940). Handbook of Economic Entomology for South India. Madras Government Press. Pp. 520.

Berger, A. (1992). Larval movements of *Chilo partellus* (Lepidoptera: Pyralidae) within and between plants: timing, density responses and survival. *Bull Entomol Res.* 82(04): 441-8.

Bhutani, D.K. (1961). Insect pests of maize and their control. *Indian Farm.* 11(4): 7-11.

Boxall, R. (2002). Damage and loss caused by the larger grain borer *Prostephanus truncatus*. *Int Pest Manag Rev.* 7: 105-121.

Chandra, A., Kumar, P., Shashi, B.S., Asharani, S.S. and Singh, S. (2012). Evaluation of maize germplasm and susceptibility, and DIMBOA role against stemborer (*Chilo partellus*). *Maize J.* 1(2): 126-130.

Chandra, A., Singh, S.B. and Kumar, P. (2013). Effect of light on DIMBOA synthesis in maize leaves as revealed by modified cost-effective extraction method. *Environ Monit Assess.* 185(12): 9917-9924.

Chatterji, S.N., Young, W.R., Sharma, G.C., Sayi, I.V., Chahal, B.S., Khare, B.P., Rathore, Y.S., Panwar, V.P.S. and Siddiqui, K.H. (1969). Estimation of loss in yield of maize due to insect pests with special reference to borers. *Indian J Entomol.* 31(2): 109-115.

Derera, J., Pixley, K.V. and Giga, P.D. (2001). Resistance of maize to the maize weevil antibiosis. *African Crop Sci J*. 9: 431- 440.

Dhanya K.M. (2009) Potential of biopesticides for the management of *Chiloloba acuta* (Wiedemann) (Coleoptera: Scarabaeidae). MSc. Thesis, Indian Agricultural Institute, New Delhi, India.

Dicke, F.F. and Guthrie, W.D. (1988). The most important corn insects. In: Corn and Corn Improvement.American Society of Agronomy, Madison, W.I. Pp. 767-867.

Dobie, P. (1974). The laboratory assessment of the inherent susceptibility of maize varieties to post harvest infestation by *Sitophilus zeamais* Motsch. (Coleoptera: Curculionidea) infesting field corn. *J Entomol Sci.* 21: 367-375.

Dobie, P. (1977). The contribution of the tropical stored products center to the study of insect resistance in stored maize. *Trop Stored Prod Res.* 34: 7-22.

Durbey, S.L. and Sarup, P. (1984). Biological parameters related to antibiosis mechanism of resistance in maize varieties to *Chilo partellus* (Swinhoe). *J Entomol Res* 8(2): 140-147.

Fletcher, T.B. (1917). Maize Pests, Proceedings 2nd Entomological Meeting, Pusa. Pp. 188-193.

Fletcher, T.B. (1914). Some South Indian Insects and Other Animals of Importance Considered Especially from Economic Point of Views. Government Printing Press, Madras. Pp. 565.

Gundappa. (2012). Role of Phenolic acids (Ferulic acid and p-coumaric acid) in imparting resistance against maize stemborer *Chilo partellus*, PhD thesis, Indian Agricultural Research Institute, New Delhi-12.

Gundappa, Kumar, P. and Suby, S.B. (2013). Antibiosis effect of phenolic acids (ferulic acid and P-Coumaric acid) on maize spotted stemborer, *Chilo partellus* (Swinhoe) (lepidoptera: pyralidae). *Indian J Entomol.* 75(3): 247-250.

Gutierrez, C., Castanera, P. and Torres, V. (1988). Wound induced changes in DIMBOA (2, 4 dihydroxy 7 methoxy 2H 1, 4 benzoxazin 3 (4H) one) concentration in maize plants caused by Sesamia nonagrioides (Lepidoptera: Noctuidae). *Ann Appl Biol.* 113(3): 447-54.